Ecofriendly Frontiers

Olubukola Oluranti Babalola
Olukayode Oladipo Amund
Akinlolu Olalekan Akanmu
Editors

Ecofriendly Frontiers

Harnessing Microbial Applications for Food Security

Springer

Editors
Olubukola Oluranti Babalola
Department of Life Sciences
Imperial College London
Ascot, Berkshire, UK

Food Security and Safety Focus Area
Faculty of Natural and Agricultural
Sciences, North-West University
Mmabatho, South Africa

Olukayode Oladipo Amund
Department of Microbiology
Faculty of Science, University of Lagos
Lagos, Nigeria

Akinlolu Olalekan Akanmu
Food Security and Safety Focus Area
Faculty of Natural and Agricultural
Sciences, North-West University
Mmabatho, South Africa

ISBN 978-3-031-98699-4 ISBN 978-3-031-98700-7 (eBook)
https://doi.org/10.1007/978-3-031-98700-7

This Springer imprint is published by the registered company Springer Nature Switzerland AG
The registered company address is: Gewerbestrasse 11, 6330 Cham, Switzerland

Preface

Agriculture is currently at a pivotal stage in the history of humankind, as it faces the pressing challenges of unsustainable farming practices, increasing food requirements by the rapidly growing population, and environmental degradation. While conventional agricultural systems are heavily reliant on chemical inputs resulting in biodiversity loss, soil depletion, and environmental contamination, there arises the need for more sustainable and eco-friendly farming approaches as alternatives to the existing systems, toward managing the present odds against sustainable agriculture. However, harnessing the benefits of microbial applications is one of the most promising strategies for attaining sustainability in agriculture. The relationship of plant and microbe has been observed in diverse ecosystems, and the microbes are found to play some essential roles in the agroecosystems, such as the facilitation of nutrient cycling, suppression of plant pathogens, and enhancement of the plant resilience to environmental stresses. Furthermore, microbial communities also aid plants significantly in optimizing their nutrient absorption, improving drought and disease tolerance, and enhancing overall plant health. Therefore, leveraging these important microbial processes will readily transform agriculture from the chemical-intensive practices into a more sustainable model.

Ecofriendly Frontiers: Harnessing Microbial Applications for Food Security provides a comprehensive evaluation of the current advances and practical applications of microbes in agriculture and food production. It also explores the mechanisms of operation of the plant growth-promoting microbes and their contributions to modulating plant hormone production, enhancing soil fertility, and inducing systemic resistance to diseases and pests. Furthermore, the book delves into cutting-edge management strategies entailing synthetic microbial communities, microbial inoculants, and plant breeding for microbiome compatibility. While these approaches are promising for enhancing agricultural productivity, they are also effective in mitigating climate change effects.

A broader outlook of the food system beyond the plant–microbe relationships was made; it explores the benefits of microbes in food production, preservation, and safety. Microorganisms have been employed in these roles for centuries, where microbial fermentation has been taken huge advantage of in enhancing the nutritional value, quality, and extension of the shelf life of food. Impressively, the advancement in biotechnology now offers the opportunity to optimize the fermentation processes toward improving food safety and developing innovative microbial-based food solutions.

This book further elucidates the importance of microbial applications in addressing post-harvest losses and food waste, as the million tons of food being wasted yearly cumulates to significant environmental, economic, and social consequences. Thus, the microbial bioconversion of food waste into valuable products such as biofuels, biofertilizer, enzymes, and bioplastics presents appropriate solution to the problem of food wastage. Furthermore, the roles of microbial bioremediation techniques in the detoxification of food products that are contaminated with heavy metals and other harmful substances, toward attaining food safety, are also reported. In line with the advances in microbial biotechnology in recent times, the integration of state-of-the-art techniques such as next-generation sequencing, synthetic biology, and artificial intelligence have been harnessed in transforming the study and applications of microbial communities in agriculture. These technologies have enabled the precise manipulation of microbial ecosystems to enhance soil health, agricultural productivity, and food safety. However, some associated challenges with this development, such as regulatory frameworks, public perception, and farmer adoption, need to be addressed to effectively maximize the benefits of microbial innovations.

Ecofriendly Frontiers: Harnessing Microbial Applications for Food Security aims to bridge the gap between scientific research and real-world implementation. It provides policymakers, academics, researchers, and agricultural professionals with valuable insights into the potential of microbial solutions for a more resilient and sustainable food system. By fostering interdisciplinary collaboration and innovation, this book underscores the transformative role of microorganisms in securing the future of global food production while preserving ecological balance.

Mmabatho, South Africa — Olubukola Oluranti Babalola
Lagos, Nigeria — Olukayode Oladipo Amund
Mmabatho, South Africa — Akinlolu Olalekan Akanmu

Contents

Editors and Contributors

About the Editors

Olubukola Oluranti Babalola (Pr.Sci.Nat, MASSAF, FASLP, FTWAS, FAS, FAAS, FISC), the Vice President of both the Organization for Women in Science for the Developing World (OWSD) and The World Academy of Sciences (TWAS), is an NRF-rated B2 internationally acclaimed scientist with over 21 years of research experience focusing on rhizosphere metagenomics, coupled with an MBA in General Management. Without reservation, she is the Research Director of the Food Security & Safety Focus Area at North-West University (NWU), South Africa, leading as the Head and Principal Investigator of a Microbial Ecology Research laboratory. She is a member of the editorial board for *Microbiology Spectrum*, *Applied and Environmental Microbiology* (American Society for Microbiology, USA), *BMC Microbiology* (Elsevier), *Transgenic Research* (Springer), and *Biochemistry and Biophysics Reports* (Elsevier). She is an International Advisor to F1000Research Plant Science Gateway, Nano-Horizon and Heliyon (Cell Press). She has received many awards, including being the finalist, GenderInSite 2020 and Clarivate highly cited research top 1% for field years 2022 and 2023. Recently, she received the 2024 South African Women in Science Awards (SAWiSA). She joined NWU in 2009; at NWU, she is passionate about capacity building and has graduated 37 doctoral fellows, 25 masters, and numerous Honors students. Several of her PhD products are postdocs or staff in America, Australia, Belgium, Canada, Italy, Sweden, and the UK, also scattered over Africa. She is a prolific author with over 400 publications. She is an AAAS-TWAS Science diplomacy alumnus, impacting the world after the eye-opening awareness of policymaking, diplomacy, and diplomatic engagement. She is taking action using the gender lens and forging collaborations with diplomats. She is a fellow of The Nigerian Academy of Science (NAS), The World Academy of Sciences (TWAS), and the Academy of Science of South Africa (ASSAF) and an honorary fellow of the International Science Council (ISC). Her wealth of international experience spans the Americas, Asia, Europe, and Oceania,

64 countries to be precise. She enjoys international collaborations globally, research grants, and many awards. She has an H-index of 83, with over 32,000 citations. She is #1 in Africa for Soil Science and Plant Nutrition (https://www.adscientificindex.com/scientist.php?id=96341). Her recent Springer-edited book is *Food Security and Safety: Africa's Perspective*. In 2018, she delivered her professorial inaugural lecture.

Olukayode Oladipo Amund is a distinguished microbiologist specializing in Petroleum and Environmental Microbiology. He earned a First Class (Hons.) degree in Biology from the University of Lagos in 1977 and a PhD in Biotechnology from Cranfield University, UK, in 1982. He began his academic career at the University of Lagos in 1978, rising to the rank of Professor in 1998. He has held several leadership roles, including Dean of the Faculty of Science, Head of the Department of Botany and Microbiology, and Director of the Distance Learning Institute. Before his appointment as Director of Distance Learning, he served as the Dean of Student Affairs. Additionally, he was a Member of the Governing Council of the University of Lagos. He was the Foundation Vice-Chancellor of Wesley University of Science and Technology and later served as the third substantive Vice-Chancellor of Elizade University. A Fellow of the Nigerian Academy of Science (FAS), he is also a fellow of the African Academy of Sciences, the Royal Society of Biology (UK), the Institute of Policy Management Development (FIPMD), and the African Scientific Institute (FASI). He remains an active member of several professional societies. With over 100 research articles published in international journals, he has made significant contributions to his field. His research is widely cited, with 36 articles featured on the United States' National Center for Biotechnology Information (NCBI). He has contributed book chapters, presented at numerous conferences, and secured multiple research grants. He has also collaborated extensively with international academics and provided consultancy services to government and industry. He has received numerous awards in recognition of his contributions to science and education. He is happily married with children and grandchildren.

Akinlolu Olalekan Akanmu is a postdoctoral researcher at North-West University, South Africa. He earned his Bachelor of Agriculture degree from the University of Agriculture, Abeokuta, Nigeria, followed by both master's and PhD degrees from the Department of Botany University of Ibadan, Nigeria. His research interest is centered on soil-host-microbe interaction, toward ensuring food security and safety. Akinlolu is an active member of several professional bodies, including the American Phytopathological Society (APS) and International Society for Microbial Ecology (ISME). He has served as a co-guest editor of "the Proceedings of Nigeria Academy of Science," he has also authored and co-authored more than 50 research papers in journal articles published in reputable national and international journals.

Contributors

Ajayi Abraham Molecular Biology and Biotechnology Department, Nigerian Institute of Medical Research Yaba, Lagos, Nigeria

Ismail Abiola Adebayo Department of Medical Biochemistry, Molecular Biology, and Genetics, School of Medicine and Pharmacy, College of Medicine and Health Sciences, University of Rwanda, Butare, Rwanda

Oyetayo Olaoluwa Adefiranye Centre for Bioresources and Nanotechnology, University of Lagos, Akoka, Lagos State, Nigeria

Taofeek Tope Adegboyega Department of Microbiology and Parasitology, School of Medicine and Pharmacy, College of Medicine and Health Sciences, University of Rwanda, Butare, Rwanda

A. O. Adesemoye USDA-ARS National Soil Dynamics Laboratory, Auburn, AL, USA

Rasheed Omotayo Adeyemo Department of Microbiology and Parasitology, School of Medicine and Pharmacy, College of Medicine and Health Sciences, University of Rwanda, Butare, Rwanda

Akinlolu Olalekan Akanmu Food Security and Safety Focus Area, Faculty of Natural and Agricultural Sciences, North-West University, Mmabatho, South Africa

Saheed Adekunle Akinola Department of Microbiology and Parasitology, School of Medicine and Pharmacy, College of Medicine and Health Sciences, University of Rwanda, Butare, Rwanda

Food Security and Safety Focus Area, Faculty of Natural and Agricultural Sciences, North-West University, Mmabatho, South Africa

Nwanneka Marie Akinyemi Centre for Bioresources and Nanotechnology, University of Lagos, Akoka, Lagos State, Nigeria

Adenike Eunice Amoo Department of Biology, Edge Hill University, Ormskirk, UK

Olukayode Oladipo Amund Department of Microbiology, Faculty of Science, University of Lagos, Akoka, Lagos State, Nigeria

Bukola Rhoda Aremu Food Security and Safety Focus Area, Faculty of Natural and Agricultural Sciences, North-West University, Mmabatho, South Africa

Department of Chemistry and Biochemistry, University of Windsor, Windsor, ON, Canada

Anjana J. Atapattu Agronomy Division, Coconut Research Institute, Lunuwila, Sri Lanka

Modupe Stella Ayilara Food Security and Safety Focus Area, Faculty of Natural and Agricultural Sciences, North-West University, Mmabatho, South Africa

Olubukola Oluranti Babalola Department of Life Sciences, Imperial College London, Ascot, Berkshire, UK

Food Security and Safety Focus Area, Faculty of Natural and Agricultural Sciences, North-West University, Mmabatho, South Africa

Muritala Issa Bale Department of Microbiology and Parasitology, School of Medicine and Pharmacy, College of Medicine and Health Sciences, University of Rwanda, Butare, Rwanda

Michael Olubusayo Binuyo Department of Microbiology and Parasitology, School of Medicine and Pharmacy, College of Medicine and Health Sciences, University of Rwanda, Butare, Rwanda

M. Buba Department of Agronomy, Faculty of Agriculture, Federal University of Kashere, Kashere, Gombe State, Nigeria

Chioma Blaise Chikere Department of Microbiology, Faculty of Science, University of Port Harcourt, Port Harcourt, Rivers State, Nigeria

Department of Environmental Sciences, College of Agriculture and Environmental Sciences (CAES), Florida Science Campus, University of South Africa, Florida, South Africa

Oyinade David Department of Plant Science and Biotechnology, Federal University, Oye Ekiti, Ekiti, Nigeria

D. M. N. S. Dissanayaka Agronomy Division, Coconut Research Institute, Lunuwila, Sri Lanka

Israel Edamkue Department of Microbiology, Faculty of Science, University of Port Harcourt, Port Harcourt, Rivers State, Nigeria

Department of Biotechnology, Institute of Parasitology, Faculty of Agriculture and Environmental Sciences, McGill University, Ste-Anne-de-Bellevue, QC, Canada

S. A. Ganiyu Department of Agronomy, Faculty of Agriculture, Federal University of Kashere, Kashere, Gombe State, Nigeria

Smith Stella Ifeanyi Molecular Biology and Biotechnology Department, Nigerian Institute of Medical Research Yaba, Lagos, Nigeria

Rebaona Reaobaka Molefe Food Security and Safety Focus Area, Faculty of Natural and Agricultural Sciences, North-West University, Mmabatho, South Africa

Priyanka More Department of Biology, Edge Hill University, Ormskirk, UK

T. D. Nuwarapaksha Agronomy Division, Coconut Research Institute, Lunuwila, Sri Lanka

O. I. Nwachukwu Department of Soil Science and Land Resources Management, Michael Okpara University of Agriculture, Umudike, Umuahia, Nigeria

Adewale Olusegun Obadina Department of Food Science and Technology, Federal University of Agriculture, Abeokuta, Nigeria

Oluwafemi Sunday Obayori Department of Microbiology, Faculty of Science, Lagos State University, Ojo, Lagos, Nigeria

Chioma Christiana Obi Department of Biological Sciences, Bells University of Technology, Ota, Nigeria

Adeyinka Odebode Department of Basic Science, Faculty of Science and Technology, Kampala International University, Kampala, Uganda

Elijah Ige Ohimain Department of Microbiology, Niger Delta University, Wilberforce Island, Bayelsa State, Nigeria

I. G. Okoro Department of Soil Science and Land Resources Management, Michael Okpara University of Agriculture, Umudike, Umuahia, Nigeria

Oluwaseun Oyetunji Olasunkanmi Department of Microbiology, Faculty of Science, Lagos State University, Ojo, Lagos, Nigeria

Olumayowa Mary Olowe Plant Pathology Unit, Department of Botany, University of Ibadan, Ibadan, Nigeria

Ganiyu Oladunjoye Oyetibo Centre for Bioresources and Nanotechnology, University of Lagos, Akoka, Lagos State, Nigeria

Geomicrobiology and Environmental Biotechnology Group, Department of Microbiology, Faculty of Science, University of Lagos, Akoka, Lagos State, Nigeria

Olubukola Blessing Oyetibo Department of Microbiology, Lagos State University, Ojoo, Lagos State, Nigeria

Olusola Bandele Oyewole Department of Food Science and Technology, Federal University of Agriculture, Abeokuta, Nigeria

Lateef Babatunde Salam Department of Biological Sciences, Microbiology Unit, Elizade University, Ilara-Mokin, Ondo State, Nigeria

Jean Paul Sinumvayo Department of Microbiology and Parasitology, School of Medicine and Pharmacy, College of Medicine and Health Sciences, University of Rwanda, Butare, Rwanda

H. A. Torbert USDA-ARS National Soil Dynamics Laboratory, Auburn, AL, USA

S. S. Udumann Agronomy Division, Coconut Research Institute, Lunuwila, Sri Lanka

Edmond Osi Uguomore Department of Microbiology, Faculty of Science, University of Port Harcourt, Port Harcourt, Rivers State, Nigeria

D. B. Watts USDA-ARS National Soil Dynamics Laboratory, Auburn, AL, USA

Abbreviations

AMF	Arbuscular mycorrhiza fungi
AOB	Ammonia-oxidizing bacteria
BT	*Bacillus thuringiensis*
CCS	Circular consensus sequencing
CRISPR	Clustered Regularly Interspaced Palindromic Repeat
CUE	Microbial Carbon Use Efficiency
DAPG	2,4-Diacetylphloroglucinol
ECM	Ectomycorrhizal
EPNs	Entomopathogenic nematodes
FDA	Food and Drug Administration
GEB	Genetically engineered bacterium
IEA	International Energy Agency
ISR	Induced Systemic Resistance
JA	Jasmonic acid
LAB	Lactic Acid Bacteria
LPs	Cyclic lipopeptides
NGS	Next-generation sequencing
PGPR	Plant Growth-Promoting Rhizobacteria
PSMs	Phosphate Solubilizing Microorganisms
QIIME	Quantitative Insight into Microbial Ecology
SA	Salicylic acid
SDGs	Sustainable Development Goals
SIP	Stable Isotope Probing
SOC	Soil organic carbon
SOM	Soil Organic Matter
SSF	Solid-state fermentation
TGS	Third-generation sequencing era
VOCs	Volatile organic compounds
WGS	Whole genome sequencing

List of Figures

List of Tables

Part I
Microbial Roles in Sustainable Agriculture and Food Systems

Chapter 1
Unraveling the Role of Plant-Microbe Interactions in Sustainable Agriculture

Anjana J. Atapattu, T. D. Nuwarapaksha, D. M. N. S. Dissanayaka, and S. S. Udumann

Abstract Current agricultural systems have led to widespread environmental degradation due to practices heavily reliant on agrochemical inputs. Urgent action is needed to transition to sustainable farming methods that nourish soils, reduce external inputs, and increase productivity. A rapidly emerging solution involves harnessing beneficial plant-microbe partnerships. In natural settings, plants thrive in association with microbial communities that enhance nutrient access, improve disease resilience, and increase stress tolerance. Translating these services to agriculture holds immense potential for reducing environmental impacts. This chapter gives a conclusive overview of the current scientific understanding and future outlooks for optimizing plant-microbe interactions across diverse cropping systems. It describes key mechanisms through which plant growth-promoting microbes enhance productivity by increasing nutrient availability, modulating phytohormones, and inducing systemic resistance to pests and diseases. These mechanisms demonstrate the potential to reduce reliance on agrochemical inputs. The chapter places particular emphasis on management strategies, including microbial inoculant development, plant breeding, and synthetic community design to engineer plant microbiomes for human benefit. Translational examples highlight enhanced yields, stress resilience, and reductions in fertilizer and pesticide needs across various crops, illustrating the potential for more sustainable agriculture. Despite these promising aspects, significant obstacles remain for successful implementation at commercial scales. The future calls for policies, regulations, and ecosystem valuation metrics that account for services provided by healthy plant-soil-microbe interactions. Managing and enhancing beneficial plant-microbe symbioses holds exceptional promise as a foundation for the future of sustainable agriculture globally. Ultimately, the chapter delves into the future directions current state of scientific advancements, the practical applications of translating research findings, real-world potential, and prospects in targeted plant-microbe biofertilization and biocontrol. The chapter's objective is to demonstrate how plant-microbe interactions can foster sustainable agriculture by enhancing plant health, nutrient

A. J. Atapattu (✉) · T. D. Nuwarapaksha · D. M. N. S. Dissanayaka · S. S. Udumann
Agronomy Division, Coconut Research Institute, Lunuwila, Sri Lanka

O. O. Babalola et al. (eds.), *Ecofriendly Frontiers*,
https://doi.org/10.1007/978-3-031-98700-7_1

availability, and resilience while minimizing chemical inputs. To achieve this, it recommends integrating microbial partnerships through strategies such as developing targeted microbial inoculants, breeding plants that support beneficial microbiomes, and managing environmental factors to optimize these interactions. This approach aims to build a sustainable agricultural framework by promoting soil health, conserving natural resources, and addressing global food security challenges in an environmentally conscious manner.

Keywords Agricultural systems · Management strategies · Microbial communities · Nutrient availability · Productivity increase

1.1 Introduction

Conventional agriculture has become the mainstay for decades in the space of food production supported by the massive application of synthetic fertilizers, pesticides, and monoculture practices (Rosset and Altieri 1997). This model in turn has elevated crop yield but with detrimental outcomes on the environment and health. Major problems are, amongst others, soil fertility decrease from excessive fertilization with synthetic ones, biodiversity losses due to overuse of pesticides, low resilience of monoculture systems, water pollution by chemical runoff, and human health issues associated with pesticide exposure and low-nutrient foods (Tripathi et al. 2020). The solutions to these problems will rely on environmental conservation activities that are centered on sustainable agricultural practices that balance the need for food production, environmental conservation, and social welfare. Sustainable agriculture emphasizes preserving soil fertility and health through natural methods such as incorporating organic matter, growing cover crops, and practicing crop rotation (Dissanayaka et al. 2024b; Tahat et al. 2020). Its emphasis lies in sustaining productivity without over-reliance on synthetic or chemical inputs. It promotes diversification by incorporating mixed crops, livestock, and agroforestry components, mirroring natural systems to enhance the ecosystem's resilience against stressors (Dissanayaka et al. 2023c; Udumann et al. 2023). At its core, sustainable practices work to eliminate the use of synthetic fertilizers and pesticides through the application of integrated pest management, organic amendments, and the long relationship between plants and the microbial community that make them (Dissanayaka et al. 2023a; Gupta et al. 2022). This plant-microbe interaction introduces transformative characteristics that can overcome the perennial challenges of organic agriculture.

Plants and microorganisms have developed close connections over millions of years (Rosenberg et al. 2010). Altruistic and mutually compliant interaction, developed through a lengthy time, is the result. The community of microbiomes in the soil and associated with plants is pivotal for several key aspects of sustainable farming practices. Notably, the microbiome enhances soil fertility and enables recycling of nutrients. It stimulates and promotes plant growth through beneficial interactions. The microbiome provides biological control against pests and diseases afflicting crops. Additionally, it bolsters the plant's natural defenses against environmental stresses like drought conditions and high soil salinity. The proper functioning of this microbial ecosystem is central to the ecological principles behind sustainable agriculture (Suman et al. 2022). For example, bacteria facilitate mineral absorption, produce hormones boosting growth, trigger plant protections, inhibit pathogens, and lead to additional resistance. Innovative methods such as biofertilizers, biopesticides, and microbial inoculants formulated based on these positive interactions not only assist the farmers in using agrochemicals less but also allow them to operate sustainable and highly productive agricultural systems. Additionally, there is also the involvement of the plant microbiome in the stability of agricultural ecosystems and intricate biodiversity. Microbiome hence promotes resilience with in-depth research of these refined plant-microbe relationships and their applications in the agricultural sphere, a transformation towards a more sustainable agriculture system that promotes ecological balance, economic stability, and social justice through advances in science is highly likely (Nadarajah and Abdul Rahman 2023).

1.1.1 Current Challenges in Conventional Agriculture

Most agricultural systems have so far relied significantly on synthetic fertilizers, pesticides, as well as monoculture cultivation to produce food for many years (Altieri et al. 2017; Nuwarapaksha et al. 2024). This strategy has greatly added to the produce output and responded to the continually rising global food demand, however, it has also taken a toll on the environment in addition to human health. Another big scope of challenge that modern agriculture faces is the erosion of soil fertility as a result of the poor agricultural policy. Overusing chemical fertilizers can deplete soil organic matter and harm soil microbes, which are needed for soil health and nutrient cycling (Srivastava 2021). This has consequently elevated the demand for massive amounts of fertilizers to keep the same yield levels thereat creating a vicious spiral of soil accrued depreciation and dependency on external inputs. Also, there are great consequences of the run-of-the-mill of synthetic pesticides on the flora and fauna that are in those areas. The blind use of chemicals by farmers has led to the destruction of beneficial insects, a pollinator, and a predator for pests (Oberemok et al. 2023). The rise of pest resistance is another factor driving unsustainable agricultural practices. The threat of crops being damaged by resistant pests has led farmers to reduce crop diversity and rotate crops less frequently. This has necessitated using stronger and more harmful pesticide chemicals to combat the resistant pests. However, the increased use of these potent chemical pesticides has further degraded the environment and posed health risks to people living in agricultural regions. The vicious cycle of pest resistance followed by stronger pesticide use continues to harm the environment and human health.

Monoculture farming has an undeniable role as well, where a single crop is farmed over significant areas, to the challenges typically encountered in conventional agriculture (Sarkar et al. 2012; Udumann et al. 2024). Imposing the monoculture of crops leads to inadequate crop diversity, which agricultural systems more vulnerable to threats such as pests, diseases, and the effects of climate change, and natural resource utilization is very low (Dissanayaka et al. 2023a; Nuwarapaksha et al. 2022). Moreover, monoculture practices have resulted in a degeneration of agrobiodiversity, which runs in parallel with a lack of traditional crop products and the correlated loss of traditional ecological knowledge (Thrupp 2000). The environmental damages associated with conventional agricultural production are not limited to farming activity but extend these impacts also to the larger ecosystem. Due to rainfall or snowmelt, synthetic phosphates and pesticides make their way into rivers and lakes, thus causing eutrophication of aquatic ecosystems and consequentially poor water quality and enhanced biodiversity loss. Apart from this, the greenhouse gas emissions arising due to the production and usage of artificial fertilizers, in addition to energy-consuming agriculture means like modern farming which releases a lot of greenhouse gas into the atmosphere are great contributors to climate change (Kabange et al. 2023). Besides mentioned above, it is noted that conventional agriculture has also been correlated with some human health concerns. Residues of pesticides are responsible for various forms of diseases, among which are endocrine disruptors, neurological disorders, and diseases related to an increased risk of 2eveloping cancer (Sabarwal et al. 2018).

1.1.2 The Need for Sustainable Agriculture Practices

Considering that today much viability of conventional agriculture practices is questioned, the transition to sustainable agriculture become the priority. The way of agriculture that is sustainable is designed for food production and security but without the destruction of the natural ecosystems and the wellness of people and all living things (Viana et al. 2022). Maintaining healthy soil is vital for sustainable agriculture. Instead of solely relying on chemical inputs, emphasis should be placed on improving and preserving soil fertility through the combined use of natural and chemical methods. This includes incorporating organic matter into the soil, planting cover crops, and practicing crop rotation. By adopting these practices, the overall fertility and productivity of the soil can be enhanced while ensuring long-term sustainability (Dissanayaka et al. 2022; Nuwarapaksha et al. 2023b; Tully and McAskill 2020). This also aids in restoring the natural nutrient cycling processes and reduces plant diseases that affect the nutrient uptake system. Moreover, it increases the biodiversity and abundance of beneficial soil microbes that are vital in crop disease management and increases agricultural productivity. In addition to diversification, stable agriculture is another important part. Rather than joining what is known as monoculture farming practices, components of sustainable agriculture systems involve planting a wider range of crops, crop rotations, organic farming, livestock,

Fig. 1.1 Components of sustainable agriculture

intercropping, and ideas of agroforestry elements (Fig. 1.1). The diversity mimics natural ecosystems, enhancing agriculture's resilience by making the system robust against pests, diseases, and climate change impacts (Swift et al. 2004). Moreover, this combination promotes efficient nutrient cycling and reduces harmful inputs, benefiting crops, livestock, and the environment.

The sustainable agriculture practice also involves the reduction of other synthetic inputs, like fertilizer and pesticides (Wezel et al. 2014). Integrated pest management (IPM), cover cropping, and organic amendments used by sustainable agriculture will lead to multiple chemical dependence reductions as well as raise the productivity of efficient and useful farmlands. Moreover, the conservation process will not only have positive impacts on the environment but also decrease the possibility of the recurrence of farm workers' diseases. Also, even the chance for consumers to acquire illnesses caused by toxic substances released by chemical fertilizers will be reduced. Added to this is the notion of sustainable agriculture that gives preference to the preservation of traditional environmental knowledge as well as agrobiodiversity keeping (Negi and Maikhuri 2013). Traditional farming systems can be enhanced by incorporating sustainable practices and by keeping related varieties of crops. Transitioning to sustainable agriculture is not only beneficial for the environment but also has the potential these positive implications concerning human health conditions and approved (Atapattu et al. 2024b; Wezel et al. 2020).

1.1.3 The Potential of Plant-Microbe Interactions in Sustainable Agriculture

Sustainable agriculture hinges on the intricate, dynamic interactions between plants and their highly diverse microbial environments (Eviner and Hawkes 2008). We discover these complex plant-microbe relationships offering us the possibility of a fundamental change to the faced agriculture problems and building the ground for a more sustainable prosperity. Through the constant interaction with plants and microbes which took ages of co-evolution, the latter developed complex and mutually beneficial relations. Microorganisms, for example, bacteria, fungi, and archaea, populate the plant's root system, thus the symbiotic associations are among the most crucial factors for the plants' growth, healthy development, and resistance (Odelade and Babalola 2019). The microbe containing symbionts that reside in plant tissues is the subject matter known as plant microbiome, and this group has important functions that benefit the agriculturally sustainable developments. One of the most notable functions of the plant microbiome is that it works in favor of soil nutrient cycle and fertility (Dubey et al. 2019). Cultivation of microorganisms like the nitrogen-fixing bacteria and phosphate-dependent fungi would prove instrumental in reducing the demand for man-made fertilizers (Dissanayaka et al. 2024a; Jacoby et al. 2017). Additionally, growth compounds from these microbiomes can boost plant growth, development, and yields without the negative environmental impacts of conventional fertilizers. Plants harboring beneficial microbes gain effective defenses against pests and diseases (Syed Ab Rahman et al. 2018). Some microorganisms induce plant's innate defense mechanisms, while others antagonize or outcompete pathogens and pests directly. Biological control involves harnessing these living organisms to combat plant diseases and pests, ultimately decreasing reliance on synthetic pesticides. This approach safeguards the environment, promoting more sustainable and eco-friendly agriculture (Ayilara et al. 2023).

Additionally, the plant microbiome can enhance crop resilience by mitigating abiotic stresses like drought, salinity, and high temperatures. Certain microbes provide plant tolerance capabilities, such as improved water and mineral uptake or synthesis of protective compounds against stress damage (Kumar and Verma 2018). Researchers, scientists, and microbiologists are keen on the role microbiome plays in the environment, mainly, the ability for these microbes to perform necessary functions, giving rise to a rapid interest and development in microbial-based products and approaches for sustainable agriculture. Biological control agents, microbial stimulants, and organic inoculants such as biofertilizers, and biopesticides are just a few of the newly-developed approaches that exploit the power of plant-microbe relationships for sustainable and productive agriculture (Pirttilä et al. 2021). By supporting microbial abundance and diversity, sustainable farming practices promote soil function, natural balance, and overall agroecosystem resilience (Bhaduri et al. 2022). As our brains become more meaningful in the ways that plants and other microbes communicate, it becomes apparent that the applications for these relationships will continue to be discovered as we deepen our understanding of how plants associate with microbiomes. This approach, through plant-microbe interactions, chips in to bring about a future for agriculture that goes beyond environmentalism, can be self-sustaining economically and sees all stakeholders provided for. This chapter gives an inclusive overview of the current scientific understanding and future outlooks for optimizing plant-microbe interactions across diverse cropping systems. It also, highlights the potential of harnessing beneficial plant-microbe interactions to reduce reliance on agrochemical inputs and promote sustainable agriculture practices that nourish soils, increase productivity, and reduce environmental impacts.

1.2 Plant Microbe Interactions in Natural Ecosystems

Interactions between plants and microbes are crucial in natural ecosystems, as they play a key role in nutrient cycling and enhancing total ecosystem productivity (Aqeel et al. 2023). The equilibrium of the systems and the biodiversity within them is also supported by the plant-microbe linkages. The microbes are the most crucial players in the transit of nutrients from inorganic to organic being responsible for a wide range of chemical processes in the biogeochemical culmination (Gachara et al. 2024). Meanwhile, within sustainable agronomy, these associations afford farmers with chances to improve plant health, and reduce chemical applications while at the same time developing environment-friendly strategies to control pests and diseases. Beneficial root bacteria and mycorrhizal fungi can trigger systemic resistance in plants, activating defense pathways and generating antimicrobial compounds to combat pathogens and pests (Alamoudi 2023). Biological control agents possess the ability not only to combat plant pathogens but also to act as antibiotic producers, compete for resources, and parasitize pathogens themselves. Plant-microbe interactions are not limited only by studying the impacts on ease of digestibility, but also on stresses of abiotic conditions like drought, salinity, and nutrient deficiencies (Hossain et al. 2022). Microorganisms can release phytohormones that control stresses and the formation of available nitrogen and nutrients. Besides they can make the competitors weak and resistant against diseases. These useful plant-microbe circumstance holds the strong prospect for creating green practices of agriculture culture, preserving ecosystem health, and upgrading crop productivity while leaving behind the use of easily destructible methods of chemical inputs and reducing environmental damages (Dewi et al. 2023).

1.2.1 Ecosystem Nutrient Cycling

Nutrient cycling is an essential process that facilitates the movement of nutrients throughout the ecosystem, ensuring their availability across various parts of the ecosystem to sustain life. The plant-microbe coexistence is a crucial aspect of the success of natural systems and nutrient recycling process, respectively (Ma et al. 2016). This incorporation allows for the cycling of nitrogen, carbon, phosphorus, and other vital elements among abiotic components thus they remain available for support of productivity and biodiversity of ecosystems. In maintaining such intricate relationships, natural ecosystems can be carried along a sustainable way and be supportive of a wide range of life options (Wu 2013). Knowing and preserving these tenderminded plant-microbe interactions is essential for ecosystem management, conservation, and the general realization of a happy place for the ecosystems of our rather unique planet.

1.2.1.1 Plant-Microbe Interactions for the Nitrogen Cycle

The nitrogen cycle is an intricate biogeochemical process that regulates the movement of nitrogen, a fundamental element for life, through different reservoirs within the environment (Zerkle and Mikhail 2017). The nitrogen cycle comprises various crucial stages facilitated by diverse microbial communities (Fig. 1.2). Nitrogen fixation, conducted by diazotrophs such as rhizobia and cyanobacteria, transforms atmospheric nitrogen (N_2) into ammonia (NH_3), a form accessible to living organisms. This fixed nitrogen then undergoes nitrification, a two-step process where ammonia is converted to nitrites (NO_2^-) and then to nitrates (NO_3^-) by nitrifying bacteria and archaea (Cáceres et al. 2018). Nitrates can be readily taken up by plants. However, excessive nitrification can lead to nitrogen loss through leaching and denitrification. Denitrification, performed by facultative anaerobic bacteria like *Pseudomonas* and *Bacillus*, converts nitrates back to gaseous forms of nitrogen, such as nitrous oxide (N_2O) and N_2, completing the cycle by returning nitrogen to the atmosphere (Aryal et al. 2022). Ammonification, another critical process, involves the breakdown of organic nitrogen compounds by heterotrophic bacteria and fungi, releasing ammonia back into the soil for recycling (Grzyb et al. 2021).

Nitrogen Fixation

Nitrogen fixation is the process of converting atmospheric N_2 into NH_3 or other compounds that can be readily utilized by living organisms (Mansingh et al. 2021). This is the entry point of the nitrogen cycle and is carried out by a diverse group of microorganisms known as diazotrophs. The major microbial players in nitrogen fixation include symbiotic bacteria (*Rhizobium, Bradyrhizobium*, and *Sinorhizobium* species) that form a mutualistic relationship with leguminous plants, residing in root

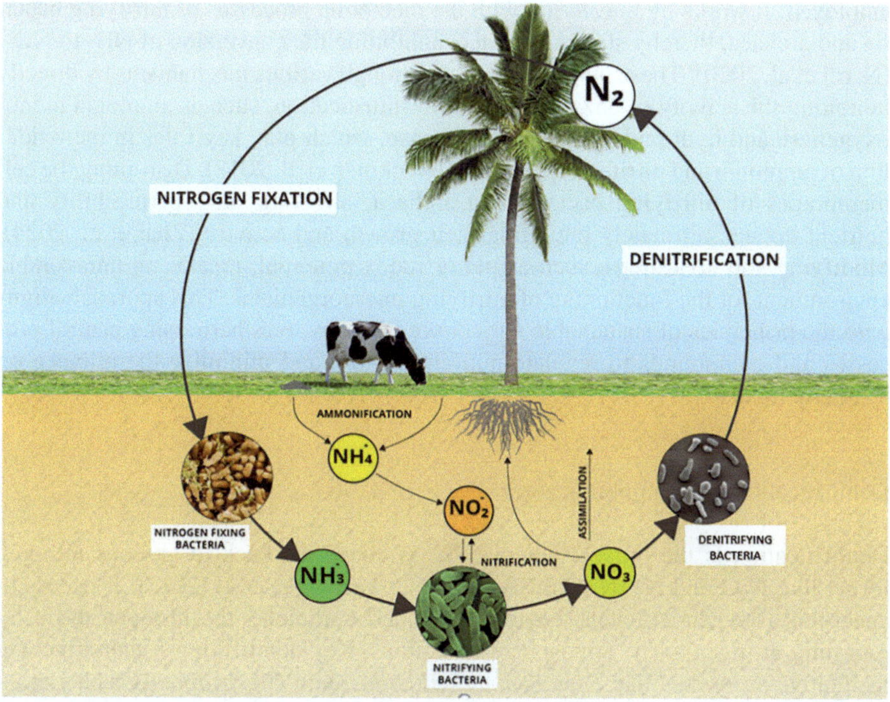

Fig. 1.2 Steps of nitrogen cycle (Source: Figures by authors)

nodules and fixing nitrogen in exchange for carbohydrates from the host plant (Coba de la Peña et al. 2018; Nuwarapaksha et al. 2023a; Udumann et al. 2023). Free-living bacteria *Azotobacter, Clostridium*, and *Klebsiella* species are free-living diazotrophs found in various terrestrial and aquatic environments (Khan et al. 2021). *Cyanobacteria, Anabaena, Nostoc*, and *Trichodesmium* are examples of cyanobacteria that can fix nitrogen through specialized heterocyst cells (Fujita and Uesaka 2022). These diazotrophs possess the nitrogenase enzyme complex, which catalyzes the energy-intensive process of breaking the triple bond of the N_2 molecule and converting it into ammonia (Gu and Milton 2020).

Nitrification

Nitrification is a two-step process in the soil where NH_3 undergoes transformation. In the first step, ammonia-oxidizing bacteria and archaea convert NH_3 to NO_2^- (Mohanty et al. 2018). Subsequently, nitrite-oxidizing bacteria further convert NO_2^- into NO_3^-, which is a form of nitrogen easily taken up by plants. However, excessive nitrification can lead to nitrogen loss through leaching and denitrification, resulting in issues like groundwater contamination and greenhouse gas emissions. To mitigate these problems, a process called "biological nitrification inhibition" can be employed. It works by interfering with the metabolic processes of nitrifying bacteria and archaea, thereby slowing down or inhibiting the conversion of NH_3 to NO_3- (Nardi et al. 2020). These inhibitors can act through various mechanisms by directly inhibiting the activity of enzymes crucial for nitrification, such as ammonia monooxygenase and hydroxylamine oxidoreductase, which play key roles in the oxidation of ammonia and nitrite, respectively (Beeckman et al. 2018). Disrupting the cell membranes of nitrifying bacteria and archaea, affecting their permeability and nutrient uptake, ultimately inhibiting their growth and activity (Zhai et al. 2024). Modifying soil conditions, such as pH or redox potential, creates an unfavorable environment for the functioning of nitrifying microorganisms. This approach aligns with the principles of sustainable agriculture, as it involves harnessing natural processes and compounds to regulate nutrient cycling and minimize the reliance on synthetic inputs.

Denitrification and Ammonification

Denitrification is the process where NO_3^- is converted back to gaseous nitrogen forms like N_2O and N_2 by denitrifying bacteria (Saggar et al. 2013). It occurs in anaerobic (oxygen-deficient) environments and completes the nitrogen cycle by returning nitrogen to the atmosphere conditions. Key denitrifying bacteria include facultative anaerobes like *Pseudomonas, Bacillus,* and *Clostridium* (Pashaei et al. 2022). Ammonification is another crucial process where organic nitrogen compounds (e.g. proteins, nucleic acids) are broken down by heterotrophic bacteria and fungi, releasing NH_3 as a byproduct (Deb et al. 2024) This process recycles organic nitrogen back into the soil nitrogen pool.

1.2.1.2 Plant-Microbe Interactions for Phosphorus Cycle

While phosphorus is essential for plant growth as a macro element, a significant portion of it exists in the soil in forms that are immobile and inaccessible to plants. In typical soils a meager 0.1% of total soil phosphorus is present in the solution as useful ions namely $H_2PO_4^-$ and HPO_4^{2-} (Ibrahim et al. 2022). With that in mind, soil-borne bacteria known as phosphate-solubilizing microorganisms can dissolve and make available phosphorus sources that are soluble, therefore benefiting plants in the process. Formation of this group is caused by photosynthetic activity of bacteria and fungi that synthesize nutritional products for plants (1–50% of soil microbial biomass) and lower amounts of these products (0.1–0.5%) (Paśmionka et al. 2021). These microbes, through their solubilization of insoluble phosphates, make available phosphorus in the form of a liquid nutrient, which is more easily absorbed by roots for the better development of the plants and higher yields. Moreover, they precipitate the excess of phosphorus, thus limiting the chances of eutrophication which is also enhancing the bioremediation (Wang and Jiang 2016). Proper use of phosphatase-producing microorganisms is consistent with sustainable agriculture because it doesn't require large applications of chemical phosphorus fertilizers which often step on environmental aspects and degrade resources.

Phosphate-Solubilizing Bacteria

Numerous bacterial genera, including *Bacillus, Pseudomonas, Enterobacter, Rhizobium,* and *Burkholderia*, are known for their efficiency in solubilizing phosphates. Additionally, *Rhodococcus, Arthrobacter, Serratia, Chryseobacterium, Xanthomonas, Klebsiella, Agrobacterium, Azotobacter, Erwinia, Kushneria,* and *Pantoea* have also been identified as potential bacteria for this purpose (Kumar and Shastri 2017). These bacteria can solubilize phosphorus through various mechanisms producing organic acids like gluconic, citric, oxalic, and lactic acids, which can chelate and solubilize insoluble phosphorus compounds. Producing enzymes such as phosphatases and phytases, which can hydrolyze organic phosphorus compounds and release inorganic phosphates and producing protons (H^+) or CO_2, to reduce the pH in the root zone, increasing the solubility of phosphorus compounds (Rawat et al. 2021).

Phosphate-Solubilizing Fungi

Certain fungal genera, such as *Aspergillus, Penicillium, Acremonium, Hymenella, Fusarium*, and *Neosartorya*, are known for their ability to solubilize phosphorus (Rawat et al. 2021). Similar to phosphate-solubilizing bacteria, these fungi can solubilize phosphorus through mechanisms like producing organic acids that chelate and solubilize insoluble phosphorus compounds, producing enzymes that hydrolyze organic phosphorus compounds and release inorganic phosphates, and producing protons or CO_2 to reduce the pH in the root zone, thereby increasing the solubility of phosphorus compounds (Kalayu 2019; Rawat et al. 2021).

Arbuscular Mycorrhizal Fungi

These fungi represent a distinct strain, highly evolved to exclusively collaborate with plant roots in a symbiotic relationship. It's estimated that approximately 80% of plants are engaged in mycorrhizal associations (Hirsch et al. 2001). These fungi establish a symbiotic relationship not only with phosphate-solubilizing bacteria but also with the plant host, enabling them to acquire both organic and inorganic phosphorus from the soil, alongside the plant. As a result, members of this soil fungi group's hyphae encourage the colonization of the rhizosphere by phosphate-solubilizing microorganisms by securing access to the plant's nutrient flux, and producing carbohydrates, while, at the same time, helping the acquisition of microelements as well as the formation of a proper soil structure. Additionally, this biomass can amplify the number and type of bacterial community within a community and also the nitrogen-fixing and the recycling process of the nutrient absorbed by plants to attenuate the nutrients leaching (Ren et al. 2020).

1.2.1.3 Plant-Microbe Interactions for Carbon Cycle

Microbial interactions are the main drivers of the multiple processes of the carbon cycle which is the very complicated mechanism controlling the movement and transformation of the molecular compounds of carbon from one reservoir of the Earth to the other (Schimel and Schaeffer 2012). The carbon cycle consists of a communication between the atmosphere, terrestrial biosphere, oceans, and the sediments (Mackenzie et al. 2004). Bacteria, archaea, fungi, and protists are vital microbes facilitating carbon cycling through multi-step reactions driving this cyclical flow.

Photosynthesis and Carbon Fixation

Process of photosynthesis, primary producers use the energy from sunlight to capture the CO_2 from the atmosphere and turn it into organic compounds that they need to live and grow (Anemaet et al. 2010). Oxygenic photosynthesis, which is facilitated by microorganisms such as cyanobacteria, is supposed to be the process that ushered in the presence of oxygen on earth. Cyanobacteria, also known as blue-green algae, are arguably one of the most ancient and widely distributed groups of phototrophic microorganisms (Sciuto and Moro 2015). These possess specific structures known as carboxysomes which compartmentalize the carbon dioxide and lead to the enhancement of the carbon fixation process. Bacteria, like sulfur-oxidizing and methane-oxidizing bacteria, further supplement the carbon fixation ability of the ecosystem (Paul et al. 2017). They do this by converting inorganic carbon sources into living substances These microbes utilize the energy that is obtained from the oxidation of those reduced inorganic compounds, e.g. hydrogen sulfide or methane, to perform the fixation of carbon. Some chemosynthetic bacteria, known as Calvin cycle autotrophs, employed the identical biochemical mechanism as both photosynthetic organisms and plants to fix carbon. The Calvin-Benson-Bassham cycle, converting inorganic carbon to organic compounds for cellular processes like growth and maintenance, is a central carbon cycle process (Wang et al. 2022).

Decomposition and Respiration

Aerobic heterotrophic microbes like bacteria and fungi are key players, breaking down organic matter and releasing CO_2 through respiration, driving the carbon cycle (Plante 2007). These actors discharge some enzymes, herewith cellulases, proteases, and lignin peroxidases, that allow them to digest complex carbohydrates like cellulose, proteins, and lignin, and obtain simple forms of carbon to be used in their processes. Microbe respiration, as well, is a great contributor to the carbon cycle since, along with CO_2 utilized by plants during photosynthesis, it completes the cycle (Raza et al. 2023). Environmental factors like temperature, moisture, and nutrient availability govern the rate of microbial decomposition of organic matter. Altering the mentioned parameters results in slowing or speeding up the procedure of decomposition, thus leading to changing the general landscape of the carbon cycle. As far as fungi are concerned, their role is vital when it comes to the decomposition of extremely recalcitrant organic matter such as lignin and humic substances, due to being capable of using many oxidative enzymes (Tunlid et al. 2016).

Methanogenesis and Methane Oxidation

Methanogenic archaea, found in anaerobic environments like wetlands, rice paddies, and the digestive systems of ruminants, are responsible for methane (CH_4) production as part of their metabolic processes (Yuan et al. 2019). The archaea are anaerobic stricteners and they belong to the Archaea domain, which differs from bacteria. Their energy source is inorganic compounds such as acetate or hydrogen and carbon dioxide, the by-product is methane. Methane is a strong greenhouse gas that is released by methanogen during the C-cycle in the world (Dean et al. 2018). Wetlands, especially, tend to be major hot spots for methane emission because there is a group of methanogenic archaea in these wet areas. On the bright side, methanotrophic bacteria can oxidise methane using it to convert the methane back to CO_2 and water, thus taming the methane release effectively (Sahoo et al. 2021). Methanotrophic bacteria encompass diverse species that can solely utilize methane oxidation for their carbon and energy needs. Bacteria possess enzyme specialties such as methane monooxygenases which are responsible for the oxidation of methane. The imbalance of methanogenesis and methane oxidation plays a vital role in regulating atmospheric methane levels and carries great implications for the global carbon cycle and climate change (Schaefer 2019).

1.2.2 Plant-Microbe Interactions for Pest and Disease Resistant

Plant-microbe interactions have become a focal point in sustainable agriculture, with experts and growers collaborating to redefine safe and resilient strategies for pest and disease management (Thakur et al. 2023). These images offer an opportunity to promote plant health, decrease chemical inputs, and thereby contribute to sustainability efforts. In this comprehensive review, we will describe in detail the relationships of plants and microbes, their roles in insect and disease tolerance, and the potential applications of these practices in sustainable agriculture.

1.2.2.1 Plant Growth-Promoting Rhizobacteria (PGPR)

Plant Growth-Promoting Rhizobacteria (PGPR) comprise a group of bacterial strains that inhabit the rhizosphere, the soil area near plant roots (Santoyo et al. 2021a, b). These bacteria possess the ability to confer resistance against various categories of disease-causing pathogens or invasive insects in plants through a mechanism known as Induced Systemic Resistance (ISR). During this process, genetic regulation is activated, triggering the synthesis of antimicrobial substances that spread throughout the plant, providing it with a natural form of broad-spectrum antimicrobial protection. Inoculation with beneficial microbes primarily operates through the jasmonic acid and ethylene signaling pathways, resulting in the synthesis of phytoalexins, pathogenesis-related proteins, and other natural defense compounds (Khoshru et al. 2023). In their competition with pathogens and pests for nutrients and agendas, PGPRs are often outwitted and thus become ineffective in an area where the pests are pests. Moreover, certain PGPRs can secrete antibiotics, siderophores (which might well be characterized as iron-chelating compounds), as well as other metabolites adding to the broad-spectrum pathogen and pest control. In addition, PGPRs affect the hydrolase activity of plant hormones like auxins, cytokines, and gibberellins whose levels are responsible for plant growth and improvement of its vigor and resistance against biotic stress factors (Ivo Ganchev 2021).

1.2.2.2 Mycorrhizal Fungi

It is a fungus among the most common always prefers association with root-based plants and hence, is often found in them. By nature, these microorganisms do not only produce antibiotics against unwanted pests or plant diseases that destroy plants, but the microbes play other useful roles such as improving plants' abilities to acquire water and nutrients ever better (Timmis and Ramos 2021). Plants with enhanced health can potentially be more resistant to biotic stresses since they can be able to do more with their available resources to use these for defensive mechanisms and these can recover quickly from any pathogen or pest attacks. The mycorrhizal colonization can be a trigger that fastens the systemic resistance ISR in plants which brings out the defense-associated genes and the production of anti-microbial compounds mainly mediated by jasmonic acid and ethylene signaling paths (Yu et al. 2022). Furthermore, mycorrhizal fungi can participate in vegetation structure creation through the formation of physical boundaries like thicker cell walls and higher lignification which reduce pathogen penetration and that of the insects on the plants. On the other hand, some types of fungi forming mycorrhizae also synthesize antimicrobial substances that can suppress the microbial organisms and dissuade the pests.

1.2.2.3 Biological Control Agents

Certain microbes can play the role of biological control agents, they demonstrate plant pathogens and pests directly by antagonism and suppression (Lahlali et al. 2022). These microbial species could be versatile in the fact that some are known to produce hormones that will hamper the growth and development of plant pathogens by interrupting the cell membrane integrity, intrusion of metabolic processes, or cell suicide. Such mycopathogenic fungi like *Trichoderma* species can parasitize other fungi including plant pathogenic pathogens and therefore, the only enzymes that they can excrete are chitinases and glucanases (Dutta et al. 2023). In addition, they can share the nutrient resources, space, and habitat with pathogens and pests, thus, preventing their proliferation. Additionally, the biological control agents may elicit ISR and Systemic Acquired Resistance (SAR) in plants through which the defenses are activated, the compounds adjust, and plants respond to environmental threats.

1.2.2.4 Modulation of Plant Defense Pathways

The microorganisms can affect the plants' actions such as a defense pathway affects the expression of defense-related genes and lead to the production of defensive substances. The salicylic acid (SA) signal pathway is particularly significant in bringing plants to the resistance state against biotrophic fungal, bacterial, and viral pathogens as described by the increased accumulation of pathogenesis-related (PR) proteins, phytoalexins, PR proteins, other antimicrobial compounds (Klessig et al. 2018). The jasmonic acid (JA) pathway is mainly tailored for plants defending against necrotrophic pathogens and insect predators, where the plant makes protease inhibitors volatiles, and other compounds that prevent feeding by herbivorous insects and stop the growth of the pathogens ET pathway links to SA and JA pathways and gives defense system responsibility that actives defensive genes and their products like phytoalexins (Schweiger et al. 2014). The result is the fortification of physical barriers like cell wall thickening. Reactive oxygen species (ROS) are signaling molecules in plant's defensive responses that trigger genes related to defense. Some microbes can induce or regulate this ROS signaling, thus, leading to defense gene expression, the production of antimicrobial compounds, and the hypersensitive response (localized cell death) to suppress microbial spread (Huang et al. 2019).

1.2.3 Plant-Microbe Interactions for Stress Tolerance

Plants engage in complex interactions with microorganisms, which can enhance their ability to withstand various stresses, forming the foundation for sustainable agriculture. This relationship explicitly involves the misty tie between plants and a prominent clot of microbes and therefore, it plays a key role in plants' growth and development in the face of either biotic or abiotic factors. These subtle diseases on plants caused by interactions between plants and microbes may encourage green techniques in farming (Solomon et al. 2024). The understanding and employing of these interactions are key factors of sustainable agriculture, which heavily decreases input use, increases the productivity of crops, and ensures environmental safety. Furthermore, plants can experience stress from the accumulation of heavy metals in their tissues, but the presence of beneficial soil microbes can help mitigate this stress by enhancing plant tolerance and promoting phytoremediation processes. Beneficial microbes enhance plant tolerance and facilitate phytoremediation by employing various mechanisms that reduce the toxicity of heavy metals in the soil. Many microbes, such as Acinetobacter sp. and Bacillus subtilis, have high sorption efficiencies, allowing them to absorb and accumulate heavy metals, thereby decreasing the concentration of free metal ions available to plants (Pradhan et al. 2017). Additionally, some microbes, like Pseudomonas aeruginosa and Aspergillus sp., produce enzymes that transform heavy metals into less toxic forms, making them easier for plants to uptake and sequester safely (Yaashikaa et al. 2022). Certain microbes also produce siderophores, molecules that bind to heavy metals and prevent them from interfering with plant cellular functions, enhancing plant growth in contaminated soils. Furthermore, microbes like Bacillus subtilis form biofilms around plant roots, immobilizing heavy metals and aiding in phytostabilization (Sharma 2022). Many of these microbes also generate antioxidants that help plants cope with oxidative stress caused by heavy metal exposure, increasing their resilience in polluted environments. Saccharomyces cerevisiae helps plants tolerate chromium stress by mitigating oxidative damage (Li et al. 2019). Through these combined actions, these beneficial microbes play a critical role in reducing metal toxicity and boosting plant tolerance, making them invaluable partners in phytoremediation efforts (Table 1.1).

1.2.3.1 Phytohormone Production and Modulation

Phytochemicals are responsible for signaling required for the regulation of plant growth, development, and stress responses. The group of Plant Growth Promoting Microorganisms (PGPM) that produce or regulate levels of plant hormones are auxins, cytokinins, gibberellins, abscisic acid (ABA), ethylene, and jasmonate (Akhtar et al. 2022). Such phytohormone regulation can provide the plant with a significant boost to not only cope but also thrive even in hostile environments. Under the banner of abiotic stresses, plant ABA is a very essential hormone that is mainly responsible for drought and salinity stresses (Abhinandan et al. 2018). These microbes are referred to as ABA producers and could help the plants to tailor to the stresses by controlling the stomovas, water uptake, and stress gene process. The fungus Piriformospora indica has been proven to induce more ABA content and responsive genes by activating them in Arabidopsis thaliana, which directly contributes to its increased drought tolerance (Sherameti et al. 2008). Decrease in auxins and cytokinins concentration has their main function of root architecture control, thus they enable plants to discover and access large soil territories under drought stress

Table 1.1 Beneficial Microbes Involve in Phytoremediation

Heavy metal	Micro-organism	Average sorption efficiency (%)
Chromium (Cr)	*Acinetobacter* sp.	87
	Sporosarcina saromensis	82.5
	Bacillus cereeus	81
	Bacillus circulans	96
	Bacillus cereus	78
	Bacillus subtilis	99.6
	Pseudomonas aeruginosa	72
	Aspergillus sp.	92
	Saccharomyces cerevisiae	95
Lead (Pb)	*Cellulosimicrobium* sp.	99.3–84.6
	Methylobacterium organophilum	62.28
	Bacillus firmus	98.3
	Staphylococcus	82.6
Mercery (Hg)	*Enterobacter cloacae*	28.6
	Klebsiellap neumoniae	29.8
	Pseudomonas aeruginosa	90
	Bacillus licheniformis	70
	Candidapara psilosis	80
Copper (Cu)	*Micrococcussp*	55
	Desulfovibrio desulfuricans	90.3–90.1
	Bacillu sfirmus	74.9

Source: Adopted from Medfu Tarekegn et al. (2020)

conditions in search of water and nutrients. This bacterium reduces the effects of drought in tomatoes by the regulation of auxin and cytokinin, which results in increased the ability of roots to develop. Ethylene is responsible for reaction processes in plant responses, but having high concentrations can be toxic (Iqbal et al. 2013). Some of the bugs may contain a special enzyme called ACC deaminase that drives the breakdown of ACC which is the parent of ethylene molecules. The by-product of this process mitigates the ethylene on the cranium during high-stress situations.

1.2.3.2 Nutrient Acquisition and Solubilization

The availability of nutrients and the process of uptake of nutrients are major factors that contribute to plant growth and stress tolerance. Some bacteria can synergistically be involved in nutrient scavenging and solubilization thus helping overcome nutrient deficiencies or stresses that are often the worsening factor (Hamid et al. 2021). Another specialist group was the rhizobacteria and mycorrhizal fungi. They excreted some organic acids and activated some phosphatase and other enzymes, making the transformation of insoluble phosphate into soluble phosphate which was taken up by the root of plants. Improving phosphorus utilization can indeed enhance plant growth and increase stress tolerance, particularly under conditions where phosphorus availability is limited (Vance et al. 2003). For N_2 fixations, diazotrophs like *Azospirillum, Azotobacter,* and *Rhizobium* species will do the job and feed the plant with complete plant available nitrogen source which will boost the plant growth and improvement of plants to a stronger ability to fight the stress (Dobbelaere et al. 2003). Others can generate siderophores, siderophore-like molecules that bind iron—an important component in stress response mechanisms. They can solubilize and transport iron making it dilute and hence, available to plants that are incapable of coping with conditions of iron deficiency and other stress conditions.

1.2.3.3 Induction of Systemic Resistance

The advantageous microorganisms result in the term of the increased systemic resistance (ISR) of the plants arousing their protective implement of all the stresses (biotic and abiotic). ISR is a state of physiological mobilization of complex defense capabilities that could be functionally contributed by an array of genes and pathways under activation (Borah et al. 2023). At this time, microbes bestow certain signaling paths on the plants, and the plants, correspondingly, begin accruing pathogenesis-related proteins, phytoalexins, and other defense compounds. Such antifungal, antibacterial, and abiotic stress tolerance compounds are universally present and inherit diverse adaptive attributes including drought, salinity, and temperature resistance. Signaling through ISR is indeed complex, akin to the Mitogen-Activated Protein Kinase (MAPK) pathways. It involves the activation of transcription factors and the production of signaling molecules such as salicylic acid, jasmonic acid, and ethylene (Bukhat et al. 2020). The release of signaling molecules triggered by the initial stress condition primes the cell to activate its defense mechanisms, leading to a faster and more effective response to subsequent stress events. Studies have demonstrated that *Pseudomonas fluorescens*, a species of rhizobacterium, induces ISR in Arabidopsis thaliana, offering protection against fungal pathogens and enhancing drought tolerance. Similarly, the fungus *Trichoderma virens* has been shown to elicit an ISR response in crop plants, improving their antibiotic activity and enhancing tolerance to both salinity and drought (Fazeli-Nasab et al. 2022).

1.2.3.4 Biofilm formation and Stress Tolerance

Certain plant-borne microbes form biofilms, complex columnette-like bacterial communities covered with an extracellular polysaccharide matrix. Figure 1.3 illustrates the beneficial relationship between plant roots and soil microorganisms.

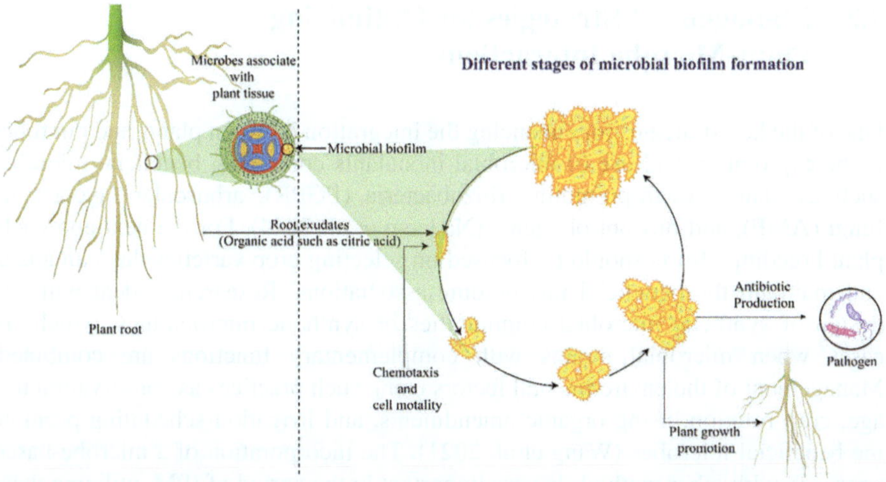

Fig. 1.3 Role of microbial biofilm on plant growth (Source: Figures by authors)

When plant roots release organic compounds (exudates) like citric acid into the soil, they attract helpful microorganisms that form biofilms on the root surface. These microorganisms then support plant health in multiple ways. These biofilms make the surroundings around the roots and the leaves of the plants so favorable that plants stay protected from stresses. Biofilm is responsible for water retention in the rhizosphere environment hence preventing the loss of water by plants especially during drought (Khan et al. 2020). Biofilms regulate nutrient availability by dissolving and incorporating nutrients into the biofilm matrix, which will guarantee a stably nutrition supply to plants (Ciccazzo et al. 2016). Microbes comprising the biofilm can secrete enzymes as well as metabolites that are capable of solubilizing these insoluble, nutrient forms, providing plants with the ability to uptake these nutrients. The ability of biofilms to sequester and store nutrients is observed as another way of preventing runaway nutrient loss from the rhizosphere. Lastly, biofilms prevent pathogens and regulate stress in plants by also producing substances that combat these conditions. Biofilm matrix has a physical barrier that prohibits any pathogenic entry, colonization, and invasion (Zarzosa-Moreno et al. 2020). Biofilm-forming microbes together with non-pathogenic microbes also contribute to antimicrobial production, e.g., antibiotics, enzymes, and volatile organic compounds, that repel the pathogens. As biofilms secrete certain chemical compounds that immediate plant stress relief, for example, exopolysaccharides that act as osmoprotectants, the impact of drought- or salinity-induced osmotic stress is reduced (Enebe and Babalola 2018). These food chain climbers stabilize plant cell membranes and ensure that the enzymes don't get denatured by carrying out their functions under harsh conditions.

1.3 Management Strategies for Optimizing Plant-Microbe Interactions

One of the key strategies for enhancing the integration between plants and microbes is the exploration and use of microbial inoculants containing beneficial microbes such as plant growth-promoting rhizobacteria (PGPR), arbuscular mycorrhizal fungi (AMF), and biocontrol agents (Nadeem et al. 2014). For the next approach, plant breeding efforts should be focused on selecting crop varieties that can attach and maintain the beneficial microbiome associations. Researchers deal with the design of synthetic microbial communities or synthetic microbiomes, which are cases when microbial strains with complementary functions are combined. Management of the environmental factors using such practices as conservation tillage, crop rotation, using organic amendments, and irrigation scheduling promote the beneficial microbes (Wang et al. 2021). The incorporation of a microbe-based approach with other methods is very important in the period of IPM, utilizing them as biocontrol agents together with cultural practices that favor them, but careful use of pesticides will minimize negative impacts (Baker et al. 2020). Ultimately, this approach exploits the efficacy of the microbial population to improve plant growth, nutrient uptake, stress tolerance, pest and disease resistance, and sustainability of crop production by optimizing these to micro-organism interactions (Fig. 1.4).

1.3.1 Development of Microbial Inoculants

Microbial inoculants, also known as bioinoculants or biofertilizers, involve the incorporation of living or inactivated microorganisms that support healthy plant growth, improve nutrient uptake, and assist plants in coping with both abiotic and biotic stresses (Mącik et al. 2020). These inoculants often contain Plant Growth-Promoting Rhizobacteria (PGPR), Arbuscular Mycorrhizal Fungi (AMF), or biocontrol agents. PGPR, for instance, can enhance nutrient uptake by fixing nitrogen, solubilizing phosphate, producing plant hormones that combat pathogens, and inducing plant disease resistance against pathogens (Sağlam et al. 2024). AMF managing close to the plant roots, assists with the claiming of nutrients, phosphorus, and other several minerals, from the soil mainly. One example is the biocontrol agents including the species which are called *Trichoderma* with *Bacillus*. This can prevent plant pathogens through antagonism, competition, and induce systemic resistance. The design of powerful microbial inoculants is narrowed down into several points, among which are strain selection, carrier application, and delivery mechanism (Pavlicevic et al. 2022). Both dry pellet and liquid formulations are used for inoculant production, and the different application approaches may include seed coating, soaking the soil with fermenter extract, spraying on foliage, or simply mixing the fermenting product with the soil or growth media.

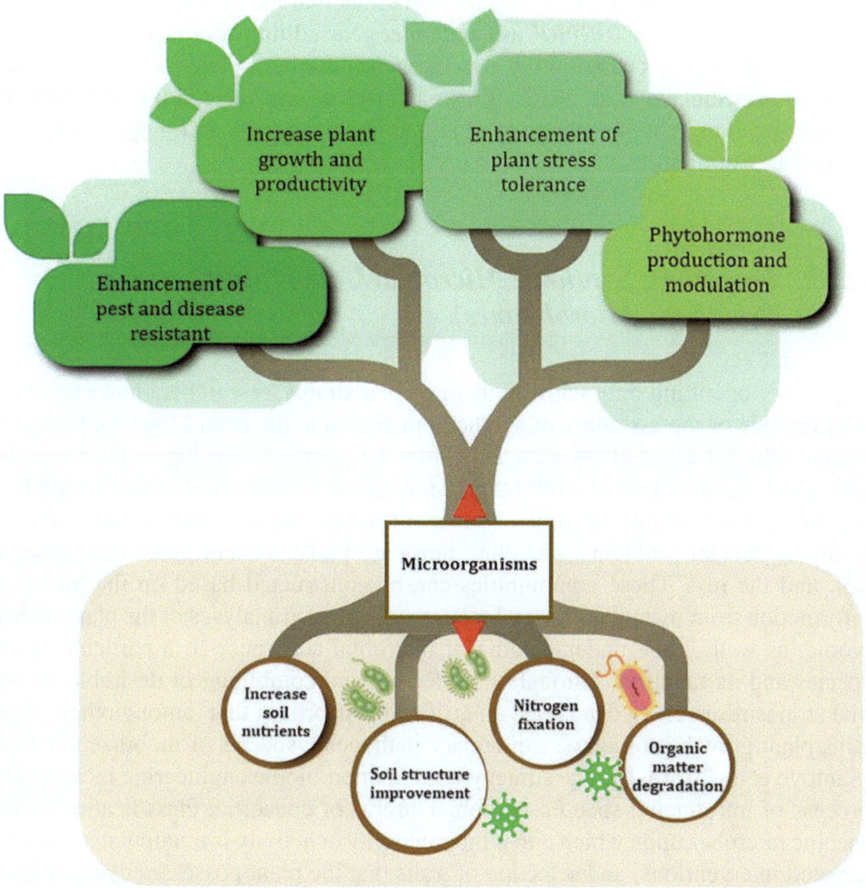

Fig. 1.4 Benefits of plant-microbe interactions in sustainable agriculture (Source: Figures by authors)

1.3.2 Plant Breeding for Beneficial Microbiome Associations

Genotypic selection in plant breeding may thus target the establishment and maintenance of symbiotic associations with their microbial communities, which make up the plant microbiome (Wille et al. 2019). This mechanism however is focused on employing the movement of the characteristics of plant-microbe interactions and therefore enriches the harvest and adaptability of crops. By screening wide germplasm collections, both cultivated and wild species, for certain microbiome-associated characteristics, such as root exudate profile, microbe recruitment patterns, and plant responses to specific species microbiota, researchers can discover new associating mechanisms and biochemical pathways of interactions (Tosi et al. 2020). Polymer markers and genomics-guided breeding could contribute to the discovery and utilization of those gene variations that are beneficial for the plant-microbe interactions. CRISPR/Cas9 and other gene editing approaches can be used to modify plants by changing the specific genes that are related to plant-microbe interactions (Shelake et al. 2019). They can also be used for building the forms of the new generation of plants with increased abilities to recruit and support beneficial microbial communities.

1.3.3 Designing Synthetic Microbial Communities (Synthetic Microbiomes)

In place of operating only with single microbial strains, researchers now can apply the principle of the assembly of synthetic microbes in the form of synthetic microbiomes which are aimed to enhance the microbe-plant relationship and increase the crop yield (Larrainzar et al. 2005). Synthetic microbiomes can be collected employing integrating multiple microbial strains or species that complement each other in resolving several problems, including nutrient uptake, system protection, biocontrol, and the rest. These communities can be constructed based on the intake of information from metagenomic and metatranscriptomic analyses of the plant microbiome, as well as the understanding of microbial adherence to a particular plant species and its function. Through the selection and combining of desirable microbial strains researchers can create an artificial consortium that, among others, contains plant growth promoters, can replace pathogenic species or mobilize nutrients (Santoyo et al. 2021a, b). The strategy used in microbiome engineering includes the process of introducing specific microbial strains or consortia; classification of the specific microbes upon which either their integrity or activity is manipulated through targeted interventions; and selection of traits that the plant possesses depending on the type of microbes which are to be recruited and supported.

1.3.4 Managing Environmental Factors

The environment influences the plant microbiome both through physical characteristics and through patterns of interaction. Implementation of these strategies can be the key to creating good conditions, where beneficial microorganisms can be assisted and boost their positive influence on plant growth and health. Certain soil management practices, such as tillage, crop rotation, organic matter amendment and biochar application, serve as remarkable tools for enhancing the soil microbiome to a higher level (Dissanayaka et al. 2023b; Ekanayaka et al. 2023; Wipf et al. 2021). Emphasis on the protection tillage methods and inclusion of organic matter can increase microbial density and make the microorganisms that are beneficial to farm lives. Also, crop rotation makes plant life cycles dissimilar to beneficial neutralizers such as bacteria and introduces new microorganisms to the soil (Fageria and Baligar 2008). Management of water, from watering and drainage schedules, can exploit the microbiome of plants consequently to modifications of the levels of the soil moisture and oxygen. The successful management of water helps build environments that cultivate microbial species that may benefit the host plant but destroy the ones deemed harmful (Timmis and Ramos 2021). There are other environmental elements such as humidity, light intensity, and temperature that can affect the plant microbiome. Traditional greenhouse management techniques, such as regulating temperature and humidity, can be indicative of a habitat that is conducive to desirable microbial symbionts and that promotes the activities and associations of which they are part of plant growth and health. IPM techniques aim to foster the coordination of measures like culture, biology, and chemical systems to keep pests and diseases under control, avoiding influencing the environment and human safety adversely (Sekabira et al. 2022). PGPR, *Trichoderma,* and *Bacillus* are included in control agents to be used in IPM programs so that the plant pathogens and pests can be suppressed by them.

1.4 Obstacles and Challenges in Commercially Optimizing Plant-Microbe Interactions

Impartation of desirable plant-microbiome interactions is a challenge to attain consistently and reproducibly in diverse urban settings (Trivedi et al. 2021). An example of these factors includes variations in soil properties, environmental conditions, and plant genotypes that can hence have significant effects and outcomes on the microbial inoculants' efficacy and performance. Ensuring the viability and functionality of microbes is challenging during storage, transportation, and subsequent application because of the difficulty of environmental requirements (O'Callaghan 2016). Transparency, open communication, and education as critical matters for dealing with these concerns. As the production technology is scaling from the bench scale to large-scale, increasing the efficiency of production and maintaining the quality, purity, and consistency of the product is the most pertinent challenge. This might involve considerable capital investment, an optimized process, quality control and assurance items, and successful viability to compete.

1.4.1 Consistency and Reproducibility

Conditionally, it is one of the most difficult aspects to reliably apply the resulting of plant-microbe interactions in diverse agricultural settings and environment conditions. Characteristics of soil may vary secondary to each other due to the difference in pH, nutrient availability, texture, and microbial community composition, which may occur across different geographic regions and even within one field (Xia et al. 2020). This variability in fact may be a source of the lack of success of the microbial inoculant introduction due to poor performance. Overall, getting predictable effects might be problematic. The role of environmental factors in the propagation and operation of beneficial microorganisms, which includes temperature, moisture, and seasonal happening, cannot be over-emphasized (Marasco et al. 2022). Microbe adapting to survive in a variety of conditions is a key factor in its effectiveness. Selection of a specific plant genotype and matching that with the introduced microorganisms can determine the level and kind of plant-microbe interactions that occur (Gamalero et al. 2022). The process of microbial inoculants increasing their effectiveness across a more extensive range of crop varieties is a must for their practical implementation.

1.4.2 Microbial Inoculant Formulation and Delivery

Designing effective microbial inoculants or growing helpful bacteria in sufficient quantity is one of the most important points for the establishment of plant-microbe relationships (Vishwakarma et al. 2020). Achieving multipurpose suspensions that have lasting activity and event glint without harming the colonies is a very difficult task. Consideration of both the processes of delivery and the targeting of location is required, whether that location is soil, seeds, or tissue, without jeopardizing the viability and functionality of the inoculants (Koskey et al. 2021). Multiple modes of delivery, for instance, coating the seeds, application through soil drenching, or surface area application, vary in the ways and restrictions they have. Concerning inoculants and their involved application tools, it is vital to make them compatible with conventional agricultural tools like tillage, irrigation, and crop protection measures to make the integration and utilization simple by the farmers (O'Callaghan et al. 2022).

1.4.3 Regulatory Frameworks and Biosafety Considerations

The commercialization of microbial inoculants that would be used for agricultural applications needs to ensure the fulfillment of regulatory requirements and the view of biosafety issues (Kaur and Vishnu. 2022). Among the nations that have already constructed federal programs for microbial products approval and registration for use in agriculture are included. These regulations may differ depending on region, and such differences together with the fact they can drain both precious time and resources are the more reason for small and medium enterprises to dedicate more time and resources to them. All-round biosafety evaluation is as important as it analyses the challenges involved in the bringing of non-native microorganisms or genetically edited organisms to the agricultural ecosystem (Ghimire et al. 2023).

The safety of the microorganisms that are introduced to humans, animals, and the environment is a very crucial factor that has to be addressed for the successful commercialization of a particular product. The public perspective and receptions of microbial inoculants have a huge influence on the commercial success of the same (Mącik et al. 2020). Resolving worries connected with risks and effects that are unexpected by the integration of foreign microorganisms is a key factor, both for accepting those innovations by the public and their worldwide implementation.

1.4.4 Scaling Up Production

Commercialization of plant microbial inoculants presents a noteworthy trouble as scaling up production is one of the main challenges (Timmusk et al. 2017). The transition from laboratory or small-scale production to large, low-cost manufacturing is as necessary as possible to reach commercial viability on a large scale. Although it provides various benefits, this procedure includes certain challenges. Keeping the distribution and deepness of cultures of microorganisms in correspondence with the production or processing is a very complicated task. Microbes of different species exhibit high degrees of microbial diversity whereby certain strains may have specialized requirements for growth while others are sensitive to environmental factors, thus making replication of the conditions for optimality at a larger scale difficult (Gonzalez and Aranda 2023). In addition to that, the development of the pertinent downstream processing methods, i.e. separation, concentration, and formulation. These processes should result in keeping the attribute of microbial agents plus an acceptable shelf-life for storage and transportation is necessary. Large-scale production quality is the main decisive factor in having consistent product quality and performance. Setting up control and standardization protocols plays a crucial role in this aspect. Secondly, scaling up manufacturing production often involves proportionate large capital investment in specialized apparatus, shed, and utility amenities. It can be a huge obstacle for small and medium enterprises like them if they have no means to access it, it will be difficult for them to compete because they don't have the market advantage.

1.5 Future Directions

The current issue of the persistent agricultural problem is most critically seen in the situation where climate change, population growth, and attractive resources are coming in an excessive amount. The intricate relationship between plants and microbes offers a very efficient method for innovative means that could help address the problem of feeding the ever-increasing population with a decreased ecological footprint (Chen and Zhang 2015). The research on plant-microbe associations could entirely change the crop management paradigm and enable enormous development prospects for crops, along with the alignment of agriculture with nature. Investigating the subtle interactions involving plants and their symbiotic microbes has already brought a good deal of important information about those mechanisms that regulate plant growth, development, and resilience (Farrar et al. 2014). Microbes, as wide-ranging as the benign bacteria, fungi, endophytes, and symbiotics, have major roles such as nutrient acquisition, stress tolerance, and disease resistance for plants. Through recognizing, exploiting, and utilizing these minuscule relations, researchers can effectively develop innovative ways to improve crop yields, decrease the use of harmful inputs, and combat biotic and abiotic stresses.

A good area of study for the future is the creation of microbial inoculants specifically for certain species of crops and prevailing seasons (Farrar et al. 2014). These adjuncts can constitute strains of selected PGPR, mycorrhizal fungi, or endophytes, which would increase the rate of moisture, nutrient absorption, stress tolerance, and disease resistance. Through the development of improved and environmentally responsible application methods and delivery systems, these beneficial microbes can become a basis for affordable and ecologically sound agriculture (Naik et al. 2019). Besides, the discovery of the plant microbiome, which is composed of the whole community of microbes living on and inside plant tissues, has grown the field of plant health and productivity research to a new dimension. The previously unforeseen way omics technologies, including metagenomics, metatranscriptomics, and metabolomics, were applied allowed researchers to untie the intricate web of plant microbiome interactions (Lucaciu et al. 2019). With the help of delving into such intricate relationships, researchers can detect microbial players as well as their roles in the process. Subsequently, this data can serve as the basis for the design of targeted microbiome interventions.

Besides, the union between plant-microbe interactions with other new technologies like precision agriculture and controlled environment agriculture can have a very suggestive nature of sustaining farming (Atapattu et al. 2024a; Weekley et al. 2012). The precision agriculture approach aims to increase the effectiveness of applying microbial inoculants and choose the most suitable agricultural methods based on the local elements of soil like nutrients and environmental conditions, thus achieving the positive influences of beneficial plant-microbe interaction. Controlled environment agriculture in the form of vertical farming and greenhouse systems is the situation where the optimal conditions both for initiation and success of the beneficial microbial community are guaranteed which therefore enhances crop performance and efficiency levels of the used resources (Lastochkina et al. 2022). Interdisciplinarity including plant biology, microbiology, genetics, ecology and agricultural sciences as the respective areas will be a remarkable avenue in also advancing our comprehension of plant-microbe biological interactions and their application in practice. Through the establishment of transdisciplinary dialogues and amalgamating the ideals from different sectors, researchers can bring forth multilevel strategies that will integrate the plants, microbes, and the environment equally.

Finally, the transformation of research findings to practical applications requires the collaboration of different stakeholders like farmers and policymakers, and the players in industry through joint effort (Cronin et al. 2021). Proper communication and knowledge transfer shall be a critical aspect in solving the gap between science-based discoveries and their implementation in agricultural field activities. It is possible to get agriculture practices that are more sustainable by using collaboration among stakeholders as well as integrating traditional knowledge and indigenous procedures. Plant-microbe interactions hold great promise for sustainable agriculture's future. Unveiling the intricate connections between microorganisms and plants could reveal novel ways to boost plant productivity, reduce environmental impacts, and advance sustainable agricultural practices (Cronin et al. 2021). Considering thorough research, technology breakthroughs, and inclusive collaboration, we are going to be the ones who build a reproductive, productive, and eco-friendly system of food production that defines their future as being safe and sustainable.

1.6 Conclusion

The chapter emphasizes the essential role of plant-microbe interactions in fostering sustainable agriculture, highlighting how microbial communities can enhance plant resilience, nutrient access, and growth. Through biological processes like nutrient cycling, pest and disease resistance, and stress tolerance, these interactions offer a viable alternative to conventional agrochemical-reliant practices. Sustainable agriculture seeks to reduce environmental impacts while boosting productivity, and microbial partnerships play a pivotal role in this transformation. Microbial inoculants, biofertilizers, and biopesticides harness the natural benefits of microbes, thereby promoting healthier soil ecosystems, enhancing plant vigor, and reducing dependency on chemical inputs. Key strategies for optimizing these interactions include the development of microbial inoculants tailored to specific crops, breeding plants that foster beneficial microbiomes, and designing synthetic microbial communities. Environmental management, such as conservation tillage and crop rotation, supports these efforts by creating favorable conditions for beneficial microbes to thrive. Challenges persist, including scalability, regulatory requirements, and achieving consistent results across diverse agricultural landscapes. Overcoming these obstacles requires innovations in microbial formulation, delivery systems, and regulatory frameworks to support safe, effective applications. Future research will continue to unravel the complex interactions between plants and their microbiomes, utilizing advanced technologies like genomics and precision agriculture to maximize the benefits of microbial associations. By integrating scientific advancements with traditional knowledge and fostering interdisciplinary collaboration, the agricultural sector can pave the way for more resilient, ecologically balanced, and economically sustainable farming systems. Ultimately, the chapter underscores the transformative potential of plant-microbe interactions in building a sustainable agricultural framework, promoting soil health, conserving natural resources, and addressing the global food security challenge in an environmentally conscious manner.

References

Abhinandan K, Skori L, Stanic M, Hickerson NMN, Jamshed M, Samuel MA (2018) Abiotic stress signaling in wheat—an inclusive overview of hormonal interactions during abiotic stress responses in wheat. Front Plant Sci 9:734. https://doi.org/10.3389/fpls.2018.00734

Akhtar N, Rashid MM, Perween S, Kumar G, Nanda S (2022) Mode of action of different microbial products in plant growth promotion. In: New and future developments in microbial biotechnology and bioengineering. Elsevier, pp 85–120. https://doi.org/10.1016/B978-0-323-85577-8.00016-0

Alamoudi SA (2023) Using some microorganisms as biocontrol agents to manage phytopathogenic fungi: a comprehensive review. J Plant Pathol 106(1):3–21. https://doi.org/10.1007/s42161-023-01542-7

Altieri M, Nicholls C, Montalba R (2017) Technological approaches to sustainable agriculture at a crossroads: an agroecological perspective. Sustainability 9(3):349. https://doi.org/10.3390/su9030349

Anemaet IG, Bekker M, Hellingwerf KJ (2010) Algal photosynthesis as the primary driver for a sustainable development in energy, feed, and food production. Mar Biotechnol 12(6):619–629. https://doi.org/10.1007/s10126-010-9311-1

Aqeel M, Ran J, Hu W, Irshad MK, Dong L, Akram MA, Eldesoky GE, Aljuwayid AM, Chuah LF, Deng J (2023) Plant-soil-microbe interactions in maintaining ecosystem stability and coordinated turnover under changing environmental conditions. Chemosphere 318:137924. https://doi.org/10.1016/j.chemosphere.2023.137924

Aryal B, Gurung R, Camargo AF, Fongaro G, Treichel H, Mainali B, Angove MJ, Ngo HH, Guo W, Puadel SR (2022) Nitrous oxide emission in altered nitrogen cycle and implications for climate change. Environ Pollut 314:120272. https://doi.org/10.1016/j.envpol.2022.120272

Atapattu AJ, Perera LK, Nuwarapaksha TD, Udumann SS, Dissanayaka NS (2024a) Challenges in achieving artificial intelligence in agriculture. In: Chouhan SS, Saxena A, Singh UP, Jain S (eds) Artificial intelligence techniques in smart agriculture. Springer, Singapore. https://doi.org/10.1007/978-981-97-5878-4_2

Atapattu AJ, Ranasinghe CS, Nuwarapaksha TD, Udumann SS, Dissanayaka NS (2024b) Sustainable agriculture and sustainable development goals (SDGs). (pp. 1–27). https://doi.org/10.4018/979-8-3693-4864-2.ch001

Ayilara MS, Adeleke BS, Akinola SA, Fayose CA, Adeyemi UT, Gbadegesin LA, Omole RK, Johnson RM, Uthman QO, Babalola OO (2023) Biopesticides as a promising alternative to synthetic pesticides: A case for microbial pesticides, phytopesticides, and nanobiopesticides. Front Microbiol 14. https://doi.org/10.3389/fmicb.2023.1040901

Baker BP, Green TA, Loker AJ (2020) Biological control and integrated pest management in organic and conventional systems. Biol Control 140:104095. https://doi.org/10.1016/j.biocontrol.2019.104095

Beeckman F, Motte H, Beeckman T (2018) Nitrification in agricultural soils: impact, actors and mitigation. Curr Opin Biotechnol 50:166–173. https://doi.org/10.1016/j.copbio.2018.01.014

Bhaduri D, Sihi D, Bhowmik A, Verma BC, Munda S, Dari B (2022) A review on effective soil health bio-indicators for ecosystem restoration and sustainability. Front Microbiol 13. https://doi.org/10.3389/fmicb.2022.938481

Borah P, Gogoi N, Asad SA, Rabha AJ, Farooq M (2023) An insight into plant growth-promoting Rhizobacteria-mediated mitigation of stresses in plant. J Plant Growth Regul 42(5):3229–3256. https://doi.org/10.1007/s00344-022-10787-y

Bukhat S, Imran A, Javaid S, Shahid M, Majeed A, Naqqash T (2020) Communication of plants with microbial world: exploring the regulatory networks for PGPR mediated defense signaling. Microbiol Res 238:126486. https://doi.org/10.1016/j.micres.2020.126486

Cáceres R, Malińska K, Marfà O (2018) Nitrification within composting: A review. Waste Manag 72:119–137. https://doi.org/10.1016/j.wasman.2017.10.049

Chen H-G, Zhang Y-HP (2015) New biorefineries and sustainable agriculture: increased food, biofuels, and ecosystem security. Renew Sust Energ Rev 47:117–132. https://doi.org/10.1016/j.rser.2015.02.048

Ciccazzo S, Esposito A, Borruso L, Brusetti L (2016) Microbial communities and primary succession in high altitude mountain environments. Ann Microbiol 66(1):43–60. https://doi.org/10.1007/s13213-015-1130-1

Coba de la Peña T, Fedorova E, Pueyo JJ, Lucas MM (2018) The symbiosome: legume and rhizobia co-evolution toward a nitrogen-fixing organelle? Front Plant Sci 8. https://doi.org/10.3389/fpls.2017.02229

Cronin E, Fosselle S, Rogge E, Home R (2021) An analytical framework to study multi-actor partnerships engaged in interactive innovation processes in the agriculture, forestry, and rural development sector. Sustainability 13(11):6428. https://doi.org/10.3390/su13116428

Dean JF, Middelburg JJ, Röckmann T, Aerts R, Blauw LG, Egger M, Jetten MSM, de Jong AEE, Meisel OH, Rasigraf O, Slomp CP, in't Zandt MH, Dolman AJ (2018) Methane feedbacks to the global climate system in a warmer world. Rev Geophys 56(1):207–250. https://doi.org/10.1002/2017RG000559

Deb S, Lewicka-Szczebak D, Rohe L (2024) Microbial nitrogen transformations tracked by natural abundance isotope studies and microbiological methods: a review. Sci Total Environ 926:172073. https://doi.org/10.1016/j.scitotenv.2024.172073

Dewi WS, Amalina DD, Romadhon MR (2023) Microbial biofilm for soil health, plant growth, and productivity under multi stress. A review. IOP Conference Series: Earth and Environmental Science 1162(1):012008. https://doi.org/10.1088/1755-1315/1162/1/012008

Dissanayaka D, Nuwarapaksha T, Udumann S, Dissanayake D, Atapattu AJ (2022) A sustainable way of increasing productivity of coconut cultivation using cover crops: A review. Circular Agric Syst 2(1):1–9. https://doi.org/10.48130/CAS-2022-0007

Dissanayaka DMNS, Dissanayake DKRPL, Udumann SS, Nuwarapaksha TD, Atapattu AJ (2023a) Agroforestry—a key tool in the climate-smart agriculture context: a review on coconut cultivation in Sri Lanka. Front Agron 5:1162750. https://doi.org/10.3389/fagro.2023.1162750

Dissanayaka DMNS, Udumann SS, Nuwarapaksha TD, Atapattu AJ (2023b) Effects of pyrolysis temperature on chemical composition of coconut-husk biochar for agricultural applications: a characterization study. Technol Agron 3(1):10. https://doi.org/10.48130/TIA-2023-0013

Dissanayaka NS, Udumann SS, Nuwarapaksha TD, Atapattu AJ (2023c) Agroforestry: an avenue for resilient and productive farming through integrated crops and livestock production. In: Transitioning to zero hunger, Delwendé Innocent Kiba (ed). MDPI Books, pp 115–136. https://doi.org/10.3390/books978-3-03897-863-3-5

Dissanayaka DMNS, Udumann SS, Atapattu AJ (2024a) In: Raj A, Jhariya MK, Banerjee A, Jha RK, Singh KP (eds) Synergies between tree crops and ecosystems in tropical agroforestry. In Agroforestry, pp 49–87. https://doi.org/10.1002/9781394231164.ch3

Dissanayaka DMNS, Udumann SS, Nuwarapaksha TD, Atapattu AJ (2024b) Harnessing the potential of Mucuna cover cropping: a comprehensive review of its agronomic and environmental benefits. Circular Agric Syst 4(1):10. https://doi.org/10.48130/cas-0024-0001

Dobbelaere S, Vanderleyden J, Okon Y (2003) Plant growth-promoting effects of diazotrophs in the rhizosphere. Crit Rev Plant Sci 22(2):107–149. https://doi.org/10.1080/713610853

Dubey A, Malla MA, Khan F, Chowdhary K, Yadav S, Kumar A, Sharma S, Khare PK, Khan ML (2019) Soil microbiome: a key player for conservation of soil health under changing climate. Biodivers Conserv 28(8–9):2405–2429. https://doi.org/10.1007/s10531-019-01760-5

Dutta P, Mahanta M, Singh SB, Thakuria D, Deb L, Kumari A, Upamanya GK, Boruah S, Dey U, Mishra AK, Vanlaltani L, Vijay Reddy D, Heisnam P, Pandey AK (2023) Molecular interaction between plants and Trichoderma species against soil-borne plant pathogens. Front Plant Sci 14. https://doi.org/10.3389/fpls.2023.1145715

Ekanayaka EMGN, Dissanayake DKRPL, Udumann SS, Dissanayaka DMNS, Nuwarapaksha TD, Herath HMSK, Atapattu AJ (2023) Sustainable utilization of king coconut husk as a feedstock in biochar production with the highest conversion efficiency and desirable properties.

In IOP Conference Series: Earth and Environmental Science 1235(10):012009. https://doi.org/10.1088/1755-1315/1235/1/012009
Enebe MC, Babalola OO (2018) The influence of plant growth-promoting rhizobacteria in plant tolerance to abiotic stress: a survival strategy. Appl Microbiol Biotechnol 102(18):7821–7835. https://doi.org/10.1007/s00253-018-9214-z
Eviner VT, Hawkes CV (2008) Embracing variability in the application of plant–soil interactions to the restoration of communities and ecosystems. Restor Ecol 16(4):713–729. https://doi.org/10.1111/j.1526-100X.2008.00482.x
Fageria NK, Baligar VC (2008) Chapter 7 ameliorating soil acidity of tropical Oxisols by liming for sustainable crop production (pp. 345–399). https://doi.org/10.1016/S0065-2113(08)00407-0
Farrar K, Bryant D, Cope-Selby N (2014) Understanding and engineering beneficial plant–microbe interactions: plant growth promotion in energy crops. Plant Biotechnol J 12(9):1193–1206. https://doi.org/10.1111/pbi.12279
Fazeli-Nasab B, Shahraki-Mojahed L, Piri R, Sobhanizadeh A (2022) Trichoderma: improving growth and tolerance to biotic and abiotic stresses in plants. In: Trends of applied microbiology for sustainable economy. Elsevier, pp 525–564. https://doi.org/10.1016/B978-0-323-91595-3.00004-5
Fujita Y, Uesaka K (2022) Nitrogen fixation in cyanobacteria. In: Cyanobacterial physiology. Elsevier, pp 29–45. https://doi.org/10.1016/B978-0-323-96106-6.00007-1
Gachara G, Kenfaoui J, Suleiman R, Kilima B, Taoussi M, Aberkani K, Belabess Z, Meddich A, Handaq N, Laasli S-E, Barka EA, Lahlali R (2024) The role of soil microbiome in driving plant performance: an overview based on ecological and ecosystem advantages to the plant community. J Crop Health 76(1):3–17. https://doi.org/10.1007/s10343-023-00935-z
Gamalero E, Bona E, Glick BR (2022) Current techniques to study beneficial plant-microbe interactions. Microorganisms 10(7):1380. https://doi.org/10.3390/microorganisms10071380
Ganchev I (2021) Role of multispecies biofilms with a dominance of Bacillus subtilis in the rhizosphere. Biol Bull 48(S2):S47–S64. https://doi.org/10.1134/S1062359021150061
Ghimire BK, Yu CY, Kim W-R, Moon H-S, Lee J, Kim SH, Chung IM (2023) Assessment of benefits and risk of genetically modified plants and products: current controversies and perspective. Sustainability 15(2):1722. https://doi.org/10.3390/su15021722
Gonzalez JM, Aranda B (2023) Microbial growth under limiting conditions-future perspectives. Microorganisms 11(7):1641. https://doi.org/10.3390/microorganisms11071641
Grzyb A, Wolna-Maruwka A, Niewiadomska A (2021) The significance of microbial transformation of nitrogen compounds in the light of integrated crop management. Agronomy 11(7):1415. https://doi.org/10.3390/agronomy11071415
Gu W, Milton RD (2020) Natural and engineered electron transfer of nitrogenase. Chemistry 2(2):322–346. https://doi.org/10.3390/chemistry2020021
Gupta A, Singh UB, Sahu PK, Paul S, Kumar A, Malviya D, Singh S, Kuppusamy P, Singh P, Paul D, Rai JP, Singh HV, Manna MC, Crusberg TC, Kumar A, Saxena AK (2022) Linking soil microbial diversity to modern agriculture practices: a review. Int J Environ Res Public Health 19(5):3141. https://doi.org/10.3390/ijerph19053141
Hamid B, Zaman M, Farooq S, Fatima S, Sayyed RZ, Baba ZA, Sheikh TA, Reddy MS, El Enshasy H, Gafur A, Suriani NL (2021) Bacterial plant biostimulants: a sustainable way towards improving growth, productivity, and health of crops. Sustainability 13(5):2856. https://doi.org/10.3390/su13052856
Hirsch AM, Lum MR, Downie JA (2001) What makes the rhizobia-legume symbiosis so special? Plant Physiol 127(4):1484–1492. https://doi.org/10.1104/pp.010866
Hossain A, Ali ME, Maitra S, Bhadra P, Rahman MME, Ali S, Aftab T (2022) The role of soil microorganisms in plant adaptation to abiotic stresses: current scenario and future perspectives. In: Plant perspectives to global climate changes. Elsevier, pp 233–278. https://doi.org/10.1016/B978-0-323-85665-2.00001-7
Huang H, Ullah F, Zhou D-X, Yi M, Zhao Y (2019) Mechanisms of ROS regulation of plant development and stress responses. Front Plant Sci 10. https://doi.org/10.3389/fpls.2019.00800

Ibrahim, M., Iqbal, M., Tang, Y.-T., Khan, S., Guan, D.-X., & Li, G. (2022). Phosphorus Mobilization in Plant–Soil Environments and Inspired Strategies for Managing Phosphorus: A Review. Agronomy, 12(10), 2539. https://doi.org/10.3390/agronomy12102539

Iqbal N, Trivellini A, Masood A, Ferrante A, Khan NA (2013) Current understanding on ethylene signaling in plants: the influence of nutrient availability. Plant Physiol Biochem 73:128–138. https://doi.org/10.1016/j.plaphy.2013.09.011

Jacoby R, Peukert M, Succurro A, Koprivova A, Kopriva S (2017) The role of soil microorganisms in plant mineral nutrition—current knowledge and future directions. Front Plant Sci 8. https://doi.org/10.3389/fpls.2017.01617

Kabange NR, Kwon Y, Lee S-M, Kang J-W, Cha J-K, Park H, Dzorkpe GD, Shin D, Oh K-W, Lee J-H (2023) Mitigating greenhouse gas emissions from crop production and management practices, and livestock: a review. Sustainability 15(22):15889. https://doi.org/10.3390/su152215889

Kalayu G (2019) Phosphate solubilizing microorganisms: promising approach as biofertilizers. Int J Agron 2019:1–7. https://doi.org/10.1155/2019/4917256

Kaur J, Vishnu. (2022) Bacterial inoculants for rhizosphere engineering: applications, current aspects, and challenges. In: Rhizosphere engineering. Elsevier, pp 129–150. https://doi.org/10.1016/B978-0-323-89973-4.00004-1

Khan N, Ali S, Tariq H, Latif S, Yasmin H, Mehmood A, Shahid MA (2020) Water conservation and plant survival strategies of rhizobacteria under drought stress. Agronomy 10(11):1683. https://doi.org/10.3390/agronomy10111683

Khan S, Nadir S, Iqbal S, Xu J, Gui H, Khan A, Ye L (2021) Towards a comprehensive understanding of free-living nitrogen fixation. Circul Agric Syst 1(1):1–11. https://doi.org/10.48130/CAS-2021-0013

Khoshru B, Mitra D, Joshi K, Adhikari P, Rion MSI, Fadiji AE, Alizadeh M, Priyadarshini A, Senapati A, Sarikhani MR, Panneerselvam P, Mohapatra PK, Das, Sushkova S, Minkina T, Keswani C (2023) Decrypting the multi-functional biological activators and inducers of defense responses against biotic stresses in plants. Heliyon 9(3):e13825. https://doi.org/10.1016/j.heliyon.2023.e13825

Klessig DF, Choi HW, Dempsey DA (2018) Systemic acquired resistance and salicylic acid: past, present, and future. Mol Plant-Microbe Interact 31(9):871–888. https://doi.org/10.1094/MPMI-03-18-0067-CR

Koskey G, Mburu SW, Awino R, Njeru EM, Maingi JM (2021) Potential use of beneficial microorganisms for soil amelioration, Phytopathogen biocontrol, and sustainable crop production in smallholder agroecosystems. Front Sustain Food Syst 5. https://doi.org/10.3389/fsufs.2021.606308

Kumar R, Shastri B (2017) Role of phosphate-solubilising microorganisms in sustainable agricultural development. In: Agro-environmental sustainability. Springer International Publishing, pp 271–303. https://doi.org/10.1007/978-3-319-49724-2_13

Kumar A, Verma JP (2018) Does plant—microbe interaction confer stress tolerance in plants: a review? Microbiol Res 207:41–52. https://doi.org/10.1016/j.micres.2017.11.004

Lahlali R, Ezrari S, Radouane N, Kenfaoui J, Esmaeel Q, El Hamss H, Belabess Z, Barka EA (2022) Biological control of plant pathogens: a global perspective. Microorganisms 10(3):596. https://doi.org/10.3390/microorganisms10030596

Larrainzar E, O'Gara F, Morrissey JP (2005) Applications of autofluorescent proteins for in situ studies in microbial ecology. Ann Rev Microbiol 59(1):257–277. https://doi.org/10.1146/annurev.micro.59.030804.121350

Lastochkina O, Aliniaeifard S, SeifiKalhor M, Bosacchi M, Maslennikova D, Lubyanova A (2022) Novel approaches for sustainable horticultural crop production: advances and prospects. Horticulturae 8(10):910. https://doi.org/10.3390/horticulturae8100910

Li C, Xu Y, Li L, Yang X, Wang Y (2019) Acid stress induces cross-protection for cadmium tolerance of multi-stress-tolerant Pichia kudriavzevii by regulating cadmium transport and antioxidant defense system. J Hazard Mater 366:151–159. https://doi.org/10.1016/j.jhazmat.2018.11.101

Lucaciu R, Pelikan C, Gerner SM, Zioutis C, Köstlbacher S, Marx H, Herbold CW, Schmidt H, Rattei T (2019) A bioinformatics guide to plant microbiome analysis. Front Plant Sci 10. https://doi.org/10.3389/fpls.2019.01313

Ma Y, Oliveira RS, Freitas H, Zhang C (2016) Biochemical and molecular mechanisms of plant-microbe-metal interactions: relevance for phytoremediation. Front Plant Sci 7. https://doi.org/10.3389/fpls.2016.00918

Mącik M, Gryta A, Frąc M (2020) Biofertilizers in agriculture: an overview on concepts, strategies and effects on soil microorganisms (pp. 31–87). https://doi.org/10.1016/bs.agron.2020.02.001

Mackenzie FT, Lerman A, Andersson AJ (2004) Past and present of sediment and carbon biogeochemical cycling models. Biogeosciences 1(1):11–32. https://doi.org/10.5194/bg-1-11-2004

Mansingh S, Das KK, Sultana S, Parida K (2021) Recent advances in wireless photofixation of dinitrogen to ammonia under the ambient condition: A review. J Photochem Photobiol C: Photochem Rev 47:100402. https://doi.org/10.1016/j.jphotochemrev.2021.100402

Marasco R, Fusi M, Mosqueira M, Booth JM, Rossi F, Cardinale M, Michoud G, Rolli E, Mugnai G, Vergani L, Borin S, De Philippis R, Cherif A, Daffonchio D (2022) Rhizosheath–root system changes exopolysaccharide content but stabilizes bacterial community across contrasting seasons in a desert environment. Environ Microb 17(1):14. https://doi.org/10.1186/s40793-022-00407-3

Medfu Tarekegn M, Zewdu Salilih F, Ishetu AI (2020) Microbes used as a tool for bioremediation of heavy metal from the environment. Cogent Food Agric 6(1):1783174. https://doi.org/10.1080/23311932.2020.1783174

Mohanty SR, Yadav R, Dubey G, Ahirwar U, Ahirwar N, Aparna K, Rao DLN, Kollah B (2018) How sequential reduction of terminal electron acceptors modulates nitrification and dynamics of nitrifying bacteria and archaea in a tropical vertisol. J Agric Sci 156(2):215–224. https://doi.org/10.1017/S0021859618000266

Nadarajah K, Abdul Rahman NSN (2023) The microbial connection to sustainable agriculture. Plants 12(12):2307. https://doi.org/10.3390/plants12122307

Nadeem SM, Ahmad M, Zahir ZA, Javaid A, Ashraf M (2014) The role of mycorrhizae and plant growth promoting rhizobacteria (PGPR) in improving crop productivity under stressful environments. Biotechnol Adv 32(2):429–448. https://doi.org/10.1016/j.biotechadv.2013.12.005

Naik K, Mishra S, Srichandan H, Singh PK, Sarangi PK (2019) Plant growth promoting microbes: potential link to sustainable agriculture and environment. Biocatal Agric Biotechnol 21:101326. https://doi.org/10.1016/j.bcab.2019.101326

Nardi P, Laanbroek HJ, Nicol GW, Renella G, Cardinale M, Pietramellara G, Weckwerth W, Trinchera A, Ghatak A, Nannipieri P (2020) Biological nitrification inhibition in the rhizosphere: determining interactions and impact on microbially mediated processes and potential applications. FEMS Microbiol Rev 44(6):874–908. https://doi.org/10.1093/femsre/fuaa037

Negi VS, Maikhuri RK (2013) Socio-ecological and religious perspective of agrobiodiversity conservation: issues, concern and priority for sustainable agriculture, central Himalaya. J Agric Environ Ethics 26(2):491–512. https://doi.org/10.1007/s10806-012-9386-y

Nuwarapaksha T, Udumann S, Dissanayaka D, Dissanayake D, Atapattu AJ (2022) Coconut based multiple cropping systems: an analytical review in Sri Lankan coconut cultivations. Circ Agric Syst 2(1):1–7. https://doi.org/10.48130/CAS-2022-0008

Nuwarapaksha TD, Dissanayaka NS, Udumann SS, Atapattu AJ (2023a) Gliricidia as a beneficial crop in resource-limiting agroforestry systems in Sri Lanka. Ind J Agroforest 25(1):12–18

Nuwarapaksha TD, Udumann SS, Dissanayaka NS, Atapattu AJ (2023b) Coconut-based livestock farming: a sustainable approach to enhancing food security in Sri Lanka. In transitioning to zero hunger. Delwendé Innocent Kiba (ed), MDPI Books, pp 197–213. https://doi.org/10.3390/books978-3-03897-863-3-8

Nuwarapaksha TD, Udumann SS, Atapattu AJ (2024) In: Raj A, Jhariya MK, Banerjee A, Jha RK, Singh KP (eds) Fostering food and nutritional security through agroforestry practices. In Agroforestry, pp 285–318. https://doi.org/10.1002/9781394231164.ch10

O'Callaghan M (2016) Microbial inoculation of seed for improved crop performance: issues and opportunities. Appl Microbiol Biotechnol 100(13):5729–5746. https://doi.org/10.1007/s00253-016-7590-9

O'Callaghan M, Ballard RA, Wright D (2022) Soil microbial inoculants for sustainable agriculture: limitations and opportunities. Soil Use Manag 38(3):1340–1369. https://doi.org/10.1111/sum.12811

Oberemok VV, Gal'chinsky NV, Useinov RZ, Novikov IA, Puzanova YV, Filatov RI, Kouakou NJ, Kouame KF, Kra KD, Laikova KV (2023) Four Most pathogenic superfamilies of insect pests of suborder sternorrhyncha: invisible superplunderers of plant vitality. Insects 14(5):462. https://doi.org/10.3390/insects14050462

Odelade K, Babalola O (2019) Bacteria, fungi and archaea domains in rhizospheric soil and their effects in enhancing agricultural productivity. Int J Environ Res Public Health 16(20):3873. https://doi.org/10.3390/ijerph16203873

Pashaei R, Zahedipour-Sheshglani P, Dzingelevičienė R, Abbasi S, Rees RM (2022) Effects of pharmaceuticals on the nitrogen cycle in water and soil: a review. Environ Monit Assess 194(2):105. https://doi.org/10.1007/s10661-022-09754-7

Paśmionka IB, Bulski K, Boligłowa E (2021) The participation of microbiota in the transformation of nitrogen compounds in the soil—a review. Agronomy 11(5):977. https://doi.org/10.3390/agronomy11050977

Paul BG, Ding H, Bagby SC, Kellermann MY, Redmond MC, Andersen GL, Valentine DL (2017) Methane-oxidizing bacteria shunt carbon to microbial mats at a marine hydrocarbon seep. Front Microbiol 8. https://doi.org/10.3389/fmicb.2017.00186

Pavlicevic M, Abdelraheem W, Zuverza-Mena N, O'Keefe T, Mukhtar S, Ridge G, Ranciato J, Haynes C, Elmer W, Pignatello J, Pagano L, Caldara M, Marmiroli M, Maestri E, Marmiroli N, White JC (2022) Engineered nanoparticles, natural nanoclay and biochar, as carriers of plant-growth promoting bacteria. Nano 12(24):4474. https://doi.org/10.3390/nano12244474

Pirttilä AM, Mohammad Parast Tabas H, Baruah N, Koskimäki JJ (2021) Biofertilizers and biocontrol agents for agriculture: how to identify and develop new potent microbial strains and traits. Microorganisms 9(4):817. https://doi.org/10.3390/microorganisms9040817

Plante AF (2007) Soil biogeochemical cycling of inorganic nutrients and metals. In: Soil microbiology, ecology and biochemistry. Elsevier, pp 389–432. https://doi.org/10.1016/B978-0-08-047514-1.50019-6

Pradhan D, Sukla LB, Sawyer M, Rahman PKSM (2017) Recent bioreduction of hexavalent chromium in wastewater treatment: a review. J Ind Eng Chem 55:1–20. https://doi.org/10.1016/j.jiec.2017.06.040

Rawat P, Das S, Shankhdhar D, Shankhdhar SC (2021) Phosphate-solubilizing microorganisms: mechanism and their role in phosphate solubilization and uptake. J Soil Sci Plant Nutr 21(1):49–68. https://doi.org/10.1007/s42729-020-00342-7

Raza T, Qadir MF, Khan KS, Eash NS, Yousuf M, Chatterjee S, Manzoor R, ur Rehman S, Oetting JN (2023) Unraveling the potential of microbes in decomposition of organic matter and release of carbon in the ecosystem. J Environ Manag 344:118529. https://doi.org/10.1016/j.jenvman.2023.118529

Ren A-T, Abbott LK, Chen Y, Xiong Y-C, Mickan BS (2020) Nutrient recovery from anaerobic digestion of food waste: impacts of digestate on plant growth and rhizosphere bacterial community composition and potential function in ryegrass. Biol Fertil Soils 56(7):973–989. https://doi.org/10.1007/s00374-020-01477-6

Rosenberg E, Sharon G, Atad I, Zilber-Rosenberg I (2010) The evolution of animals and plants via symbiosis with microorganisms. Environ Microbiol Rep 2(4):500–506. https://doi.org/10.1111/j.1758-2229.2010.00177.x

Rosset PM, Altieri MA (1997) Agroecology versus input substitution: a fundamental contradiction of sustainable agriculture. Soc Nat Resour 10(3):283–295. https://doi.org/10.1080/08941929709381027

Sabarwal A, Kumar K, Singh RP (2018) Hazardous effects of chemical pesticides on human health–cancer and other associated disorders. Environ Toxicol Pharmacol 63:103–114. https://doi.org/10.1016/j.etap.2018.08.018

Saggar S, Jha N, Deslippe J, Bolan NS, Luo J, Giltrap DL, Kim D-G, Zaman M, Tillman RW (2013) Denitrification and N2O:N2 production in temperate grasslands: processes, measurements, modelling and mitigating negative impacts. Sci Total Environ 465:173–195. https://doi.org/10.1016/j.scitotenv.2012.11.050

Sağlam NG, Rachid NA, Güngör ND (2024) Secondary metabolites and biological compounds of actinomycetes and their applications. In: Bacterial secondary metabolites. Elsevier, pp 123–145. https://doi.org/10.1016/B978-0-323-95251-4.00013-2

Sahoo KK, Goswami G, Das D (2021) Biotransformation of methane and carbon dioxide into high-value products by Methanotrophs: current state of art and future prospects. Front Microbiol 12. https://doi.org/10.3389/fmicb.2021.636486

Santoyo G, Guzmán-Guzmán P, Parra-Cota FI, de Santos-Villalobos S, los Orozco-Mosqueda Ma. del C, Glick BR (2021a) Plant growth stimulation by microbial consortia. Agronomy 11(2):219. https://doi.org/10.3390/agronomy11020219

Santoyo G, Urtis-Flores CA, Loeza-Lara PD, Orozco-Mosqueda M, del C, Glick BR (2021b) Rhizosphere colonization determinants by plant growth-promoting Rhizobacteria (PGPR). Biology 10(6):475. https://doi.org/10.3390/biology10060475

Sarkar A, Aronson KJ, Patil S, Hugar LB, van Loon GW (2012) Emerging health risks associated with modern agriculture practices: a comprehensive study in India. Environ Res 115:37–50. https://doi.org/10.1016/j.envres.2012.03.005

Schaefer H (2019) On the causes and consequences of recent trends in atmospheric methane. Curr Clim Chang Rep 5(4):259–274. https://doi.org/10.1007/s40641-019-00140-z

Schimel JP, Schaeffer SM (2012) Microbial control over carbon cycling in soil. Front Microbiol 3. https://doi.org/10.3389/fmicb.2012.00348

Schweiger R, Heise A-M, Persicke M, Müller C (2014) Interactions between the jasmonic and salicylic acid pathway modulate the plant metabolome and affect herbivores of different feeding types. Plant Cell Environ 37(7):1574–1585. https://doi.org/10.1111/pce.12257

Sciuto K, Moro I (2015) Cyanobacteria: the bright and dark sides of a charming group. Biodivers Conserv 24(4):711–738. https://doi.org/10.1007/s10531-015-0898-4

Sekabira H, Tepa-Yotto GT, Djouaka R, Clottey V, Gaitu C, Tamò M, Kaweesa Y, Ddungu SP (2022) Determinants for deployment of climate-smart integrated Pest management practices: a meta-analysis approach. Agriculture 12(7):1052. https://doi.org/10.3390/agriculture12071052

Sharma P (2022) Role and significance of biofilm-forming microbes in phytoremediation—a review. Environ Technol Innov 25:102182. https://doi.org/10.1016/j.eti.2021.102182

Shelake RM, Pramanik D, Kim J-Y (2019) Exploration of plant-microbe interactions for sustainable agriculture in CRISPR era. Microorganisms 7(8):269. https://doi.org/10.3390/microorganisms7080269

Sherameti I, Tripathi S, Varma A, Oelmüller R (2008) The root-colonizing endophyte pirifomospora indica confers drought tolerance in arabidopsis by stimulating the expression of drought stress–related genes in leaves. Mol Plant-Microbe Interact 21(6):799–807. https://doi.org/10.1094/MPMI-21-6-0799

Solomon W, Janda T, Molnár Z (2024) Unveiling the significance of rhizosphere: implications for plant growth, stress response, and sustainable agriculture. Plant Physiol Biochem 206:108290. https://doi.org/10.1016/j.plaphy.2023.108290

Srivastava RK (2021) Biofertilizers application in agriculture: A viable option to chemical fertilizers. In: Biofertilizers. Wiley, pp 393–411. https://doi.org/10.1002/9781119724995.ch13

Suman J, Rakshit A, Ogireddy SD, Singh S, Gupta C, Chandrakala J (2022) Microbiome as a key player in sustainable agriculture and human health. Front Soil Sci 2. https://doi.org/10.3389/fsoil.2022.821589

Swift MJ, Izac A-MN, van Noordwijk M (2004) Biodiversity and ecosystem services in agricultural landscapes—are we asking the right questions? Agric Ecosyst Environ 104(1):113–134. https://doi.org/10.1016/j.agee.2004.01.013

Syed Ab Rahman SF, Singh E, Pieterse CMJ, Schenk PM (2018) Emerging microbial biocontrol strategies for plant pathogens. Plant Sci 267:102–111. https://doi.org/10.1016/j.plantsci.2017.11.012

Tahat M, Alananbeh KM, Othman YA, Leskovar DI (2020) Soil health and sustainable agriculture. Sustainability 12(12):4859. https://doi.org/10.3390/su12124859

Thakur N, Nigam M, Mann NA, Gupta S, Hussain CM, Shukla SK, Shah AA, Casini R, Elansary HO, Khan SA (2023) Host-mediated gene engineering and microbiome-based technology optimization for sustainable agriculture and environment. Funct Integr Genomics 23(1):57. https://doi.org/10.1007/s10142-023-00982-9

Thrupp LA (2000) Linking agricultural biodiversity and food security: the valuable role of agrobiodiversity for sustainable agriculture. Int Aff 76(2):265–281. https://doi.org/10.1111/1468-2346.00133

Timmis K, Ramos JL (2021) The soil crisis: the need to treat as a global health problem and the pivotal role of microbes in prophylaxis and therapy. Microb Biotechnol 14(3):769–797. https://doi.org/10.1111/1751-7915.13771

Timmusk S, Behers L, Muthoni J, Muraya A, Aronsson A-C (2017) Perspectives and challenges of microbial application for crop improvement. Front Plant Sci 8. https://doi.org/10.3389/fpls.2017.00049

Tosi M, Mitter EK, Gaiero J, Dunfield K (2020) It takes three to tango: the importance of microbes, host plant, and soil management to elucidate manipulation strategies for the plant microbiome. Can J Microbiol 66(7):413–433. https://doi.org/10.1139/cjm-2020-0085

Tripathi S, Srivastava P, Devi RS, Bhadouria R (2020) Influence of synthetic fertilizers and pesticides on soil health and soil microbiology. In: Agrochemicals detection, treatment and remediation. Elsevier, pp 25–54. https://doi.org/10.1016/B978-0-08-103017-2.00002-7

Trivedi P, Mattupalli C, Eversole K, Leach JE (2021) Enabling sustainable agriculture through understanding and enhancement of microbiomes. New Phytol 230(6):2129–2147. https://doi.org/10.1111/nph.17319

Tully KL, McAskill C (2020) Promoting soil health in organically managed systems: a review. Org Agric 10(3):339–358. https://doi.org/10.1007/s13165-019-00275-1

Tunlid A, Floudas D, Koide R, Rineau F (2016) Soil organic matter decomposition mechanisms in ectomycorrhizal fungi. In: Molecular mycorrhizal symbiosis. Wiley, pp 257–275. https://doi.org/10.1002/9781118951446.ch15

Udumann SS, Dissanayaka NS, Nuwarapaksha TD, Thelwadana EP, Atapattu AJ (2023) Assessing the growth potential of Sunn hemp (Crotalaria juncea L.) as a cover crop for major coconut-growing soils. *Trends*. Horticulture 6(2):3579. https://doi.org/10.24294/th.v6i2.3579

Udumann SS, Ranasinghe CS, Karunarathna LKNG, Kaliyadasa PE, Nuwarapaksha TD, Premathilaka UGAT, Atapattu AJ (2024) Optimizing intercropping selection for coconut plantations based on PAR and agro-climatic zones. Agrofor Syst 1:13–2859. https://doi.org/10.1007/s10457-024-00977-w

Vance CP, Uhde-Stone C, Allan DL (2003) Phosphorus acquisition and use: critical adaptations by plants for securing a nonrenewable resource. New Phytol 157(3):423–447. https://doi.org/10.1046/j.1469-8137.2003.00695.x

Viana CM, Freire D, Abrantes P, Rocha J, Pereira P (2022) Agricultural land systems importance for supporting food security and sustainable development goals: a systematic review. Sci Total Environ 806:150718. https://doi.org/10.1016/j.scitotenv.2021.150718

Vishwakarma K, Kumar N, Shandilya C, Mohapatra S, Bhayana S, Varma A (2020) Revisiting plant–microbe interactions and microbial consortia application for enhancing sustainable agriculture: a review. Front Microbiol 11. https://doi.org/10.3389/fmicb.2020.560406

Wang C, Zhao J, Feng Y, Shang M, Bo X, Gao Z, Chen F, Chu Q (2021) Optimizing tillage method and irrigation schedule for greenhouse gas mitigation, yield improvement, and water conservation in wheat–maize cropping systems. Agric Water Manag 248:106762. https://doi.org/10.1016/j.agwat.2021.106762

Wang Q, Zhang Q, Han Y, Zhang D, Zhang C-C, Hu C (2022) Carbon cycle in the microbial ecosystems of biological soil crusts. Soil Biol Biochem 171:108729. https://doi.org/10.1016/j.soilbio.2022.108729

Wang C, Jiang HL (2016) Chemicals used for in situ immobilization to reduce the internal phosphorus loading from lake sediments for eutrophication control. Crit Rev Environ Sci Technol 46(10):947–997. https://doi.org/10.1080/10643389.2016.1200330

Weekley J, Gabbard J, Nowak J (2012) Micro-level management of agricultural inputs: emerging approaches. Agronomy 2(4):321–357. https://doi.org/10.3390/agronomy2040321

Wezel A, Casagrande M, Celette F, Vian J-F, Ferrer A, Peigné J (2014) Agroecological practices for sustainable agriculture. A review. Agron Sustain Develop 34(1):1–20. https://doi.org/10.1007/s13593-013-0180-7

Wezel A, Herren BG, Kerr RB, Barrios E, Gonçalves ALR, Sinclair F (2020) Agroecological principles and elements and their implications for transitioning to sustainable food systems. A review. Agron Sustain Develop 40(6):40. https://doi.org/10.1007/s13593-020-00646-z

Wille L, Messmer MM, Studer B, Hohmann P (2019) Insights to plant–microbe interactions provide opportunities to improve resistance breeding against root diseases in grain legumes. Plant Cell Environ 42(1):20–40. https://doi.org/10.1111/pce.13214

Wipf HM-L, Xu L, Gao C, Spinner HB, Taylor J, Lemaux P, Mitchell J, Coleman-Derr D (2021) Agricultural soil management practices differentially shape the bacterial and fungal microbiomes of *Sorghum bicolor*. Appl Environ Microbiol 87(5). https://doi.org/10.1128/AEM.02345-20

Wu J (2013) Landscape sustainability science: ecosystem services and human well-being in changing landscapes. Landsc Ecol 28(6):999–1023. https://doi.org/10.1007/s10980-013-9894-9

Xia Q, Rufty T, Shi W (2020) Soil microbial diversity and composition: links to soil texture and associated properties. Soil Biol Biochem 149:107953. https://doi.org/10.1016/j.soilbio.2020.107953

Yaashikaa PR, Kumar PS, Jeevanantham S, Saravanan R (2022) A review on bioremediation approach for heavy metal detoxification and accumulation in plants. Environ Pollut 301:119035. https://doi.org/10.1016/j.envpol.2022.119035

Yu Y, Gui Y, Li Z, Jiang C, Guo J, Niu D (2022) Induced systemic resistance for improving plant immunity by beneficial microbes. Plants 11(3):386. https://doi.org/10.3390/plants11030386

Yuan J, Yi X, Cao L (2019) Three-source partitioning of methane emissions from paddy soil: linkage to methanogenic community structure. Int J Mol Sci 20(7):1586. https://doi.org/10.3390/ijms20071586

Zarzosa-Moreno D, Avalos-Gómez C, Ramírez-Texcalco LS, Torres-López E, Ramírez-Mondragón R, Hernández-Ramírez JO, Serrano-Luna J, de la Garza M (2020) Lactoferrin and its derived peptides: an alternative for combating virulence mechanisms developed by pathogens. Molecules 25(24):5763. https://doi.org/10.3390/molecules25245763

Zerkle AL, Mikhail S (2017) The geobiological nitrogen cycle: from microbes to the mantle. Geobiology 15(3):343–352. https://doi.org/10.1111/gbi.12228

Zhai Y, Bai J, Chang P, Liu Z, Wang Y, Liu G, Cui B, Peijnenburg W, Vijver MG (2024) Microplastics in the terrestrial ecosystem: exploring the menace to the soil-plant-microbe interactions. TrAC Trends Anal Chem 174:117667. https://doi.org/10.1016/j.trac.2024.117667

Chapter 2
Microbial Contributions to Nutrient Cycling and Soil Fertility

Chioma Blaise Chikere, Israel Edamkue, and Edmond Osi Uguomore

Abstract The ecosystem is a dynamic network in which important materials are constantly consumed and recycled, with biogeochemical cycles governing their systemic movement in order to support the environment. Microorganisms play a major role in this process, driving the conversion and biotransformation of these components, from the ubiquitous carbon and nitrogen cycles to the more minute elements: sulphur, phosphorus, and potassium cycles, which are handled by specially adapted microorganisms. These minute organisms use various methods for the conversion of these elements: Cyanobacteria help plants fix atmospheric CO_2 during photosynthesis; *Rhizobium* fix nitrogen in the root nodules of plants; as well as the reductive activities (assimilatory and dissimilatory fixation) of microbes in these cycles and others. It has become necessary to study the roles of these microbes in ensuring the continuation of these cycles, as their processes are important for scientific developments in various areas, including for sustainable agricultural practice to achieve sustainable development goals (SDGs) such as that of "Zero Hunger". Next-generation sequencing has emerged as a valuable tool for researching microbial activities, especially with the role of whole-genome sequencing aided by machine learning—allowing tremendous depth of research of entire microbiomes. Finally,

C. B. Chikere (✉)
Department of Microbiology, Faculty of Science, University of Port Harcourt, Port Harcourt, Rivers State, Nigeria

Department of Environmental Sciences, College of Agriculture and Environmental Sciences (CAES), Florida Science Campus, University of South Africa, Florida, South Africa
e-mail: chioma.chikere@uniport.edu.ng

I. Edamkue
Department of Microbiology, Faculty of Science, University of Port Harcourt, Port Harcourt, Rivers State, Nigeria

Department of Biotechnology, Institute of Parasitology, Faculty of Agriculture and Environmental Sciences, McGill University, Ste-Anne-de-Bellevue, QC, Canada

E. O. Uguomore
Department of Microbiology, Faculty of Science, University of Port Harcourt, Port Harcourt, Rivers State, Nigeria

O. O. Babalola et al. (eds.), *Ecofriendly Frontiers*,
https://doi.org/10.1007/978-3-031-98700-7_2

the recycling of nutrients throughout the ecosystem depends on microbial communities' activities in reservoirs and sinks. Beyond ecological applications, these microbial capabilities have the potential for novel applications, such as enhancing medicine and drug discovery advancements.

Keywords Ecosystem · Machine learning · Next-generation sequencing · Sustainable development goals

2.1 Introduction

The ecosystem has a designated flow or movement for materials (vital chemicals) within the system. Plants and the microorganisms living on or around them have been intertwined for over 450 million years, forming a complex unit called a holobiont. The pressures of evolution have shaped the makeup of these holobionts, favouring microbes that benefit the plants and help them thrive (Hassani et al. 2018). These microorganisms break down dead plants and animals, releasing nutrients that plants need for growth. These microorganisms are important for all living life forms on land, from the smallest plants to the biggest animals. This chapter will explore how these amazing creatures (microorganisms) help plants get the food they need and keep the earth healthy and green. Nutrient cycling entails the process where nutrients get absorbed, transferred, released, and reabsorbed within the environment. All these occur through the help of organisms (microorganisms), their communities, and their interactions in the ecosystem. Microbes form the backbone of ecological communities by aiding in these nutrient conversions. The decomposition of organic matter, especially when carbon is readily available, leads to the release of nutrients that plants and microbes can take up. This process is mediated by diverse microbial communities with different functional capabilities. This nutrient conversion occurs via these microbes' decomposition of organic matter and the apparent biogeochemical cycles/chemical pathways involved (Fierer et al. 2023).

This decomposition provides them with energy for growth and the essential element, carbon, for constructing new cells, contributing significantly to the overall cycle of nutrients. The soil organic matter is a mix of the living aspect, which consists of the microbes; the dead, which are fresh residues and the humus, which is completely spent organic matter. Microbes are known to be ubiquitous, and many of them yet to be cultured or even discovered, most of which reside in the soil. Many nutrient reservoirs are found within soil and plants, with other nutrient reservoirs being present within bodies like the oceans, waiting majorly for the action of microbes for the release of their nutrients. Every ecosystem has multiple biochemical cycles occurring, and in the soil, nutrients such as nitrogen, carbon, and sulfur are constantly recycled due to the high presence of organic matter deposited within the soil by decomposing organisms (Zhao et al. 2022).

Microbial contributions to nutrient cycling and soil fertility play a crucial role in sustaining terrestrial ecosystems. The intricate web of interactions between microorganisms, plants, and the surrounding environment forms the foundation of nutrient cycling, which is essential for maintaining soil fertility and supporting plant growth. In this comprehensive overview, we will explore the various ways in which microbes influence nutrient cycling and contribute to soil fertility (Iqbal et al. 2023).

2.2 Microbial Involvement in Nutrient Cycling

Microbial involvement in nutrient cycling is pivotal for maintaining soil fertility and promoting plant growth. Bacteria and fungi are the primary drivers of these processes. Bacteria excel in breaking down organic matter, converting it into plant-available nutrients. For instance, nitrogen-fixing bacteria, such as those in the genus *Rhizobium*, convert atmospheric nitrogen into ammonia, a form usable by plants, thus enriching soil nitrogen levels (Hashim et al. 2022). They also participate in nitrification and denitrification, processes vital for nitrogen transformation and release back into the atmosphere. Fungi, on the other hand, decompose complex organic compounds and extract minerals, enhancing nutrient availability. These microbial activities not only feed plants but also improve soil structure and suppress pathogens, fostering a healthy soil ecosystem. A systematic review analyzing 120 articles found that 31.4% of studies indicate microbial activity directly regulates nutrient cycling, while 68.6% suggest it enhances soil fertility and crop productivity (Marzouk et al. 2024). This reveals the important role of microbes in sustainable agriculture by reducing reliance on chemical fertilizers and promoting natural nutrient cycling processes. The byproducts of microbial decomposition, like humus, improve soil structure, aeration, and water-holding capacity. This, in turn, promotes plant growth, increases crop yields, and reduces reliance on chemical fertilizers. Furthermore, this microbial community acts as a natural defence system, suppressing soil-borne pathogens and fostering beneficial symbioses with plant roots (Ptaszek et al. 2023) (Fig. 2.1).

Nutrient cycling is the process where essential elements like carbon, nitrogen, phosphorus, and sulfur move through various compartments of the ecosystem. Microorganisms, including bacteria, fungi, and archaea, actively participate in these cycles, mediating the transformation and availability of nutrients. The presence of carbon from organic matter enables most microbes concerned with the nutrient cycle to have an energy source. The largest nutrient cycling surface exists in the soil, which contains a rich quantity and variety of species that aid the chemical breakdown and recycling of component material. Human activities on the soil can either increase or decrease the natural progression of the cycling processes. In soil, there is a large quantity of carbon, which in turn allows plants to have more nutrient availability, seeing as carbon is an energy source for the bacteria to metabolize (Freschet et al. 2021).

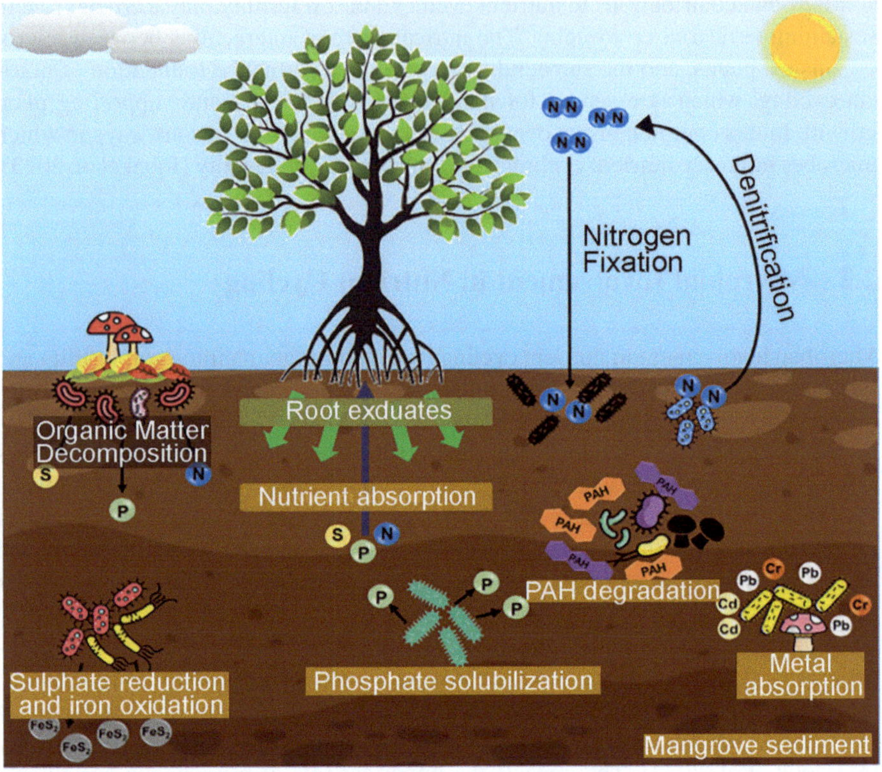

Fig. 2.1 This diagram shows the microbes interacting to release nutrients such as phosphorus (P), nitrogen (N), Sulphur (S), and iron sulphide (FeS_2), aiding the absorption of metals like cadmium (Cd), chromium (Cr); and aiding degradation of Polycyclic aromatic hydrocarbons (PAH). Microbes are involved in almost every nutrient cycle within the soil and environment. They play different roles, from mineralizing to solubilizing and even cleaning up atmospheric elements. This interplay of microbes in nutrient cycling also affects healthy crop growth and agriculture directly through microbial interactions with plants and plant roots. Image Credit: (Palit et al. 2022)

Biogeochemical cycling necessitates the interaction of biological, geological, and chemical processes. As stated earlier, microbes are the major enablers of the biological component of nutrient cycles. On a larger scale, the absence of microbial metabolic actions in these processes would tip the balance of the cycles, adversely impacting the various ecosystems they are found within.

2.2.1 Carbon Cycling and Soil Organic Matter Decomposition

Carbon cycling is a vital process in ecosystems, and soil organic matter (SOM) decomposition plays a crucial role in this cycle. Soil organic carbon (SOC) is a key component of soil fertility and agricultural sustainability, influencing plant productivity and soil health. The decomposition of SOM is primarily driven by soil microbes, which break down complex organic materials into simpler compounds, releasing carbon dioxide (CO_2) back into the atmosphere and making nutrients available for plant uptake (Timmis and Ramos 2021). Certain nutrient cycles involve microbes directly, such cycling pathways that take place in the soil due to soil organic matter, water, and air utilization by microbes (Fig. 2.2).

Soil is the second largest reservoir of carbon on the earth, in the form of soil organic carbon (SOC) (Wu et al. 2024). In the soil, three types of organic matter help in the creation of nutrients in the soil cycle. The types of organic matter found are living organisms, fresh organic residue, and humus. Soil comprises a living microbial community, recently deposited organic residues with high

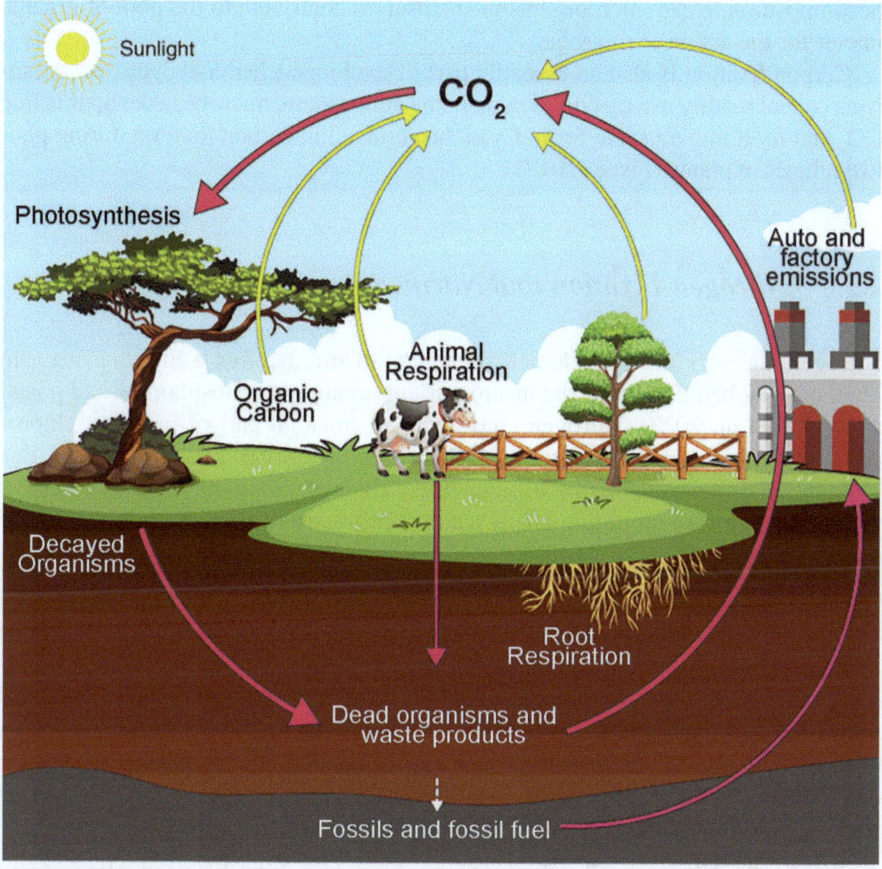

Fig. 2.2 The carbon cycle—The illustration depicts the movement of carbon in the environment. CO_2 from the air is extracted by the aid of cyanobacteria in plants during photosynthesis. Within the soil, microbes act on organic matter as a source of energy while changing the organic carbon forms to CO_2, once again released into the atmosphere. Image credit: (https://byjus.com/neet/nutrient-cycle/)

decomposition potential for nutrient release, and humus, a stable organic fraction that is more resistant to microbial and chemical degradation. Activities such as tilling and other disturbances cause the release of pathway byproducts.

The biogeochemical cycles of carbon and soil organic matter decomposition are particularly important to chemical pathways, as carbon is an important factor for energy and the growth of microbes while ensuring the creation of carbon dioxide sinks. Soil microorganisms obtain energy through dissimilation during electron transfer in the form of respiration or fermentation to liberate carbon and synthesize organic matter for assimilation and fixation. Soil microorganisms obtain energy through dissimilation during electron transfer in the form of respiration or fermentation to liberate carbon and synthesize organic matter for assimilation and fixation (Wu et al. 2024). Organic carbon from dead and decaying organisms is acted on by microorganisms, which is then released into the atmosphere upon completion of decomposition. Dead microorganisms themselves contribute to the pool of organic matter for the action of microbes.

Carbon fixation is also an essential part of the biogeochemical cycle. As organisms cannot readily use carbon dioxide in the atmosphere, microbes take up this free CO_2 and fix it into a usable form. Cyanobacteria aid in carbon fixation during photosynthesis in plants (Fisher 2017).

2.2.2 Nitrogen Fixation and Nitrification

Nitrogen, the very basic fundamental element of life, is often a limiting factor in plant growth, being a part of the macronutrients along with phosphorous and potassium (Liu et al. 2023). Nitrogen constitutes an essential part of proteins, chlorophyll, and DNA, undergoing a series of transformations to become parts of this biological machinery, ensuring life can thrive. Understanding the necessary steps of the nitrogen cycle within the soil is paramount to sustainable agriculture and the health of our planet.

Nitrogen Fixation

Nitrogen fixation is the process by which atmospheric nitrogen (N_2) is converted into ammonia (NH_3) or ammonium (NH_4^+). Certain bacteria and archaea primarily carry out this process. The enzyme nitrogenase, present in these organisms, breaks the strong triple bond in N_2 and combines nitrogen with hydrogen to form NH_3. This reaction requires significant energy input, often derived from ATP hydrolysis (Lee et al. 2014).

There are three main types of nitrogen fixation:

1. **Symbiotic Nitrogen Fixation:** Occurs in partnerships between plants (e.g., legumes) and bacteria like *Rhizobium* or *Bradyrhizobium.* The plant provides carbohydrates, while the bacteria fix nitrogen into a usable form (Zahran 1999).
2. **Associative Nitrogen Fixation:** Involves close but non-symbiotic associations between plants and nitrogen-fixing bacteria (Mus et al. 2016).

3. **Free-living Nitrogen Fixation:** Performed by soil bacteria that fix nitrogen independently without plant partnerships (Smercina et al. 2019).

Nitrification

Nitrification is the biological oxidation of ammonia (NH_3) to nitrate (NO_3^-). This process is carried out by specialized bacteria and archaea and involves two main steps:

1. **Ammonium Oxidation**: Ammonium (NH_4^+) is oxidized to nitrite (NO_2^-) by ammonia-oxidizing bacteria (AOB) or archaea (AOA).
2. **Nitrite Oxidation**: Nitrite (NO_2^-) is further oxidized to nitrate (NO_3^-) by nitrite-oxidizing bacteria (NOB) (Ni et al. 2023).
3. Nitrification is essential for converting ammonia into nitrate, which is more soluble and can be easily absorbed by plants. Factors like oxygen availability, pH, and temperature significantly influence the efficiency of this process (Zayed et al. 2023).

These microbial nitrogen fixers transform the inert atmospheric nitrogen into the elemental currency of life, fueling verdant landscapes and bountiful harvests. This ammonia, along with other organic nitrogen compounds released by decomposing organisms, forms the "organic nitrogen pool" within the soil. This pool acts as a reservoir, slowly releasing plant-available forms like ammonium and nitrate through the actions of soil microbes (Begum et al. 2021). Nitrification, performed by specialized bacteria called nitrifying bacteria, transforms this ammonia into nitrate, the preferred form of nitrogen for most plants (Hossain et al. 2022). Plants then readily absorb the nitrate through their roots, incorporating it into the form of amino acids, the building blocks of life. This nitrogen stimulates their growth, allowing them to produce healthy leaves, sturdy stems, and a plethora of healthy fruits. As herbivores graze on these nitrogen-rich tissues, this element continues its journey through the food web, nourishing consumers at each trophic level. When organisms die and decompose, their nitrogen-laden remains return to the soil, where the decomposing bacteria, fungi, and other soil denizens break down these organic materials to liberate nitrogen compounds like ammonium back into the organic pool. This vital process is referred to as mineralization, ensuring the continuous recycling of nitrogen within the soil (Tauro et al. 2021) (Fig. 2.3).

However, not all the nitrogen available is utilized by plants. The climax of the nitrogen cycle is the denitrification process, which is performed by another specialized group of bacteria that converts nitrate back into gaseous form, which escapes into the atmosphere (Xiang et al. 2020). While seemingly a loss, denitrification plays a crucial role in regulating nitrogen levels in the soil, preventing excessive buildup and potential harm to organisms and the ecosystem (Almaraz et al. 2020).

Soil fertility, crucial for sustainable agriculture, relies heavily on the efficient cycling of nitrogen. Depleted soils lacking organic matter and microbial activity struggle to provide plants with the necessary nitrogen for healthy growth. This can lead to decreased crop yields, which can impact food security and individual livelihoods (Jahanzad et al. 2020) (Ramond et al. 2022). The nitrogen cycle is vital to

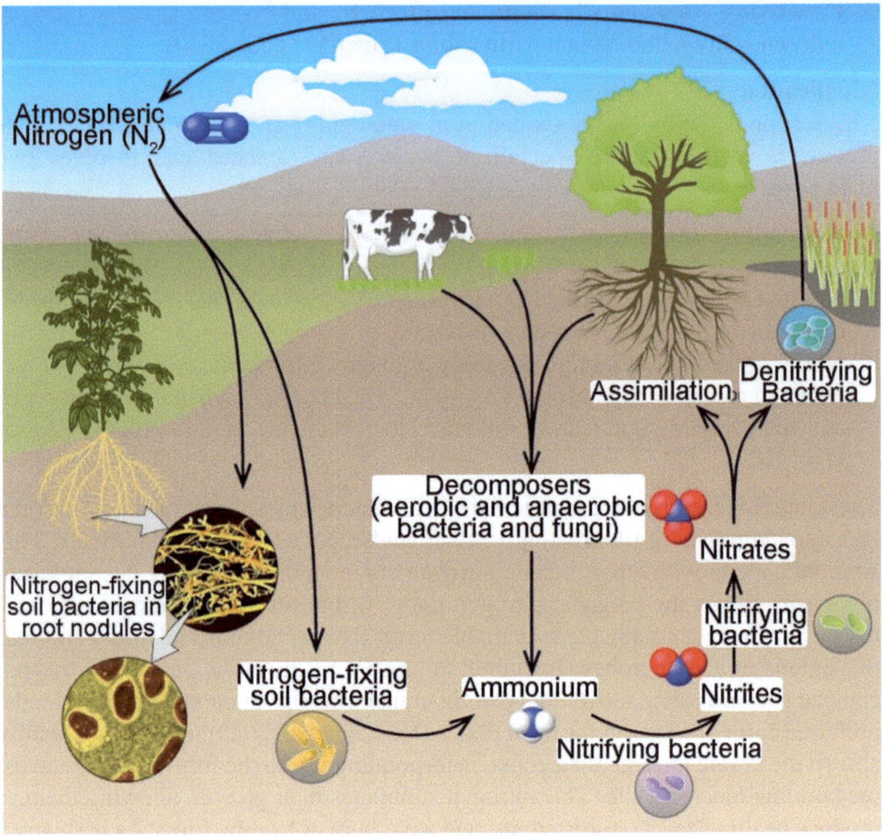

Fig. 2.3 The Nitrogen Cycle—Bacteria play a significant role in nitrogen cycling from the atmosphere and within the soil. Bacteria convert nitrogen from its various forms to others and fix nitrogen to plant root nodules. These transformations are termed either assimilatory or dissimilatory reactions of nitrogen (Image credit: https://commons.wikimedia.org/wiki/File:CNX_Chem_18_07_Nitrogen.png)

life. Through a series of transformations, this crucial element nourishes plants, sustains food webs, and ensures the continued vibrancy of our planet.

2.2.3 *Phosphorus Solubilization and Nutrient Recycling*

Phosphorus is another element that plays a crucial role in the nutrient-recycling process. This element is vital for plant life, and understanding the microbial contribution to the phosphorus cycle in the soil will help in recognizing their essential role in maintaining fertile ecosystems and ensuring food security for future generations (Constantinescu-Aruxandei and Oancea 2023). Unlike carbon and nitrogen, phosphorus has a limited terrestrial reservoir, and its availability is limited by its tendency to bind to soil particles, rendering it inaccessible to plants. This limited availability presents a unique challenge for its utility. Luckily, microbes have evolved ingenious ways to make this vital element readily available for consumption by plant populations. The microorganisms involved in the phosphorus cycle are diverse, with each member playing a specific role. Bacteria, the most abundant and diverse group, play an active role. Some, like the phytate-mineralizing bacteria, secrete enzymes that break down organic phosphorus, releasing the element in a form plants can readily absorb. Others, like acid-producing bacteria, lower the soil pH, making insoluble phosphate minerals more soluble and bioavailable (Kaur and Kaur 2022) (Fig. 2.4).

Fungi, on the other hand, play a crucial role as support. They form symbiotic relationships with plants, extending their root reach and tapping phosphorus from organic matter and minerals. Additionally, Fungal threads can directly take in and use phosphate, helping to recycle it in the soil, while tiny organisms like protozoa and nematodes act as consumers, feeding on bacteria and fungi. This grazing process releases nutrients, including phosphorus, back into the soil, making it available for other microbes and plants (Koehle et al. 2023).

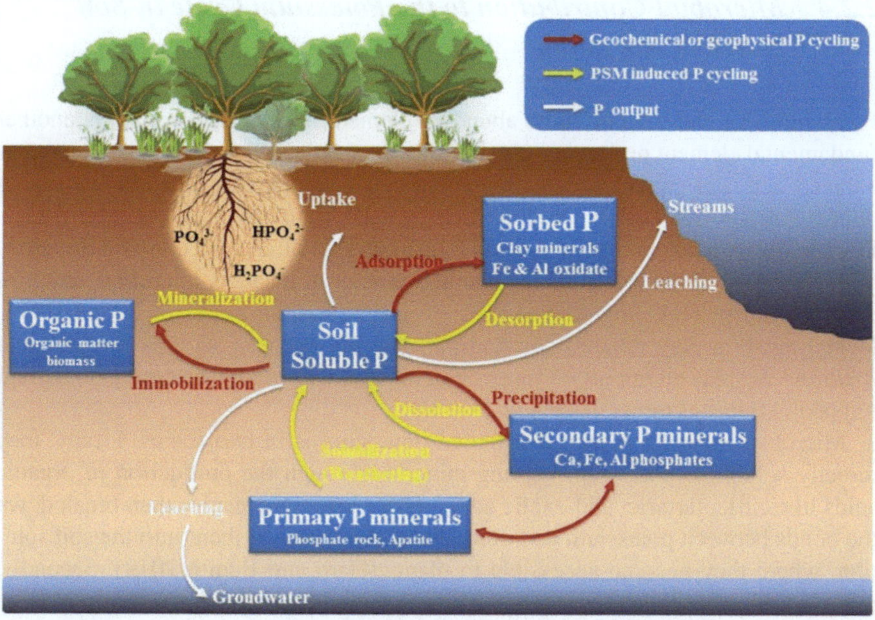

Fig. 2.4 Phosphorous solubilization—Phosphorous is predominantly cycled by climatic action as it is easily bound in soil particles. As such, phosphorous solubilizing bacteria are specially adapted to secrete chemicals to release bound particles. In the diagram, microbe-mediated changes in the cycle are identified by the yellow-coloured arrows (Image credit: https://microbenotes.com/phosphorus-cycle/#:~:text=Different%20microorganisms%20like%20Actinomycetes%2C%20Pseudomonas,to%20plants%20and%20other%20animals)

The microbial contribution to the phosphorus cycle goes beyond simply making it available. They drive the continuous conversion and remodification of phosphorus between organic and inorganic forms. During **mineralization**, microbes decompose organic matter, releasing phosphate ions. This process makes phosphorus readily available for plants and other organisms. Conversely, during **immobilization**, microbes incorporate phosphorus into their biomass, temporarily binding it and making it unavailable for immediate uptake by plants.

The balance between these two processes determines the overall availability of phosphorus in the soil. Factors like soil temperature, moisture, and pH influence microbial activity and, consequently, the rate of mineralization and immobilization (Muhammad 2022). The microbial contribution to the phosphorus cycle is not confined to individual soil particles; they play an important role in maintaining the health and fertility of entire ecosystems. Microbial activity helps build and maintain soil organic matter, a vital reservoir of nutrients, including phosphorus. Healthy soils with diverse microbial communities are more resilient to environmental disturbances, such as erosion and climate change, and are better able to support plant growth and food production (Timmis and Ramos 2021).

2.2.4 Microbial Contribution to the Potassium Cycle in Soil Nutrient Recycling

Potassium (K), the seventh most abundant element in the earth's crust, is another fundamental element promoting plant health, as stated previously. It regulates water balance, activates enzymes, and strengthens cell walls, playing a paramount role in photosynthesis, respiration, and nutrient uptake (Sardans and Peñuelas 2021). Despite its abundance, readily available potassium is often scarce in soil due to its high solubility. However, in their relentless pursuit of balance, microbial activities ensure efficient recycling. Up to 98% of soil potassium exists in bound forms, locked away in clay minerals or feldspar, but soil bacteria such as *Bacillus* and *Arthrobacter* play an important function in mineral weathering, releasing potassium trapped within soil minerals (Fig. 2.5).

Microbial interaction with this bound-potassium is in two stages. Firstly, they directly solubilize potassium-bearing minerals through the production of organic acids like citric, tartaric, and oxalic acids. These acidic molecules then break down the bonds between potassium and other elements, releasing them into the soil solution, where they become accessible to plants (Haro and Benito 2019). Secondly, microbes adopt a subtler approach, forming symbiotic partnerships with plants, e.g., *rhizobacteria*, colonizing the rhizosphere (the zone around plant roots), interacting with root hairs, and stimulating the release of enzymes and organic compounds that further liberate potassium. This mutualistic relationship benefits both parties—the

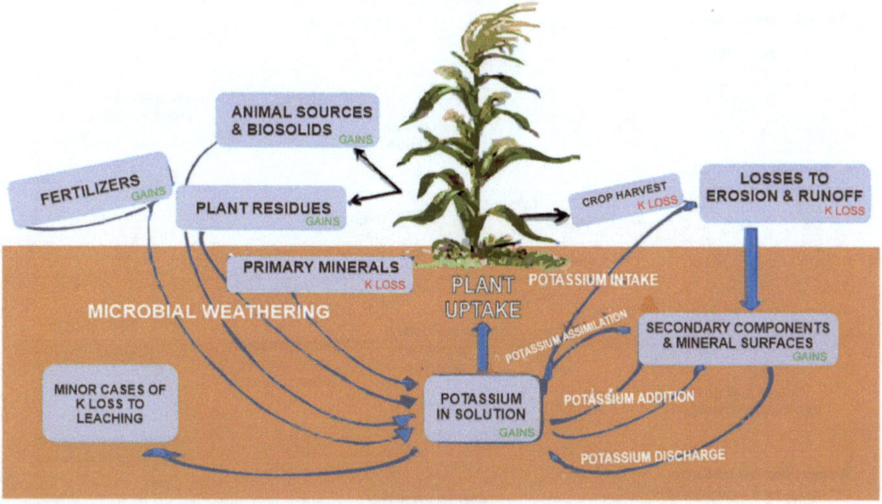

Fig. 2.5 The Cycle of Potassium in the Environment—Potassium is an important element for plant growth. Microbes act on mineralized potassium to release a form usable by plants. Hence, microbes that solubilize potassium interact closely with plant root nodules. The actions presented by the microbes are a symbiotic relationship, beneficial to both plants and microbes (Modified from Tian et al. (2021))

plants gain essential potassium, while the microbes receive sugars and organic matter from the roots.

The diversity of microbial potassium solubilizers is astounding. Bacteria like *Bacillus*, *Pseudomonas*, *Serrati*a, and fungi like *Penicillium* and *Aspergillus* contribute to this process by employing their unique mode of action of the potassium-liberating method. Some even secrete biosurfactants, molecules that lower the surface tension of minerals, increasing potassium accessibility (Kushwaha et al. 2020). This activity contributes to the microbial interaction of not only ensuring adequate potassium supply for plants but also contributes to other vital soil functions. Organic acids produced by solubilizers can chelate other nutrients, making them more available.

The activities of the microbial contribution to the potassium cycle play a crucial role in maintaining soil health and fertility. Studies have shown that soils rich in potassium-solubilizing microbes exhibit enhanced plant growth, yield, and resilience to stress (Jini et al. 2024; Nawaz et al. 2023; Olaniyan et al. 2022; Pattnaiket al. 2021; Zhang et al. 2024). This has profound implications for sustainable agriculture, offering a natural and eco-friendly alternative to chemical fertilizers, which can lead to soil degradation and environmental pollution (Wang et al. 2022).

Therefore, promoting microbial communities through cover cropping, organic compost addition, and judicious fertilizer management is essential for maintaining a vibrant potassium cycle and sustaining soil health (Hartman et al. 2018).

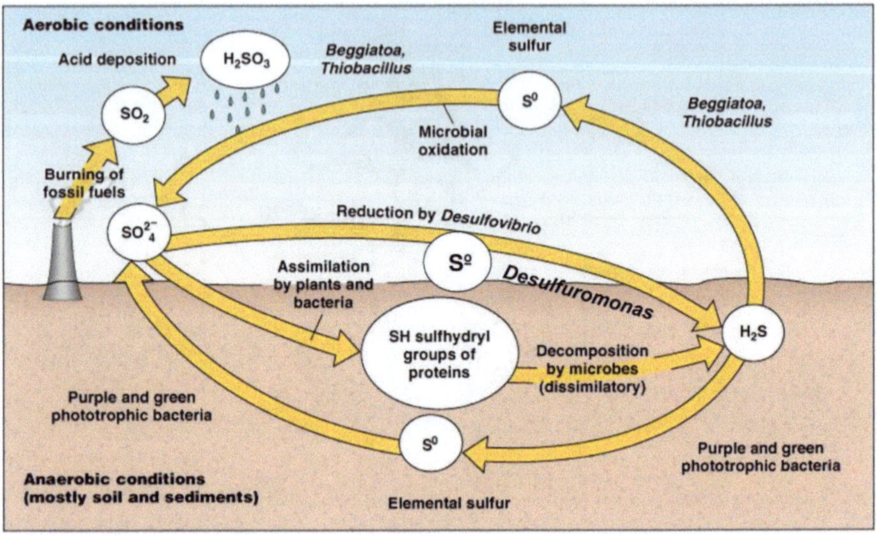

Fig. 2.6 The Sulfur Cycle—Sulfur is sourced from rocks by weathering action and then assimilated by microbes and plants from which it enters the food chain. Sulfur is released into the air to form sulfates by combination with oxygen. Assimilatory sulfate reduction occurs in cases where bacteria reduce small amounts of sulfates to synthesize minute sulfate-bound cell components. Dissimilatory sulfate reduction has sulfate-reducing bacteria reducing sulfates in large quantities as an energy source, releasing sulfide as a by-product of the respiratory action. Many of these are in archea genera such as Desulfuromonas and other general sulfate-reducers (Proteus, Campylobacter, Pseudomonas and Salmonella), which are not sulfate-dependent. (Image credit: https://www.slideserve.com/mattox/sulfur-cycle-facts)

2.2.5 *Sulfur Cycling and Microbial Transformation*

Sulfur is another essential nutrient for plant growth and is a part of nutrient cycling, playing essential roles in protein synthesis, enzyme and coenzyme formation, and various metabolic processes (Fike et al. 2023). The sulfur cycle is a dynamic process that moves sulfur through the biosphere, involving organic and inorganic transformations mediated by microorganisms. Sulfate-reducing bacteria convert sulfate to hydrogen sulfide, while other bacteria oxidize sulphur compounds. These processes influence sulfur availability for plants and contribute to soil fertility (Fig. 2.6).

2.2.5.1 Microbial Transformations within the Sulfur Cycle

Microorganisms facilitate crucial steps of the sulfur cycle through oxidation, reduction, and mineralization processes:

- **Oxidation:** Sulfur-oxidizing bacteria (SOB), such as those belonging to the *Thiobacillus* genus, are critical in converting reduced sulfur compounds (e.g., hydrogen sulfide - H_2S, elemental sulfur—SO) into more oxidized states like sulfate (SO_4^{2-}) (Nguyen et al. 2022). This oxidation makes sulfur available for plant uptake.

- **Reduction:** On the other hand, sulfate-reducing bacteria (SRB), including species like *Desulfovibrio*, convert sulphates back into reduced forms like sulphide (Nguyen et al. 2022). This reduction process is significant in anaerobic environments and can contribute to the release of H_2S.
- **Mineralization and Immobilization**: Diverse soil microorganisms break down organic sulfur-containing compounds like amino acids and release them as inorganic sulfates, a process known as **mineralization** (Fike et al. 2023). Conversely, **immobilization** describes the assimilation of inorganic sulfur into the organic matter of microbes.

2.3 The Interplay in Nutrient Recycling

The sulfur cycle is not isolated but significantly impacts the recycling of other key nutrients:

- **Nitrogen and Sulfur Interaction:** Microbial transformations of sulfur compounds often influence the nitrogen cycle (Vitousek et al. 2002). For instance, nitrification (conversion of ammonium to nitrate) can be affected by sulfide oxidation, potentially leading to acidification.
- **Carbon and Sulfur Linkage:** The decomposition of organic matter involves both carbon and sulfur cycling. Microbes utilizing organic matter for energy and carbon release or incorporate sulfur compounds during these processes (Fike et al. 2023).

2.4 Environmental and Agricultural Significance of the Sulfur Cycle

- **Soil Health and Plant Nutrition:** Sulfur cycling directly influences soil fertility and plant nutrient availability, as sulfate is a primary form of sulfur plants can absorb (Fike et al. 2023). An effective sulfur cycle ensures an adequate supply of this essential nutrient.
- **Environmental Acidification:** Anthropogenic activities like fossil fuel burning release sulfur dioxide (SO_2) into the atmosphere. Microbial sulfur oxidation can contribute to acid rain formation, causing ecological damage (Vitousek et al. 2002).

2.5 Microbial Interactions and Soil Nutrient Dynamics

The productivity arising from soil fertility lies not in the individual mode of action of this microbial activity but in the harmonious collaboration of this diverse microbial community (polymicrobial interaction) (Kalinitchenko et al. 2024). A healthy soil microbiome, rich in bacterial, fungal, and archaeal populations, offers resilience and adaptability to environmental fluctuations. Each microbial group contributes its unique function to the nutrient recycling process, ensuring a balanced and efficient performance. Disrupting this diversity through excessive chemical use or monoculture practices can lead to discordant performances, with detrimental consequences for soil health and plant productivity.

Microbial communities exhibit complex interactions with each other and with plants, influencing nutrient cycling dynamics. Competition, mutualism, and predation among microorganisms shape the availability of nutrients in the soil. Understanding these interactions is crucial for managing soil fertility effectively.

Discovering the Untapped Nutrients One of the key roles of soil microbes lies in the mineralization of organic matter. Bacteria and fungi, through their enzymatic activities, break down complex organic compounds like dead leaves and plant roots, releasing locked-away nutrients such as nitrogen, phosphorus, and potassium into forms readily accessible to plants (Hameeda et al. 2019). This process replenishes the soil's nutrient pool and makes it bioavailable for plant uptake, fostering healthy plant growth.

Symbiotic Partnerships Beyond mere decomposition, microbes often engage in symbiotic relationships with plants, forming mutually beneficial partnerships. Arbuscular mycorrhizal (AM) fungi, for example, form a network with plant roots, extending their reach and facilitating the acquisition of nutrients, particularly phosphorus, from the soil in exchange for plant-derived sugars (Chen et al. 2021). This intricate dance highlights the interdependence between plants and microbes in maintaining soil fertility and plant health.

Competition and Cooperation The stage of soil is not always one of harmony. Competition for resources like nutrients can fuel antagonistic interactions between different microbial groups. This competition can influence the overall efficiency of nutrient cycling and potentially limit nutrient availability for plants (Bongiorno et al. 2019). However, this relationship does not end there. Microbes can also exhibit cooperative behaviours, like the exchange of nutrients or signalling molecules, creating a complex web of interactions that ultimately shapes the overall nutrient dynamics in the soil (Cordovez-Cedeño et al. 2019).

2.6 Impact of Land Use and Management Practices on Soil Fertility

Pollution and contamination pose a significant threat to soil fertility and plant health (Feng and Balkcom 2017). These activities often lead to a decline in microbial diversity and activity and ultimately impact soil fertility. Bacteria like *Rhodococcus* and *Pseudomonas* employ their metabolic machinery to break down complex hydrocarbons, while others, like *Stenotrophomonas*, tackle heavy metals and pesticides (Kumari et al. 2023). Activities of these microbes contribute to the continued preservation of fertile soil for plant life.

2.7 Technological Advancements in Studying Microbial Nutrient Cycle Contributions

2.7.1 Advances in Molecular Biology and DNA Sequencing Technologies Have Revolutionized the Study of Microbial Communities in Soils

The past decade has witnessed the rise of metagenomics and Next Generation sequencing (NGS), a powerful technique that allows scientists to analyze the entire genetic material of a microbial community within an environmental sample. By analyzing the genes associated with specific metabolic pathways and sequencing, scientists can gain valuable insights into the functional capabilities of these microbial communities and their potential contributions to the nutrient-cycling process (He et al. 2020; Roux et al. 2018; Nayfach et al. 2016). Metagenomics has revealed previously unknown microbial species and their roles in nutrient transformation processes (Jia et al. 2021). Other methods include stable isotope probing (Radl et al. 2018; Chen et al. 2019), microfluidics (Zhu et al. 2022), and in the now, artificial intelligence and machine learning (Surekha et al. 2024).

These advancements and ongoing research provide an increasingly detailed picture of the diverse and complex world of microbial contributions to nutrient recycling. By harnessing these technologies, scientists can develop strategies to optimize nutrient cycling in various ecological settings, promoting sustainable agricultural practices and environmental health.

2.8 Challenges and Future Directions

2.8.1 Challenges in Soil and Nutrient Cycling Improvements

Soil management practices greatly impact soil microbes and nutrient cycling. Unsustainable land practices, such as overgrazing, bush burning, and continuous cropping, as well as the application of mineral fertilizer, chemical pesticides, and herbicides, disrupt the natural cycle by reducing plant dependency on certain microbial populations present in the soil (Fall et al. 2022). Studies may be successful in vitro, but on-field applications appear ineffective with the continuation of the current norms in agriculture.

Socio-economic realities in areas such as sub-Saharan Africa impede efforts to ensure and improve soil fertility, creating barriers to knowledge transfer and scientific advancement (Stewart et al. 2020). Issues within the region, such as the lack of funding for science, make it impossible to implement modern agricultural practices on a larger scale.

Despite significant progress, fully understanding the complexity of microbial contributions to nutrient cycling remains challenging (Ge et al. 2022). Climate change, pollution, and other environmental stressors threaten microbial communities and soil fertility. Future research must focus on sustainable land management practices and innovative technologies to address these challenges.

2.8.2 Future Research Directions for Improvement

Understanding the interactions between microbes and the continued progress of the nutrient cycle is sacrosanct for developing sustainable land-management practices. Promoting soil organic matter build-up, encouraging plant-microbial symbioses, and reducing reliance on chemical fertilizers can create conditions that favour diverse and active microbial communities, leading to increased soil fertility and efficient nutrient recycling.

Furthermore, research into harnessing the power of specific microbes, like phosphate-solubilizing bacteria, could lead to the development of biofertilizers and bioremediation techniques that enhance phosphorus availability and minimize environmental impacts (Vyas et al. 2022). Understanding the complexities of the microbial potassium cycle offers exciting possibilities for improving agricultural practices.

Research into isolating and optimizing potassium-solubilizing microbes holds the potential for developing highly effective biofertilizers, promoting plant growth, and reducing reliance on chemical fertilizers. Further exploration of plant-microbe interactions can also help tailor crop systems to specific soil conditions, maximizing nutrient efficiency and minimizing environmental impact (Kumar et al. 2021).

Investing in soil microbial health is not just about improving crop yields; it is also about securing a sustainable future for future generations. By fostering a thriving microbial community, we can build resilient food systems, mitigate climate change, and protect the very foundation of life on Earth. By recognizing the invisible hand of microbes in shaping soil fertility, we can rewrite the narrative of food production, one microscopic step at a time.

Nurturing the activities of the microbial world requires a shift in our agricultural practices and techniques. There is a need to move beyond the confines of chemical-intensive practices and embrace a holistic approach that prioritizes microbial well-being. Such techniques to improve microbial activity in the soil may be seen in other chapters of this textbook.

Understanding and nurturing this microbial world is therefore essential for achieving the pivotal Sustainable Development Goals (SDGs) set by the United Nations, particularly those related to food security, sustainable agriculture, and climate action. The contributions of soil microbes align perfectly with the Sustainable Development Goals (SDGs) aspirations. SDG-2, aiming for "Zero Hunger," can be significantly advanced by harnessing the power of microbes to improve soil fertility and increase food production. Sustainable agricultural practices like organic farming and reduced tillage, which prioritize enhancing microbial activity, can lead to resilient food systems and enhanced food security. SDG-13, which focuses on "Climate Action," also benefits from the carbon sequestration potential of healthy soils teeming with microbes. Microbes contribute to the storage of atmospheric carbon in soil organic matter, mitigating climate change and promoting soil health. (Gil et al. 2019).

2.9 Conclusion

The microbial contribution to nutrient recycling and soil fertility is a story of hidden and unseen heroes essential to supporting life's continual existence on Earth. By appreciating their role in nurturing healthy soil ecosystems, it is possible to ensure a future where the elements of life continue to sustain life for generations to come.

Furthermore, microbial contributions to nutrient cycling and soil fertility are fundamental for the health and productivity of terrestrial ecosystems, thus helping the attainment of the SDGs of "Zero Hunger", "Climate Action," and sustainable agriculture as earlier described. Recognizing the delicate relationships between microorganisms and nutrient dynamics is crucial for developing sustainable agricultural practices and conserving natural ecosystems in the face of global environmental changes. The power provided by NGS and machine learning are tools available to aid the achievement of new heights in understanding our environment. Additional modern techniques, such as microfluidics and SIP, may be utilized to produce a more robust approach to the research in the field. Continued research and interdisciplinary collaboration will further enhance our knowledge and guide effective soil management strategies with the input of microbes.

References

Almaraz M, Wong MY, Yang WH (2020) Looking back to look ahead: a vision for soil denitrification research. Ecology 101(1):e02917. https://doi.org/10.1002/ecy.2917

Begum R, Jahangir MMR, Jahiruddin M, Islam MR, Rahman MT, Rahman ML, Ali MY, Hossain MB, Islam KR (2021) Nitrogen fertilization impact on soil carbon pools and their stratification and lability in subtropical wheat-mungbean-rice agroecosystems. PLoS One 16(10):e0256397. https://doi.org/10.1371/journal.pone.0256397

Bongiorno G, De Marco A, Giribaldi V, Anzuay MS (2019) Competition for nutrients shapes the bacterial community structure of the rhizosphere. Microbiology Open 8(11):e00920. https://www.frontiersin.org/articles/10.3389/ffgc.2023.1135692

Chen Y, Lu H, Sun G, Liu X (2019) Stable isotope probing (SIP) for investigating nutrient transformation in the environment: a review. Appl Microbiol Biotechnol 103(18–19):6821–6834. https://www.ncbi.nlm.nih.gov/pmc/articles/PMC10477051/

Chen M, Wang F, Wu X, Feng G, Lin X (2021) Arbuscular mycorrhizal fungi enhance the utilization efficiency of fertilizer nitrogen by maize plants under drought stress. Plant Soil 463(1–2):339–352. https://www.ncbi.nlm.nih.gov/pmc/articles/PMC9298553/

Constantinescu-Aruxandei D, Oancea F (2023) Closing the nutrient loop-the new approaches to recovering biomass minerals during the biorefinery processes. Int J Environ Res Public Health 20(3):2096. https://doi.org/10.3390/ijerph20032096

Cordovez-Cedeño R, Jaramillo-Soto S, Orjuela-Mora P (2019) Co-occurrence patterns of soil bacteria reveal potential cooperation for nutrient cycling. Microbiol Res 228:126348. https://www.frontiersin.org/journals/microbiology/articles/10.3389/fmicb.2023.1160683/full

Fall AF, Nakabonge G, Ssekandi J, Founoune-Mboup H, Apori SO, Ndiaye A, Badji A, Ngom K (2022) Roles of arbuscular mycorrhizal fungi on soil fertility: contribution in the improvement of physical, chemical, and biological properties of the soil. Front Fungal Biol 3:723892. https://doi.org/10.3389/FFUNB.2022.723892/BIBTEX

Feng Y, Balkcom KS (2017) Nutrient cycling and soil biology in row crop systems under intensive tillage. Soil health and intensification of agroecosystems. Elsevier, pp 231–255. https://doi.org/10.1016/B978-0-12-805317-1.00011-7

Fierer N, Walsh CM (2023) Can we manipulate the soil microbiome to promote carbon sequestration in croplands? PLoS Biol 21(7):e3002207. https://doi.org/10.1371/journal.pbio.3002207

Fike DA, Sulc RM, Bender A, Dick WA, Watson ME (2023) Sulfur as part of a fertility program in the North-Central United States. Agron J 13(1):1–26

Fisher MR (2017) 3.2 biogeochemical cycles. Open Oregon Educational Resources

Freschet GT, Pagès L, Iversen CM, Comas LH, Rewald B, Roumet C, Klimešová J, Zadworny M, Poorter H, Postma JA, Adams TS, Bagniewska-Zadworna A, Bengough AG, Blancaflor EB, Brunner I, Cornelissen JHC, Garnier E, Gessler A, Hobbie SE, Meier IC, McCormack ML (2021) A starting guide to root ecology: strengthening ecological concepts and standardising root classification, sampling, processing and trait measurements. New Phytol 232(3):973–1122. https://doi.org/10.1111/nph.17572

Ge H, Ding K, Guo F, Wu X, Zhai N, Wang W (2022) Green and superior adsorbents derived from natural plant gums for removal of contaminants: a review. Materials (Basel) 16(1):179. https://doi.org/10.3390/ma16010179

Gil JDB, Reidsma P, Giller K, Todman L, Whitmore A, van Ittersum M (2019) Sustainable development goal 2: improved targets and indicators for agriculture and food security. Ambio 48(7):685–698. https://doi.org/10.1007/s13280-018-1101-4

Hameeda B, Akhter MS, Iqbal J (2019) Mineralization of organic matter and potential for plant growth promotion by novel bacterial strains. Front Life Sci 12(3):232–243. https://www.sciencedirect.com/science/article/pii/S0038071707002891

Haro R, Benito B (2019) The role of soil fungi in K^+ plant nutrition. Int J Mol Sci 20(13):3169. https://doi.org/10.3390/ijms20133169

Hartman K, van der Heijden MGA, Wittwer RA, Banerjee S, Walser JC, Schlaeppi K (2018) Cropping practices manipulate abundance patterns of root and soil microbiome members paving the way to smart farming. Microbiome 6(1):14. https://doi.org/10.1186/s40168-017-0389-9

Hashim S, Waqas M, Rudra RP, Khan AA, Mirani AA, Sultan T, Ehsan F, Abid M, Saifullah M (2022) On-farm composting of agricultural waste materials for sustainable agriculture in Pakistan. Scientifica 2022:5831832–5831812. https://doi.org/10.1155/2022/5831832

Hassani MA, Durán P, Hacquard S (2018) Microbial interactions within the plant holobiont. Microbiome. NLM (Medline) 6. https://doi.org/10.1186/s40168-018-0445-0

He J, Wang X, Liu S, Lv J, Wang Y, Li X, Yu Y (2020) Next-generation sequencing in microbial ecology: application and future prospects. Microbiome 8(1):1–13. https://www.ncbi.nlm.nih.gov/books/NBK513764/

Hossain S, Chow CWK, Cook D, Sawade E, Hewa GA (2022) Review of nitrification monitoring and control strategies in drinking water system. Int J Environ Res Public Health 19(7):4003. https://doi.org/10.3390/ijerph19074003

Iqbal B, Li G, Alabbosh KF, Hussain H, Khan I, Tariq M, Ahmad N (2023) Advancing environmental sustainability through microbial reprogramming in growth improvement, stress alleviation, and phytoremediation. Elsevier B.V, Plant Stress. https://doi.org/10.1016/j.stress.2023.100283

Jahanzad E, Holtz BA, Zuber CA, Doll D, Brewer KM, Hogan S, Gaudin ACM (2020) Orchard recycling improves climate change adaptation and mitigation potential of almond production systems. PLoS One 15(3):e0229588. https://doi.org/10.1371/journal.pone.0229588

Jia Z, Wang X, Ma Y, Liu A (2021) Metagenomic insights into microbial community diversity and metabolic potential in the rhizosphere of wheat under long-term fertilization regimes. Front Microbiol 12:660438. https://pubmed.ncbi.nlm.nih.gov/38224956/

Jini D, Ganga VS, Greeshma MB, Sivashankar R, Thirunavukkarasu A (2024) Sustainable agricultural practices using potassium-solubilizing microorganisms (KSMs) in coastal regions: a critical review on the challenges and opportunities. Environ Dev Sustain 26(6):13641–13664. https://doi.org/10.1007/S10668-023-03199-9

Kalinitchenko VP, Swidsinski AV, Glinushkin AP, Overcash MR, Meshalkin VP, Minkina TM et al (2024) New approach to soil fertility management: biogeosystem technique methodology (critical review). Land Degrad Dev 35(7):2367–2381

Kaur R, Kaur S (2022) Exploration of phytate-mineralizing bacteria with multifarious plant growth-promoting traits. Biotechnologia 103(2):99–112. https://doi.org/10.5114/bta.2022.116204

Koehle AP, Brumwell SL, Seto EP, Lynch AM, Urbaniak C (2023) Microbial applications for sustainable space exploration beyond low earth orbit. NPJ microgravity 9(1):47. https://doi.org/10.1038/s41526-023-00285-0

Kumar S, Diksha, Sindhu SS, Kumar R (2021) Biofertilizers: An ecofriendly technology for nutrient recycling and environmental sustainability. Curr Res Microb Sci 3:100094. https://doi.org/10.1016/j.crmicr.2021.100094

Kumari S, Das S (2023) Bacterial enzymatic degradation of recalcitrant organic pollutants: catabolic pathways and genetic regulations. Environ Sci Pollut Res 30(33):79676–79705

Kushwaha P, Kashyap PL, Srivastava AK, Tiwari RK (2020) Plant growth promoting and antifungal activity in endophytic bacillus strains from pearl millet (Pennisetum glaucum). Brazil J Microbiol 51(1):229–241. https://doi.org/10.1007/s42770-019-00172-5

Lee CC, Ribbe MW, Hu Y (2014) Cleaving the n,n triple bond: the transformation of dinitrogen to ammonia by nitrogenases. Met Ions Life Sci 14:147–176. https://doi.org/10.1007/978-94-017-9269-1_7

Liu Q, Cheng L, Nian H, Jin J, Lian T (2023) Linking plant functional genes to rhizosphere microbes: a review. Plant Biotechnol J 21(5):902–917. https://doi.org/10.1111/pbi.13950

Marzouk SH, Kwaslema DR, Omar MM, Mohamed SH (2024) Harnessing the power of soil microbes: their dual impact in integrated nutrient management and mediating climate stress for sustainable rice crop production. A systematic review. Heliyon 11(1):e41158. https://doi.org/10.1016/j.heliyon.2024.e41158

Muhammad I, Lv JZ, Wang J, Ahmad S, Farooq S, Ali S, Zhou XB (2022) Regulation of soil microbial community structure and biomass to mitigate soil greenhouse gas emission. Front Microbiol 13:868862. https://doi.org/10.3389/fmicb.2022.868862

Mus F, Crook MB, Garcia K, Garcia Costas A, Geddes BA, Kouri ED, Paramasivan P, Ryu MH, Oldroyd GED, Poole PS, Udvardi MK, Voigt CA, Ané JM, Peters JW (2016) Symbiotic nitrogen fixation and the challenges to its extension to nonlegumes. Appl Environ Microbiol 82(13):3698–3710. https://doi.org/10.1128/AEM.01055-16

Nawaz A, Qamar ZU, Marghoob MU, Imtiaz M, Imran A, Mubeen F (2023) Contribution of potassium solubilizing bacteria in improved potassium assimilation and cytosolic K+/Na+ ratio in rice (Oryza sativa L.) under saline-sodic conditions. Front Microbiol 14. https://doi.org/10.3389/FMICB.2023.1196024/FULL

Nayfach S, Albrecht L, Williams RN, Cuadros-Rodríguez J, Schulz PA, Leah ND et al (2016) Integrated metagenomic and metabolic analysis of human intestinal microbial communities. Cell Syst 3(4):382–392. https://www.frontiersin.org/journals/microbiology/articles/10.3389/fmicb.2017.01606/full

Nguyen PM, Do PT, Pham YB, Doan TO, Nguyen XC, Lee WK, Nguyen DD, Vadiveloo A, Um MJ, Ngo HH (2022) Roles, mechanism of action, and potential applications of sulfur-oxidizing bacteria for environmental bioremediation. Sci Total Environ 852:158203. https://doi.org/10.1016/j.scitotenv.2022.158203

Ni G, Leung PM, Daebeler A, Guo J, Hu S, Cook P, Nicol GW, Daims H, Greening C (2023) Nitrification in acidic and alkaline environments. Essays Biochem 67(4):753–768. https://doi.org/10.1042/EBC20220194

Olaniyan FT, Alori ET, Adekiya AO, Ayorinde BB, Daramola FY, Osemwegie OO, Babalola OO (2022) The use of soil microbial potassium solubilizers in potassium nutrient availability in soil and its dynamics. Ann Microbiol 72(1). https://doi.org/10.1186/S13213-022-01701-8

Palit K, Rath S, Chatterjee S, Das S (2022) Microbial diversity and ecological interactions of microorganisms in the mangrove ecosystem: threats, vulnerability, and adaptations. Environ Sci Pollut Res 29:32467–32512

Pattnaik S, Mohapatra B, Gupta A (2021) Plant growth-promoting microbe mediated uptake of essential nutrients (Fe, P, K) for crop stress management: microbe–soil–plant continuum. Front Agron 3. https://doi.org/10.3389/FAGRO.2021.689972/FULL

Ptaszek M, Canfora L, Pugliese M, Pinzari F, Gilardi G, Trzciński P, Malusà E (2023) Microbial-based products to control soil-borne pathogens: methods to improve efficacy and to assess impacts on microbiome. Microorganisms 11(1):224. https://doi.org/10.3390/microorganisms11010224

Radl V, Richter A, Šimek M, Ačková-Tesárková M (2018) Stable isotope probing reveals distinct microbial communities utilizing dissolved organic matter of different origin in a freshwater reservoir. ISME J 12(1):204–215

Ramond JB, Jordaan K, Díez B, Heinzelmann SM, Cowan DA (2022) Microbial biogeochemical cycling of nitrogen in arid ecosystems. Microbiol Mol Biol Rev 86(2):e0010921. https://doi.org/10.1128/mmbr.00109-21

Roux S, Emerson JB, Eloe-Fadrosh JA, Freeman KM (2018) 16S rRNA gene sequencing and its applications in microbial ecology. Cold Spring Harb Perspect Med 8(6):a034080

Sardans J, Peñuelas J (2021) Potassium control of plant functions: ecological and agricultural implications. Plants (Basel, Switz) 10(2):419. https://doi.org/10.3390/plants10020419

Smercina DN, Evans SE, Friesen ML, Tiemann LK (2019) To fix or not to fix: controls on free-living nitrogen fixation in the rhizosphere. Appl Environ Microbiol 85(6):e02546–e02518. https://doi.org/10.1128/AEM.02546-18

Stewart ZP, Pierzynski GM, Middendorf BJ, Vara Prasad PV (2020) Approaches to improve soil fertility in sub-Saharan Africa. J Exp Bot 71(2):632–641. https://doi.org/10.1093/JXB/ERZ446

Surekha S, Kumar D, Reddy AD, Gupta S (2024) Functional profiling of microbial communities through AI approaches. Genomic Intelligence: Metagenomics and Artificial Intelligence, pp 104–118. https://doi.org/10.1201/9781003570233-7/functional-profiling-microbial-communities-ai-approaches-surekha-dhirendra-kumar-alla-devivaraprasad-reddy-sheetanshu-gupta

Tauro TP, Mtambanengwe F, Mpepereki S, Mapfumo P (2021) Soil fungal community structure and seasonal diversity following application of organic amendments of different quality under maize cropping in Zimbabwe. PLoS One 16(10):e0258227. https://doi.org/10.1371/journal.pone.0258227

Tian J, Ge F, Zhang D, Deng S, Liu X (2021) Roles of phosphate solubilizing microorganisms from managing soil phosphorus deficiency to mediating biogeochemical P cycle. Biology 10(2):158. https://doi.org/10.3390/BIOLOGY10020158

Timmis K, Ramos JL (2021) The soil crisis: the need to treat as a global health problem and the pivotal role of microbes in prophylaxis and therapy. Microb Biotechnol 14(3):769–797. https://doi.org/10.1111/1751-7915.13771

Vitousek PM, Hättenschwiler S, Olander L, Allison S (2002) Nitrogen and nature. Ambio: J Hum Environ 31(2):97–101

Vyas P, Sharma S, Gupta J (2022) Vermicomposting with microbial amendment: implications for bioremediation of industrial and agricultural waste. Biotechnologia 103(2):203–215. https://doi.org/10.5114/bta.2022.116213

Wang G, Ren Y, Bai X, Su Y, Han J (2022) Contributions of beneficial microorganisms in soil remediation and quality improvement of medicinal plants. Plants (Basel, Switz) 11(23):3200. https://doi.org/10.3390/plants11233200

Wu H, Cui H, Fu C, Li R, Qi F, Liu Z, Yang G, Xiao K, Qiao M (2024) Unveiling the crucial role of soil microorganisms in carbon cycling: a review. Sci Total Environ 909:168627. https://doi.org/10.1016/j.scitotenv.2023.168627

Xiang Y, Shao Z, Chai H, Ji F, He Q (2020) Functional microorganisms and enzymes related nitrogen cycle in the biofilm performing simultaneous nitrification and denitrification. Bioresour Technol 314:123697

Zahran HH (1999) Rhizobium-legume symbiosis and nitrogen fixation under severe conditions and in an arid climate. Microbiol Mol Biol Rev 63(4):968–989. https://doi.org/10.1128/MMBR.63.4.968-989.1999

Zayed O, Hewedy OA, Abdelmoteleb A, Ali M, Youssef MS, Roumia AF, Seymour D, Yuan ZC (2023) Nitrogen journey in plants: from uptake to metabolism, stress response, and microbe interaction. Biomolecules 13(10):1443. https://doi.org/10.3390/biom13101443

Zhang L, Tan C, Li W, Lin L, Liao T, Fan X, Peng H, An Q, Liang Y (2024) Phosphorus-, potassium-, and silicon-solubilizing bacteria from forest soils can mobilize soil minerals to promote the growth of rice (Oryza sativa L.). Chem Biol Technol Agriculture 11(1). https://doi.org/10.1186/S40538-024-00622-9

Zhao Y, Tian Y, Li X, Song M, Fang X, Jiang Y, Xu X (2022) Nitro-gen fixation and transfer between legumes and cereals under various cropping regimes. Rhizosphere 22:100546. https://doi.org/10.1016/j.rhisph.2022.100546

Zhu X, Wang K, Yan H, Liu C, Zhu X, Chen B (2022) Microfluidics as an emerging platform for exploring soil environmental processes: a critical review. Environ Sci Technol 56(2):711–731

Chapter 3
Microbial Diversity in Humid Tropical Agricultural Ecosystem: Impact on Food Security

Adeyinka Odebode and Oyinade David

Abstract Soil microbiome plays a critical role in the sustenance of plant health and ecological balance. Microbial diversity is essential for increasing food yield through its effects on plant growth, soil health, and food processing. By promoting effective nutrient uptake and strengthening soil structure, a diversified microbial population increases soil resilience and productivity, which in turn promotes plant growth. Such as, the symbiotic associations that mycorrhizal fungi form with plant roots which aids in boosting crop yields and support sustainable farming methods. Also, ability of land to maintain food production is quickly depleted, agricultural production methods in humid tropical environment are typically unsustainable. Therefore, beneficial soil bacteria have been employed in crop production as phytostimulators, biopesticides, and biofertilizers. These microbes also help plants become more resilient. Furthermore, it is recognised that some microbes, referred to as plant growth-promoting rhizobacteria (PGPR), directly promote plant growth by suppressing plant pathogens, solubilizing phosphate, and generating phytohormones. Additionally, a healthy microbial environment can create antibiotic chemicals and outcompete pathogenic species, which lowers the prevalence of diseases. By reducing the need for chemical pesticides, this biological control also helps to produce food in a safer and healthier manner.

Keywords Microbes · Soil · Ecosystem · Biocontrol · Environment

A. Odebode (✉)
Department of Basic Science, Faculty of Science and Technology, Kampala International University, Kampala, Uganda
e-mail: adeyinkabode@kiu.ac.ug

O. David
Department of Plant Science and Biotechnology, Federal University, Oye Ekiti, Ekiti, Nigeria
e-mail: oyinade.dedeke@fuoye.edu.ng

O. O. Babalola et al. (eds.), *Ecofriendly Frontiers*,
https://doi.org/10.1007/978-3-031-98700-7_3

3.1 Introduction

Soil is a vital natural resource that is essential for the sustenance of life on Earth. It provides a multitude of ecological services, which can be attributed to the expansion of associated niches through biological diversity and soil activity. In particular, the provision of plant health and biomass yield is contingent upon the soil microflora. The capacity of soil microorganisms to degrade the majority of natural organic molecules and naturally occurring polyphenolic compounds, in addition to persistent xenobiotics, is of vital importance (Ngosong et al. 2010). This also regulates the mechanism by which plants are productive in terrestrial ecosystems and thereby sustaining biogeochemical cycles. The soil is home to four distinct types of living organisms which are microflora, macrofauna, mesofauna and microfauna which play various activities directly or indirectly thereby affecting food security through their roles. The decomposition of organic matter is initiated and carried out primarily by soil micro-organisms (Odaro-Junior 2020). Furthermore, these organisms influence and control the physical properties of the soil and the water regime, they determine the turnover rates of soil organic matter, ability to fix carbon in the soil, decrease the emission of greenhouse gases, offer protection against stress, immunity to infections, and enhance the health of the plant. Also, the chemical processes involved in the degradation of approximately 45,000 chemical compounds commonly used by humans are mediated by microorganisms. In addition, the vast majority of antibiotics and other pharmaceuticals, plus a select few industrial enzymes, are derived from these microorganisms (Ranveer et al. 2014).

The elimination of toxins such as heavy metals and pesticides, facilitation of plant growth, reduction of wastes, and enhancement of nutrient cycling are processes carried out by soil microbes. All of these roles are of significant importance to farmers and to society at large (Maron et al. 2011). It has been discovered that modern farming practices such as heavy tillage, the application of toxic agrochemicals, monocropping, etc., affect the composition of the soil microbial population and the sustainability of the soil.

A high diversity of soil organisms is indicative of a healthy biological community, whereas a low diversity suggests the presence of stress in the soil community (Zhang et al. 2015). Consequently, the activity of microorganisms is employed to assess soil biological quality (SBQ), given that these organisms engage in a multitude of soil metabolic processes that serve as the foundation for the Earth's ecosystem. As the understanding of the ecology of soil microbes deepens, it is imperative that land use decisions are informed by this knowledge so as to preserve their function in the ecosystem (Maron et al. 2018). The diverse array of microorganisms, which vary in type, utilise secondary extracellular chemicals to prevent wind and water erosion on the surface of the soil. It is well established that changes in microbial diversity are regarded as a critical factor in determining soil quality and health (Torsvik et al. 2002). Nannipieri et al. (2017), observed a correlation exist between the level of nutrient supply and the number, diversity, and functionality of soil microorganisms. Microbial diversity also depends on various factors such as different Agricultural practices, vegetation amongst others (Fig. 3.1).

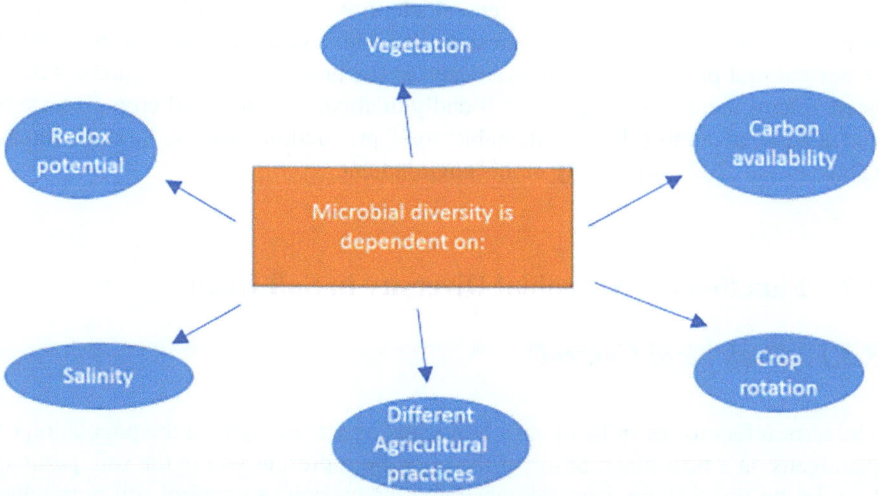

Fig. 3.1 Factors affecting microbial diversity

Pests and diseases present a significant challenge in humid tropical zones, impeding the achievement and maintenance of food security. Consequently, microbial-based solutions are presented as efficient and environmentally friendly alternatives to chemically-based pesticides, which are often perceived as posing significant environmental risks. As a consequence of the antagonistic properties exhibited by certain microbes in the context of competition for resources, it is possible for researchers to design new biocontrol strategies that both protect crops and enhance the management of resources (Schroth et al. 2000).

The impact of pests and diseases has the potential to result in significant losses in agricultural food production, particularly in regions with challenging climatic conditions, commonly referred to as humid tropical areas. The conventional approach to pest control often involve the use of chemical pesticides, which have the potential to exert adverse effects on the environment, pollinators, other wildlife, and human beings in the long term. Ultimately, microbial-based strategies offer a sustainable solution to this problem (Singh et al. 2023).

The recognition of such beneficial microorganism provides an opportunity to promote the biocontrol methods as a means of reducing losses due to pests and diseases in crops. In addition, the use of microbial-based treatments serves to protect crop production while simultaneously reducing the impact of chemical pesticides on the environment. However, microbial-based strategies of pest management are not merely a means of simple pest control. Such strategies contribute to the development of resilient and sustainable agroecosystems, which in turn reduce the likelihood of pesticide resistance emerging and enhance biodiversity. It is also noteworthy that these strategies are aligned with the sustainable agriculture practices, thereby supporting the long-term goal of food security (Elnahal et al. 2022).

It can therefore be said that the general adoption of microbial-based pest control measures represents a further progressive development, particularly in the context of agricultural practices in regions situated in the humid tropics. Researchers have been able to identify effective, eco-friendly methods of integrated crop protection and a rational approach to sustainable food production in these strategic areas (Thakur et al. 2020) especially in this environment.

3.2 Functions of Microbial Diversity in Soil Health

3.2.1 Biological Control

The term refers to the utilization of biological agents to regulate the population of pathogens on a host plant. Some of the microflora present within the soil, particularly in the rhizosphere, have demonstrated the capacity to control soil-borne diseases through the mechanisms of biological control. The microorganisms employed in this technique are referred to as bio-agents or biocontrol agents. In his report, O'Brien (2017), highlighted that bio control agents most commonly used are typically bacteria or fungi isolates isolated from the endosphere or rhizosphere. It is for this reason that fungi and bacteria are the major biological agents that influence the health of the soil. They play a vital role in plant protection against pathogenic microorganisms by breaking down organic matter and releasing nutrients that may be utilized by plants. These defensive mechanisms are biological responses that enable the maintenance of soil health and the protection of plants from disease-causing pathogens (Frac et al. 2015). The introduction of microbial inoculants into the soil may also prevent the establishment of numerous pathogens. This may result in diseased plants. At the level of the rhizosphere, microbial population prevents diseases caused by pathogens in the soil through the secretion of enzymes. Iavicoli et al. (2003) discovered that host defenses could also be stimulated by *pseudomonads* that are known to secrete the antibiotic 2,4-diacetylphloroglucinol (DAPG). The capacity of plants to attach and develop roots and obtain nutrients that facilitate growth may be in competition with pathogens and other organisms present in the rhizosphere. Bacteria present in the soil that produce siderophores can deprive fungi of iron, since fungal siderophores are less effective in binding to iron. Also, bacteria are capable of neutralising deadly viruses or, alternatively, of imposing what is known as 'induced systemic resistance' on plants. The beneficial bacteria that naturally reside in soil typically produce antibiotics that are toxic to other organisms.

3.2.2 Alteration of Soil Physical Properties

The developmental stages of plants are influenced by environmental factors such as water and soil nutrients, particularly nitrogen (N) and phosphorus (P) in natural and agro-ecosystems. The final source of carbon for the entire microbial community of the rhizosphere, which supports the macrofauna, is the vegetation cover on the soil surface (Bhatia 2008). Accordingly, the nature and composition of microbial communities at the subsurface level are contingent upon the vegetation cover on the Earth's surface (Orwin 2005). Hyphal threads and extracellular polysaccharides which binds soil particles, are produced by fungi and actinomycetes. The formation of larger amounts of soil aggregates and the infiltration of water into the soil are facilitated by these linkages. Bacteria synthesise polysaccharides which forms microaggregates and also improve the soil structure, as well as the binding of clay, silt and sand particles together (Hoorman 2011).

3.2.3 Nutrient Cycling

Microorganisms are important in the cycling of nutrients such as nitrogen, sulphur and phosphorus and in the decomposition of organic matter, as their occurrence in the soil is limited. These composts affect the carbon and nutrient cycles on a global scale (Pankhurst et al. 1997). The process of decomposition of plant and animal litter produces energy that is integrated into the soil ecosystem. In this way, organic waste is converted into biomass or into CO2, H_2O, phosphorus, mineral nitrogen and other nutrients (Bloem et al. 1997). The enzymes released into the soil by plants and other living organisms are urease, cellulase, chitinase, dehydrogenase, phosphatase and amylase (Gupta et al. 1993).

3.2.4 Carbon Sequestration

Various micro-organisms enable a process known as carbonisation, whereby carbon in organic form is eliminated to produce stable forms essential for improving soil fertility and overall structure. Improving soil organic carbon levels has two benefits: It increases the soil's fertility, and also reduce the level of greenhouse gases in the atmosphere (Lal 2004).

3.2.5 Promoting Plant Growth

Beneficial effects of PGPR on plant growth leads to the promotion of plant development through secretion of phytohormones, solubilisation of nutrients and improvement of stress tolerance. Examples of PGPR include; *Bacillus* and *Pseudomonas* species. Enhanced Plant growth gives better and increased crop yield and also improves the quality of soil in the field (Vessey 2003).

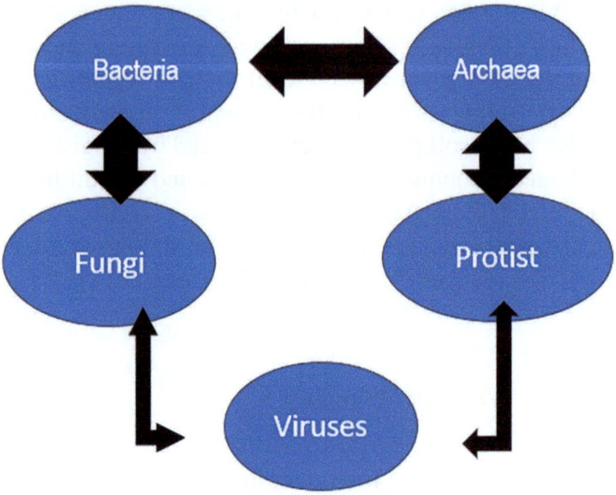

Fig. 3.2 Interplay between various microbes present in soil

Soil fertility and organic matter turnover are attributes that are positively associated with microbial diversity and activities (Fig. 3.2). Increased soil microbial communities are considered as an indication of a "healthy" organic soil, and there is a positive correlation between microbial community structure and function and ecosystem function and diversity (Nicole et al. 2016). Ecological theory also affirms that the taxonomic richness of soils, i.e. the functionalities of these soils are not completely gone even if their population becomes low, because of the compensation that depends on the species overlap at high diversity levels (Pierre-Alain et al. 2018).

3.3 Application of Microbes for Sustainable Soil Fertility

Microorganisms play a very important role in maintaining soil fertility by contributing to soil composition/structure, plant growth and nutrient availability. Some examples of how microorganisms are used to maintain soil fertility are as follows:

3.3.1 Rhizobium Species

These micro-organisms work in symbiotic partnership with leguminous plants to reduce atmospheric nitrogen and convert it to a state that can be utilised by plants. As a result of this, the need for artificial nitrogen fertilisers is reduced due to the legumes' ability to fix nitrogen in the soil. For example, the use of Rhizobium inoculants on legumes prior to planting increases the nitrogen content in the soil and encourage growth of crops (Van Hamme et al. 2003).

3.3.2 Phosphate Solubilizing Microorganisms (PSMs)

Bacillus and Pseudomonas species can solubilize insoluble forms of phosphate, making it available for plant uptake. PSM inoculants can also be added to the soil to increase the solubility of phosphate, thereby increasing plant growth and reducing the use of phosphate fertilisers (Rodríguez and Fraga 1999).

3.3.3 Arbuscular Mycorrhizal Fungi (AMF)

These fungi grow in association with plant roots and thereby improve phosphorus uptake and plant stress tolerance. By applying AMF to the soil or seedlings, it is possible to improve plant growth and development and soil health at the same time, reducing dependence on chemical fertilisers (Smith and Read 2010).

3.3.4 Bio-Remediation

Bacteria, especially those in the genera *Alcanivorax* and *Pseudomonas* and also certain fungi are capable of remediating hydrocarbon-contaminated soils by breaking them down into usable soil nutrients. These microbes can be introduced into polluted soils to assist in the process of bioremediation and thus rehabilitate the soil for agriculture (Das and Chandran 2011).

3.3.5 Disease Suppression

Through competition, predation and other means, the antibiotics produced by different microbial populations can reduce the rate of spread of soil-borne diseases. Healthy plants require little or no chemical pesticides (Mendes et al. 2013).

3.4 Importance of Microbial Diversity in Tropical Agricultural Ecosystem

Microbial diversity is an important factor in contributing to a safe and healthy ecosystem for tropical agriculture. Microorganisms also help to improve soil nutrient availability, plant growth and health, and are ecologically essential for ecosystem balance. The following are some key ideas that highlight the importance of microbial diversity. The following are some key points that highlight the importance of microbial diversity:

3.4.1 *Environmental Resilience*

Microbial diversity helps to withstand environmental challenges such as salt, drought and heavy metal pollution. Some microorganisms have the ability to detoxify contaminants, improve soil water-holding capacity and soil structure (van der Heijden et al. 2008).

3.4.2 *Biodiversity Conservation*

The overall level of biodiversity in tropical agricultural environments depends on microbial diversity. For example, microorganisms are critical to ecosystem processes in support of plants and animals (Giller et al. 1997).

It is therefore important to understand that the health and longevity of tropical agricultural ecosystems is largely determined by the diversity of their microbial population. Farming practices that increase crop productivity, strengthen the resilience of agricultural systems to a variety of biotic and abiotic factors, and overall improve soil quality by cultivating healthy microbial communities. The role of microbes in humid tropical environment is further elaborated in Table 3.1.

3.5 Microbial Species and Ecotypes in Humid Tropical Ecosystem

The humid tropical ecosystem is characterized by high biodiversity, mainly due to the high microbial richness and the presence of different ecotypes that provide considerable stability and functionality to the ecosystem. These are the bacteria, fungi, archaea and protozoa that have different functions in the process of decomposition, nutrient cycling and the well-being of plants (Yang et al. 2022). The following are important microbial species commonly observed in humid tropical habitats:

3.5.1 *Bacteria*

3.5.1.1 Rhizobia

A nitrogen-fixing bacteria that works in mutual cooperation with leguminous plants, i.e. the bacteria help the plant by converting atmospheric nitrogen into a form that the plant can use.

Table 3.1 Micro-organism roles in tropical environment

Factors	Effect on food security	Results	Micro-organisms involved	References
Soil health and fertility	Crop yield increase	The increase in crop production and the fertility of the soil is given a boost by a variety of enhanced and nitrogen-fixing bacteria that are found in the soil.	Nitrogen fixing bacteria, Trrichoderma	Liang et al. (2019)
Plant growth promotion	Improve general plant health and growth.	Mycorrhizal fungi and PGPRs are two types of microorganisms that can improve plant growth by increasing water uptake rates, producing plant hormones and making compounds in the soil available for plant uptake.	Fungi	Hossain et al. (2017)
Disease suppression	Crop losses reduced	This is the case when biocontrol agents in the form of a diverse microbial community build resistance to pathogens, suppress the pathogens, produce antimicrobial substances and thus reduce the incidence of plant diseases and consequently the use of pesticides.	Fungi, bacteria	Singh,et al. (2019)
Biological nitrogen fixation	Reduced use of chemical fertilisers	The need for synthetic fertilisers is partly reduced because nitrogen-fixing bacteria convert atmospheric nitrogen into forms that plants can absorb, improving soil fertility.	Azotobacter sp	Dal Cortivo et al. (2017)
Stress tolerance	Increased resilience to environmental stress	Most microbes are used to increase plant's tolerance to salinity, drought and high temp, ensuring consistent crop production regardless of environmental conditions.	Rhizosphere bacteria	Liu et al. (2022)

(continued)

Table 3.1 (continued)

Factors	Effect on food security	Results	Micro-organisms involved	References
Food processing and fermentation	Improved food quality and shelf life	Microbes are important in fermentation because they improve the taste, texture and composition of food. They also improve the preservation of food so that it stays fresh for longer, a process known as biopreservation.	Bacteria	Lee et al. (2020)
Biocontrol agents	Thorough and deliberate management of pests and diseases	Some of the desirable micro-organisms also help to reduce the use of chemical pesticides in agriculture by acting as biocontrols of viruses and insects.	Pseudomonas fluorescens	David et al. (2018)
Enhancement of Soil structure	Enhanced root branching and water uptake	The actions of microorganisms contribute to the improvement of soil structure, which is necessary for healthy root development, soil aeration, and water retention—All of which are critical components of crop productivity.	Fungi, bacteria	Gupta (2011)
Nutrient recycling	Persistent fertility of the soil	The organic matter is broken down by the microorganisms, which also replenish the nutrients needed to keep the soil fertile and conducive to consistent and regular food production.	Nitrogen-fixing bacteria	Yuan et al. (2021)
Phytoremediation	Detoxification of contaminated land	Depending on their ability to break down contaminants and make the soil suitable for farming, several types of microorganisms can help expand the amount of arable land.	Pseudomonas	Alves et al. (2022)

(continued)

Table 3.1 (continued)

Factors	Effect on food security	Results	Micro-organisms involved	References
Biofortification	Increased nutrient density in the food crops that are produced	Microbial exchange can improve the bioavailable nutrients in crops, including biofortified crops with increased vitamin and mineral contents.	Yeast	Obafemi et al. (2022)
Symbiotic relationships	Increased effectiveness in the use of resources	It is possible for plants and bacteria to cohabit. One such example is the legume rhizobia, which improves resource utilisation and boosts crop productivity.	Azotobacter	Gupta (2011)
Environmental sustainability	Long-term agricultural productivity	Both production and the microbial load can be sustained by implementing environmentally friendly agricultural methods that lead to sustainable farming practices.	Fungi, bacteria	David et al. (2018)

3.5.1.2 Pseudomonas Species

These bacteria produce antibiotics and other bioactive compounds that are known to enhance plant growth and prevent soil-borne diseases.

3.5.1.3 Burkholderia Spp.

Members of this genus are involved in the breakdown and fixation of nitrogen in complex organic compounds. Some species are plant growth promoting rhizobacteria or PGPR.

3.5.2 Fungi

3.5.2.1 Mycorrhizal Fungi

These fungi form mutualistic associations with plant roots in which, in exchange for nutrients, mostly phosphorus. Plants gain the ability to resist unwanted stress factors.

3.5.2.2 Trichoderma

Cyanobacteria

3.5.2.3 Nostoc Species

A nitrogen-fixing cyanobacteria that can live synergistically with lichens, fungi and other plants to improve soil quality (David et al. 2018).

3.5.3 *Protozoa*

3.5.3.1 Amoeba

They are important in nutrient cycling, by regulating microbial populations and also feeds on bacteria and other microbes.

3.5.4 *Archaea*

In particular, methanogenic archaea are microorganisms that decompose organic matter and produce methane under anaerobic conditions, in moist tropical soils to produce methane (Dal Cortivo et al. 2017).

3.6 Microbial Sustainability in Humid Tropical Area and Future Directions

Understanding the link between ecosystem performance and organization of microbial diversity explains the importance of microbial diversity in sustaining tropical agro-ecosystems. This understanding proves to be essential in predicting the readiness of ecosystems to respond to changes in environmental conditions (Logue et al. 2015). Plant diversity influences biomass production, soil carbon retention, suppression of fungal pathogens, and control of insect herbivores maintained by ecosystems (Cardinale et al. 2012). Therefore, understanding soil microbial communities is essential, as they are involved in ecosystem functions such as decomposition (Glassman et al. 2018) and regulation of greenhouse gas emissions in soil (Domeignoz-Horta et al. 2018). The effects of these communities on ecosystem function could be manifested through changes in the plant community or changes in soil structure, as shown in Fig. 3.3. Consequently, changes in soil structure affect the movement of water and oxygen in the soil, which in one way or another affects

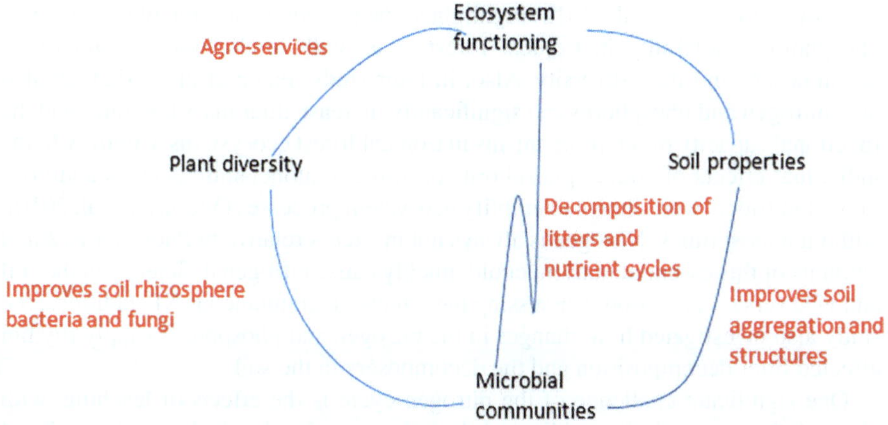

Fig. 3.3 Cyclic Stability of plant ecosystem and microbial community

plant growth. Their influences are reflected in the positive relationship between the level of microbial richness and the productivity of soil aggregation (Lehmann et al. 2017), microbial growth efficiency, (Domeignoz-Horta et al. 2020), as well as the formation of soil organic matter for decomposition (Kallenbach et al. 2016).

Furthermore, soil microbial communities serve as vital reservoirs for plant and animal biodiversity, contributing significantly to global climate regulation and biogeochemical cycling (Townsend et al. 2011). However, the conversion of tropical ecosystems to other uses poses a significant threat to these communities (Stork et al. 2007). Given the context of global change, it is imperative to grasp how environmental factors, biogeographic patterns, and land use changes synergistically influence the structure and function of microbial communities in these ecosystems.

The distribution of microbes, which is a key indicator of soil health, is not only intricately linked to environmental factors, but also exhibits variation across temporal and spatial scales (Shade et al. 2012). Physical properties, soil organic matter (SOM) content, and depth play a key role in determining microbial diversity and overall soil health. Biological indicators such as microbial biomass, soil respiration, enzyme activities, and microbial diversity serve as reliable measures of soil health (Nielsen and Winding 2002). Understanding these dynamics is essential in the face of ongoing global change, by providing insights into how environmental factors and land-use change interact to influence microbial communities in tropical ecosystems.

Therefore, in tropical terrestrial ecosystems, both agricultural and natural, the major Nitrogen pool is in Soil Organic Matter and not in biomass as is commonly believed (Jordan 1985). Through microbial activity such as ammonification and nitrification, organic N in the soil is converted to a soluble form, making it readily available to plants. It can then either be recycled back into the soil through root decay, leaves, branches, root secretions, canopy droppings and frass, etc. However, there is a lapse in this cycle, with denitrification and leaching as major concerns, especially in the humid tropics.

A study by Fanin et al. (2012) highlighted the temporal variation of nitrogen and phosphorus availability in tropical forest soils, with implications for microbial decomposers and their diversity. Also, in their study, Fanin et al. (2012) detailed how nitrogen and phosphorus can significantly increase litter mass loss rates and the functional capacity of microorganisms in tropical forest ecosystems. Contrary to the individual effects of adding phosphorus or nitrogen alone, simultaneous additions of both nutrients were shown to amplify ecosystem processes (Vitousek et al. 2010). Although most times, Nitrogen is always not in excess relative to Phosphorus, small amounts of this essential nutrient could quickly cause nitrogen deficiency in the soil and vice versa. In addition to assessing the effects of simultaneous NP additions, the study also investigated how changes in the nitrogen and phosphorus supply rhythm affected litter decomposition and the decomposers in the soil.

One significant challenge of the nitrogen cycle is the effects of leaching, with nitrate being particularly mobile and the primary anion leached from agricultural soils in the humid tropics (Greenland and Hayes 1981; Bibi et al. 2016). Adding to the complexity, each leached nitrate ion is most often associated with an equivalent cation, usually Potassium, which is a macroelement essential for the nutrition of agricultural and natural ecosystems. Thus, actively growing plant communities can effectively protect against nitrate leaching, even in soils with low anion exchange capacity (Marrs et al. 1983).

Phosphorus, on the other hand, is highly immobile and is therefore easily fixed in the soil, hence the low leaching rates reported by Wild (1981) and Mwende, (2019). Specific types of soils found in tropical regions are known to fix very large amounts of phosphorus in chemically inactive forms. It should be noted that despite the total phosphorus presence, it may not be available for use by plants.

On average, up to 50% of the total phosphorus concentration in the surface horizons of tropical soils can be associated with organic matter; in the Oxisols, Ultisols, and Alfisols, these figures are higher, reaching 60–80%. Some microorganisms contribute directly to making many forms of organic phosphorus available to plants (Stewart and McKercher 1982; Gargallo-Garriga et al. 2024). Some microorganisms live in association with roots which secrete phytohormones that promote plant growth; as a result, these microorganisms can mobilise phosphorus more efficiently. It is therefore expected that phosphorus, nitrogen, and phosphorus will be depleted most rapidly in tropical ecosystems. Nitrate concentrations are generally high after deforestation, but phosphorus, which is known to be immobile, soon becomes scarce. Nitrate levels decrease rapidly over the weeks due to ash erosion, leaching, and plant uptake (Ewe et al. 1981; Bian et al. 2020).

During early succession stages, nitrogen requirement is high, which exhausts the available forms of nitrogen and further increases the competition between roots and microorganisms. The mycorrhizal root systems spread out to a greater extent in the soil thereby enabling the plant to access the available phosphorus reserve. At this point, the rate of plant growth may be limited by nitrogen or phosphorus. Also, plants recover and remobilize a large portion of extractable phosphorus. An increase in soil acidity causes unabsorbed inorganic phosphorus to bind with iron and aluminum oxides consequently, reducing the availability of phosphorus (Sanchez 1976;

Johan et al. 2021). At the same time, the demand for nitrogen decreases, and the nitrogen content in the system increases due to atmospheric nitrogen fixation and retention of mineral nitrogen from rain.

It is established that the addition of nitrogen influences phosphorus dynamics in the soil. Some of the microorganisms that solubilize P include; *Pseudomonas* spp., and *Bacillus* spp., and *Erwinia* spp. These microbes play a vital role in promoting phosphorus release. In the long term, the increase of soil N may result in microbial P limitation since it affects the availability of soil P through control of mineralization, phosphatase, and soil PH. Nitrogen addition can also affect soil P availability by reducing Phosphate Solubilizing Microorganisms and alkaline phosphatase activities thereby reducing P solubility. There is a correlation between the availability of phosphorus, which depends on the level of Nitrogen addition, and soil pH (Morris et al. 2022).

3.6.1 Microbes for Pests and Disease Control

The increasing problems with the toxicity of synthetic chemical pesticides, while the role of insect pests and pathogens is becoming more important due to the need to feed the growing world population, the banning of some chemical pesticides, the appearance of new invasive species, and the evolution of pesticide-resistant pests in line with impact of climate change and specialized monoculture, all lead to the expansion of the field of biological control in plant protection as part of sustainable agriculture. Biological control proves to be a cheaper, environmentally friendly, and more sustainable way of protecting crops from biotic challenges compared to other methods (Rajwinder et al. 2024).

Many microorganisms have shown efficacy in controlling soil-borne diseases and nematodes. The processes involved in the biocontrol of plant diseases by microbial biocontrol agents include antibiosis, hyperparasitism, induction of resistance and competition for space and nutrients (Köhl et al. 2019).

At present, only a few genera, species and strains of Biological Control Agents (BCA) are registered against targeted Soil-borne pathogens such as *Coniothyrium minitans*, *Gliocladium catenulatum*, *Pseudomonas chlororaphis*, *Streptomyces griseovirides*, *Streptomyces lydicus* and the fungal strains of *Trichoderma asperellum*, *T. atroviride* and *T. harzianum*. Similarly, *Bacillus firmus and Purpureocillium lilacinum* are the only BCAs approved for nematode control (Pertot et al. 2015).

Some microorganisms including *Bacillus* spp., *Pseudomonas* spp., *Acinetobacter calcoaceticus*, *Azotobacter* spp., *Azospirillum* spp. Strains of the genera Rhizobium, *Mesorhizobium, Bradyrhizobium, Burkholderia* are involved in biostimulation through the production of indoleacetic acid, facilitating nitrogen fixation, Phosphate solubilization, siderophore production, Hydrogen Cyanide (HCN) production and 1-aminocyclopropane-1-carboxylate (ACC) deaminase. Some of these activities result in an increase in crop yield, disease and pest control, and improved soil health (Kang et al. 2019).

3.6.2 Microbes for Soil Amendments/Remediation

The presence of many types of microorganisms in mixed contaminated soils is one such aspect critical to the success of phytoremediation, which induces shifts in the microbial environment. This practice directly or indirectly influences the regulation of metabolic processes and influences the density and distribution of microbiota in soils (Mierzwa-Hersztek et al. 2017).

Today, bioaugmentation is considered one of the newest and most popular approaches in phytoremediation. Bioaugmentation is the process of adding microbial cultures to enhance the natural degradation of contaminants. Microbial and fungal enzymes used for bioaugmentation are produced via catabolic processes (Hai et al. 2012).

Microbes also show the ability to adsorb or absorb heavy metals, which is the general method of increasing ionic metals in soils (Alloway 1995). Some of the primary forms of metal removal and/or immobilisation include precipitation, chemical adsorption or ion exchange, surface precipitation, stable organic ligand complexation, and redox reactions. However, the exact nature of immobilization remains unclear due to the problem of current analytical techniques and the nature of the soil matrix of the contaminant (Wuana and Okieimen 2011). Adsorption involves the chelation of the heavy metals to the cell surface and then absorption by the cell (Danis et al. 2008). The cell surface consists of the cell wall and the mucus layer, which makes it relatively easy for heavy metals to be adsorbed and absorbed. Various functional groups on the cell surface can interact with metal ions, of which nitrogen, oxygen, sulphur, and phosphorus are the most common coordinating atoms (Brady and Duncan 1994). In addition, phosphoric acid anions and carboxyl anionic groups on the microbial cell wall surface carry a negative charge that attracts the cationic surface charge that is most commonly found in heavy metal bodies. This interaction allows transport across the cell membrane or binding of metal ions (Sarret et al. 1998). For example, at a pH of 7.2, *Bacillus* was found to be capable of adsorbing 60% of the Cu2+ concentration within one minute, with a maximum adsorption time of ten minutes (He and Tebo 1998).

In addition, facultative anaerobic chemoorganotrophic microorganisms such as *Bacillus*, *Clostridium*, *Pseudomonas* and *Geobacter* are involved in the processes of mobilization and reduction of Fe oxides (Petrie et al. 2003). In addition, certain fungi have been found to play an important role in the reduction of Fe3+ oxides (Bonneville 2014). Fe in the oxidised state becomes an alternative electron acceptor for the energy-rich electron donors to fulfill respiratory obligations, especially from easily mineralizable organic roots. It is important to understand that the availability of Fe-reducing microorganisms is also significantly higher near the rhizosphere than in the bulk soil (Sutton-Grier et al. 2011).

Microbial-dependent redox processes play a pivotal role in making Heavy Metals less toxic and completely harmless. Outer membrane c-type cytochromes (OM c-Cyts) and trans outer membrane porin-cytochrome proteins of the *Geobacter* and *Shewanella* genus are essential for the microbial reduction of HMs (Shi et al. 2016).

Specific enzymes plus multicopper oxidases such as CueO, CuiD, CopR and chromate reductase (ChroA) are involved in the cellular efflux and reduction of copper and chromium. MerA protein also aid in the detoxification of mercury (Hg) to elemental Hg (Hg0). Bacteria can assist in the accumulation of HMs by plants and also facilitate the detoxification of metals. For example, *Geobacillus* has been reported to convert the hazardous As^{3+} to the less toxic As^{5+}, and some other bacteria found in contaminated soils were also capable of similar transformations (Majumder et al. 2013).

Proteins, nucleic acids, complex carbohydrates and lipids, which are the components of Extracellular Polymeric Substances (EPS), are particularly important in the adsorption of HM ions. Different bacterial species such as *Sulfobacillus*, *Thiobacillus denitrificans*, *T. thioparus*, *Acidiphilium cryptum*, *Cronobacter sp*., *Ferroplasma acidiphilum*, *Acidithiobacillus caldus*, *A. ferrooxidans*, *A. thiooxidans* and *Ferribacterium limneticum* are all involved in the remediation process (Anawar 2015). Intracellular sequestration and extracellular sequestration are some of the ways in which Fungi assist in the removal of HM. *A. niger* has been shown to be a good biosorbent for the removal of Pb (II), while *Termitomyces clypeatus* biomass has previously been shown to remove Cr (VI) through various popular functional groups including imidazole, sulfhydryl, phosphate, hydroxyl and carboxyl (Iram et al. 2015). Cellular biomass such as *Saccharomyces cerevisiae* cell biomass has been studied to extract Cu (II) from wastewater and *Rhizopus* and *Thiobacillus* also show ability for absorption of multiple heavy metal ions (Amirnia et al. 2015).

3.7 Microbial Diversity in Humid Tropical Areas: A Foundation for Sustainability

Microbial diversity is the foundation of ecosystems in humid tropics and therefore plays an essential role in the ability of such areas to sustain themselves. The intricacies of these synergistic relationships are critical to the Earth's dynamic processes such as nutrient cycling and decomposition of organic matter, including disease control. Environmental conditions in the humid tropics are conducive to the growth of diverse microbial communities, but these ecosystems are currently at great risk from climate change (Onen et al. 2020).

Climate change is altering patterns of microbial function, particularly in the high humidity regions of the tropical belt. Changes in precipitation regimes, temperature, and other forms of extreme weather events affect the existing microbial ecosystems (Hutchins et al. 2019). Such disruptions inevitably lead to changes in species distributions and abundance, with negative impacts on processes, which include nutrient cycling, among many others. These impacts require adequate knowledge of how climate change affects microbial diversity to set the right pace for conservation and management strategies (Weiskopf et al. 2020).

The call for the development and application of appropriate measures to protect microbial diversity in the humid tropics is urgent. For this reason, any conservation action should ensure that microbial habitats are protected, as microbes and their habitats are so closely linked. Alongside such initiatives, restoration programmes can be of great importance in rehabilitating degraded microbial ecosystems (Sharma et al. 2016).

As microbial habitats are also sensitive to similar challenges such as land-use change and climatic disturbances, there is a need to take a preventative approach to conservation. The pristine nature of these ecosystems must be maintained to ensure the microbial diversity that underpins ecosystem processes (Reinold et al. 2019). This involves the process of identifying areas that contain important microbes such as soil, water and vegetation, and then aiming to conserve the different ecological niches.

Thus, restoration initiatives are emerging as key elements in mitigation efforts where microbial populations have been negatively impacted by human activities or other environmental challenges. Such activities include the deliberate inoculation of beneficial microbes, habitat restoration, and the application of effective measures to enhance microbial diversity. Consequently, restoration positively influences the overall condition of different humid tropical habitats by stimulating the natural recovery processes of microbial ecosystems (Reinold et al. 2019; Simonson et al. 2021).

Furthermore, the adoption of sustainable land management practices based on scientific knowledge of microbial ecosystems is a way solution to counteract the negative impacts of climate change. Once microbial function in processes such as carbon and nutrient cycling and soil structure is understood, it will be easier to introduce practices that favour microorganisms. These include approaches such as agroecology, precision agriculture, and organic farming, which allow microbial habitats to be protected and coexist with microscopic species (Kolapo et al. 2022).

Therefore, a timely and complex strategy to conserve microbial diversity in the humid tropics involves habitat preservation, restoration efforts/projects and the introduction of sustainable management. By focusing on the relationships between microorganisms and their environment, these approaches/strategies promote the sustainability and resilience of ecosystems in the context of contemporary environmental concerns (Sayer et al. 1995).

There is a need to identify and promote microbial species that are adapted to climate change, in order to maintain the stability of ecosystems. Further research should focus on investigating microbial coping strategies in the presence of stressors. This involves investigating microbial genetic diversity to find valuable adaptive traits in the face of climatic change dynamics. In this way, climate-resilient microbial communities provide a basis for bio-sustainability in Humid Tropical Areas (Hutchins et al. 2019).

3.7.1 *Harnessing Microbial Functional Traits for Ecosystem Services*

The diversity of functional groups of microorganisms in the humid tropics is crucial for supporting ecosystem services. Aspects of traits such as nitrogen fixation, decomposition of organic matter and disease control are some of the traits that contribute to the maintenance and productivity of these ecosystems. An important area of microbial research that will continue to be relevant and advance in future is to determine and expand more on the functional characteristics of microbes and their relationship to the enhancement of ecosystem services (Delgado-Baquerizo et al. 2016).

Bioprospecting can be defined as a methodical search of microbial ecosystems to find specific novel functional properties. There is a need to focus on sustainable use of natural resources, especially in humid tropical regions, because the potential of bioprospecting is likely to help discover more microorganisms that can be used to improve agriculture and efficient environmental management. It is possible for scientists to select microbial traits that increase soil fertility, enhance the rate of nutrient cycling, and also the ability to withstand pests and diseases (Vuong et al. 2022).

Beneficial microorganisms are a better solution for improving soil productivity. These inoculants/beneficial microorganisms can increase the efficiency of nutrient absorption from soil to plants, improve plant health, and enhance plant production. The overall goal is to benefit from beneficial microbes which can lead to sustainable agricultural practices, and improve soil fertility in humid tropical environments (Sammauria et al. 2020).

3.8 Microbial Technology for Ensuring Food Security

Among the various approaches to achieving food security in humid tropical areas, microbial technology appears to be one of the most important. This technology includes various applications such as the improvement of measurable productivity of sustainable agriculture, bioremediation of contaminated soils, and microbial pest and disease control, all of which are critical to agricultural productivity (Rehman et al. 2022).

Soil sustainability is a key component of microbial technology in humid tropical areas. The presence of microorganisms such as bacteria and fungi plays a specific role in soil fertility, nutrient cycling, and overall soil health. Therefore, the positive impact of microbial inoculants and the enhancement of microbial diversity are part of the objectives of ecological measures to manage soil resources. This, in turn, creates the best environment that can support plant growth and withstand most of the stress factors in the environment (de Moura et al. 2016).

Biological control of pests and diseases is another sustainable innovative strategy besides the use of chemicals. Due to the antagonistic properties of certain microorganisms, bio control techniques can be developed to control pests and diseases that affect crops. This not only ensures agricultural productivity but also helps to reduce the use of chemical pesticides, which also helps to protect the environment and support environmental preservation and promote the establishment of a more resilient agro ecosystem (Thakur et al. 2020).

In essence, microbial technology provides food security in Humid Tropical regions in direct and dynamic ways that are engaging. Through efficient and sustainable use of soil, bioremediation, microbial pest and disease control, the resilience and sustainability of agricultural production in these areas is enhanced. Thus, the use of these microbial technologies is a progressive approach that is in line with the concept of environmental conservation and the practice of modern agriculture (Singh and Trivedi 2017).

Agricultural yields depend on soil health, and microbial technology is at the forefront of successful soil management. Measures such as biofertilisers, which use microbes to fix nitrogen, can be used in place of traditional, highly hazardous fertilisers. Precision agriculture using microbial insight is also helping farmers use resources more efficiently and reduce pollution (Hartmann and Six 2023).

Another important area of microbial technology is the bioremediation of contaminated soil. About agricultural fields, some of the problems that could be encountered in countries located in the humid tropical regions are mainly due to soil contamination from a variety of sources, including agrochemicals. Bioremediation is the modern technique of using microbes that are capable of degrading or immobilizing contaminants, thereby improving soil quality and reducing hazards to crops, ecosystems and human health (Mishra and Mohan 2017).

Contaminated soils represent a multifaceted problem, ranging from human economic activities such as industrial processes, waste disposal or the use of agrochemicals. In addition to the negative impact on the yields of the agricultural land they occupy, they are also associated with likely outcomes such as the degradation of the surrounding ecosystem and adverse effects on the health of people living in the area (Steffan et al. 2018).

These and other challenges explain why bioremediation uses the metabolic versatility of microorganisms to metabolise, neutralise or immobilise contaminants in the soil. This process is not only environmentally friendly, but also offers the possibility of reducing the toxicity of waste to harmless levels. Various bacteria, fungi and archaea with different functions in the remediation process have been identified, forming a complex microbial community that modifies a range of contaminants (Bala et al. 2022).

Thus, bioremediation is not just about the ecological aspect. This approach is proactive in improving and maintaining soil fertility for the purpose of sustainable food production on existing agricultural land. Beneficial microorganisms in the processes help in nutrient cycling, amendments, and improved stability in the ecosystem (Pande et al. 2020).

The use of microbial diversity for bioremediation offers a mutually integrated balance between ecological restoration and agricultural sustainability. With regard to the problem of contaminated soils addressed by this approach, the health of the environment is protected, as is the integrity of agricultural land, so that the production of safe and nutritious food can continue ad infinitum (Paul et al. 2005).

3.8.1 Challenges and Opportunities in the Integration of Microbial Technology

While acknowledging the substantial potential benefits that microbial technology holds for ensuring global food security, it is imperative to recognize and address various challenges inherent in its adoption. The successful implementation of microbial technology at scale necessitates a comprehensive examination of several key factors, including economic considerations and the potential ecological impacts associated with its deployment. On the operational front, transitioning from laboratory-scale success to large-scale implementation poses a significant challenge. The intricacies of scaling up microbial technology require meticulous planning, infrastructure development, and technological optimization. Economic considerations, such as the cost-effectiveness of large-scale implementation, financial feasibility, and the creation of sustainable business models, play a pivotal role in determining the success of widespread adoption (Timmis et al. 2017).

Moreover, the potential ecological impacts of deploying microbial technology on a large-scale demand scrutiny. Unintended consequences, such as alterations in soil microbiota or broader ecosystem dynamics, must be thoroughly evaluated to mitigate any adverse effects on biodiversity and environmental sustainability. In addressing these challenges, collaborative efforts emerge as a cornerstone for success. The involvement of multidisciplinary teams, comprising researchers, policymakers, and local communities, is crucial in navigating the complexities of implementing microbial technology. Researchers play a pivotal role in advancing scientific understanding, policymakers contribute by developing supportive regulatory frameworks, and local communities provide valuable insights into the economic socioeconomic dynamics and cultural considerations (Iqbal et al. 2023).

In essence, a holistic and collaborative approach is essential to overcoming the challenges associated with implementing microbial technology at scale. By fostering synergy among stakeholders, we can unlock the full potential of microbial technology, ensuring not only food security but also sustainable and resilient agricultural practices for the future (McLaney et al. 2022).

3.9 Conclusion and Future Perspectives

Microbial diversity is crucial for sustainable food production and security because it supports the productivity and health of agricultural systems.. It enables stable and sustainable agricultural practices by boosting nutrient cycling, enhancing soil structure, encouraging plant development, reducing illnesses, and increasing resilience to environmental shocks. Thus, supporting and preserving microbial diversity is a crucial tactic for attaining global food security.

The trajectory of humid tropical areas in the coming years hinges significantly on our capacity to both harness and safeguard microbial diversity within these ecosystems. A holistic strategy for sustainability in these regions necessitates a multifaceted approach, incorporating measures to address the impacts of climate change, promote functional traits, and leverage microbial technology. Climate change poses a profound threat to humid tropical areas, with rising temperatures, altered precipitation patterns, and extreme weather events challenging the stability of ecosystems. Harnessing microbial diversity becomes crucial in this context, as certain microorganisms can play pivotal roles in enhancing soil fertility, water retention, and overall ecosystem resilience. Additionally, understanding and promoting functional traits within microbial communities can contribute to the adaptation and survival of flora and fauna in the face of changing environmental conditions (Tabari 2020).

The integration of microbial technology emerges as a key element in this comprehensive strategy. Leveraging advancements in biotechnology, such as precision agriculture and microbial-based solutions, can enhance agricultural productivity, mitigate the impacts of climate change, and promote sustainable land management practices. As research continues to advance and technology evolves, our understanding of the intricate relationships between microorganisms and their environments deepens. This knowledge is instrumental in shaping effective strategies for ecosystem management in humid tropical regions. Ongoing interdisciplinary collaboration, involving scientists, policymakers, and local communities, is essential in this pursuit. Researchers contribute by expanding our understanding of microbial diversity and ecosystem dynamics, policymakers play a role in developing supportive policies, and local communities provide valuable insights into the socio-economic context (French et al. 2021).

By committing to sustainable practices and fostering interdisciplinary collaboration, we can build resilient ecosystems in humid tropical areas. These ecosystems, equipped to withstand current challenges and adapt to future uncertainties, will play a critical role in preserving biodiversity, maintaining ecological balance, and ensuring the long-term sustainability of these vital regisons (Crabbé 2019).

References

Alloway BJ (1995) Heavy metals in soils, 2nd edn. Springer, Dordrecht

Alves AR, Yin Q, Oliveira RS, Silva EF, Novo LA (2022) Plant growth-promoting bacteria in phytoremediation of metal-polluted soils: current knowledge and future directions. Sci Total Environ 838:156435

Amirnia S, Ray MB, Margaritis A (2015) HMs removal from aqueous solutions using Saccharomyces cerevisiae in a novel continuous bioreactor–biosorption system. Chem Eng J 264:863–872. https://doi.org/10.1016/j.cej.2014.12.016

Anawar HM (2015) Sustainable rehabilitation of mining waste and acid mine drainage using geochemistry, mine type, mineralogy, texture, ore extraction and climate knowledge. J Environ Manag 158:111–121. https://doi.org/10.1016/j.jenvman.2015.04.045

Bala S, Garg D, Thirumalesh BV, Sharma M, Sridhar K, Inbaraj BS, Tripathi M (2022) Recent strategies for bioremediation of emerging pollutants: a review for a green and sustainable environment. Toxics 10(8). https://doi.org/10.3390/toxics10080484

Bhatia CR (2008) Role of microbial diversity for soil, health nutrition. In: Molecular and plant mechanisms of plant and microbe coexistence. Springer, Berlin, Heidelberg, p 5374

Bian Z, Wang Y, Zhang X, Li T, Grundy S, Yang Q, Cheng R (2020) A review of environment effects on nitrate accumulation in leafy vegetables grown in controlled environments. Foods 9(6):732. https://doi.org/10.3390/foods9060732. PMID: 32503134; PMCID: PMC7353485

Bibi S, Saifullah, Naeem A, Dahlawi S (2016) Environmental impacts of nitrogen use in agriculture, nitrate leaching and mitigation strategies. In: Hakeem K, Akhtar J, Sabir M (eds) Soil science: agricultural and environmental prospectives. Springer, Cham. https://doi.org/10.1007/978-3-319-34451-5_6

Bloem J, de Ruiter PC, Bouwman LA (1997) Soil food webs and nutrient cycling in agro-ecosystems. In: van Elsas JD, Trevors JT, Wellington HME (eds) *Modern soil microbiology*. New York, Marcel Dekker, pp 245–278

Bonneville, S. (2014) The life cycle of Fe Fe (III) oxide: impact of fungi and bacteria. EGUGA:7252

Brady D, Duncan JR (1994) Cation loss during accumulation of heavy metal cations by Saccharomyces cerevisiae. Biotechnol Lett 16:543–548

Cardinale BJ et al (2012) Biodiversity loss and its impact on humanity. Nature 486:59–67

Crabbé A (2019) In: Leal Filho W (ed) Interdisciplinarity and sustainable development BT—encyclopedia of sustainability in higher education. Springer International Publishing, pp 998–1006. https://doi.org/10.1007/978-3-030-11352-0_215

Dal Cortivo C, Barion G, Visioli G, Mattarozzi M, Mosca G, Vamerali T (2017) Increased root growth and nitrogen accumulation in common wheat following PGPR inoculation: assessment of plant-microbe interactions by ESEM. Agric Ecosyst Environ 247:396–408

Danis U, Nuhoglu A, Demirbas A (2008) Ferrous ion-oxidizing in Thiobacillus ferrooxidans batch cultures: influence of pH, temperature and initial concentration of Fe^{2+}. Fresenius Environ Bull 17:371–377

Das N, Chandran P (2011) Microbial degradation of petroleum hydrocarbon contaminants: an overview. Biotechnol Res Int 2011:941810. https://doi.org/10.4061/2011/941810

David BV, Chandrasehar G, Selvam PN (2018) Pseudomonas fluorescens: a plant-growth-promoting rhizobacterium (PGPR) with potential role in biocontrol of pests of crops. In: Crop improvement through microbial biotechnology. Elsevier, pp 221–243

de Moura EG, Gehring C, Braun H, Ferraz Junior ADSL, Reis FDO, Aguiar ADCF (2016) Improving farming practices for sustainable Soil use in the humid tropics and rainforest ecosystem health. Sustainability 8(9). https://doi.org/10.3390/su8090841

Delgado-Baquerizo M, Maestre FT, Reich PB, Jeffries TC, Gaitan JJ, Encinar D, Berdugo M, Campbell CD, Singh BK (2016) Microbial diversity drives multifunctionality in terrestrial ecosystems. Nat Commun 7:10541. https://doi.org/10.1038/ncomms10541

Domeignoz-Horta LA et al (2018) Peaks of in situ N_2O emissions are influenced by N_2O-producing and reducing microbial communities across arable soils. Glob Change Biol 24:360–370

Domeignoz-Horta LA et al (2020) Microbial diversity drives carbon use efficiency in a model soil. Nat Commun 11:3684

Elnahal ASM, El-Saadony MT, Saad AM, Desoky E-SM, El-Tahan AM, Rady MM, Abu Qamar SF, El-Tarabily KA (2022) The use of microbial inoculants for biological control, plant growth promotion, and sustainable agriculture: A review. Eur J Plant Pathol 162(4):759–792. https://doi.org/10.1007/s10658-021-02393-7

Ewe JJ, Berish CW, Brown BJ, Price N, Raich JW (1981) Slash and burn impacts on a Costa Rican wet forest site. Ecology 62:816–829

Fanin N, Barantal S, Fromin N, Schimann H, Schevin P, Hättenschwiler S (2012) Distinct microbial limitations in litter and underlying soil revealed by carbon and nutrient fertilization in a tropical rainforest. PLoS ONE 7:e49990

French E, Kaplan I, Iyer-Pascuzzi A, Nakatsu CH, Enders L (2021) Emerging strategies for precision microbiome management in diverse agroecosystems. Nat Plants 7(3):256–267. https://doi.org/10.1038/s41477-020-00830-9

Fungicide Resistance Action Committee (2015) Minutes of the 2015 SDHI Working Group meeting:Recommendations for 2016. FRAC

Gargallo-Garriga A, Sardans J, Llusià J et al (2024) Different profiles of soil phosphorous compounds depending on tree species and availability of soil phosphorus in a tropical rainforest in French Guiana. BMC Plant Biol 24:278. https://doi.org/10.1186/s12870-024-04907-x

Giller KE, Beare MH, Lavelle P, Izac AMN, Swift MJ (1997) Agricultural intensification, soil biodiversity and agroecosystem function. Appl Soil Ecol 6(1):3–16

Glassman SI et al (2018) Decomposition responses to climate depend on microbial community composition. Proc Natl Acad Sci USA 115:11994–11999

Gupta V (2011) Microbes and soil structure. In: Gliński J, Horabik J, Lipiec J (eds) Encyclopedia of agrophysics. Encyclopedia of earth sciences series. Springer, Dordrecht. https://doi.org/10.1007/978-90-481-3585-1_91

Gupta VVSR, Farrell RE, Germida JJ (1993) Activity of arylsulfatase in Saskatchewan soils. Can J Soil Sci 73(3):341–347

Hai FI, Modin O, Yamamoto K, Fukushi K, Nakajima F, Nghiem LD (2012) Pesticide removal by a mixed culture of bacteria and white-rot fungi. J Taiwan Inst Chem Eng 43:459–462

Hartmann M, Six J (2023) Soil structure and microbiome functions in agroecosystems. Nat Rev Earth Environ 4(1):4–18. https://doi.org/10.1038/s43017-022-00366-w

He LM, Tebo BM (1998) Surface charge properties of and cu(II) adsorption by spores of the marine bacillus sp. strain SG-1. Appl Environ Microbiol 64:1123–1129. [PubMed]

Hoorman JJ (2011) The Role of Soil Bacteria. Fact sheet agriculture and natural resources. The Ohio State University extension. SAG-13-11 pp 1–4

Hossain MM, Sultana F, Islam S (2017) Plant growth-promoting fungi (PGPF): phytostimulation and induced systemic resistance. Plant-Microbe Interactions in Agro-Ecological Perspectives: Volume 2: Microbial Interactions and Agro-Ecological Impacts, pp 135–191

Hutchins DA, Jansson JK, Remais JV, Rich VI, Singh BK, Trivedi P (2019) Climate change microbiology — problems and perspectives. Nat Rev Microbiol 17(6):391–396. https://doi.org/10.1038/s41579-019-0178-5

Iavicoli A, Boutet E, Buchala A, and Métraux JP (2003) Induced systemic resistance in Arabidopsis thaliana in response to root inoculation with Pseudomonas fluorescens CHA0. Mol Plant-Microbe Interact 16:851–858

Iqbal B, Li G, Alabbosh KF, Hussain H, Khan I, Tariq M, Javed Q, Naeem M, Ahmad N (2023) Advancing environmental sustainability through microbial reprogramming in growth improvement, stress alleviation, and phytoremediation. Plant Stress 10:100283. https://doi.org/10.1016/j.stress.2023.100283

Iram S, Shabbir R, Zafar H, Javaid M (2015) Biosorption and bioaccumulation of copper and lead by heavy metal-resistant fungal isolates. Arab J Sci Eng 40:1867–1873. https://doi.org/10.1007/s13369-015-1702-1

Johan PD, Ahmed OH, Omar L, Hasbullah NA (2021) Phosphorus transformation in soils following co-application of charcoal and wood ash. Agronomy 11(10):2010. https://doi.org/10.3390/agronomy11102010

Jordan CF (1985) Nutrient cycling in tropical Forest ecosystems. Wiley, Chichester, p 190

Kallenbach C et al (2016) Direct evidence for microbialderived soil organic matter formation and its ecophysiological controls. Nat Commun 7:13630

Kang SM, Shahzad R, Bilal S, Khan AL, Park YG, Lee KE et al (2019) Indole-3-acetic-acid and ACC deaminase producing Leclercia adecarboxylata MO1 improves Solanum lycopersicum

L. growth and salinity stress tolerance by endogenous secondary metabolites regulation. BMC Microbiol 19(1):1–14
Köhl J, Kolnaar R, Ravensberg WJ (2019) Mode of action of microbial biological control agents against plant diseases: relevance beyond efficacy. Front Plant Sci 10:845. https://doi.org/10.3389/fpls.2019.00845
Kolapo A, Didunyemi AJ, Aniyi OJ, Obembe OE (2022) Adoption of multiple sustainable land management practices and its effects on productivity of smallholder maize farmers in Nigeria. Resourc, Environ Sustain 10:100084. https://doi.org/10.1016/j.resenv.2022.100084
Lal R (2004) Soil carbon sequestration impacts on global climate change and food security. Science 304(5677):1623–1627. https://doi.org/10.1126/science.1097396
Lee SH, Whon TW, Roh SW, Jeon CO (2020) Unraveling microbial fermentation features in kimchi: from classical to meta-omics approaches. Appl Microbiol Biotechnol 104:7731–7744. https://doi.org/10.1007/s00253-020-10804-8
Lehmann A et al (2017) Soil biota contributions to soil aggregation. Nat Ecol Evol 1:1828–1835
Logue JB, Findlay SEG, Comte J (2015) Editorial: microbial responses to environmental changes. Front Microbiol 6:1364
Liang C, Amelung W, Lehmann J, Kästner M (2019) Quantitative assessment of microbial necromass contribution to soil organic matter. Glob Chang Biol 25(11):3578–3590
Liu S, García-Palacios P, Tedersoo L, Guirado E, van der Heijden MG, Wagg C, Chen D, Wang Q, Wang J, Singh BK, Delgado-Baquerizo M (2022) Phylotype diversity within soil fungal functional groups drives ecosystem stability. Nat Ecol Evol 6(7):900–909
Majumder A, Bhattacharyya K, Bhattacharyya S, Kole SC (2013) Arsenictolerant, arsenite-oxidising bacterial strains in the contaminated soils of West Bengal, India. Sci Total Environ 46:1006–1014. https://doi.org/10.1016/j.scitotenv.2013.06.068084-5
Maron PA, Mougel C, Ranjard L (2011) Soil microbial diversity: methodological strategy, spatial overview and functional interest. C R Biol 334(5–6):403–411. https://doi.org/10.1016/j.crvi.2010.12.003
Maron PA, Sarr A, Kaisermann A, Lévêque J, Mathieu O, Guigue J, Karimi B, Bernard L, Dequiedt S, Terrat S, Chabbi A (2018) High microbial diversity promotes soil ecosystem functioning. Appl Environ Microbiol 84(9):e02738–e02717
Marrs RH, Roberts RD, Skeffington RA, Bradshaw AD (1983) Nitrogen and the development of ecosystems. See Ref 135(6):113–136
McLaney E, Morassaei S, Hughes L, Davies R, Campbell M, Di Prospero L (2022) A framework for interprofessional team collaboration in a hospital setting: advancing team competencies and behaviours. Healthc Manage Forum 35(2):112–117. https://doi.org/10.1177/08404704211063584
Mendes R, Garbeva P, Raaijmakers JM (2013) The rhizosphere microbiome: significance of plant beneficial, plant pathogenic, and human pathogenic microorganisms. FEMS Microbiol Rev 37(5):634–663. https://doi.org/10.1111/1574-6976.12028
Mierzwa-Hersztek M, Klimkowicz-Pawlas A, Gondek K (2017) Influence of poultry litter and poultry litter biochar on soil microbial respiration and nitrifying bacteria activity. Waste Biomass Valoriz J 9:1–11
Mishra M, Mohan D (2017) In: Rakshit A, Abhilash PC, Singh HB, Ghosh S (eds) Bioremediation of contaminated soils: an overview BT—adaptive soil management: from theory to practices. Springer Singapore, pp 323–337. https://doi.org/10.1007/978-981-10-3638-5_16
Morris KA, Hornum S, Crystal-Ornelas R, Pennington SC, Bond-Lamberty B (2022) Soil respiration response to simulated precipitation change depends on ecosystem type and study duration. J Geophys Res Biogeosciences 127(11):e2022JG006887
Mwende Muindi E (2019) Understanding soil phosphorus. Int J Plant Soil Sci 31(2):1–18. https://doi.org/10.9734/ijpss/2019/v31i230208
Nannipieri P, Ascher J, Ceccherini MT, Landi L, Pietramellara G, Renella G (2017) Microbial divers

Ngosong C, Jarosch M, Raupp J, Neumann E, Ruess L (2010) The impact of farming practice on soil microorganisms and arbuscular mycorrhizal fungi: crop type versus long-term mineral and organic fertilization. Appl Soil Ecol 46:134–142. https://doi.org/10.1016/j.apsoil.2010.07.004
Nicole E. Tautges, Sullivan TS, Reardon CL, Burke IC (2016) Soil microbial diversity and activity linked to crop yield and quality in a dryland organic wheat production system. Appl Soil Ecol 108:258–268. https://doi.org/10.1016/j.apsoil.2016.09.003
Nielsen MN, Winding A (2002) Microorganisms as Indicators of soil Health. Technical Report 388 National Environmental Research Institute, Denmark, (http://www.dmu.dk/1_viden/2_Publikationer/3_fagrapporter/rapporter/FR388.pdf)
O'Brien PA (2017) Biological control of plant diseases. Australas Plant Pathol 46(4):293–304
Obafemi YD, Oranusi SU, Ajanaku KO, Akinduti PA, Leech J, Cotter PD (2022) African fermented foods: overview, emerging benefits, and novel approaches to microbiome profiling. npj Science of Food 6(1):15
Odaro-Junior UB (2020) Tropical crops and microbes. IntechOpen. https://doi.org/10.5772/intechopen.89531
Onen O, Aboh A, Akor M (2020) Microbial diversity: values and roles in ecosystems. Asian J Biol:10–22. https://doi.org/10.9734/AJOB/2020/v9i130075
Orwin KH, Wardle DA (2005) Plant species composition effects on belowground properties and the resistance and resilience of the soil microflora to a drying disturbance. Plant Soil 278(1):205–221
Pande V, Pandey SC, Sati D, Pande V, Samant M (2020) Bioremediation: an emerging effective approach towards environment restoration. Environ Sustain 3(1):91–103. https://doi.org/10.1007/s42398-020-00099-w
Pankhurst CE, Doube BM, Gupta VVSR (1997) Biological indicators of soil health: synthesis. In: Pankhurst CE, Doube BM, Gupta VVSR (eds) Biological indicators of SOIL health. CAB International, pp 419–435
Paul D, Pandey G, Pandey J, Jain RK (2005) Accessing microbial diversity for bioremediation and environmental restoration. Trends Biotechnol 23(3):135–142. https://doi.org/10.1016/j.tibtech.2005.01.001
Pertot I, Alabouvette C, Esteve EH, Franca S (2015). The use of microbial biocontrol agents against soil-borne diseases. Retrieved from European Commission EIPAGRI Focus Group. 2015. https://ec.europa.eu/eip/agriculture/sites/agrieip/fles/8_eip_sbd_mp_biocontrol_fnal.pdf
Petrie L, North NN, Dollhopf SL, Balkwill DL, Kostka JE (2003) Enumeration and characterization of Fe (III)-reducing microbial communities from acidic subsurface sediments contaminated with uranium (VI). Appl Environ Microbiol 69(12):7467–7747
Pierre-Alain, Amadou MS, Aurore K, Jean L, Olivier M, Julien G, Battle K, Laetitia B, Samuel D, Sébastien T, Abad C, Lionel R (2018) High microbial diversity promotes soil ecosystem functioning. Appl Environ Microbiol 84(2018). https://doi.org/10.1128/AEM.02738-17
Rajwinder Kaur, Choudhary D, Bali S, Bandral SS, Varinder Singh M, Ahmad A, Rani N, Singh TG, Chandrasekaran B (2024) Pesticides: an alarming detrimental to health and environment. Sci Total Environ 915:170113
Ranveer K, Yogendra SG, Vivek K (2014) Interaction and symbiosis of AM fungi, actinomycetes and plant growth promoting rhizobacteria with plants: strategies for the improvement of plants health and defense system. Int J Curr Microbiol App Sci 3(7):564–585
Rehman A, Farooq M, Lee D-J, Siddique KHM (2022) Sustainable agricultural practices for food security and ecosystem services. Environ Sci Pollut Res 29(56):84076–84095. https://doi.org/10.1007/s11356-022-23635-z
Reinold M, Wong HL, MacLeod FI, Meltzer J, Thompson A, Burns BP (2019) The vulnerability of microbial ecosystems in A changing climate: potential impact in Shark Bay, vol 9(3), Life (Basel). https://doi.org/10.3390/life9030071
Rodríguez H, Fraga R (1999) Phosphate solubilizing bacteria and their role in plant growth promotion. Biotechnol Adv 17(4–5):319–339. https://doi.org/10.1016/S0734-9750(99)00014-2
Sammauria R, Kumawat S, Kumawat P, Singh J, Jatwa TK (2020) Microbial inoculants: potential tool for sustainability of agricultural production systems. Arch Microbiol 202(4):677–693. https://doi.org/10.1007/s00203-019-01795-w

Sanchez PA (1976) Properties and management of soils in the tropics. Wiley, New York, p 618
Sarret G, Manceau A, Spadini L, Roux JC, Hazemann JL, Soldo Y, Eybert-Bérard L, Menthonnex J (1998) Structural determination of Zn and Pb binding sites in Penicillium chrysogenum cell walls by EXAFS spectroscopy. Environ Sci Technol 32:1648–1655
Sayer J, Zuidema P, Rijks M (1995) Managing for biodiversity in humid tropical forests. Int For Rev 74:282
Schroth G, Krauss U, Gasparotto L, Aguilar J, Vohland K (2000) Pests and diseases in agroforestry systems of the humid tropics. Agrofor Syst 50:199–241. https://doi.org/10.1023/A:1006468103914
Shade A, Peter H, Allison SD, Baho DL, Berga M, Bürgmann H, Huber DH, Langenheder S, Lennon JT, Martiny JBH, Matulich KL, Schmidt TM, Handelsman J (2012) Fundamentals of microbial community resistance and resilience. Front Microbio 3:417
Sharma S, Gupta A, Shukla A, Ahmad E, Sharma M, Ramesh A (2016) Microbial conservation strategies and methodologies: status and challenges. Ind J Plant Genetic Res 29:340. https://doi.org/10.5958/0976-1926.2016.00062.0
Shi L, Dong H, Reguera G, Beyenal H, Lu A, Liu J et al (2016) Extracellular electron transfer mechanisms between microorganisms and minerals. Nat Rev Microbiol 14:651–662. https://doi.org/10.1038/nrmicro.2016.93
Simonson WD, Miller E, Jones A, García-Rangel S, Thornton H, McOwen C (2021) Enhancing climate change resilience of ecological restoration — a framework for action. Perspect Ecol Conserv 19(3):300–310. https://doi.org/10.1016/j.pecon.2021.05.002
Singh BK, Trivedi P (2017) Microbiome and the future for food and nutrient security. Microb Biotechnol 10(1):50–53. https://doi.org/10.1111/1751-7915.12592
Singh M, Singh D, Gupta A, Pandey KD, Singh PK, Kumar A (2019) Plant growth promoting rhizobacteria: application in biofertilizers and biocontrol of phytopathogens. In: PGPR amelioration in sustainable agriculture. Woodhead Publishing, pp 41–66
Singh BK, Delgado-Baquerizo M, Egidi E, Guirado E, Leach JE, Liu H, Trivedi P (2023) Climate change impacts on plant pathogens, food security and paths forward. Nat Rev Microbiol 21(10):640–656. https://doi.org/10.1038/s41579-023-00900-7
Smith SE, Read DJ (2010) Mycorrhizal symbiosis. Academic press
Steffan JJ, Brevik EC, Burgess LC, Cerdà A (2018) The effect of soil on human health: an overview. Eur J Soil Sci 69(1):159–171. https://doi.org/10.1111/ejss.12451
Stewart JWB, McKercher RB (1982) In: Bums RG, Slater JH (eds) Phosphorus cycle. Experimental microbial ecology. Blackwells, Oxford, pp 221–238. 683 pp
Stork NE, Balston J, Farquhar GD, Franks PJ, Holtum JAM, Liddell MJ (2007) Tropical rainforest canopies and climate change. Austral Ecol 32:105–112
Sutton-Grier AE, Keller JK, Koch R, Gilmour C, Megonigal JP (2011) Electron donors and acceptors influence anaerobic soil organic matter mineralization in tidal marshes. Soil Biol Biochem 43(7):1576–1583
Tabari H (2020) Climate change impact on flood and extreme precipitation increases with water availability. Sci Rep 10(1):13768. https://doi.org/10.1038/s41598-020-70816-2
Thakur, N., Kaur, S., Tomar, P., Thakur, S., & Yadav, A. N. (2020). Chapter 15—microbial biopesticides: current status and advancement for sustainable agriculture and environment A. A. Rastegari, A. N. Yadav, & N. B. T.-N. F. D. in M. B. B. Yadav (eds.); pp. 243–282). Elsevier. https://doi.org/10.1016/B978-0-12-820526-6.00016-6
Timmis K, de Vos WM, Ramos JL, Vlaeminck SE, Prieto A, Danchin A, Verstraete W, de Lorenzo V, Lee SY, Brüssow H, Timmis JK, Singh BK (2017) The contribution of microbial biotechnology to sustainable development goals. Microb Biotechnol 10(5):984–987. https://doi.org/10.1111/1751-7915.12818
Torsvik V, Øvreås L, Thingstad TF (2002) Prokaryotic Diversity—Magnitude, Dynamics, and Controlling Factors. Science 296(5570):1064–1066
Townsend AR, Cleveland CC, Houlton BZ, Alden CB, White JWC (2011) Multi-element regulation of the tropical forest carbon cycle. Front Ecol Environ 9:9–17

van der Heijden MGA, Bardgett RD, van Straalen NM (2008) The unseen majority: soil microbes as drivers of plant diversity and productivity in terrestrial ecosystems. Ecol Lett 11(3):296–310

Van Hamme JD, Singh A, Ward OP (2003) Recent advances in petroleum microbiology. Microbiol Mol Biol Rev 67(4):503–549. https://doi.org/10.1128/MMBR.67.4.503-549.2003

Vessey JK (2003) Plant growth promoting rhizobacteria as biofertilizers. Plant Soil 255(2):571–586. https://doi.org/10.1023/A:1026037216893

Vitousek PM, Porder S, Houlton BZ, Chadwick OA (2010) Terrestrial phosphorus limitation: mechanisms, implications, and nitrogen-phosphorus interactions. Ecol Appl 20:5–15

Vuong P, Chong S, Kaur P (2022) The little things that matter: how bioprospecting microbial biodiversity can build towards the realization of United Nations sustainable development goals. Npj Biodiversity 1(1):4. https://doi.org/10.1038/s44185-022-00006-y

Weiskopf SR, Rubenstein MA, Crozier LG, Gaichas S, Griffis R, Halofsky JE, Hyde KJW, Morelli TL, Morisette JT, Muñoz RC, Pershing AJ, Peterson DL, Poudel R, Staudinger MD, Sutton-Grier AE, Thompson L, Vose J, Weltzin JF, Whyte KP (2020) Climate change effects on biodiversity, ecosystems, ecosystem services, and natural resource management in the United States. Sci Total Environ 733:137782. https://doi.org/10.1016/j.scitotenv.2020.137782

Wild A (1981) Mass flow and diffusion. See Ref. 85 2:37–80

Wuana RA, Okieimen FE (2011) Heavy metals in contaminated soils: a review of sources, chemistry, risks and best available strategies for remediation. ISRN Ecol 2011:402647

Yang G, Ryo M, Roy J, Lammel DR, Ballhausen MB, Jing X, Zhu X, Rillig MC (2022) Multiple anthropogenic pressures eliminate the effects of soil microbial diversity on ecosystem functions in experimental microcosms. Nat Commun 13(1):4260

Yuan MM, Guo X, Wu LW, Zhang Y, Xiao NJ, Ning DL et al (2021) Climate warming enhances microbial network complexity and stability. Nat Clim Chang 11:343–348. https://doi.org/10.1038/s41558-021-00989-9

Zhang S, Li Q, Lü Y, Sun X, Jia S, Zhang X (2015) Conservation tillage positively influences the micro flora and microfauna in the black soil of Northeast China. Soil Tillage Res 2015(149):46–52. https://doi.org/10.1016/j.still.2015.01.001

Chapter 4
Microbial Approaches: A Paradigm Shift for Nutrient Use Efficiency

A. O. Adesemoye, D. B. Watts, and H. A. Torbert

Abstract Emerging technologies in the development of biostimulants have opened the door for new opportunities to improve agricultural sustainability. The technological development has led to the promotion of microbial inoculant use as an alternative or supplement to traditional synthetic fertilizers and the reduction of agro-environmental associated problems such as leaching, pollution and greenhouse gas (GHG) emissions. These emerging products are being developed using beneficial fungi such as Arbuscular Mycorrhizal Fungi and free-living bacteria, especially Plant Growth-Promoting Rhizobacteria (PGPR) such as *Bacillus* species. Although, changes in anthropogenic activities are responsible for increasing atmospheric concentrations of GHGs, these bioinoculants as drivers of biochemical processes in soil, could cause a shift in ecosystem functions, reducing the production of CO_2, CH_4, and N_2O. Various names have been given to the products e.g., biofertilizer, fertilizer helper, inoculants, biopesticide, etc. Regardless of which name, development of this biological approach has brought about a paradigm shift in nutrient use efficiency. However, the technology must be further improved because farmers' acceptance and product use has not matched the growth of the subsector. This review examines the ongoing developments in the subsector, interrogates the identifiable impediments preventing broad acceptance, and discusses what is needed for improving the efficacy of these new approaches to better fit into an integrated nutrient management (INM) system.

Keywords Inoculant · Fertilizer · PGPR · Nutrient management · Biologicals · Greenhouse gas

A. O. Adesemoye (✉) · D. B. Watts · H. A. Torbert
USDA-ARS National Soil Dynamics Laboratory, Auburn, AL, USA
e-mail: tony.adesemoye@usda.gov

O. O. Babalola et al. (eds.), *Ecofriendly Frontiers*,
https://doi.org/10.1007/978-3-031-98700-7_4

4.1 Introduction

Use of microbial inoculants as an alternative or supplement to synthetic chemical inputs has opened the door for new opportunities to sustain agriculture. Before now, agricultural sustainability has been challenged by continuous and indiscriminate use of chemical inputs that contribute to environmental pollution, accumulation of greenhouse gases (GHGs), and the unprecedented challenge of managing environmental issues worldwide. Intensive crop production driven by increased food demand and agricultural profitability over the last several decades utilizes massive amounts of chemical inputs including, herbicides, pesticides, and synthetic fertilizers, which has led to negative impacts on the environment. Also, the application of fertilizers and pesticides may cause changes in the mineral content of food crops and has the potential to leave residues in food; for instance, a deviation from the expected range of aluminum (Al), cobalt (Co), and nickel (Ni) was reported in the rice plants (Qian et al. 2019). Some of these chemical inputs may also have carcinogenic properties which can negatively impact human health (Alori and Babalola 2018). To overcome nutrient stress during a growing season, crops naturally form alliances with soil microbes to mitigate stresses, survive, and maintain homeostasis (Bouain et al. 2019; Fan et al. 2021). These microbes can directly or indirectly influence ecosystem functions and potentially have a lasting effect on improving cropping systems. Nutrient cycling may be altered to minimize stress by leveraging these natural alliances and improving the concentration of such beneficial microbes that can solubilize macronutrients, enhance root nutrient uptake, and improve foliar nutrient content (Moore et al. 2022). These organisms frequently contribute to soil organic matter through biosynthesized compounds, e.g., extracellular enzymes, which can sorb to minerals and lead to the production of mineral-associated organic matter (Cotrufo et al. 2013; Watts et al. 2024).

In recent decades, research have identified various categories of soil microorganisms including plant growth promoting rhizobacteria (PGPR) which can enhance nutrient uptake and improve crop production. The use of these microorganisms is not associated with any known negative environmental concern like the types associated with chemical fertilizers or pesticides (Adesemoye et al. 2008). Microbial-enhanced uptake of macro and micronutrients has also been reported (Adesemoye et al. 2009; Bashan and de-Bashan 2010; Calvo et al. 2014; Sheng and He 2006). At least two mechanisms of action, including augmentation and displacement have been suggested for the establishment and persistence of these microbes in soil (Kurkjian et al. 2021). In turn, the activities that are triggered in plants may cause a shift in microbial populations and ecosystem functions. Carefully planned application strategy of these soil microbes can enhance their key roles in maintaining soil health and sustainability by regulating the rate of plant and animal residue decomposition and nutrient availability. Employing practices that improve the function of belowground biological processes in the rhizosphere will promote soil health, environmental quality, and overall sustainability in intensive crop production systems (He and Zhang 2014; Watts et al. 2024).

Consequently, interests in biorational or biological based materials are growing and the development and improvement in microbial approaches is a paradigm shift for nutrient use efficiency. Microbial inoculants are a class of products that has emerged consisting of different microbial taxa that include bacteria, fungi, and algae. Microbial inoculant is under the broad category called biostimulants that include multiple classes of substances/products (Calvo et al. 2014; Parikh et al. 2018). Though interest in these products is growing, and the market is expanding with new products, enthusiasm of farmers has not matched the growth of the sector mainly because of some inconsistencies reported when the products are used. The reason for the inconsistencies is not yet understood but efficacy of these approaches must improve to strengthen integrated nutrient management (INM) systems. As previously explained, the INM system promotes low chemical input and improved nutrient use efficiency by combining both natural and synthetic nutrient sources efficiently without compromising crop productivity (Adesemoye et al. 2008).

Identifying strains with many desired traits, combining multiple strains into a consortium, and using carrier materials may help achieve improved efficacy enabling the technology to spread more quickly across the world. Effective identification and utility of appropriate genera of bacteria and fungi in INM is an important factor that can help achieve the desired improvement in product efficacy for agricultural productions (Akinrinlola et al. 2018). Beneficial microorganisms, if effectively applied, could be the approach agriculture needs to sustainably improve productivity with less or no chemical input. However, this may just remain an aspiration unless proper actions are taken locally and globally. The Food and Agriculture Organization (FAO) suggested that the development of local food production in many countries is one of the most important factors that determine progress or failure in improving food security (Adesemoye and Egamberdieva 2013). This review examines the ongoing developments in the biorational/biological sector of agricultural industry and interrogates the identifiable impediments preventing broad acceptance as well as discusses what is needed for improving the efficacy of these new approaches to better fit into the INM systems.

4.2 Different Groups of Soil Microbes Beneficial to Plants

4.2.1 Plant Growth-Promoting Rhizobacteria (PGPR)

Plant growth-promoting rhizobacteria (PGPR) are free-living beneficial bacteria in the soil that can easily establish relationships with plant roots. PGPR belong to several genera such as *Azospirillum*, *Azotobacter*, *Bacillus*, *Pseudomonas* (Adesemoye et al. 2008; Akinrinlola et al. 2018) but does not include those with associative symbiotic relationship such as in legumes. They play important roles in crop production which include (i) stimulation of root growth, (ii) elevation of soil fertility by improving nutrient availability for plant roots and/or fixing nitrogen, (iii) suppression of soil-borne pathogens and (iv) alleviation of abiotic stresses in plants (Egamberdieva and Adesemoye 2016). *Bacillus* species is a frequently used taxonomic group utilized in bacteria-based agricultural products. This group is effective in enhancing nutrient uptake (Adesemoye et al. 2008; Lin et al. 2019), functioning as a biological control agent against plant pathogens (Parikh et al. 2018) and pests (Disi et al. 2018), and can tolerate high temperatures, pH, and osmotic conditions. Also, *Bacillus* sp. is easily adaptable to the arid- and salt-affected environments (Egamberdieva and Adesemoye 2016).

PGPR activities in plants occur through direct or indirect mechanisms of action. The direct mechanisms are nitrogen fixation, mineral solubilization of phosphorus, potassium, sulphur, etc., siderophores and phytohormones production, including auxin e.g., indole acetic acid (IAA), gibberellin, and 1-aminocyclopropane-1-carboxylate (ACC) deaminase. PGPR-induced plant growth promotion via nutrient uptake occurs through enhanced plant rooting, which leads to greater surface area and increased root hairs (Adesemoye and Kloepper 2009). Indirect mechanisms occur through competition with pathogens, induced systemic resistance (ISR) or other antagonistic activity against phytopathogenic microorganisms through production of antimicrobials, cyanide, volatile compounds, cell wall-degrading enzymes, and lytic enzymes such as chitinases, proteases, cellulases, and lipases. These mechanisms have been described previously by various authors (Calvo et al. 2014; Adesemoye and Kloepper 2009; Glick et al. 2007), so we would not elaborate on them further in this review.

4.2.2 Actinomycetes, Lactic Acid Bacteria, and Other Beneficial Bacteria

Actinomycetes play major roles in organic matter cycling, nitrogen fixation, and decomposition of complex remains of dead plants, animals, and fungi. They are potent biological control agents against plant pathogens, help improve the availability of nutrients and minerals, enhance metabolites production, and promote plant growth regulators (Bhatti et al. 2017). Especially with the known ability to form stable humus, these organisms could potentially be combined with other beneficial microorganisms to improve nutrient use efficiency.

Lactic acid bacteria (LAB) are a significant group of beneficial Gram-positive bacteria widely used as probiotic organisms for fermented foods in the food and pharmaceutical industry, which can regulate soil organic matter and biochemical cycles (Raman et al. 2022). They also have potential to promote soil health and fertility, stimulate plant growth, alleviate various abiotic stress, and produce organic acid and metabolites that can protect plants against fungal and bacterial pathogens. However, these properties have not been well explored, and this might be partly because of the phylogenetic complexity and diversity of the group. Lactic acid bacteria (LAB) are also used as direct fed microbials and as commercial silage inoculants for reducing methane production in ruminant livestock (Doyle et al. 2019). As natural inhabitants of the intestinal tract of mammals, LAB have high potential use as a methane mitigation agent because enteric fermentation in ruminants is the largest anthropogenic source of agricultural methane, a greenhouse gas with a significant role in global warming (Doyle et al. 2019). Methane is frequently produced as a fermentation end-product in the rumen, exhaled from the lungs via respiration from the animal and released to the atmosphere. Previous reports showed that silage treated with lactic acid bacteria inoculants would increase ruminal microbial biomass when tested in vitro. Therefore, Monteiro et al. (2021) evaluated the inoculation of alfalfa silage with *Lactobacillus plantarum* (LP) and reported that cows receiving the LP treatment had greater milk production and minor changes in total microbial non-ammonia N flow in the omasal canal compared to the control treatment. However, the mechanisms used by LAB are not yet well understood, so they have been underutilized for the needed methane mitigation.

4.2.3 Beneficial Fungi: Species of *Mycorrhiza* and *Trichoderma*

Mycorrhizal fungi and *Trichoderma* species have been reported as potential inoculants that may significantly increase plant height, dry weight, number of nodules, pods, or yield of chickpea (Garg and Baher 2013; Patel et al. 2012). It was demonstrated by Adesemoye et al. (2008) that, when applied singly or in combination with PGPR, mycorrhizae improved nutrient uptake in corn (Fig. 4.1). *Trichoderma* is an important fungal genus used in biological products for the control of pathogens and to enhance nutrient uptake (Parikh et al. 2018). For instance, five strains of

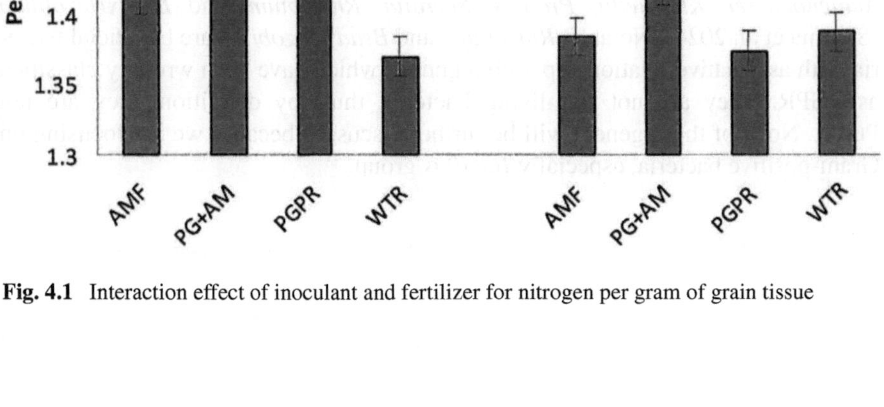

Fig. 4.1 Interaction effect of inoculant and fertilizer for nitrogen per gram of grain tissue

Trichoderma (one strain of *T. velutinum*, two strains of *T. gamsii*, and two strains of *T. virens*) were evaluated by Parikh et al. (2018) and shown to have excellent inhibition levels against six tested pathogenic *Fusarium* isolates. The Trichodermal treatments resulted in reductions of in-vitro growth of the pathogens and seedling diseases in corn, soybean, and wheat (Parikh et al. 2018).

Shown are arbuscular mycorrhiza fungi (AMF), plant growth promoting rhizobacteria (PGPR), co-inoculation of PGPR and AMF, and water control (WTR), with either poultry litter (PL) or ammonium nitrate (NH_4). This figure was previously published in Adesemoye et al. (2008).

4.2.4 Other Beneficial Organism Groups

Evaluations of soil suppressiveness have provided insight that there are many groups of beneficial organisms in the soil that are yet to be identified (Adesemoye et al. 2023). Through soil suppressiveness, diseases caused by soil-borne pathogens are reduced over time and could become negligible without the use of chemical pesticides. However, the situation remains very difficult to be fully explained because only a fraction of the organisms involved have been isolated and characterized as evidenced from microbiome analyses (Adesemoye et al. 2023; Carrion et al. 2019). Thus, it is unclear how many among these diverse organisms could potentially be harnessed as plant helpers to shore up nutrient use efficiency. In a recent study of the diversity of microbial communities in soils suppressive to soilborne pathogens in Nebraska, relatively higher operational taxonomic units (OTUs) were observed for potential biocontrol agents such as *Bacillus*, *Paenibacillus*, *Massilia*, *Pseudomonas*, and *Sphingomonas* and Adesemoye et al. (2023) suggested that these groups played important roles in the development of soil suppressive properties.

Streptomyces species are another important group that have plant growth-promoting potentials and biological control properties against pests. Wang et al. (2022) reported that *Streptomyces aureoverticillatus* HN6 has disease prevention and growth promotive effects, which can improve the rhizosphere fertility of plants and modulate the rhizosphere microbial community of plants. Other groups of organisms that have been investigated or explored as potential biostimulants include but not limited to species of *Stenotrophomonas, Enterobacter, Staphylococus, Acinenobacter, Klebsiella, Proteus, Serratia, Rhizobium,* and *Bradyrhizobium* (Sharma et al. 2024). Notably, *Rhizobium* and *Bradyrhizobium* are beneficial bacteria with associative relationship with legumes, which have been wrongly classified as PGPR. They are not free-living bacteria; thus, by definition, they are not PGPR. None of these genera will be further discussed because we are focusing on Gram-positive bacteria, especially *Bacillus* group.

4.2.5 Using Microbes as a Consortium or in Combination with Other Materials

Changes in precipitation, water levels, availability of water for irrigation, salinity, and infection/infestation from pathogens and pests has major impact on agricultural production (Egambergieva and Adesemoye 2016). Using microbial inoculants in a consortium or with other materials is a promising approach to help alleviate the impact of environmental changes on agriculture as well as the impact of agriculture on those changes, which occur in a cycle. Numerous studies have evaluated the synergist effects of two or more combined strains of microbes on crop production in the last two decades (Mpanga et al. 2019). These studies have shown that a consortium of PGPR is better in promoting plant growth as compared to individual strains (Singh et al. 2014; Wang et al. 2020). This is likely because the individual strains have different mechanisms that are working together leading to a synergy for improved nutrient uptake, plant growth, and overall plant health (Khatoon et al. 2020). Inoculation of PGPR with urea, calcium-ammonium nitrate, or urea-ammonium nitrate has been used to achieve increased nutrient uptake of N, P, and K during the early growth stage of corn plants. Also, combinations of microbial inoculants with seaweed extracts or humic substances could enhance crop production (Calvo et al. 2014). Evidently, there are many areas where benefits from these combinations can be utilized but are not yet properly explored.

4.3 Enhanced Uptake of Macro and Micronutrients

Essential nutrients (macro nutrients) such as nitrogen (N), phosphorus (P), potassium (K), sulfur (S), magnesium (Mg), and calcium (Ca), are needed in relatively large quantities for plant nutrition, while micronutrients such as iron (Fe), manganese (Mn), copper (Cu), zinc (Zn), boron (B), molybdenum (Mo), chlorine (Cl), and nickel (Ni), are required in comparatively smaller amounts (Andresen et al. 2018; Gong et al. 2020). Deficiency of any one of these macros or micronutrients can have negative consequences at the plants molecular and phenotypic level, effecting development and yield (Gong et al. 2020). Thus, given that nutrient deficiency is common in most soils, the production and application of synthetic fertilizers are often used to meet global food demand in most agricultural systems. As indicated previously, improper or long-term use of these synthetic fertilizers can pose serious problems to the environment and one of the major issues is that N fertilizer manufacturing consumes 1–2% of the world's fossil fuel energy output (Chen et al. 2018). Also, plants only take up 30–50% of the N fertilizer that is applied to soil, leaving the remaining N susceptible to gaseous loss or leaching, which may cause negative impacts to the environment (Lehnert et al. 2018). Similarly, between 15% and 30% of the P (Syers et al. 2008) and about 40% of the potassium (Zou and Lu 2010) are utilized by the crop, demonstrating the importance of improving crop nutrient uptake. Although most fertilized soils have enough plant nutrients, they are often present in insoluble forms; thereby, rendering them unavailable for uptake by plants (Shafi et al. 2017). If plants can more efficiently absorb and use these soil nutrients, this could improve the efficiency of applied fertilizers; thereby reducing cost of inputs and preventing nutrient loss to the environment (Baligar et al. 2001).

The importance of research on the use of plant growth promoting bacteria to improve nutrient acquisition (Dobbaelere et al. 2003) from soil cannot be overemphasized. Nutrient uptake from the soil is the most important function in the process of plant accumulation before translocation to seeds and PGPR inoculants can help plants in the process, leading to increased growth, mobility, and nutrient uptake (Lucas et al. 2004; Cakmakci et al. 2006). One of the main mechanisms of action is by modifying the root morphology. This results in greater root surface for greater exploration and nutrient uptake to occur. Hernandez and Chailloux (2004) and Adesemoye et al. (2008) are two examples among other studies, which have shown that PGPR-elicited plant growth promotion can result in enhanced N uptake by plant roots. In these studies, it was suggested that the increase in plant N content might result from increased fertilizer N utilization efficiency, suggesting that PGPR should be useful in integrated nutrient management systems (Adesemoye et al. 2008). There have been reports of PGPR contributing 30–50% of the N needs for crop production. For example, in an intensive wheat rotation system at Avon, South Australia, free-living non symbiotic nitrogen fixation accounted for 20 kg ha^{-1} $yr.^{-1}$, (30–50% of crop needs) (Vadakattu and Patterson 2006).

Presently, from a biofertilizer standpoint, the most common microbial inoculants being researched and marketed are nitrogen-fixing microorganisms. This is not surprising, given that N is the most common limited plant nutrient needed in the greatest amount (O'Callaghan et al. 2022). Additionally, PGPR have been shown to improve uptake of other nutrients and many plant parameters (Calvo et al. 2017; Herman et al. 2008; Podile and Kishore 2007). Calvo et al. (2017) reported increased nutrient uptake of N, P, and K from PGPR-inoculated corn plants grown with urea, calcium-ammonium nitrate, and urea-ammonium nitrate during the early growth stage under controlled conditions in a glasshouse. Many other authors have also shown higher plant biomass, significantly increased yield, plant height, head diameter, chlorophyll content, and/or accumulation of macro and micronutrient in broccoli (Yildirim et al. 2011), leaves of squash (Tchiaze et al. 2016), chickpea (Elkoca et al. 2008), barley (Cakmakci et al. 2007), tomato (Adesemoye et al. 2009), lettuce (Lai et al. 2008), and strawberry (Gunes et al. 2009).

Specifically, bacterial inoculations of *Bacillus cereus*, *Rhizobium rubi*, and *Brevibacillus reuszeri* with manure increased yield 17.0%, 20.2%, and 24.3% and chlorophyll content 14.7%, 14.0%, and 13.7% over control, respectively (Yildirim et al. 2011) and O'Callaghan et al. (2022) showed that microbial inoculants increased yield by approximately 22.25% compared to treatments with no-microbial inoculants. Several other studies have confirmed that microbial inoculants improved crop yield, nutrient uptake, and chlorophyll content (Gontia-Mishra et al. 2016; Hahm et al. 2017; Agegnehu et al. 2016; Talebi 2011). O'Callaghan et al. (2022) suggested that *Bacillus* was one of the most abundant genera reported in the rhizosphere that can easily colonize and proliferate in soil and on plant surfaces. Members of the genus often possess multifunctional plant growth-promoting traits including N-fixation, P-solubilisation, phytohormone production (Swarnalakshmi et al. 2020), biofilm development (Ansari and Ahmad 2018), and plant exudates utilization as nutrients (Tan et al. 2013). Consequently, *Bacillus* species are one of the most promising biofertilizers in sustainable agriculture. While several microbial inoculants have been shown to improve crop yield and nutrient uptake, it is important to note that these microbial species may not be effective in all ecosystems. Microbial species may provide specific functions to certain plants. As a result, microbial inoculants containing diverse species may provide multiple benefits. However, the lasting effect of the microbes in the soil over multiple growing seasons is not well understood and the effect on modulating the microbial community structure over multiple seasons should be investigated further.

4.4 Greenhouse Gas Emissions and Microbe-Based Mitigation

Greenhouse gases (GHG) are naturally occurring gases that keep the earth warm by trapping heat in the atmosphere. Since the industrial revolution, the earth has warmed by approximately 1.5% (IPCC 2018). This global warming effect is primarily due to changes in anthropogenic activities. Studies have shown that agricultural activities are the primary source of GHG (Jacinthe et al. 2014; Smith et al. 2021; Watts et al. 2023). If mitigation practices are not implemented to reduce these GHG emissions, projections suggest that the Earth could warm by as much as 5 °C over the next century (IPCC 2018). This global warming awareness has led to increased pressure from society to minimize atmospheric concentrations of greenhouse gases. Carbon dioxide (CO_2), methane (CH_4), and nitrous oxide (N_2O) have been identified as the primary atmospheric gases contributing to warming of the earth (USEPA 2020). Carbon dioxide is the predominant GHG from agricultural activities and is predicted to account for more than 80% of all GHGs emissions in the US. While N_2O is emitted in a lower quantity, it is regarded as the most potent greenhouse gas emitted to the atmosphere, having a radiative forcing 300 times that of CO_2 (Myhre et al. 2013). Although changes in anthropogenic activities are largely responsible for the increasing atmospheric concentrations of GHGs, soil microbes are the main drivers of biochemical reactions (cycling of organic C, N and other nutrients) in soil, thereby affecting the production and consumption of CO_2, CH_4, and N_2O (Smith et al. 2021; Watts et al. 2023). Thus, modulating these soil microbes can potentially slow or accelerate gas efflux with implications for global warming.

Nitrous oxide reduction strategies and other technologies aimed to reduce agriculture's footprint on global warming, are limited. A very important reduction approach is the use of control release fertilizers and nitrification inhibitors (Mosier et al. 1998; Singh and Verma 2007) and has remained the most common approach as of the time of this publication. However, this approach has had mixed results and may not be as effective in regions with high rainfall events (Dell et al. 2014; Parkin and Hatfield 2014; Watts et al. 2015). Improving nitrogen use efficiency (NUE) of agricultural systems and promoting more efficient N use by plants have been proposed for reducing N_2O emissions (Calvo et al. 2016). This approach relies on the premise of applying fertilizer N or manure to crops at the right time (period of rapid plant nutrient uptake) based on soil test recommendations and realistic yield goals (Snyder et al. 2009), to maximize fertilizer N acquisition by crops. While this had led to the promotion of better fertilization timing that increase NUE; there is still a lack of sustainable management practices that could help achieve efficient use of N (Calvo et al. 2016). Given the important role of soil microbes in N transformations, manipulating native microbial communities in soil by inoculation of specific microorganisms could potentially reduce N_2O emissions from soil. In this context, microbial-based inoculants known to improve plant growth could potentially be that novel approach to improve NUE.

Presently, there is limited information on the influence of microbial inoculants on N_2O, CO_2, and CH_4 emissions from soils. Microbial-based inoculants could help accomplish N_2O reductions using two approaches: by decreasing N_2O efflux from N-fertilized soils (Calvo et al. 2013) and/or by increasing (directly or indirectly) crop N-uptake, thereby reducing N losses from the agricultural system (Adesemoye et al. 2008). Previous researchers have evaluated the replacement of chemical N fertilizers (e.g. ammonium nitrate) with Bacillus-based PGPR (Adesemoye et al. 2008, 2010) on crop production and have shown that the amount of N needed for crop production can be reduced. This implies that nitrous oxide production can be reduced by a reduction in N fertilizer use. Additionally, Wang et al. (2014) observed from a study conducted on tea fields that, the fermentation products of a *Trichoderma viride* mixture with straw could significantly mitigate N_2O emissions by 33.3–71.8%. Calvo et al. (2013) demonstrated the potential of inoculants containing a Bacillus mixture (BM), SoilBuilder™(SB), and SoilBuilder™ metabolites (SBM) for decreasing N_2O emissions in the presence of UAN-32 and a nonfertilized, control under laboratory incubations using only soil, while no differences were observed with urea (Fig. 4.2).

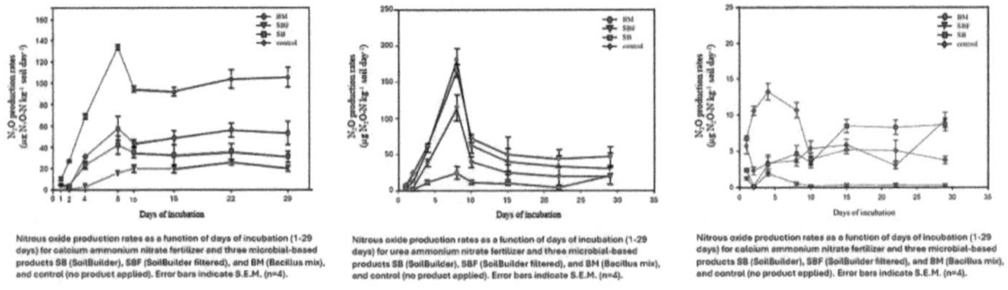

Fig. 4.2 Nitrous oxide production rates after incubation of microbial blend products compared to calcium ammonium nitrate fertilizer

When the effects of the microbial inoculants and metabolite were revisited in a container study, under greenhouse conditions in the presence of a plant, cumulative N_2O fluxes from the 41-day study showed significant reductions in N_2O of 15% (SB), 41% (BM), and 28% (SBM) with CAN-17 fertilizer. When UAN-32 was used, reductions of 34% (SB), 35% (SBM), and 49% (BM) were obtained, but still, no reductions in N_2O emissions occurred with urea (Calvo et al. 2013). These results suggest that reduction of N_2O emissions in soil inoculated with PGPR may be fertilizer dependent. In addition to increased N uptake from soil by plants, Wu et al. (2018) observed that cumulative N_2O emissions over a 45-day period were significantly reduced (452.57%) with the application of *Bacillus* in a containerized vegetable production system with irrigation under greenhouse conditions. Based on data from quantitative PCR and the network analysis of DNA sequencing, Wu et al. (2018) demonstrated that the *Bacillus* inoculants partially inhibited the nitrification process by significantly reducing ammonia-oxidizing bacteria. At the same time, the study showed that Bacillus-based inoculants enhanced the denitrification process, mainly by increasing the abundance of N_2O-reducing bacteria. These results suggest that microbial inoculants can play a role in reducing greenhouse gas emissions. More work is needed to expand upon this novel phenomenon and to increase awareness within the scientific community about the potential use of microbial inoculants to reduce GHG emission. Microbial inoculant use for nitrous oxide reduction may be a tool of the future to reduce agriculture's footprint on greenhouse gas emissions.

4.5 Biorational/Biological Products and the Market

Biorationals are low risk materials obtained from animals, plants, bacteria, and some minerals, which are useful for nutrient management, pathogen control, and pest management, to prevent large quantities of synthetic agricultural inputs across the globe. Commercial biological products for nutrient use and plant pathogens and pests control continue to expand, and it is currently among the fastest-growing areas in the agricultural sector (Murugesan et al. 2024; Parikh and Adesemoye 2018). Different components and types of biorationals will be discussed in this section.

4.5.1 Biostimulants and Inoculants

The terms 'biostimulant' and 'inoculant' are sometimes used interchangeably in error. Here, we will attempt to differentiate them. There is not yet a universally accepted definition or agreement for the term "biostimulant" among countries and even among U.S States or Federal agencies. We have considered definitions proposed for biostimulant by the following reports or groups: (i) the U.S Agriculture Improvement Act of 2018 (a.k.a. Farm Bill), (ii) report submitted by the USDA to

Congress on Plant Biostimulants in December 2019, (iii) a guideline by the Biostimulant industry and (iv) The European biostimulants industry council (EBIC). Therefore, here we define plant biostimulants as a broad category of products that can improve plant nutritional processes, nutrient availability, uptake, or use efficiency, leading to improved plant health and tolerance to abiotic stresses, and enhanced plant growth, quality, and yield. As per previous descriptions, biostimulants include microbial inoculants/their metabolites; plant and algal extracts; carbon-based deposits e.g., humic and fulvic acids; protein hydrolysates; and purified single molecules from natural or synthetic sources (Anon 2022; Calvo et al. 2014).

As an emerging area, understanding biostimulants and their classification are still unfolding as well. Thus, global regulations of plant biostimulants are variable from one country to the other. Within the United States, it varies from one state to the other, making production and registration sometimes convoluted whereby manufacturers may be confused about under which category to register their products in the state where they intend to sell the product. Based on the situation, the U.S Environmental Protection Agency, EPA has acknowledged that products being sold as plant biostimulants may or may not trigger regulation under the Federal Insecticide, Fungicide, and Rodenticide Act (FIFRA), 7 U.S.C. 136–136y as plant regulators. Meanwhile, plant biostimulant is a very fast-growing sector of the agriculture industry, according to a 2021 Dunham Trimmer® Global Biostimulant Report, and it was projected to become approximately a 3-billion-dollar market by 2025 (Anon 2022).

A variety of distinct product labels and categories under biostimulants has included soil amendment, plant amendment, plant inoculant, plant regulator, and beneficial substances. These products were previously divided into eight categories by du Jardin in 2012 (Calvo et al. 2014) and has been more recently divided into five compositional categories: (i) Microbial-based (inoculants), (ii) algal or plant extract, (iii) complex carbon-based, (iv) protein hydrolysate derived from plant, animal, or microbial protein feedstock, and (v) defined molecules purified from minerals, plants, animals, microbes, or obtained by synthesis (Anon 2022). The United States Biostimulant Industry Workgroup listed three methods for the verification of plant biostimulant efficacy claims especially before they could be registered: (i) Relevant published literature must be associated with efficacy claims, where product ingredients and/or properties are stated with recommended rates of application; (ii) Research studies with evidence about the product must be conducted using methods and designs that are scientifically recognized with proper control treatments, while the test crops, study locations, and method of statistical analysis are clearly stated; and (iii) a combination of product-specific scientific results and relevant published literature must support product claim (Anon 2022). For instance, an acceptable standard study layout with a picture such as the example shown in Fig. 4.3 could be required as evidence.

Inoculants are a subset of biostimulants, which are called biofertilizer, or microbial inoculants. Microbial inoculants or the active ingredient in an inoculant product are usually free-living bacteria (especially PGPR) and fungi, e.g., *Trichoderma* sp.

Fig. 4.3 Hay field study with inoculants being conducted by the authors at the Auburn University' E. V. Smith Research Center, Shorter, AL, USA

and arbuscular mycorrhizal fungi (AMF). Many of these microbial-based biostimulants are sold as a wettable formulation and may be prepared through a carrier-based system such as biochar and peat that support long-term storage. The preparation further supports the microorganisms' ability to enhance nutrient availability for plant roots through solubilization of minerals (Alori and Babalola 2018; Calvo et al. 2014; Egambergieva and Adesemoye 2016) and they affect ecosystem functions such as accumulation and decomposition of organic matter, through their interactions with the resident microbiome (Adesemoye et al. 2023).

4.5.2 Dual Role of Biofertilizer and Biopesticides

There are certain products among biorationals and biologicals that are used effectively as both fertilizer and pesticide and they should be discussed separately because of this uniqueness. Some governments rightly have a different handling for this group of input. For instance, Minnesota law (Minn. Stat. 18c.411) requires that any grade of fertilizer that contains a pesticide be registered as a specialty fertilizer in addition to registration as a pesticide except if they are excepted pesticides under section 25b of FIFRA. Although, all pesticides are regulated by the Environmental Protection Agency based on FIFRA, each pesticide must be registered in the state and additional rules will apply if it plays the dual role of fertilizer and biopesticide. Additionally, if the product can be considered a standalone fertilizer i.e., it contains significant source of plant nutrients, a fertilizer registration may also be required in the state.

4.5.3 Other Biorational Materials or Product Categories

Three other product categories of interest include (i) Humic substances such as humic acids, fulvic acids, and humins; (ii) protein hydrolysates and amino acids; and (iii) seaweed extracts. We do not intend to delve into these areas because an in-depth review was conducted previously by Calvo et al. (2014).

4.6 Progress with Biorationals and Biologicals

The major categories of biorationals include microbials (biologicals), botanicals, minerals/oils, and synthetic materials. The biologicals market has continued to expand through new products and a growing interest among farmers/growers, but enthusiasm has not matched up with the growth of the products. The identifiable major impediments against broad acceptance include (i) inadequate knowledge or low appreciation of the importance of biorationals and biological products by farmers/producers; (ii) low or inconsistent efficacy observed by farmers; and (iii) fear of potential losses by changing from conventional chemicals which efficacy is known to a new technology which little is known about the efficacy.

4.7 Benefits from Microbial Approach

Products across the biological agriculture input market are good for sustainable crop production and the more farmers accept the new technologies, the more the positive environmental effects can be realized on a global scale. Meanwhile, more farmers will appreciate the benefits if products were developed with high level efficacy, ease of integration into farmer's existing equipment and program, and with adaptability to diverse agro-ecologies including severe environmental conditions.

4.7.1 Improvement in Efficacy

The first author has heard concerns expressed by farmers about low efficacy, variability of efficacy, or low colonization of biological products when used in different conditions or cropping systems. Many factors can affect the efficacy of strains and microbial products, including collection strategy, screening process, materials used for product development (whether live cells, e.g., as obtainable from Fig. 4.4, spores, dead cells, or cell products/metabolites), delivery method to plants, local abiotic (environmental) factors, and the resident microbial communities where products were applied. The collection strategy and screening process for strains are

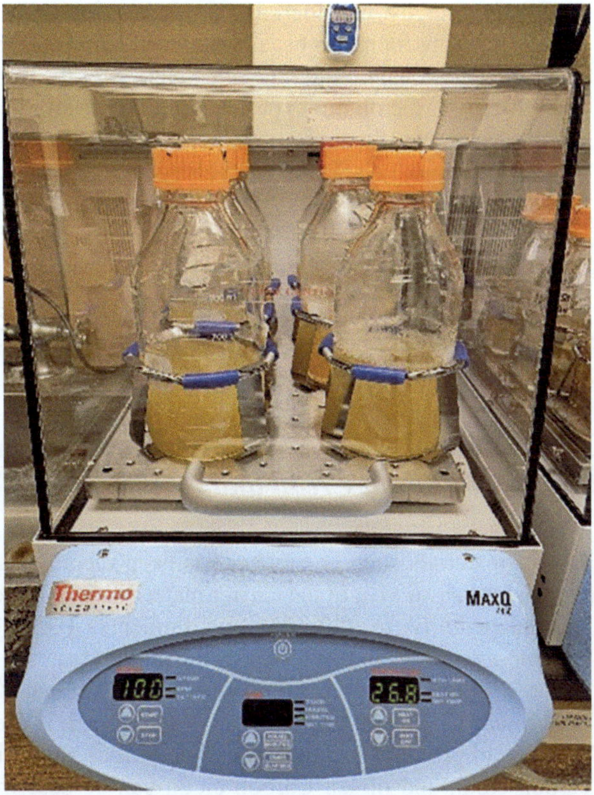

Fig. 4.4 Broth in a shaker to produce bacteria live or vegetative cells

crucial for the efficacy of biological products (Parikh and Adesemoye 2018) and a standard screening flowchart reported by Egamberdieva and Adesemoye (2016) could be adopted to avoid garbage-in garbage out. Also, knowledge of the mechanisms of action of microbials are essential for the development of the technology for improved product efficacy.

Co-inoculation of multiple organisms from different genera or species, i.e., the use of a consortium has been shown consistently to improve efficacy whether for nutrient use efficacy or for pest management (Parikh et al. 2018). Combination of strains from different genera with different modes of action has the potential for broad utility to promote plant growth, improve root system growth and architecture, enhance nutrient availability and uptake by roots, alleviate abiotic stress, and help manage biotic stresses. The delivery method of biologicals has great effects on viability and colonization in the plant host.

Environmental factors e.g., temperature, soil type, N content, salt concentration, and soil moisture content are important considerations. For example, abrupt changes in temperature can have major impacts on efficacy. Generally, when the environmental condition is hostile, microbial growth is hindered, populations will be low, and the microbial activity will be greatly reduced (Egamberdieva and Adesemoye

2016). Importantly, consideration must be given to the resident microbial populations in new environments because they will compete with any introduced strain and this competition may hinder the effectiveness of the introduced strains (Adesemoye et al. 2023). Based on our experience, resident soil microbes are hardly considered while commercial inoculants are applied in many locations different from where they were originally isolated or developed. However, the introduced population will have to compete with the resident or autochtonous organisms in each local environment where products are applied. Efforts to improve efficacy must continue and for the efforts to be successful, each of the factors identified above that can hinder efficacy must be targeted. Better knowledge of the mechanisms of actions used by microbial strains is also important for improved efficacy (Akinrinlola et al. 2018).

4.7.2 Interactions of Microbes with Biochar and their Roles in Carbon Credit and Offset

Carrier materials such as biochar has been recommended to promote viability. If organisms can survive in the carrier materials, it will greatly enhance their viability and adaptability to different environments. For instance, the population of *Enterobacter cloacae* UW5 was significantly greater in soil four weeks after application with biochar (pine wood) as a carrier, compared to a direct soil inoculation with liquid inoculums (Hale et al. 2015). Thus, it is useful in carbon capture technologies to promote sustainable carbon sequestration in the soil, and help achieve carbon credit and offset. Carbon credit can be described as a marketable unit of carbon dioxide or other greenhouse gas (GHG) emissions reduced or removed from the atmosphere while carbon offset is the process of financing other projects as per climate mitigation to balance out GHG emissions footprint (Trouwloon et al. 2023).

In recent times, biochar has been identified as a potential carbon credit and offset solutions that could help in carbon sequestration and be integrated into the Carbon (Emission) Trading System (Salma et al. 2024). Biochar, a form of charcoal, has been described as a versatile and scalable technology for carbon dioxide removal and storing carbon in the soil and could contribute significantly to carbon credits and sustainable development goals. When used as a soil amendment, biochar has been reported to improve soil physical (e.g., surface area, porosity, ion exchange, and water-holding capacity) and chemical (e.g., pH, nutrient exchange, functional groups, and carbon sequestration) properties, and impact the soil microbiome for increased plant growth, nutrient uptake, reduced greenhouse gas emissions, reduced plant diseases, and help to remediate heavy metal soil contamination (Haider et al. 2022). Owing to surface area, negative charge, and charge density, biochar easily improves soil cation exchange capacity (CEC), which makes it a unique carbon compound and ensures its usability for many above listed functions in crop production, including reducing the leaching of phosphate, ammonium, and nitrate (Haider et al. 2022; Lehmann 2007).

The potentials of combining microbial technology with biochar to manage agro-pollution, reduce soil erosion, and enhance fertility in crop production has not been well explored. Although biochar has been used as a carrier material for bacterial inoculants where it helps to reduce the effect of environmental stress factors and improve shelf life (Egamberdieva and Adesemoye 2016; Gul et al. 2023), this interaction has a lot more to offer. The combination of Biochar and PGPR have been reported to enhance plant fertility, including N, P, K, and development under drought conditions (Gul et al. 2023). In another study, a combination of *Bacillus subtilis* and green waste biochar (GWB) was reported to achieve up to 80% suppression of the pathogen, *Alternaria solani* both *in vitro* and *in vivo* on tomato, *Solanum lycopersicum* L. (Rasool et al. 2021). There is no doubt that microbial inoculants can help in the efficiency of fertilizers (Adesemoye et al. 2008, 2009), and that, physiochemical properties of biochar could alter the biomass and activity of microbes and soil enzymatic activity with consequent great effect on the elemental cycles in soil (Haider et al. 2022; Lehmann 2007). Therefore, the combined use of microbial inoculants and biochar is promising to enhance efficacy. It has been shown that the interactions of biochar and microorganisms improved safflower yield (Heidari et al. 2020), the growth of Chinese cabbage (Luan et al. 2023), and vegetation in a degraded alpine grassland (Li et al. 2023). Specifically, in a two-year split plot experiment, Heidari et al. (2020) evaluated three biochar types and four microbial treatments against a non-treated control. The highest levels of dehydrogenase, urease activity, and microbial biomass carbon was reported from the cow manure containing biochar with phosphate solubilizing microorganisms and the authors recommended these combinations for improving soil biological traits and safflower yield traits. The technology of combining inoculants and biochar needs further development to maximize the benefit of carbon capture in the soil.

4.8 Conclusion and Recommendations

This review discusses how microbial inoculants are transforming plant nutrient use efficiency, and establishes that, improvements in this progressive technology will enhance ecosystem functions in soil; thereby reducing the amount of synthetic fertilizer inputs needed for crop production. These ongoing developments in the biorational/biological subsector of the agricultural industry are exciting. Meanwhile as the technology is optimized for improved efficacy of products, there must be clarity about what constitute biorational and biological products and specifically for plant nutrition, there must be a clear definition of a biostimulant. A non-ambiguous global or regional standard must also be established for verifying the efficacy of biostimulants because the importance of biostimulant is global and not limited to any region. Governments and global institutions should work together as part of global environmental and agricultural plans to carry out the necessary legal reforms or reviews because laws related to biostimulants are not yet clear in most places and are not well implemented. Clarity and intentionality from high levels of national and global discourse will help to maximize the benefits. Additionally, this review suggests that exploring the broad spectrum of microbial groups and their utility in different combinations i.e., microbial consortia or their use with carriers such as biochar could improve efficacy and consequently, engender more widespread acceptance of the technology by farmers. Many areas for future research and development to harness unexplored potential benefits of the technology were identified in this chapter.

References

Adesemoye AO, Egamberdieva D (2013) Beneficial effects of plant growth promoting rhizobacteria on improved crop production: the prospects for developing economies. In: Maheshwari DK, Saraf M, Aeron A (eds) Bacteria in agrobiology: crop productivity. Springer, Berlin/Heidelberg, pp 45–63

Adesemoye AO, Kloepper JW (2009) Plant-microbe interactions in enhanced fertilizer use efficiency. Appl Microbiol Biotechnol 85:1–12

Adesemoye AO, Kodati S, Watts DB, Maharjan B (2023) Differential shifts in microbiome and pathogen populations associated with suppressive soil in long-term continuous corn field compared to rotation corn field. Appl Soil Ecol 192:105093

Adesemoye AO, Torbert HA, Kloepper JW (2008) Enhanced plant nutrient use efficiency with PGPR and AMF in an integrated nutrient management system. Can J Microbiol 54:876–886

Adesemoye AO, Torbert HA, Kloepper JW (2009) Plant growth-promoting rhizobacteria allow reduced application rates of chemical fertilizers. Microb Ecol 58:921–929

Adesemoye AO, Torbert HA, Kloepper JW (2010) Increased plant uptake of nitrogen from ^{15}N-depleted fertilizer using plant growth-promoting rhizobacteria. Appl Soil Ecol 46:54–58

Agegnehu G, Bass AM, Nelson PN, Bird MI (2016) Benefits of biochar, compost, and biochar–compost for soil quality, maize yield and greenhouse gas emissions in a tropical agricultural soil. Sci Total Environ 543:295–306

Akinrinlola RJ, Yuen GY, Adesemoye AO (2018) Evaluation of bacillus strains for plant growth promotion and predictability of efficacy by in vitro physiological traits. Int J of Microbiol 5686874:1–11

Alori ET, Babalola OO (2018) Microbial inoculants for improving crop quality and human health in Africa. Front Microbiol 9:403230

Andresen E, Peiter E, Kupper H (2018) Trace metal metabolism in plants. J Exp Bot 69(5):909–954. https://doi.org/10.1093/jxb/erx465

Anonymous (2022) United States Biostimulant Industry Recommended Guidelines to Support Efficacy, Composition, and Safety of Plant Biostimulant Products. A guideline developed by the Biostimulant Industry Workgroup, a collaboration of the Biological Products Industry Alliance (BPIA) and The Fertilizer Institute (TFI)

Ansari FA, Ahmad I (2018) Biofilm development, plant growth promoting traits and rhizosphere colonization by *Pseudomonas entomophila* FAP1: a promising PGPR. Adv Microbiol 8:235–251

Baligar VC, Fageria NK, He ZL (2001) Nutrient use efficiency in plants. Commun Soil Sci Plant Anal 32:921–950

Bashan Y, De-Bashan LE (2010) How the plant growth-promoting bacterium *Azospirillum* promotes plant growth—a critical assessment. Adv Agron 108:77–136, Elsevier

Bhatti AA, Haq S, Bhat RA (2017) Actinomycetes benefaction role in soil and plant health. Microb Pathog 111:458–467

Bouain N, Krouk G, Lacombe B, Rouached H (2019) Getting to the root of plant mineral nutrition: combinatorial nutrient stresses reveal emergent properties. Trends Plant Sci 24:542–552. https://doi.org/10.1016/j.tplants.2019.03.008

Cakmakci R, Donmez F, Aydin A, Sahin F (2006) Growth promotion of plants by plant growth promoting rhizobacteria under greenhouse and two different field soil conditions. Soil Biol Biochem 38:1482–1487

Cakmakci R, Donmez MF, Erdogan U (2007) The effect of plant growth promoting rhizobacteria on barley seedling growth, nutrient uptake, some soil properties, and bacterial counts. Turk J Agr For 31:189–199

Calvo P, Watts DB, Kloepper JW, Torbert HA III (2016) Application of microbial-based inoculants for reducing N2O emissions from soil under two different ammonium nitrate-based fertilizers. Soil Sci 181:427–434

Calvo P, Watts DB, Kloepper JW, Torbert HA III (2017) Effect of microbial-based inoculants on nutrient concentrations and early root morphology of corn (Zea mays). J of Plant Nutr and Soil Sci 180:56–70

Calvo P, Watts DB, Kloepper JW, Torbert HA III (2013) Microbial-based inoculants impact nitrous oxide emissions from an incubated soil medium containing urea fertilizer. J of Env Qual 42:704–712

Calvo P, Nelson L, Kloepper JW (2014) Agricultural uses of plant biostimulants. Plant Soil 383:3–41

Carrion VJ, Perez-Jaramillo J, Cordovez V, Tracanna V, de Hollander M, Ruiz-Buck D, Mendes LW, van Ijcken WFJ, Gomez-Exposito R, Elsayed SS, Mohanraju P, Arifah A, van der Oost J, Paulson JN, Mendes R, van Wezel GP, Medema MH, Raaijmakers JM (2019) Pathogen-induced activation of disease-suppressive functions in the endophytic root microbiome. Science 366:606–612

Chen JG, Crooks RM, Seefeldt LC, Bren KL, Bullock RM, Darensbourg MY, Holland PL, Hoffman B, Janik MJ, Jones AK, Kanatzidis MG, King P, Lancaster KM, Lymar SV, Pfromm P, Schneider WF, Schrock RR (2018) Beyond fossil fuel– driven nitrogen transformations. Science 360:6391–6611. https://doi.org/10.1126/science.aar6611

Cotrufo MF, Wallenstein MD, Boot CM, Denef K, Paul E (2013) The microbial efficiency-matrix stabilization (MEMS) framework integrates plant litter decomposition with soil organic matter stabilization: do labile plant inputs form stable soil organic matter? Glob Change Biol 19:988–995

Dell CJ, Han K, Bryant RB, Schmidt JP (2014) Nitrous oxide emissions with enhanced efficiency nitrogen fertilizers in a rainfed system. Agron J 106:723–731

Disi JO, Kloepper JW, Fadamiro HY (2018) Seed treatment of maize with *Bacillus pumilus* strain INR-7 affects host location and feeding by Western corn rootworm, Diabrotica virgifera virgifera. J of Pest Sci 91:515–522

Dobbelaere S, Vanderleyden J, Okon Y (2003) Plant growth promoting effects of diazotrophs in the rhizosphere. Crit Rev Plant Sci 22:107–149

Doyle N, Mbandlwa P, Kelly WJ, Attwood G, Li Y, Ross RP, Stanton C, Leahy S (2019) Use of lactic acid bacteria to reduce methane production in ruminants, a critical review. Front Microbiol 10:2207. https://doi.org/10.3389/fmicb.2019.02207

Egamberdiev D, Adesemoye AO (2016) Improvement of crop protection and yield in hostile agroecological conditions with PGPR-based biopesticide and biofertilizer formulations. In: Bioformulations—for sustainable agriculture. Arora N.K, Mehnaz S., and Balestrini R. (Ed.), Springer. Pp. 199–211

Elkoca E, Kantar F, Sahin F (2008) Influence of nitrogen fixing and phosphorus solubilizing bacteria on the nodulation, plant growth, and yield of chickpea. J Plant Nutr 31:157–171

Fan X, Zhou X, Chen H, Tang M, Xie X (2021) Cross-talks between macro- and micronutrient uptake and signaling in plants. Front Plant Sci 12:663477. https://doi.org/10.3389/fpls.2021.663477

Garg N, Baher N (2013) Role of arbuscular mycorrhizal symbiosis in proline biosynthesis and metabolism of Cicer arietinum L. (chickpea) genotypes under salt stress. J Plant Growth Regul 32:767–778

Glick BR, Todorovic B, Czarny J, Cheng Z, Duan J, McConkey B (2007) Promotion of plant growth by bacterial ACC deaminase. Crit Rev Plant Sci 26:227–242

Gong Z, Xiong L, Shi H, Yang S, Herrera-Estrella LR, Xu G, Chao DY, Li J, Wang PY, Qin F, Li J, Ding Y, Shi Y, Wang Y, Yang Y, Guo Y, Zhu JK (2020) Plant abiotic stress response and nutrient use efficiency. Sci China Life Sci 63(5):635–674. https://doi.org/10.1007/s11427-020-1683-x

Gontia-Mishra I, Sapre S, Sharma A, Tiwari S (2016) Amelioration of drought tolerance in wheat by the interaction of plant growth-promoting rhizobacteria. Plant Biol 18:992–1000

Gul F, Khan IU, Rutherford S, Dai ZC, Li G, Du DL (2023) Plant growth promoting rhizobacteria and biochar production from *Parthenium hysterophorus* enhance seed germination and productivity in barley under drought stress. Front Plant Sci 14:1175097

Gunes A, Atatoglu N, Turan M, Esitken A, Ketterings QM (2009) Effects of phosphate-solubilizing microorganisms on strawberry yield and nutrient concentrations. J Plant Nutr Soil Sci 172:385–392

Hahm MS, Son JS, Hwang YJ, Kwon DK, Ghim SY (2017) Alleviation of salt stress in pepper (Capsicum annum L.) plants by plant growth- promoting rhizobacteria. J Microbiol Biotechnol 27:1790–1797

Haider FU, Coulter JA, Liqun CAI, Hussain S, Cheema SA, Jun WU, Zhang R (2022) An overview on biochar production, its implications, and mechanisms of biochar-induced amelioration of soil and plant characteristics. Pedosphere 32(1):107–130

Hale L, Luth M, Crowley D (2015) Biochar characteristics relate to its utility as an alternative soil inoculum carrier to peat and vermiculite. Soil Biol Bioch 81:228–235

Heidari E, Mohammadi K, Pasari B, Rokhzadi A, Sohrabi Y (2020) Combining the phosphate solubilizing microorganisms with biochar types in order to improve safflower yield and soil enzyme activity. Soil Sci and Plant Nutr 66(2):255–267

He Z, Zhang H (2014) Applied Manure and Nutrient Chemistry for Sustainable Agriculture and Environment; Springer: Amsterdam, The Netherlands, 372p

Herman MAB, Nault BA, Smart CD (2008) Effects of plant growth promoting rhizobacteria on bell pepper production and green peach aphid infestations in New York. Crop Prot 27:4363–4370

Hernandez MI, Chailloux M (2004) Las micorrizas arbusculares y las bacterias rizosfericas como alternativa a la nutricion mineral del tomate. Cultivos Tropicales 25:5–12

IPCC (2018) Global Warming of 1.5°C. Masson-Delmotte V, Zhai P, Pörtner HO, Roberts D, Skea J, Shukla PR, Pirani A, Moufouma-Okia W, Péan C, Pidcock R, Connors S, Matthews JBR, Chen Y, Zhou X, Gomis MI, Lonnoy E, Maycock T, Tignor M, Waterfield T (eds) An IPCC Special Report on the impacts of global warming of 1.5°C above pre-industrial levels and related global greenhouse gas emission pathways, in the context of strengthening the global response to the threat of climate change, sustainable development, and efforts to eradicate poverty. IPCC, Geneva. https://www.ipcc.ch/site/assets/uploads/sites/2/2019/06/SR15_Full_Report_High_Respdf

Jacinthe PA, Dick WA, Lal R, Shrestha RK, Bilen S (2014) Effect of no-till duration on the methane oxidation capacity of Alfisols. Biol Fertil Soils 50:477–486

Khatoon Z, Huang S, Rafique M, Fakhar A, Kamran MA, Santoyo G (2020) Unlocking the potential of plant growth-promoting rhizobacteria on soil health and the sustainability of agricultural systems. J Environ Manag 273:111118

Kurkjian HM, Akbari MJ, Momeni B (2021) The impact of interactions on invasion and colonization resistance in microbial communities. PLoS Comput Biol 17:1008643

Lai WA, Rekha PD, Arun AB, Young CC (2008) Effect of mineral fertilizer, pig manure, and *Azospirillum rugosum* on growth and nutrient contents of Lactuca sativa L. Biol Fertil Soils 45:155–164

Lehmann J (2007) Concepts and question: bioenergy in the black. Front Ecol Environ 5:381–387

Lehnert N, Dong HT, Harland JB, Hunt AP, White CJ (2018) Reversing nitrogen fixation. Nat Rev Chem 2(10):278–289. https://doi.org/10.1038/s41570-018-0041-7

Li J, Li H, Shang J, Liu K, He Y, Shao X (2023) The synergistic effect of biochar and microorganisms greatly improves vegetation and microbial structure of degraded Alpine Grassland on Qinghai–Tibet Plateau. Agronomy 13(9):2203

Lin Y, Watts DB, Kloepper JW, Adesemoye AO, Feng Y (2019) Effect of plant-growth promoting rhizobacteria at various nitrogen rates on corn growth. Agri Sci 10:1542–1565. https://doi.org/10.4236/as.2019.1012114

Luan J, Fu Y, Tang W, Yang F, Li X, Yu Z (2023) Impact of interaction between biochar and soil microorganisms on growth of Chinese cabbage by increasing soil fertility. Appl Sci 13(23):12545

Lucas GJA, Probanza A, Ramos B, Colon Flores JJ, Gutierrez MFJ (2004) Effect of plant growth promoting rhizobacteria PGPRs on biological nitrogen fixation, nodulation, and growth of Lupinus albus I. cv. Multolupa Eng Life Sci 7:1–77

Moore JAM, Abraham PE, Michener J, Muchero W, Cregger M (2022) Ecosystem consequences of introducing plant growth promoting rhizobacteria to managed systems and potential legacy effects. New Phytol 234:1914–1918. https://doi.org/10.1111/nph.18010

Monteiro HF, Paula EM, Muck RE, Broderick GA, Faciola AP (2021) Effects of lactic acid bacteria in a silage inoculant on ruminal nutrient digestibility, nitrogen metabolism, and lactation performance of high-producing dairy cows. J Dairy Sci 104(8):8826–8834

Mosier AC, Kroeze C, Nevison C, Oenema O, Seitzinger S, van Cleemput O (1998) Closing the global N2O budget: nitrous oxide emissions through the agricultural nitrogen cycle—OECD/IPCC/IEA phase II development of IPCC guidelines for national greenhouse gas inventory methodology. Nutr Cycl Agroecosyst 52:225–248

Mpanga IK, Nkebiwe PM, Kuhlmann M, Cozzolino V, Piccolo A, Geistlinger J, Berger N, Ludewig U, Neumann G (2019) The form of N supply determines plant growth promotion by P-solubilizing microorganisms in maize. Microorganisms 7:38. https://doi.org/10.3390/microorganisms7020038

Murugesan P, Krishanan A, Ramasamy P, Chandrasekaran K, Rajendran J, Paramasivam S, Chinnaiyan U, Panneerselvam C, Aziz AT, Alasmari A (2024) Biorational methods for effective pest control management in stored products for agricultural sustainability. Entomol Res 54(1):12697

Myhre G, Shindell D, Bréon FM, Collins W, Fuglestvedt J, Huang J, Koch D, Lamarque JF, Lee D, Mendoza B, Nakajima T, Robock A, Stephens G, Takemura T, Zhang H (2013) Anthropogenic and natural radiative forcing. In: Stocker TF, Qin D, Plattner GK, Tignor M, Allen SK, Boschung J, Nauels A, Xia Y, Bex V, Midgley PM (eds) Climate Change 2013: The physical science basis. Contribution of Working Group I to the Fifth Assessment Report of the Intergovernmental Panel on Climate Change. Cambridge University Press, Cambridge, UK, and New York

O'Callaghan M, Ballard RA, Wright D (2022) Soil microbial inoculants for sustainable agriculture: limitations and opportunities. Soil Use Manag 38:1340–1369

Parikh L, Adesemoye AO (2018) Impact of delivery method on the efficacy of biological control agents and the virulence of Fusarium root rot pathogen in the greenhouse. Biocontrol Sci Tech 28(12):1191–1202

Parikh L, Eskelson MJ, Adesemoye AO (2018) Relationship of in vitro and in planta screening: improving the selection process for biological control agents against Fusarium root rot in row crops. Arch Phytopathol Plant Protect 51:156–169

Parkin TB, Hatfield JL (2014) Enhanced efficiency fertilizers; effect on nitrous oxide emissions in Iowa. Agron J 106:694–702

Patel DP, Singh HB, Shroff S, Sahu J (2012) Studies on growth promotion activities of *Trichoderma harzianum* on chickpea. Adv Plant Sci 25:193–195

Podile AR, Kishore GK (2007) Plant growth-promoting rhizobacteria. In: Gnanamanickam SS (ed) Plant-associated bacteria. Springer, The Netherlands, pp 195–230

Qian L, Zhang C, Zuo F, Zheng L, Li D, Zhang A, Zhang D (2019) Effects of fertilizers and pesticides on the mineral elements used for the geographical origin traceability of rice. J Food Compos Anal 83:103276

Raman J, Kim JS, Choi KR, Eun H, Yang D, Ko YJ, Kim SJ (2022) Application of lactic acid bacteria (LAB) in sustainable agriculture: advantages and limitations. Int J Mol Sci 23(14):7784

Rasool M, Akhter A, Soja G, Haider MS (2021) Role of biochar, compost and plant growth promoting rhizobacteria in the management of tomato early blight disease. Sci Rep 11(1):6092

Salma A, Fryda L, Djelal H (2024) Biochar: a key player in carbon credits and climate mitigation. Resources 13(2):31

Shafi J, Tian H, Ji M (2017) Bacillus species as versatile weapons for plant pathogens: a review. Biotechnol Biotechnol Equip 31:446–459

Sharma P, Pandey R, Chauhan NS (2024) Biofertilizer and biocontrol properties of *Stenotrophomonas maltophilia* BCM emphasize its potential application for sustainable agriculture. Front Plant Sci 15:1364807. https://doi.org/10.3389/fpls.2024.1364807

Sheng XF, He LY (2006) Solubilization of potassium-bearing minerals by a wild-type strain of *Bacillus edaphicus* and its mutants and increased potassium uptake by wheat. Can J Microbiol 52:66–72

Singh A, Jain A, Sarma BK, Upadhyay RS, Singh HB (2014) Rhizosphere competent microbial consortium mediates rapid changes in phenolic profiles in chickpea during *Sclerotium rolfsii* infection. Microbiol Res 169:353–360. https://doi.org/10.1016/j.micres.2013.09.014

Singh SN, Verma A (2007) The potential of nitrification inhibitors to manage the pollution effect of nitrogen fertilizers in agricultural and other soils: a review. Environ Practice 9:266–279

Smith P, Reay D, Smith J (2021) Agricultural methane emissions and the potential for mitigation. Phil Trans R Soc A 379:20200451

Snyder CS, Bruulsema TW, Jensen TL, Fixen PE (2009) Review of greenhouse gas emissions from crop production systems and fertilizer management effects. Agric Ecosyst Environ 133:247–266

Swarnalakshmi K, Yadav V, Tyagi D, Dhar DW, Kannepalli A, Kumar S (2020) Significance of plant growth promoting rhizobacteria in grain legumes: growth promotion and crop production. Plants 9:1596

Syers JK, Johnston AE, Curtin D (2008) Efficiency of soil and fertilizer phosphorus use. FAO fertilizer and plant nutrition bulletin 18. Rome, Italy, p 122

Talebi R (2011) Evaluation of chlorophyll content and canopy temperature as indicators for drought tolerance in durum wheat (Triticum durum Desf.). Aust J Basic Appl Sci 5:1457–1462

Tan S, Yang C, Mei X, Shen S, Raza W, Shen Q, Xu Y (2013) The effect of organic acids from tomato root exudates on rhizosphere colonization of *Bacillus amyloliquefaciens* T-5. Appl Soil Ecol 64:15–22

Tchiaze AI, Taffouo VD, Fankem H, Kenne M, Baziramakenga R, Ekodeck GE, Antoun H (2016) Influence of nitrogen sources and plant growth-promoting rhizobacteria inoculation on growth, crude fiber, and nutrient uptake in squash (Cucurbita Moschata Duchesne Ex Poir.) plants. Not Bot Horti Agrobo 44:53–59

Trouwloon D, Streck C, Chagas T, Martinus G (2023) Understanding the use of carbon credits by companies: a review of the defining elements of corporate climate claims. Global Chall 7(4):2200158

USEPA (2020) Inventory of U.S. greenhouse gas emissions and sinks: 1990–2018. United States Environmental Protection Agency EPA 430-R-20-002. USEPA, Washington, DC

Vadakattu G, Paterson J (2006) Free-living bacteria lift soil nitrogen supply. Farming Ahead 169:40

Wang J, Li R, Zhang H, Wei G, Li Z (2020) Beneficial bacteria activate nutrients and promote wheat growth under conditions of reduced fertilizer application. BMC Microbiol 20:38. https://doi.org/10.1186/s12866-020-1708-z

Wang Y, Wang X, Greenfield P, Jin D, Bai Z (2014) Draft genome sequence of *Bacillus amyloliquefaciens* EBL11, a new strain of plant growth-promoting bacterium isolated from rice rhizosphere. Genome Announc 2:e00732–e00714

Wang T, Ahmad S, Yang L, Yan X, Zhang Y, Zhang S, Wang L, Luo Y (2022) Preparation, biocontrol activity and growth promotion of biofertilizer containing *Streptomyces aureoverticillatus* HN6. Front Plant Sci 13:1090689

Watts DB, He Z, Yin X, Torbert HA III, Senwo Z, Tewolde H (2024) Long-term cropping management practices affect the biochemical properties of an Alabama Ultisol. Soil Systems 8(2):41. https://doi.org/10.3390/soilsystems8020041

Watts DB, Runion GB, Dick WA, Gonzalez JM, Islam KR, Flanagan DC, VanToai NT, Batte MT, Reeder RC, Kost D, Chen L, Jacinthe PA (2023) Influence of gypsum and cover crop on greenhouse gas emissions in soybean cropping systems. J Soil Water Conserv 78:154–162

Watts DB, Runion GB, Smith Nannenga K, Torbert HA (2015) Impacts of enhanced-efficiency nitrogen fertilizers on greenhouse gas emissions in a coastal plain soil under cotton. J Environ Qual 44:1699–1710

Wu S, Zhuang G, Bai Z, Cen Y, Xu S, Sun H, Han X, Zhuang X (2018) Mitigation of nitrous oxide emissions from acidic soils by *Bacillus amyloliquefaciens*, a plant growth-promoting bacterium. Glob Change Biol 24:2352–2365

Yildirim E, Karlidag H, Turan M, Dursun A, Goktepe F (2011) Growth, nutrient uptake, and yield promotion of broccoli by plant growth promoting rhizobacteria with manure. Hort Science 46:932–936

Zou J, Lu JW (2010) Yield response of winter rapeseed to potassium fertilization, use efficiency and soil's potassium critical level in the Yangtze River Valley. Optim Crop Nutr 23:6–9

Part II
Microbial Interventions for Environmental Stress and Soil Health

Part II

[illegible]

Chapter 5
Microbial Applications for Enhancing Drought and Water-Scarcity Resilience (DWSR) in Crops

Saheed Adekunle Akinola, Taofeek Tope Adegboyega, Modupe Stella Ayilara, Michael Olubusayo Binuyo, Muritala Issa Bale, Jean Paul Sinumvayo, Rasheed Omotayo Adeyemo, and Ismail Abiola Adebayo

Abstract Drought and water scarcity pose serious challenges to agricultural sustainability, endangering both economic stability and food security. Meanwhile, commonly used conventional farming methods have failed to mitigate problems associated with environmental stresses. Recent studies have demonstrated the positive impact of microbial applications as sustainable and efficient methods to improve crop resistance to water constraint and drought. This chapter focuses on the role of microbial applications in improving drought and water scarcity resilience (DWSR) in crops. The paper also discusses the mechanisms used by beneficial microbes to enhance stress tolerance on plants to modulate the physiological attributes, hormonal regulation and the activation of stress-response pathways to DWSR. Withal, the paper explored the intricate use of microbe-based agricultural innovations such as nano-fertilizers, bio-stimulants, and bio-pesticides in improving stress resilience and crop productivity. The use of microorganisms offers promising opportunities to enhance DWSR in crops for sustainable agricultural practices.

S. A. Akinola (✉)
Department of Microbiology and Parasitology, School of Medicine and Pharmacy, College of Medicine and Health Sciences, University of Rwanda, Butare, Rwanda

Food Security and Safety Focus Area, Faculty of Natural and Agricultural Sciences, North-West University, Mmabatho, South Africa
e-mail: s.akinola@ur.ac.rw

T. T. Adegboyega · M. O. Binuyo · M. I. Bale · J. P. Sinumvayo · R. O. Adeyemo
Department of Microbiology and Parasitology, School of Medicine and Pharmacy, College of Medicine and Health Sciences, University of Rwanda, Butare, Rwanda

M. S. Ayilara
Food Security and Safety Focus Area, Faculty of Natural and Agricultural Sciences, North-West University, Mmabatho, South Africa

I. A. Adebayo
Department of Medical Biochemistry, Molecular Biology, and Genetics, School of Medicine and Pharmacy, College of Medicine and Health Sciences, University of Rwanda, Butare, Rwanda

O. O. Babalola et al. (eds.), *Ecofriendly Frontiers*,
https://doi.org/10.1007/978-3-031-98700-7_5

Keywords Agricultural sustainability · Nanoscience application · Efficiently used water · Endophytic organisms · Abiotic soil stressors

5.1 Introduction

The current period of growing environmental threats necessitates an urgent push toward establishing sustainable farming methods. The primary dangers to worldwide food security include water shortages and drought conditions which are intensified by climate change and rising population figures. The urgent need to combat water stress has driven the development of microbial applications which present innovative methods to enhance crop resilience. Science has tapped into microorganism capabilities from bacteria to fungi to develop numerous interventions that strengthen plant drought resistance while improving water efficiency and maintaining agricultural output even under harsh climate conditions. The fusion of advanced microbiological research with agricultural science creates groundbreaking opportunities for crop resilience which will minimize water scarcity effects while promoting sustainable global agricultural practices (Adeleke et al. 2022a; Akinola et al. 2022).

In the future, temperature and rainfall pattern changes will become the main obstacles for food production throughout the upcoming decades. The loss of agricultural productivity stems from non-living elements which account for about 50%, while living organisms cause less than 50% of these reductions. The agricultural industry must find ways to increase crop yields or reduce damage while operating under the constraints of changing climate conditions and financial restrictions (Adeleke et al. 2022b; Bittencourt et al. 2023).

Plant stress arises from external disruptions to their balance and manifests as harmful physiological changes. Abiotic stressors such as heavy metal exposure and environmental extremes like flooding and temperature changes interfere with plant gene expression and physiological processes which lead to decreased crop yields. Through the combination of conventional and modern breeding methodologies plant breeding has become a popular approach to select genotypes that survive different abiotic stresses while focusing on creating drought-resistant varieties. Concurrently, numerous studies have delved into unraveling the physio-morphological traits and molecular underpinnings associated with enhancing DWS tolerance (Babalola et al. 2020; Bittencourt et al. 2023). This approach encounters several challenges, including reduced genetic diversity, trade-offs in productivity, high development costs, ecological risks, difficulties in farmer adoption, and the risk of overreliance. These limitations showed the importance of exploring innovative microbial applications as alternative solutions (Adeleke et al. 2022a).

Microorganisms of agricultural interest have been suggested to alleviate environmental stresses through two main mechanisms namely; protective-recovery mechanisms and plant water conservation. Consequently, there is a growing enthusiasm for translating scientific insights regarding DWS-tolerance bacteria into agroecosystems sustainability. At present, an ecofriendly use of microorganisms stand out as an effective biological technology for meeting plant growth needs while simultaneously protecting crops against environmental constraints (Akinola et al. 2020, 2021).

In this chapter, the authors address the fundamental mechanisms employed by microbial inoculants to alleviate plant water stress. These mechanisms involve intricate molecular processes facilitated by phytohormonal signaling, the induction of enzymatic activity, and the production of metabolites aimed at improving access to water and minimizing plant loss of water in the phase of environmental challenges. Furthermore, the chapter also delves into the synergistic interplay between metabolites and catalytic proteins utilized to prevent damage to plant cells and activate repair systems that bolster tolerance to water scarcity. Additionally, the chapter also examines technological applications, current and historical strategies, and future prospects regarding the use of microbial inoculants in mitigating DWS stresses.

5.2 Mechanisms of Drought and Water-Scarcity Tolerance in Plant

Drought is the most critical threat to the productivity and quality of world agriculture, restricting the growth of plant species. It is defined as a meteorological phenomenon with recurring intervals of subnormal precipitation that ruins the required soil moisture content necessary for maximum crop yield at a point in time. Thus, in this context, reduced water availability that negatively affected plant growth and developmental phases, having control over their life cycles. Under DWS conditions, the soil suffers a decrease in water potential, causing a subsequent decrease in the water potential in plants. Remarkably, the response of plants to drought is a key factor in maintaining balance throughout the atmosphere-plant-soil continuum. It is a complex phenomenon with a series of physiological, biochemical, and molecular changes. The initial response mechanism is the closure of stomata to avoid leaf embolism and cavitation and hence inhibit water loss in the plant (Cochard et al. 2002; Nardini and Salleo 2000). In a study conducted on grape leaves, the study revealed that stomata closure occurred entirely prior to the onset of cavitation (Hochberg et al. 2017). Consequently, this simultaneous closure led to a decline in CO_2 influx, increasing photorespiration and directly impacting photosynthetic capacity. During water scarcity, the reduction in carbon assimilation into plant biomass significantly affects the energy required to facilitate cellular protection, plant growth, and restoration mechanisms in response to DWS conditions.

Enhancing plant tolerance to DWS stress needs the harmonious integration of two general approaches (plant- and microbial-based approaches). One approach is to employ a suite of morpho-physiological processes to enhance the plant water status, which is primarily controlled by a complex hormonal signaling pathway of gibberellin, cytokinin, and auxin (Bandurska 2022; Iqbal et al. 2022; Wahab et al.

2022). Asides the fundamental role in sustainable plant growth, these hormones also serve as crucial mediators in stress signaling. For instance, a stress hormone known as abscisic acid (ABA) exhibits significant presence during DWS episodes. It regulates many genes participating in dehydration tolerance and triggers stomatal closure (Lin et al. 2021; Zandalinas et al. 2016). Second, the use of microbial inoculants like plant growth-promoting rhizobacteria (PGPR) and mycorrhizal fungi to improve nutrient uptake, regulate stress-responsive signaling, and increase soil water holding capacity.

An alternative strategy for water retention by the plant is the regulation of plant cell osmolarity by altering intracellular solute concentrations like glycine, proline, sorbitol, and soluble sugars. This adjustment increases to uphold cell turgor, a phenomenon known as osmotic adjustment (Turner 2017). Besides sustaining turgor pressure, these solutes shield plant cells from the detrimental impact of toxic by-products generated during DWS conditions (Constable and Hearn 1978; Turner 2017), and early indications of osmotic shift in pea roots and sorghum. Subsequent research has revealed the process of osmotic shift that helps to preserve plant's turgor. Nevertheless, it is crucial to acknowledge that the extent of this response in osmotic adjustment varies depending on the duration of exposure to the stress and the cultivar/species of plant (Blum 2017; Pilon et al. 2019).

The use of organic compounds as foliar sprays leads to a technological advancement in osmolyte accumulation and enhanced plant DWS conditions. For instance, the application of L-ornithine has been shown to increase stress tolerance in *Raphanus sativus* L., various *Brassica* spp., *Catharanthus roseus*, and sugar beet (*Beta vulgaris* var. *saccharifera* L.) (Akram et al. 2016; Hussein et al. 2019).

Targeted mechanisms for water saving are linked with mechanisms that protect and restore plant architecture. Like other abiotic stresses, drought leads to a sharp decline in the electron transport chain (ETC) and subsequent enhanced photo-oxidative stress to the plant tissues. Consequently, the enzymatic activity of Rubisco is reduced, and the membrane complex of photosystem II (PSII) is injured, causing an inhibition of the photosynthetic process. Thus, plants have a robust photoprotective response, with energy dissipation through a process known as non-photochemical quenching (NPQ) of chlorophyll fluorescence (Sohbat 2022; Tardieu 2022; Wani et al. 2021).

Plants respond with a variety of biochemical and molecular reactions to DWS. Notably among these, ROS remains one of the well-studied mechanisms. At a high level, ROS lead to oxidative deterioration, disruption, and peroxidation of cellular functions. Plants have therefore developed an antioxidant defense system such as superoxide dismutase (SOD), peroxidase (POX), catalase (CAT), and glutathione reductase (GR) to neutralize the oxidative impairments (Hernández et al. 2012). Hence, DWS tolerance is positively correlated with antioxidant response. Plant species more tolerant to DWS have greater antioxidant responses, which result in greater activity of antioxidative enzymes. This protects the plant from oxidative damage. Conversely, plant species that are more susceptible to DWS do not demonstrate alterations in the activity of these enzyme systems (Pyngrope et al. 2013).

Apart from the physiological, biochemical, and molecular reactions elicited by DWS in plants, their engagement with microbial communities residing in within the inner tissues (endophytes), the epiphytes (shoot-root surface), and the rhizosphere can contribute to improving plant resilience amidst plant stressors (Kour and Yadav 2022). Naturally, plant's microbiome shares a close relationship, hence, investigating these interactions aids in comprehending and enhancing the fundamental mechanisms of DWS tolerance.

5.3 Plant Growth-Promotion Under Drought and Water-Scarcity Stress

The root system of plants suppresses microbes, shaping a diverse ecological community impacting sustainable plant growth. Previous studies have demonstrated the use of bacterial isolates in the treatment of pepper plants displayed greater resilience to DWS compared to the control group (Mondal et al. 2020). Inoculants help boost plant root system, enhancing the plants' capacity to absorb water. This interaction exhibits a restricted specificity, with plant growth-promoting microorganisms (PGPMs) conferring resistance to DWS in plants (Ullah et al. 2019). Microorganisms residing below ground level play a pivotal role in shaping the physiological traits of plant by mitigating the impacts of DWS stress. For instance, *Brassica rapa* cultivated in soil hosting simple microbes exhibit characteristics such as fewer and smaller flowers, along with decreased chlorophyll content, in contrast to those grown alongside complex soil microbial communities (Hartman and Tringe 2019).

Six PGPMs namely; *Paenibacillus polymyxa* Mc5Re-14, *Bacillus subtilis* Col-6, *Streptomyces subrutilus* Wbn2-11B, *Serratia plymuthica* 3Re4-14, *Stenotrophomonas rhizophila* P69, *Pseudomonas fluorescens* L13-6-12 were inoculated with Chamomile seedlings in a study by Berg et al. (2013) and Rolli et al. (2015) to enhance their resilience to DWS. Inference revealed high production of bioactive secondary metabolites namely, apigenin-7-O-glucoside. In another study combining *Pseudomonas* sp. strain AKM-P6 and *P. putida* strain AKM-P7 to wheat and sorghum, the plants showed notable resilience to high-temperature stress as a result of the synthesis of cellular metabolites and high-molecular-weight proteins. These PGPR facilitate nutrient absorption, induce phytohormone balance, and augment plant resistance against phytopathogens (Li et al. 2020; Nawaz et al. 2020). Using the ACC deaminase enzyme (ACCD), plant beneficial microbes decrease ethylene in stressed plants by enzymatically degrading the plant ethylene precursor and 1-aminocyclopropane-1-carboxylate (ACC). Additionally, under salt stress conditions, *Stenotrophomonas rhizophila* DSM14405T induces crucial gene clusters associated with stress defense mechanisms, energy generation, and cell motility (Egamberdieva et al. 2016).

Plant growth-promoting microorganism bolster plant growth and increase tolerance to various stresses through a range of direct and indirect mechanisms (Fig. 5.1).

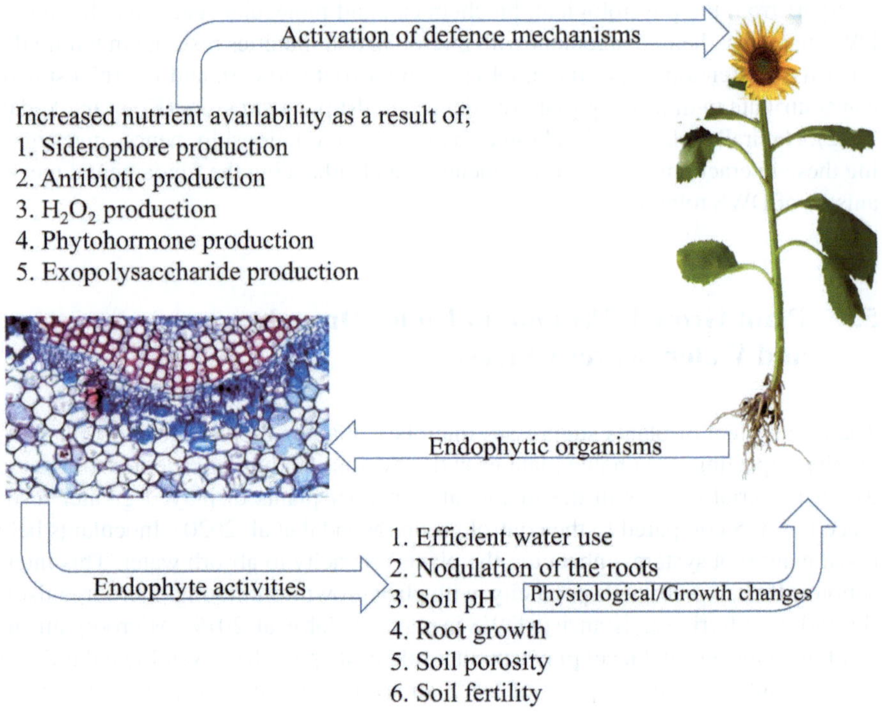

Fig. 5.1 Related strategies of PGPM on soil/plant exposed to DWS stress

5.3.1 *Plant Growth-Promoting Microorganism Alteration of Plant Antioxidant Defense*

Reactive oxygen species (ROS) such as hydrogen peroxide (H_2O_2), hydroxyl radicals (OH^-), superoxide anion radicals (O_2^-), singlet oxygen (O_2), and alkoxy radicals (RO) produced as a result of plant exposure to DWS stress disrupt normal cellular functions and cause oxidative damage by reacting with lipids, proteins, and DNA. Maintaining a lower level of ROS is crucial for initiating various defensive pathways in plants. To regulate signaling functions and oxidative stress damage effectively, it is imperative to manage ROS through coordinated control of both quenching systems and ROS generation. Plants have evolved an antioxidant system to counteract these effects, consisting of non-enzymatic and enzymatic components that prevent ROS accumulation and mitigate oxidative damage during DWS. Several PGPM have been reported to appraise this activity in plants, enhancing resilience to stressors (Kaushal and Wani 2016; Cohen et al. 2015; Kaushal 2019; Armada et al. 2014). Enzymatic components consist of glutathione reductase (GR), catalase (CAT), superoxide dismutase (SOD), and ascorbate peroxidase (APX), while the non-enzymatic component includes molecules such as cysteine, glutathione, and ascorbic acid (Akinola et al. 2023; Akinola and Babalola 2020; Ayilara et al. 2023; Kapoor et al. 2019).

In sunflower, an elevation in enzyme-defense activities and an augmentation in glutathione content were observed upon reaching a moderate level of DWS. Catalase (CAT) is a hydrogen peroxide (H_2O_2)-degrading enzyme common to both peroxisomes and mitochondria The loss of CAT activity was substituted by high peroxidases (PODs) as H_2O_2-scavenging enzymes, which were predominantly localized in the cell wall, cytosol, and chloroplasts (Sarker and Oba 2018). Also, in DWS, *Ocimum basilicum* L. (basil plants) inoculated with microbial consortia of *Pseudomonas* sp. *Azospirillum brasilense* and *Bacillus lentus* also had a noticeable hike in ascorbate peroxidase (APX) and glutathione peroxidase (GPX) activity. Cucumber plants treated with *Serratia* spp., *B. cereus*, and *B. subtilis* had lower malondialdehyde (MDA) and relative electrical conductivity of the leaves when compared to the control plants. A similar trend was observed in *A. thaliana* inoculated with *A. brasilense*, which displayed reduced MDA content during DWS conditions (Table 5.1) (Cohen et al. 2015; Kaushal and Wani 2016). The activity of ascorbate peroxidase (APX) and superoxide dismutase (SOD) increased considerably overtime when mung bean plants was treated with *Pseudomonas*, while CAT activity fluctuates (Kaushal 2019). Similar observations were made in Lavandula plants inoculated with *Bacillus thuringiensis*, where the activities of APX and glutathione reductase (GR) decreased under DWS (Armada et al. 2014).

5.3.2 *Exopolysaccharides (EPS) from PGPM Alleviate DWS Stress*

Drought and Water-Scarcity stress alters the biological and physicochemical properties of the soil, impairing soil microbial activities and crop yields. Microbes produce EPS that indirectly affect soil structure, shielding them from harsh conditions and facilitating their survival. During DWS stress, PGPM releases EPSs, comprising high molecular weight compounds consisting of a minor proportion of protein and uronic acid, intricate polysaccharides, and organic molecules. The EPSs on the surface of microbial cells serve as shields on the membrane structure during DWS stress. Intracellular synthesis of EPS occurs during the stationary phase of microbial growth. The rate of EPS production is contingent upon various factors, including drought stress (Osemwegie et al. 2020).

The capsular material of *A. brasilense* SP24 comprises carbohydrate complexes like lipopolysaccharide-protein (LP) and polysaccharide-lipid (PL) complexes, used to protect them during desiccation and augment their survival rate. Soil microbes secrete EPS into the soil in the form of capsules and slime materials, thereby creating a protective capsule structure in the soil. Capsules containing EPS can retain water and dry more slowly than the surrounding environment, thus protecting plant and immediate microcosm from DWS stress (Kaushal 2019). For

Table 5.1 Literatures showing PGPM strains used for DWS tolerance in different crop plants

PGPM	Plant crop	References
A. brasilense	*Zea mays*	Casanovas et al. (2002)
A. brasilense	*Lycopersicon esculentum*	Molina-Favero et al. (2008)
Peanibacillus polymyxa KIN-37, *Bacillus megaterium* TV-6D, 20E	*Brassica oleracea* var. *capitata*	Samancioglu et al. (2016)
Azospirillum spp.	*Triticum aestivum*	Arzanesh et al. (2011)
Phyllobacterium brassicacearum strain STM196	*Arabidopsis thaliana*	Bresson et al. (2013)
Bacillus subtilis	*Platycladus orientalis*	Liu et al. (2013)
P. putida H-2-3	*Glycine max*	Kang et al. (2014)
B. thuringiensis	*Lavandula dentata*	Ahmad et al. (2014)
Rhizobium phaseoli (MR-2), *Mesorhizobium cicero* (CR-30, CR-39), Rhizobium leguminosarum (LR-30)	*Triticum aestivum*	Hussain et al. (2014)
Pseudomonas azotoformans FAP5	*Triticum aestivum*	Ansari et al. (2021)
V. paradoxus 5C-2	*Pisum sativum*	Belimov et al. (2009)
ACC deaminase-producing rhizobacteria	*Triticum aestivum*	Shakir et al. (2012)
Consortia of *Mesorhizobium ciceris, Pseudomonas* 6-P, *Bacillus* isolate 23-B	*Cicer arietinum*	Sharma et al. (2013)
Bacillus licheniformis K11	*Capsicum annuum*	Lim and Kim (2013)
Bacillus thuringiensis AZP2	*Triticum aestivum*	Timmusk et al. (2014)
B. thuringiensis	*Lavandula dentata*	Ahmad et al. (2014)
Bacillus polymyxa	*Lycopersicon esculentum*	Shintu and Jayaram (2015)
Consortia containing *A. nitroguajacolicus* strain YB3, strain YB5, *P. synxantha* R81, *P. jessenii,* R62	*Oryza sativa*	Gusain et al. (2015)
Azospirillum lipoferum	*Zea mays*	Bano et al. (2013)
Raoultella planticola YL2, *Pseudomonas fluorescens* YX2, *Klebsiella variicola* F2	*Zea mays*	Gou et al. (2015)
Pseudomonas lini, Serratia plymuthica	*Ziziphus jujuba*	Zhang et al. (2020)
B. licheniformis K11	*Capsicum annuum*	Lim and Kim (2013)
Azospirillum brasilense NO40, *Bacillus amyloliquefaciens* 511	Triticum aestivum	Kasim et al. (2013)

(continued)

Table 5.1 (continued)

PGPM	Plant crop	References
Pseudomonas sp.	*Sorghum bicolor*	Dhawi (2020)
Enterobacter sp.16i and *Bacillus* sp. 12D6	*Triticum aestivum*	Jochum et al. (2019)
Trichoderma, Talaromyces, Rhizoctonia, Phoma spp., *Phytophthora, Penicillium, Gliocladium,* and *Fusarium*	*Zea mays* and *Oryza sativa*	Murali et al. (2021)
Serendipita indica	*Zea mays*	Tyagi et al. (2022)
Saccharomyces cerevisiae	*Arabidopsis thaliana*	Benito et al. (2022)
AMF	Medicinal and aromatic plants	Israel et al. (2022)

instance, sunflower plants treated with EPS-producing microbe exhibited a notable increase in root-adhering soil per root tissue (RAS/RT) and demonstrated enhanced resistance to DWS stress. Likewise, in the maize rhizosphere, inoculation with EPS-producing bacterial strains, such as *Proteus penneri*, *Pseudomonas aeruginosa*, and *Alcaligenes faecalis*, enhanced plant biomass, root and shoot length, leaf area, and moisture content (Adeleke et al. 2022a, b, Kaushal 2019).

5.3.3 The PGPM and ACC Deaminase Production Contribute to DWS Tolerance in Plants

Ethylene (ETH) is secreted by plants in response to environmental stresses viz. toxic metals, organic pollutants, salinity, heat, and drought. Ethylene under normal circumstances controls many facets of plant growth and development. Eth biosynthesis is controlled by abiotic and biotic stresses (Abiri et al. 2017). In Ethylene biosynthesis, ACC synthase (ACS) catalyzes S-adenosylmethionine to 1-aminocyclopropane-1-carboxylate (ACC). Eth controls plant homeostasis, and in DWS, its accumulation suppresses root and shoot growth, thus preventing the uptake of water and nutrients. ACC is secreted by plants into the rhizosphere, where it is taken up by the roots and converted to ethylene. The PGPM in the rhizosphere use ACC as a source of nitrogen and energy and break it down to ammonia and α-ketobutyrate. This process strengthens and prevents the root system and ethylene accumulation, helping plants cope with DWS (Danish et al. 2020).

ACC is excreted by the roots of plants and hydrolyzed by ACC deaminase-producing microbes. This reduces the amount of ACC in the rhizosphere, prompting the plant excretes more ACC. The interesting characteristic of the reaction catalyzed by ACC deaminase is the cyclopropane ring opening of ACC. Though the exact mechanism is not well understood, the cyclopropane bond most probably is cleaved via nucleophilic addition and elimination. 1-aminocyclopropane-1-carboxylate (ACC) deaminase is competitively inhibited by L-isomers of amino acids including

L-alanine, L-alpha-aminobutyric acid, L-homoserine, and L-serine, with L-serine and L-alanine as the strongest inhibitors (Singh et al. 2015). Structurally related compounds to ACC, such as 2-alkyl-ACC and vinyl-ACC can also act as substrates for ACC deaminase, as isolated from *Pseudomonas* sp. (Singh et al. 2015). Invariably, the reduced level of ACC lowers the ethylene production within plants. Organic acids, amino acids, and roots exude sugars, which attract more microbes to the rhizosphere (Singha and Pandey 2020). Also, microbes transform tryptophan found in root exudates into indole-3-acetic acid (IAA) for plant absorption. Indole-3-acetic acid, synthesized by either plants or microbes can function in two distinct manners. It can boost the transcription of ACC synthase or stimulate cell elongation, thereby elevating the ethylene levels in plants. It has been previously documented that microbes producing ACC deaminase-producing microbes decrease the level of the stress hormone ethylene within plant tissues. This phenomenon has been observed, for instance, in peas, beans, maize, and wheat, resulting in improved plant growth under DWS conditions (Chandra et al. 2019; Yoon et al. 2020).

1-aminocyclopropane-1-carboxylate deaminase-producing microbes breaks down plant ACC, thereby promoting plant growth and reducing the deleterious effects of ethylene (Akinola and Babalola 2020, 2021). Plant growth-promoting microorganisms like *Achromobacter piechaudii* ARV8 that produces ACC deaminase have tendency to increase wet and dry weight and decrease ethylene production in tomato and pepper seedlings grown under DWS. Pea plants inoculated with ACC deaminase-producing *Variovorax paradoxus* 5C-2 exhibited increased survival and growth rates under DWS conditions (Dodd et al. 2004). Additionally, the microorganisms producing ACC deaminase significantly restored nodulation, boosted seed number, and seed yield, which had been suppressed due to DWS conditions (Dodd et al. 2004). However, pea plants inoculated with ACC deaminase-producing *Pseudomonas fluorescens* G (ACC-5) showed enhanced nutrient and water uptake and significantly longer roots from the soil under DWS conditions (Mondani et al. 2019).

5.3.4 PGPM Produces Osmolytes to Ameliorate DWS Stress

Plants mitigate DWS stress conditions through metabolic adjustments that lead to the accumulation of various compatible solutes and osmolytes such as dehydrins, quaternary ammonium compounds, polyhydric alcohols, betaines, polyamines, amino acids, and sugars. In response to DWS stress, PGPMs secrete osmolytes that aid in stimulating plant growth. For instance, *Pseudomonas putida* GAP-P45 inoculated in maize significantly enhanced relative water content, leaf water potential, and plant biomass by accumulating proline under DWS conditions (Ghosh et al. 2019). Enhanced proline biosynthesis helps preserve cell water status and protects proteins and membranes from denaturation under stress conditions (Alamri et al. 2019). Maize plants inoculated with PGPM strains like *P. putida* GAP-P45 and *A. lipoferum* accumulate soluble sugars and free amino acids, thereby improving plant growth compared to control plants under DWS stress (Bano et al. 2013). Also, choline enhances the synthesis and accumulation of glycine betaine, which plays a crucial role in plant resistance (You et al. 2019). Previous studies have demonstrated that *P. fluorescens*, *Klebsiella variicola*, and *Raoultella planticola* in maize, as well as *B. subtilis* in Arabidopsis, significantly boost glycine betaine biosynthesis and accumulation, leading to improved dry matter weight and leaf-relative water content (Gou et al. 2015; Zhang et al. 2020).

The impact of PGPM, specifically *Bacillus megaterium*, *B. thuringiensis*, and *B. subtilis*, on the metabolomics profile of two chickpea varieties with varying drought sensitivities was studied by Khan et al. (2020). The study showed that the PGPM consortium greatly improved the accumulation of tryptophan, L-isoleucine, L-arginine, L-histidine, and proline in the leaves of chickpea under DWS stress. The following were also accumulated in the leaves of chickpea: the likes of 3-hydroxy-3-methylglutarate, nicotinamide, riboflavin, glycerol, aspartate, and L-asparagine. The reaction significantly promotes plant growth and induces DWS stress tolerance (Khan et al. 2020).

5.3.5 Plant Growth-Promoting Microorganism Modifies Phytohormonal Activity in Plants Under DWS Stress

Phytohormones such as IAA, cytokinins, gibberellins, ethylene, and abscisic acid are usually produced by plants to regulate growth and development. It is well established that these phytohormones are crucial for abiotic stress tolerance, helping plants survive under stressful conditions (Adeleke et al. 2022a). Plant growth-promoting microorganisms play a significant role in synthesizing phytohormones, thereby enhancing plant growth and development under environmental stresses (Khan et al. 2020). Plants produce IAA as the only active form of auxin regulating easy sprouting and development. The inoculation of auxin-producing microbes into various plant species significantly boosts nutrient and water uptake, promotes the development of root hairs and lateral roots, and supports plants in coping with DWS stress.

PGPMs influence the expression of genes activating the transport, synthesis and signaling of auxin. For instance, PGPM strains like *Bacillus* sp. LZR216 and *Phyllobacterium brassicacearum* STM196 boost endogenous IAA levels by activating genes involved in IAA synthesis. Additionally, inoculating *Arabidopsis thaliana* with *Burkholderia phytofirmans* PsJN enhances the production of auxin due to the increased level of tryptophan-related genes expression (Wang et al. 2012). PGPMs influence auxin transport by modifying the activity of auxin influx and efflux carriers. For example, a study by Wang and Ruan (2013) showed that *Bacillus* sp. LZR216 reduces the synthesis of auxin transporters AUX1 and PIN1, PIN2, and PIN3, while the study of Poupin et al. (2016) revealed the expression of PIN2 and PIN3 when treated with *B. phytofirmans* PsJN.

Gibberellins and ABA, a DWS alleviating phytohormone was reported to have been produced when *Azospirillum lipoferum* was introduced to maize plants in a study by Cohen et al. (2009). Under DWS conditions, cellular dehydration induces the biosynthesis of the stress hormone ABA. Abscisic acid (ABA) helps maintain water content by activating stress-signaling transduction pathways and regulating stomatal closure. Inoculation of *Arabidopsis thaliana* with *Azospirillum brasilense* Sp245 elevated ABA levels, resulting in reduced leaf transpiration (Bresson et al. 2013). Plants inoculated with ABA-producing strains, such as *Azospirillum brasilense* Sp245, *Bacillus licheniformis* RtM10, and *Pseudomonas fluorescens* RtM10, increase their internal ABA content and become more resistant to drought stress. The inoculation of ABA-producing PGPM strains namely; *Pseudomonas fluorescens* RtM10, *Bacillus licheniformis* RtM10, and *Azospirillum brasilense* Sp245 to plants increased the production of ABA and resistance to DWS stress.

5.3.6 Plant Growth-Promoting Microorganism Produces Volatile Compounds to Enhance DWS Tolerance in Plants

Under DWS conditions, PGPM produces several volatile organic compounds (VOCs) capable of alleviating DWS stress and enhancing plant growth. Volatile organic compounds produced by PGPM include; hydrocarbons, sulfur compounds, ketones, alcohols, esters, ethers, and aliphatic aldehydes. Several studies have explored the impact of VOCs secreted by PGPM, unveiling their potential to promote plant growth under DWS stress (ALKahtani et al. 2020; Chiappero et al. 2019). The VOCs produced by PGPM directly enhance plant functions in three ways which include; plant pathogen control, plant growth stimulation, and systemic plant resistance induction. Distinctively, VOCs mechanism of action does not necessitate physical interaction with plant parts or pathogen, while in contrast, other biocontrol mechanisms promote plant growth and manage phytopathogens by either physical contact or close proximity. It has been recorded that 300 fungi and bacteria are identified as sources of VOCs, while the emitted volatile database of microorganisms (DOVE-MO) contains records of 846 VOCs with 5431 synonyms (Ali and Khan 2021).

Volatile organic compounds act as signals for initiating systemic responses and priming against plant stresses. In a study priming wheat seedling with *Bacillus thuringiensis* AZP2 under DWS stress, wheat plant showed an increased survival rate and plant biomass, attributed to higher photosynthesis and VOCs produced the PGPM (Timmusk et al. 2014). In a study by Bhattacharyya et al. (2015), *A. thaliana* primed with *Alcaligenes faecalis* showed enhanced growth and the production of several VOCs. These VOCs regulate the gibberellins and auxin pathways, which are crucial for root and shoot growth under DWS conditions. Additionally, the VOCs also regulate ion transporters in a study by Bhattacharyya and Lee (2017).

Other indirect functions were also attributed to VOCs produced by PGPM. This was reported in a study by Qarni et al. (2021) on the potential of VOCs to activate plant-induced ISR and likewise control several plant pathogens (Raza et al. 2016). In DWS condition, *A. thaliana* primed with *Bacillus* sp. PGPM strains IN937a and GB03 produced acetoin and 2R, 3R-butanediol that helped mitigate the severity of *Pectobacterium carotovora*, a pathogen causing bacterial stem rot (Sampietro et al. 2020).

5.4 Concluding Remarks and Future Perspective

Utilization of microbes presents promising opportunities for the improvement of crop drought and water-scarcity resistance (DWSR). Nevertheless, through mechanisms involving modulation of plant hormones, secondary metabolite production, and nutrient uptake facilitation, symbiotic microbes perform central functions in alleviating the effects of water stress on crop growth and productivity. Microbial strategies have enormous potential for water-stressed agriculture if researched further and applied.

Withal, omics techniques have become effective instruments for improving crops' resistance to drought and water shortage. Through the integration of high-throughput technologies like proteomics, metabolomics, transcriptomics, and genomes, researchers are in a position to have a global view of the molecular mechanisms underlying plant response to water stress. Accessing environmentally friendly strains with the preferred characteristics is facilitated easily by the tools which enable researchers to map the key genes, proteins, and metabolites implicated in DWSR. In addition, advancement in multi-omics techniques offer useful resources for breeding programs, precision medicine, and the identification of novel molecular targets with biotechnological importance.

References

Abiri R et al (2017) Role of ethylene and the APETALA 2/ethylene response factor superfamily in rice under various abiotic and biotic stress conditions. Environ Exp Bot 134:33–44

Adeleke BS, Akinola SA, Adedayo AA, Glick BR, Babalola OO (2022a) Synergistic relationship of endophyte-nanomaterials to alleviate abiotic stress in plants. Front Environ Sci 10:2398

Adeleke BS, Ayilara MS, Akinola SA, Babalola OO (2022b) Biocontrol mechanisms of endophytic fungi. Egypt J Biol Pest Control 32:1–17

Ahmad I, Akhtar MJ, Zahir ZA, Naveed M, Mitter B, Sessitsch A (2014) Cadmium-tolerant bacteria induce metal stress tolerance in cereals. Environ Sci Pollut Res Int 21:11054–11065

Akinola SA, Adeyemo RO, Adebayo IA (2022) Overview and emergence of nanobiotechnology in plants. In: Diverse applications of nanotechnology in the biological sciences. Apple Academic Press, pp 173–197

Akinola SA, Ayangbenro AS, Babalola OO (2020) The diverse functional genes of maize rhizosphere microbiota assessed using shotgun metagenomics. J Sci Food Agric

Akinola SA, Ayangbenro AS, Babalola OO (2021) Metagenomic insight into the community structure of maize-rhizosphere bacteria as predicted by different environmental factors and their functioning within plant proximity. Microorganisms 9:1419

Akinola SA, Ayangbenro AS, Babalola OO (2023) Pedological factors as drivers of archaeal and fungal communities in maize rhizosphere: a shotgun metagenomic sequencing approach. SN Appl Sci 5:351

Akinola SA, Babalola OO (2020) The importance of adverse soil microbiomes in the light of omics: implications for food safety. Plant Soil Environ 66:421–430

Akinola SA, Babalola OO (2021) The fungal and archaeal community within plant rhizosphere: a review on their contribution to crop safety. J Plant Nutr 44:600–618

Akram NA, Waseem M, Ameen R, Ashraf M (2016) Trehalose pretreatment induces drought tolerance in radish (Raphanus sativus L.) plants: some key physio-biochemical traits. Acta Physiol Plant 38:1–10

Alamri SA, Siddiqui MH, Al-Khaishany MY, Khan MN, Ali HM, Alakeel KA (2019) Nitric oxide-mediated cross-talk of proline and heat shock proteins induce thermotolerance in *Vicia faba* L. Environ Exp Bot 161:290–302

Ali S, Khan N (2021) Delineation of mechanistic approaches employed by plant growth promoting microorganisms for improving drought stress tolerance in plants. Microbiol Res 249:126771

ALKahtani MD, Attia KA, Hafez YM, Khan N, Eid AM, Ali MA, Abdelaal KA (2020) Chlorophyll fluorescence parameters and antioxidant defense system can display salt tolerance of salt acclimated sweet pepper plants treated with chitosan and plant growth promoting rhizobacteria. Agronomy 10:1180

Ansari F, Jabeen M, Ahmad I (2021) Pseudomonas azotoformans FAP5, a novel biofilm-forming PGPR strain, alleviates drought stress in wheat plant. Int J Environ Sci Technol 18:3855–3870

Armada E, Portela G, Roldán A, Azcón R (2014) Combined use of beneficial soil microorganism and agrowaste residue to cope with plant water limitation under semiarid conditions. J Geod 232:640–648

Arzanesh MH, Alikhani H, Khavazi K, Rahimian H, Miransari M (2011) Wheat (*Triticum aestivum* L.) growth enhancement by *Azospirillum* sp. under drought stress. World J Microbiol Biotechnol 27:197–205

Ayilara MS, Adeleke BS, Akinola SA, Fayose CA, Adeyemi UT, Gbadegesin LA, Omole RK, Johnson RM, Uthman QO, Babalola OO (2023) Biopesticides as a promising alternative to synthetic pesticides: a case for microbial pesticides, phytopesticides, and nanobiopesticides. Front Microbiol 14

Babalola OO, Akinola SA, Ayangbenro AS (2020) Shotgun metagenomic survey of maize soil rhizobiome. Microbiol Resour Announc 9:e00860–e00820

Bandurska H (2022) Drought stress responses: coping strategy and resistance. Plants 11:922

Bano Q, Ilyas N, Bano A, Zafar N, Akram A, Hassan F (2013) Effect of *Azospirillum* inoculation on maize (*Zea mays* L.) under drought stress. Pak J Bot 45:13–20

Belimov AA, Dodd IC, Hontzeas N, Theobald JC, Safronova VI, Davies WJ (2009) Rhizosphere bacteria containing 1-aminocyclopropane-1-carboxylate deaminase increase yield of plants grown in drying soil via both local and systemic hormone signalling. New Phytol 181:413–423

Benito P, Ligorio D, Bellón J, Yenush L, Mulet JM (2022) A fast method to evaluate in a combinatorial manner the synergistic effect of different biostimulants for promoting growth or tolerance against abiotic stress. Plant Methods 18:111

Berg G, Zachow C, Müller H, Philipps J, Tilcher R (2013) Next-generation bio-products sowing the seeds of success for sustainable agriculture, vol 3, Agronomy edn. MDPI, pp 648–656

Bhattacharyya D, Lee YH (2017) A cocktail of volatile compounds emitted from *Alcaligenes faecalis* JBCS1294 induces salt tolerance in *Arabidopsis thaliana* by modulating hormonal pathways and ion transporters. J Plant Physiol 214:64–73

Bhattacharyya D, Yu S-M, Lee YH (2015) Volatile compounds from *Alcaligenes faecalis* JBCS1294 confer salt tolerance in *Arabidopsis thaliana* through the auxin and gibberellin pathways and differential modulation of gene expression in root and shoot tissues. Plant Growth Regul 75:297–306

Bittencourt PP, Alves AF, Ferreira MB, da Silva Irineu LES, Pinto VB, Olivares FL (2023) Mechanisms and applications of bacterial inoculants in plant drought stress tolerance. Microorganisms 11:502

Blum A (2017) Osmotic adjustment is a prime drought stress adaptive engine in support of plant production. Plant Cell Environ 40:4–10

Bresson J, Varoquaux F, Bontpart T, Touraine B, Vile D (2013) The PGPR strain *Phyllobacterium brassicacearum* STM 196 induces a reproductive delay and physiological changes that result in improved drought tolerance in *Arabidopsis*. New Phytol 200:558–569

Casanovas EM, Barassi CA, Sueldo RJ (2002) Azospiriflum inoculation mitigates water stress effects in maize seedlings. Cereal Res Commun 30:343–350

Chandra D, Srivastava R, Gupta VV, Franco CM, Sharma AK (2019) Evaluation of ACC-deaminase-producing rhizobacteria to alleviate water-stress impacts in wheat (Triticum aestivum L.) plants. Can J Microbiol 65:387–403

Chiappero J, del Rosario Cappellari L, Alderete LGS, Palermo TB, Banchio E (2019) Plant growth promoting rhizobacteria improve the antioxidant status in Mentha piperita grown under drought stress leading to an enhancement of plant growth and total phenolic content. Ind Crops Prod 139:111553

Cochard H, Coll L, Le Roux X, Améglio T (2002) Unraveling the effects of plant hydraulics on stomatal closure during water stress in walnut. Plant Physiol 128:282–290

Cohen AC, Bottini R, Pontin M, Berli FJ, Moreno D, Boccanlandro H, Travaglia CN, Piccoli PN (2015) Azospirillum brasilense ameliorates the response of *Arabidopsis thaliana* to drought mainly via enhancement of ABA levels. Physiol Plant 153:79–90

Cohen AC, Travaglia CN, Bottini R, Piccoli PN (2009) Participation of abscisic acid and gibberellins produced by endophytic Azospirillum in the alleviation of drought effects in maize. Botany 87:455–462

Constable G, Hearn A (1978) Agronomic and physiological responses of soybean and sorghum crops to water deficits I. Growth, development and yield. Funct Plant Biol 5:159–167

Danish S, Zafar-ul-Hye M, Fahad S, Saud S, Brtnicky M, Hammerschmiedt T, Datta R (2020) Drought stress alleviation by ACC deaminase producing *Achromobacter xylosoxidans* and *Enterobacter cloacae*, with and without timber waste biochar in maize. Sustainability 12:6286

Dhawi F (2020) Plant growth promoting Rhizobacteria (PGPR) regulated Phyto and microbial beneficial protein interactions. Open Life Sci 15:68–78

Dodd IC, Belimov A, Sobeih W, Safronova V, Grierson D, Davies W (2004) Will modifying plant ethylene status improve plant productivity in water-limited environments?

Egamberdieva D, Jabborova D, Berg G (2016) Synergistic interactions between *Bradyrhizobium japonicum* and the endophyte *Stenotrophomonas rhizophila* and their effects on growth, and nodulation of soybean under salt stress. Plant Soil 405:35–45

Ghosh D, Gupta A, Mohapatra S (2019) Dynamics of endogenous hormone regulation in plants by phytohormone secreting rhizobacteria under water-stress. Symbiosis 77:265–278

Gou W, Tian L, Ruan Z, Zheng P, Chen F, Zhang L, Cui Z, Zheng P, Li Z, Gao M (2015) Accumulation of choline and glycinebetaine and drought stress tolerance induced in maize (*Zea mays*) by three plant growth promoting rhizobacteria (PGPR) strains. Pak J Bot 47:581–586

Gusain YS, Singh U, Sharma A (2015) Bacterial mediated amelioration of drought stress in drought tolerant and susceptible cultivars of rice (*Oryza sativa* L.). Afr J Biotechnol 14:764–773

Hartman K, Tringe SG (2019) Interactions between plants and soil shaping the root microbiome under abiotic stress. Biochem J 476:2705–2724

Hernández I et al (2012) Antioxidant defenses against drought stress. In: Plant responses to drought stress: from morphological to molecular features, pp 231–258

Hochberg U, Windt CW, Ponomarenko A, Zhang Y-J, Gersony J, Rockwell FE, Holbrook NM (2017) Stomatal closure, basal leaf embolism, and shedding protect the hydraulic integrity of grape stems. Plant Physiol 174:764–775

Hussain MB, Zahir ZA, Asghar HN, Muhammad Asgher MA (2014) Can catalase and exopolysaccharides producing rhizobia ameliorate drought stress in wheat? Int J Agric Biol 16(1)
Hussein H-AA, Mekki B, Abd El-Sadek ME, El Lateef EE (2019) Effect of L-ornithine application on improving drought tolerance in sugar beet plants. Heliyon 5:e02631
Iqbal S, Wang X, Mubeen I, Kamran M, Kanwal I, Díaz GA, Abbas A, Parveen A, Atiq MN, Alshaya H (2022) Phytohormones trigger drought tolerance in crop plants: outlook and future perspectives. Front Plant Sci 12:799318
Israel A, Langrand J, Fontaine J, Lounès-Hadj Sahraoui A (2022) Significance of arbuscular mycorrhizal fungi in mitigating abiotic environmental stress in medicinal and aromatic plants: a review. Foods 11:2591
Jochum MD, McWilliams KL, Borrego EJ, Kolomiets MV, Niu G, Pierson EA, Jo Y-K (2019) Bioprospecting plant growth-promoting rhizobacteria that mitigate drought stress in grasses. Front Microbiol 10:466447
Kang S-M, Radhakrishnan R, Khan AL, Kim M-J, Park J-M, Kim B-R, Shin D-H, Lee I-J (2014) Gibberellin secreting rhizobacterium, *Pseudomonas putida* H-2-3 modulates the hormonal and stress physiology of soybean to improve the plant growth under saline and drought conditions. Plant Physiol Biochem 84:115–124
Kapoor D, Singh S, Kumar V, Romero R, Prasad R, Singh J (2019) Antioxidant enzymes regulation in plants in reference to reactive oxygen species (ROS) and reactive nitrogen species (RNS). Plant Genet Resour 19:100182
Kasim WA, Osman ME, Omar MN, Abd El-Daim IA, Bejai S, Meijer J (2013) Control of drought stress in wheat using plant-growth-promoting bacteria. J Plant Growth Regul 32:122–130
Kaushal M (2019) Portraying rhizobacterial mechanisms in drought tolerance: a way forward toward sustainable agriculture. In: PGPR amelioration in sustainable agriculture. Elsevier, pp 195–216
Kaushal M, Wani SP (2016) Plant-growth-promoting rhizobacteria: drought stress alleviators to ameliorate crop production in drylands. Ann Microbiol 66:35–42
Khan AL, Asaf S, Abed RM, Ning Chai Y, Al-Rawahi AN, Mohanta TK, Al-Rawahi A, Schachtman DP, Al-Harrasi A (2020) Rhizosphere microbiome of arid land medicinal plants and extra cellular enzymes contribute to their abundance. Microorganisms 8:213
Kour D, Yadav AN (2022) Bacterial mitigation of drought stress in plants: current perspectives and future challenges. Curr Microbiol 79:248
Li X et al (2020) A novel PGPR strain *Kocuria rhizophila* Y1 enhances salt stress tolerance in maize by regulating phytohormone levels, nutrient acquisition, redox potential, ion homeostasis, photosynthetic capacity and stress-responsive genes expression. Environ Exp Bot 174:104023
Lim J-H, Kim S-D (2013) Induction of drought stress resistance by multi-functional PGPR *Bacillus licheniformis* K11 in pepper. Plant Pathol J 29:201–208
Lin Z, Li Y, Wang Y, Liu X, Ma L, Zhang Z, Mu C, Zhang Y, Peng L, Xie S (2021) Initiation and amplification of SnRK2 activation in abscisic acid signaling. Nat Commun 12:2456
Liu F, Xing S, Ma H, Du Z, Ma B (2013) Cytokinin-producing, plant growth-promoting rhizobacteria that confer resistance to drought stress in Platycladus orientalis container seedlings. Appl Microbiol Biotechnol 97:9155–9164
Molina-Favero C, Creus CM, Simontacchi M, Puntarulo S, Lamattina L (2008) Aerobic nitric oxide production by *Azospirillum brasilense* Sp245 and its influence on root architecture in tomato. Mol Plant-Microbe Interact 21:1001–1009
Mondal S, Halder SK, Yadav AN, Mondal KC (2020) Microbial consortium with multifunctional plant growth-promoting attributes: future perspective in agriculture. In: Advances in plant microbiome sustainable agriculture: functional annotation future challenges. Springer, pp 219–258
Mondani F, Khani K, Honarmand SJ, Saeidi M (2019) Evaluating effects of plant growth-promoting rhizobacteria on the radiation use efficiency and yield of soybean (*Glycine max*) under water deficit stress condition. Agric Water Manag 213:707–713

Murali M, Naziya B, Ansari MA, Alomary MN, AlYahya S, Almatroudi A, Thriveni M, Gowtham HG, Singh SB, Aiyaz M (2021) Bioprospecting of rhizosphere-resident fungi: their role and importance in sustainable agriculture. J Fungi 7:314

Nardini A, Salleo S (2000) Limitation of stomatal conductance by hydraulic traits: sensing or preventing xylem cavitation? Trees 15:14–24

Nawaz M et al (2020) Salicylic acid improves boron toxicity tolerance by modulating the physio-biochemical characteristics of maize (*Zea mays* L.) at an early growth stage. Agron 10(12):2013

Osemwegie OO, Adetunji CO, Ayeni EA, Adejobi OI, Arise RO, Nwonuma CO, Oghenekaro AO (2020) Exopolysaccharides from bacteria and fungi: current status and perspectives in Africa. Heliyon 6:e04205

Pilon C, Loka D, Snider JL, Oosterhuis DM (2019) Drought-induced osmotic adjustment and changes in carbohydrate distribution in leaves and flowers of cotton (*Gossypium hirsutum* L.). J Agron Crop Sci 205:168–178

Poupin MJ, Greve M, Carmona V, Pinedo I (2016) A complex molecular interplay of auxin and ethylene signaling pathways is involved in Arabidopsis growth promotion by *Burkholderia phytofirmans* PsJN. Front Plant Sci 7:183261

Pyngrope S, Bhoomika K, Dubey R (2013) Oxidative stress, protein carbonylation, proteolysis and antioxidative defense system as a model for depicting water deficit tolerance in Indica rice seedlings. Plant Growth Regul 69:149–165

Qarni A, Billah M, Hussain K, Shah SH, Ahmed W, Alam S, Sheikh AA, Jafri L, Munir A, Malik KM (2021) Isolation and characterization of phosphate solubilizing microbes from rock phosphate mines and their potential effect for sustainable agriculture. Sustainability 13:2151

Raza G, Ali K, Ashraf MY, Mansoor S, Javid M, Asad S (2016) Overexpression of an H+-PPase gene from Arabidopsis in sugarcane improvesdrought tolerance, plant growth, and photosynthetic responses. Turk J Biol 40:109–119

Rolli E, Marasco R, Vigani G, Ettoumi B, Mapelli F, Deangelis ML, Gandolfi C, Casati E, Previtali F, Gerbino R (2015) Improved plant resistance to drought is promoted by the root-associated microbiome as a water stress-dependent trait. Environ Microbiol 17:316–331

Samancioglu A, Yildirim E, Turan M, Kotan R, Sahin U, Kul R (2016) Amelioration of drought stress adverse effect and mediating biochemical content of cabbage seedlings by plant growth promoting rhizobacteria. Int J Agric Biol 18(5):948–956

Sampietro DA, Bertini Sampietro MS, Vattuone MA (2020) Efficacy of Argentinean propolis extracts on control of potato soft rot caused by *Erwinia carotovora* subsp. J Sci Food Agric 100:4575–4582

Sarker U, Oba S (2018) Catalase, superoxide dismutase and ascorbate-glutathione cycle enzymes confer drought tolerance of *Amaranthus tricolor*. Sci Rep 8(1):16496

Shakir MA, Bano A, Arshad M (2012) Short Communciation Rhizosphere bacteria containing ACC-deaminase conferred drought tolerance in wheat grown under semi-arid climate. Soil Environ 31

Sharma A, Shankhdhar D, Shankhdhar S (2013) Enhancing grain iron content of rice by the application of plant growth promoting rhizobacteria. Plant Soil Environ 59:89–94

Shintu P, Jayaram K (2015) Phosphate solubilising bacteria (*Bacillus polymyxa*)-an effective approach to mitigate drought in tomato (*Lycopersicon esculentum* Mill.). Trop Plant Res 2:2349–9265

Singh RP, Shelke GM, Kumar A, Jha PN (2015) Biochemistry and genetics of ACC deaminase: a weapon to “stress ethylene” produced in plants. Front Microbiol 6:149440

Singha LP, Pandey P (2020) Rhizobacterial community of Jatropha curcas associated with pyrene biodegradation by consortium of PAH-degrading bacteria. Appl Soil Ecol 155:103685

Sohbat ZI (2022) Non-photochemical quenching of chlorophyll fluorescence and its components–recent advances. J Life Sci Biomed 4:76–86

Tardieu F (2022) Different avenues for progress apply to drought tolerance, water use efficiency and yield in dry areas. Curr Opin Biotechnol 73:128–134

Timmusk S, Abd El-Daim IA, Copolovici L, Tanilas T, Kännaste A, Behers L, Nevo E, Seisenbaeva G, Stenström E, Niinemets Ü (2014) Drought-tolerance of wheat improved by rhizosphere bacteria from harsh environments: enhanced biomass production and reduced emissions of stress volatiles. PLoS One 9:e96086

Turner NC (2017) Turgor maintenance by osmotic adjustment, an adaptive mechanism for coping with plant water deficits. Plant Cell Environ 40:1–3

Tyagi J, Chaudhary P, Mishra A, Khatwani M, Dey S, Varma A (2022) Role of endophytes in abiotic stress tolerance: with special emphasis on *Serendipita indica*. Int J Environ Res Public Health 16:62

Ullah A, Nisar M, Ali H, Hazrat A, Hayat K, Keerio AA, Ihsan M, Laiq M, Ullah S, Fahad S (2019) Drought tolerance improvement in plants: an endophytic bacterial approach. Appl Microbiol Biotechnol 103:7385–7397

Wahab A, Abdi G, Saleem MH, Ali B, Ullah S, Shah W, Mumtaz S, Yasin G, Muresan CC, Marc RA (2022) Plants' physio-biochemical and phyto-hormonal responses to alleviate the adverse effects of drought stress: a comprehensive review. Plants 11:1620

Wang C-J, Yang W, Wang C, Gu C, Niu D-D, Liu H-X, Wang Y-P, Guo J-H (2012) Induction of drought tolerance in cucumber plants by a consortium of three plant growth-promoting rhizobacterium strains. PLoS One 7:e52565

Wang L, Ruan Y-L (2013) Regulation of cell division and expansion by sugar and auxin signaling. Front Plant Sci 4:163

Wani KI, Naeem M, Castroverde CDM, Kalaji HM, Albaqami M, Aftab T (2021) Molecular mechanisms of nitric oxide (NO) signaling and reactive oxygen species (ROS) homeostasis during abiotic stresses in plants. Int J Mol Sci 22:9656

Yoon J, Cho L-H, Yang W, Pasriga R, Wu Y, Hong W-J, Bureau C, Wi SJ, Zhang T, Wang R (2020) Homeobox transcription factor OsZHD2 promotes root meristem activity in rice by inducing ethylene biosynthesis. J Exp Bot 71:5348–5364

You L, Song Q, Wu Y, Li S, Jiang C, Chang L, Yang X, Zhang J (2019) Accumulation of glycine betaine in transplastomic potato plants expressing choline oxidase confers improved drought tolerance. Planta 249:1963–1975

Zandalinas SI, Balfagón D, Arbona V, Gómez-Cadenas A, Inupakutika MA, Mittler R (2016) ABA is required for the accumulation of APX1 and MBF1c during a combination of water deficit and heat stress. J Exp Bot 67:5381–5390

Zhang R, Mu Y, Li X, Li S, Sang P, Wang X, Wu H, Xu N (2020) Response of the arbuscular mycorrhizal fungi diversity and community in maize and soybean rhizosphere soil and roots to intercropping systems with different nitrogen application rates. Sci Total Environ 740:139810

Chapter 6
Microbial Contributions to Soil Carbon Sequestration

O. I. Nwachukwu and I. G. Okoro

Abstract This review paper presented a global overview of how the environment is affected by the impact of carbon persistence; the dynamics between carbon and soil components as well as highlighting the role of microorganisms in carbon sequestration which was hitherto not accorded the rightful significance as an indispensable factor. The relevance of different microbial groups such as fungi and bacteria to significantly mitigate climate change was established, and methods of harnessing microbial applications for carbon sequestration were examined. The application of microbial inoculants in soil for the purpose of facilitating carbon sequestration is recommended as an eco- friendly and sustainable boost to traditional carbon sinks such as forests and peat bogs. It is concluded that future efforts at combating global warming must include the utilization of microbial inoculants in agricultural lands to enhance microbial efficiency for carbon sequestration in soil.

Keywords Carbon sequestration · CO_2 · Fungi · Bacteria · Soil · Environment

6.1 Introduction

Carbon is an element which occurs naturally in the environment, abounding in both living and non- living things which serve as reservoirs, and it is called the ‘element of life’ because it is a component of all living things (Jones 2020). Forming key molecules like protein, carbon has the capacity to form enormous number of compounds in the environment. Carbon does not remain in a static form as it is constantly changing relative to conditions in the soil, atmosphere, water bodies, human activities and the presence of other elements. The dynamics of how carbon moves from one reservoir in the environment to another is referred to as the carbon cycle

O. I. Nwachukwu (✉) · I. G. Okoro
Department of Soil Science and Land Resources Management, Michael Okpara University of Agriculture, Umudike, Umuahia, Nigeria
e-mail: nwachukwu.olayinka@mouau.edu.ng

O. O. Babalola et al. (eds.), *Ecofriendly Frontiers*,
https://doi.org/10.1007/978-3-031-98700-7_6

Fig. 6.1 The Carbon Cycle

(Regnier et al. 2022). A pictorial representation of the Carbon Cycle is presented in Fig. 6.1.

Our planet has faced massive upsurge in carbon emissions in the past century due to the operations of non-renewable sources of energy and industrialization (Dong et al. 2019). The fallout has been global warming due to increased atmospheric temperatures with attendant climate change which has affected all parts of our ecosystem. A notable shift in weather patterns has led to unpredictability of crop growing seasons (Fernandez-Bou et al. 2021). The potential harvest of crops has been affected on account of changes in pest populations and their life cycles, while flooding is occurring increasingly in all parts of the world due to rising sea levels (Muluneh 2021; Singh et al. 2023). Indeed, the survival of our planet is threatened because the accelerating increase in human populace has also escalated the requirement for food, clothing, health care and housing, all of which depend on the ecosystem. In order to achieve all these without further aggravating the ravages of climate, interventions must be eco-sustainable (Moreira et al. 2023; Bhattacharjee et al. 2023; Yunusa et al. 2022).

Eco-sustainability is an approach which aims to maintain a balance of natural ecosystems over the long term while meeting present needs and at the same time, making no concession for the needs of future existence (Fawzy et al. 2020; Diallo et al. 2020). Sustaining our ecosystem therefore requires an intentional networking of the key players which are climate, humans, the atmosphere, plants, wildlife, the marine environment with their organisms, and microbes.

The soil remains the major medium for crop growth, and it is also the largest sink for carbon (Lal 2008; Gomiero et al. 2011). Comprising of non-living components air, water, mineral matter and a living organic component, the soil is a richly diverse ecosystem (Sofo et al. 2022). In the face of the continued menace of global warming, two main goals must be achieved, and these are to reduce carbon emissions and also to store the carbon which has already been released into the atmosphere (Huisingh et al. 2015; Dogan et al. 2022). A remarkable ally to achieve the latter are microbes or microorganisms living beneath the ground (Singh et al. 2010). Microbes are the tiny living things which are found all around us in the air, soil, water and even our bodies (Kaur et al. 2021; National Library of Medicine 2022). The involvement of microbes in capturing and storing atmospheric carbon dioxide has not been made as visible as that of other players such as higher plants and sinks like water bodies and peat bogs. Apart from their traditional roles in the putrefaction of remains from plants and animals, indigenous microbes are known to both sequester carbon in their body cells in addition to aiding capturing and storing atmospheric carbon dioxide located in the soil (Buragohain 2019; Mitra et al. 2022; Hartmann and Six 2023).

6.2 Carbon Sequestration in Soils

The mechanism by which carbon is sequestrated necessitates encapsulating carbon(iv)dioxide (CO_2) from the atmosphere and incorporating it within soils to improve soil carbon reserve (Powlson et al. 2011; Olson 2013). This is described as the phenomenon of transporting atmospheric carbon dioxide into the earth's soil pool via plants, remains of plants and organic matter for it to be preserved as components of the soil carbon (Hartmann and Six 2023; Olson 2013). The preservation of carbon in soils can be in the form of rigid, long-conserved organic matter (Powlson et al. 2011). Improving the quantities of recalcitrant natural materials in soils would result in prolonged carbon storage resulting to depletion in the release of greenhouse gases (CO_2, N_20 etc.). The greater the measure of carbon retained down the soil and the lower the measure of carbon sequestration, and vice versa (Janzen 2004).

Activities including planting, livestock overgrazing as well as residue burning generally deplete soil carbon stocks (Bai and Cotrufo 2022). consequently, insufficient operational techniques will definitely result in deleterious consequences on soil carbon storage and thus be partly responsible for soil carbon reduction. The interplay of diverse biological processes, involving photosynthesis, respiration, and degradation alters carbon concentrations and retention in soils (Ontl and Schulte 2012). The mechanism of photosynthesis by autotrophic organisms and plants helps in sequestering carbon by way of enhancing atmospheric CO_2 intake which is consequently preserved in the terrestrial reservoir. Conversely, respiration, as well as decomposition, promotes carbon deficit from the terrestrial stockpile into the atmosphere (Janzen 2004). In recent years, carbon deficit from the terrestrial stockpile into the atmosphere due to several processes (e.g., anthropogenic activities and forest fires) has necessitated the need and urgency to find other solutions to combat the problem in addition to the natural way of photosynthesis (Huisingh et al. 2015; Dogan et al. 2022). Adoption of microbial inoculants shows a promising practice that has the aptitude to escalate the levels of carbon deposits in the soil and abate the levels of atmospheric carbon (Ahmed et al. 2019). As a result, the technique of inoculating microbial strains is beneficial in the course of this. Incorporating the process of inoculation of microorganisms in agricultural technologies may aid in enhancing soil properties. Bio-sequestration refers to the biological sequestration and preservation of CO_2 by soil microorganisms along side with photosynthetic organisms (Bhattacharjee et al. 2023; Moreira et al. 2023). Microbial inoculants may be applied to elevate carbon intake while decreasing carbon release from the soil pool (Ahmed et al. 2019; Canfora et al. 2021).

6.2.1 How Do Soil Microbes Sequester Carbon?

Soil microbes can sequester carbon through a variety of pathways, including biochemical processes that capture CO_2 from the atmosphere, such as photosynthesis when plants utilize atmospheric CO_2 and then store the carbon in their biomass or release some CO_2 into the rhizosphere during root respiration. Other pathways are the formation of recalcitrant tissues such as lignin in plants thereby storing such carbon which is a component of lignin for longer periods unlike softer plant tissues which are easily broken-down during decomposition. Also, due to their high demand for carbon, microalgae have also been known to be very efficient at capturing CO_2 from the atmosphere both in terrestrial and aquatic environments and storing same for long periods of time which can be comparable to the efficiency of forests (Guduru et al. 2022). Finally, microbes, through their metabolic processes and exudates have the tendency to aid the emergence of stable forms such as humus deposit in soil, and these act like aggregates which preserve carbon total organic components (Sosa-Hernández et al. 2019).

Soil microbes possess unique biological functions that enable them to be strong candidates for carbon sequestration as well as to reduce carbon emissions (Fang et al. 2014). Inadequate focus had been placed on microbial diversity responsible for carbon sequestration, thus it is critical to identify the suitable microorganism which possess the needed biological function that can increase the level of carbon for a particular soil type (Ferrero et al. 2016; Zhang et al. 2017). Additionally, microorganisms resident in the soil have various mechanisms that can modulate the intake and depletion of soil carbon, thus the variations in microbial communities can significantly influence the quantity of organic carbon deposit in the soil (Mawarda et al. 2022). Variability in microbial biomass compositions and metabolic functions across microbial populations can result in distinct processes and mechanisms for carbon sequestration in soils (Sosa-Hernández et al. 2019).

6.3 Microbial Applications: Microbial Mediated Soil Carbon Sequestration

Numerous studies have suggested that soil microorganisms have indispensable roles in regulating formation, breaking down and build up of soil organic matter (Mellado-Vazquez et al. 2019; Cotrufo et al. 2015; Lange et al. 2015). In *vice versa* situation, plant derived C resources largely influence the soil microbial activity apart from other factors such as soil temperature, soil humidity or pH while the impact of different photosynthetic pathways on the microbial community may be prompted by astonishing outcome of other environmental parameters (Voroney and Heck 2015).

Micro-organisms domicile in the soil has a vital function in speeding up and promoting the breakdown of the organic matter, thereby making available and cycling of stockpiled nutrients contained therein as well as the loss of gaseous to immediate environments and also impact the mechanism that regulates the absorption of these nutrients by the hyphae of plant roots (Babalola 2010; Babalola and Glick 2012). The equilibrium between the mechanism of photosynthesis and gaseous respiration is determined by the terrestrial carbon cycle. Autotrophic organisms, especially photosynthesizing plants, and also photoautotrophic and chemoautotrophic microorganisms has been reported to fix atmospheric carbon dioxide CO_2 into organic material which is then reversed to atmosphere through autotrophic and heterotrophic respiration and decomposition pathways (Gouglias et al. 2014). Organic carbon-consuming heterotrophic microorganisms are responsible for the metabolism of organic carbon, they feed from substrates of microbial source or carbon from plants or animals deposits. Certain amount of this carbon is conserved in their biomass; thus, the superfluous is discharged as metabolites or as CO_2 revert to the atmosphere (Nwachukwu and Pulford 2011). Similar to that, microorganisms domicile in the soil directly controls the organic carbon quantity deposited in soil and that which is allowed back into the atmosphere (Fang et al. 2014; Poppeliers et al. 2023). There microbes, obliquely alter the carbon storage in soils and plants thereby making available macronutrients such as Nitrogen and Phosphorus to regulate the productivity of the soil (Singh et al. 2010; Bardgett and van der Putten 2014).

6.3.1 Methods of Harnessing Microbial Applications: Microbial Solutions

A quarter of the earths biological diversity is domiciled in the soil, the most densely populated and dynamic and found everywhere on earth (Heijden 2008). The sustainability of crops and soil health is attributed to the indispensable responsibility of microorganism's domicile in the soil. (Wardle et al. 2004), with strategic roles in broader biome function and biogeochemical progression. Owing to their relevance, the ecological intensification of production systems can be improved by effective supervision and influence of identified soil and plant microbe, and also the trapping of atmospheric carbon dioxide by way of promoting soil stabilization and C storage and stabilisation (Ahmed et al. 2019; Buragohain 2019).

For centuries, the use of products emanating from microbes have been accustomed by human beings, in which time significant progress has been made in the context of soil inoculation, mostly with the purpose of enhancing plant growth/ adaptability and correlating productivity (Santos et al. 2019; Li et al. 2023; Canfora et al. 2021). Diazotrophic (microorganism that is able to grow without external sources of fixed nitrogen) bacteria such as rhizobia and Frankia are a group of microorganism having the ability to colonize the roots of numerous members of the legume family (Fabaceae), by activating a mechanism of nitrification using microbes which usually furnish the plants with the standard nitrogen requirements (Santos et al. 2019). This apparently reduces inorganic nutrient deposits. Some beneficial microbial groups such as filamentous fungi (e.g. *Paecilomyces spp., Trichoderma* sp., *Purpureocillium lilacinum*) and arbuscular mycorrhizal fungi (AMF) (e.g. *Clariodeoglomus sp., Rhizophagus* sp.) have similarly been applied in microbial products, which to a limited extent is sustaining advancement of the agricultural biotechnology sector which is anticipated by 2027 to attain about US$12.5 billion in revenue generation (O'Callaghan et al. 2022). Following the probable implementation of soil microbes in improving soil C, more advanced research on the involvement of beneficial microbes in soil carbon sequestration and retention is needed to aid in the advancement of microbial products with the clear purpose of establishing and preserving soil carbon.

6.4 Comparative Microbial Efficiency for Sequestration of Carbon in Soil

Bacteria are more populated in soil when compared with fungi. Although previous studies have compared the efficiency of fungi to sequester carbon with that of bacteria, some bacterial strains possess the capability of reducing CO_2 and hence could help with carbon sequestration and CO_2 mitigation (Yang et al. 2023). This implies that bacteria would further function in sequestering carbon through numerous metabolic mechanisms and processes.

Some groups of bacteria and the mechanisms of their activities are shown in Table 6.1.

Fungi have been confirmed to have significant involvement in sequestering carbon, biomass accumulation, and soil stability when equated with bacteria (Six et al. 2006; Bastida et al. 2021). Although fungi have a lower population compared to bacteria, they dominate soil organic matter, especially in undisturbed soils (Khan et al. 2016). This is attributed to the fact that fungi release a huge quantity of carbon into the soils, resulting in the incorporation of additional carbon into the soils together with fungal biomass, the production of recalcitrant organic matter tissues

Table 6.1 Selected groups of bacteria, mechanisms and implications of their activities

	Name of isolates	Mechanism of activities	Results	References
1	*Pseudomonas fluorescens*	They lower the plant ethylene amount by prolific generation of Acetyl-CoA decarbonylase/ synthase enzyme, a cytoplasmic deaminase that reduces the direct precusor of ethylene, 1-aminocyclopropane-1-carboxylate (ACC), into ammonium (NH_4) and α-ketobutyrate.	The activities of Acetyl-CoA decarbonylase/ synthase decreased the concentration the signaling molecules associated with stress responses in the plant.	Polko and Kieber (2019); Glick and Nascimento (2021)
2	*Bacillus mucilaginosus*	They capture atmospheric CO_2 through hydration of CO_2, and then – HCO_3 was utilized by bacterium and finally converted into the bacterial metabolism. By capturing atmospheric CO_2, this structure sparks up the production of carbonate to generate carbonated minerals.	*B. mucilaginosus* forms carbonic anhydrase that sequesters CO_2 from the atmosphere thereafter fixation process id engaged using bacterial breakdown.	Zhang et al. (2011)
3	*Escherichia coli*	It examines the capacity of bacterial producing carbonic anhydrase to enhance $CaCO_3$ precipitation	The engineered bacteria released carbonic anhydrase, which promoted the synthesis of calcium carbonate within a short time frame.	Xiao et al. (2015)
4	*Moorella thermoacetica*	Shows a self-replicating sequence in the direction of light-chemical carbon(iv)oxide decline by restrictively producing acetic acid from carbon(iv)oxide.	Increased production of acetic acid.	Sakimoto et al. (2016)
5	*E. coli BW25113*	Genetically reconfigured E. coli multiplies effectively using methanol as the sole source of carbon.	Increased amount of carbon in its cell membrane.	Chen et al. (2020)
6	Bacillus cereus	The metabolic techniques are achieved inside the cell.	CO_2 compounds produced were converted and produce other value-added byproducts.	Duport et al. (2016)
7.	Azotobacter and Bacillus subtilis strain)	It involves complex electrochemical and biological mechanisms. The principal limiting factors for the micro-organism's activity are water, light, oxygen and CO_2, the higher the water availability and light intensity, oxygen and CO_2 penetration can boost the microbial activity and retention of carbon.	Added to the soil organic matter in the form of Cx-Hy-Oz through photon energy as the support.	Naik et al. (2010)

and outgrowth, and improved carbon accumulation and stability through soil aggregate production. The amount of carbon accumulated by fungi in soils differs based on the variation in fungal strains and biomass compositions (Ekblad et al. 2013; Khan et al. 2016). This can be attributed to the fact that carbon is used to build the mycelium of fungi thus, the additional mycelia formed, the more carbon is retained in the soil. As a result, fungal mycelia might be regarded as an important carbon reservoir in soils. The vast microbiological form of stable biomass is fungal necromass, which provides soil organic matter more than in the case of bacterial necromass (Bastida et al. 2021). As a result, mycelial necromass might account for a major source of carbon accumulation in soil, as well as an outstanding source of carbon for animal and plant species. The capacity of mycelial necromass to withstand degradation varies by fungal strains as well as fungal tissue properties. As a result, some fungal species may be able to participate more in carbon sequestration than others. Mycelia emanating from fungi may advance into stable soil formations like soil aggregates and as such, fungal necromass will be preserved from losing its form for an elongated period of time and the more fungal necromass that is preserved, the additional carbon is stored in soils (Moitinho et al. 2020).

6.4.1 Microbial Inoculants for Carbon Sequestration

Heterologous expression of carbonic anhydrase genes emanating from *B. mucilaginosus* genome into *Escherichia coli* was carried out by Achal and Pan (2011) to examine the capacity of bacterial isolates in producing carbonic anhydrase to enhance $CaCO_3$ precipitation, and it was observed that the optimized bacteria released carbonic anhydrase, which promoted the synthesis of $CaCO_3$, in a limited time frame (Achal and Pan 2011). Therefore, carbonic anhydrase-producing microbial strains might be highly beneficial microbial inoculants to sequester carbon and capture CO_2 from the atmosphere. Bacterial sequestering of CO_2 into sediment carbonates might subsist as a viable mechanism for promoting lower carbon emissions and combating climate change. Glomalin is one of the root exudates and it is an inert glue-like proteinaceous material which is hydrophobic and secreted to preserve the hyphae from dehydrating from the spores and mycelia of mycorrhizal fungi as well as the genera of Glomus, Acaulospora, Entrophospora, Scutellospora, and Gigaspora (Panchal et al. 2022; Wang et al. 2023). Glomalin has been recorded to exhibit dual effects on sequestering carbon. First, glomalin synthesis helps in creation of soil aggregates, responsible in influencing carbon sequestration. The second manner is due to glomalin's resistant structure, which makes it difficult to degrade and so it is preserved in soils for extended periods of time. Bacteria and fungi provide glue-like compounds that glue soil particles in addition to organic and inorganic components to create soil aggregates. The generation and stability of aggregates are influenced by soil fauna and microbes, roots, external conditions, and binding agents (e.g. glomalin) (Panchal et al. 2022). The most significant biotic agents in soil aggregate stability are roots and fungal mycelia (Wang et al. 2023).

Soil aggregates' cementing capability creates a robust, homogenous framework that preserves the organic and inorganic components inside the soil aggregate. Aggregate stabilization aids in carbon sequestration by preventing carbon containing compounds situated in the aggregates from decomposition, allowing carbon to be retained and incorporated in soils for long periods. In maintaining and ameliorating the soil structure and fertility, soil aggregates are crucial, especially when focusing on soil's physical qualities (Hartmann and Six 2023; Blankinship et al. 2016).

6.5 Looking to the Future: Microbial Inoculation Biotechnology in Agriculture

Microbial inoculants are used as an economically viable and environmentally friendly method that tailor's microbes as renewable biofertilizers, and also as an alternative to replace inorganic fertilizers and pesticides (Babalola and Glick 2012; Li et al. 2023; Liu et al. 2022). There is need for relevant agencies of government to give priority attention to make deliberate policies which will include microbial inoculants in existing bio-fertilizers. This approach is an eco-friendly technique which assists in promoting agriculture by functioning as an organic biocontrol agent and biofertilizer, minimizing the unpleasant consequences of synthetic fertilizers and agrochemicals on agricultural fields and crop production, which often has adverse health and environmental effects (Chakraborty and Akhtar 2021). Also, including microbial inoculants in bio-fertilizers will be more affordable to farmers who would otherwise be unable to afford the high cost of inoculants nor the technology of using them. Microbial inoculants create an ideal environment in agricultural lands, ensuring good crop security and productivity (Suyal et al. 2016; Elnahal et al. 2022). Examining crop yield and health might be compromised in agricultural lands that lack healthy soil properties, whether under distressed or normal conditions, or in a situation of poor organic carbon content. Crop rate of production has been found to improve significantly in poor soils after introducing microbial inoculants (Bhattacharjee et al. 2023; Poppeliers et al. 2023).

Organic nutrients and soil respiration have been known to increase when using microbial inoculants thus, they might be a viable option for improving soil properties and productivity in vulnerable areas (Nwachukwu and Pulford 2011). In agro-ecosystems, the application of microbial inoculant has successfully improved plant health, minimizing heavy metal toxicity, improving plant resistance to drought and disease, and improving absorption of nutrients (Babalola 2010; Bashan et al. 2014; Fernandez-Bou et al. 2021). Furthermore, improved research into futuristic microbial consortia, formulation improvement, and a blueprint/strategy to commercial application remain in their infant stages. Thus, it is imperative that manufacturers are enabled to upscale the production to large scale for wider use, while governments and other agro- institutions undertake the subsidy of newly developed 'inoculant fortified bio-fertilizers' to make them affordable.

Several techniques based on biotechnological development and formulation optimization for practical use are being developed around the world to address the current limits. Metagenomics application and microbiome characterization is one of the novel emerging routes for finding prospective microbes responsible for carbon sequestration. The identification of new species and their pathways from metagenome have the prospects for commercial agricultural purposes and researchers can effectively understand their specific function using taxonomical or functional analysis of the metagenome. Identification of numerous numbers of unclassified DNA sequences from environmental metagenome holds the potential for a pedigree of microbial diversity for carbon sequestration.

6.6 Conclusion

Microbes play a significant function in sequestration of carbon in the soil. Consequently, they are indispensable in adapting to climatic variations and achieving sustainable production of food, and consequently, the continuous existence of the humans. The crucial link is the role that they play in carbon sequestration. With proper harnessing of this resource to complement natural sinks of carbon such as vegetation and peat bogs, the deleterious effects of global warming will be curtailed; threatened soils will be preserved and at the same time, soil health can be enhanced and sustainable food crop production can be achieved. Research efforts must therefore concentrate on activities that promote microbial carbon sequestration as well as carbon input in soils with a concurrent decrease in carbon emissions and outputs.

References

Achal V, Pan X (2011) Characterization of urease and carbonic anhydrase producing bacteria and their role in calcite precipitation. Curr Microbiol 62:894–902

Ahmed AAQ, Odelade KA, Babalola OO (2019) Microbial inoculants for improving carbon sequestration in agroecosystems to mitigate climate change. In: Handbook of climate change resilience

Babalola OO (2010) Beneficial bacteria of agricultural importance. Biotechnol Lett 32:1559–1570

Babalola OO, Glick BR (2012) The use of microbial inoculants in African agriculture: current practice and future prospects. J Food Agric Environ 10:540–549

Bai Y, Cotrufo MF (2022) Grassland soil carbon sequestration: current understanding, challenges, and solutions. Science 377(6606):603–608

Bardgett RD, van der Putten WH (2014) Belowground biodiversity and ecosystem functioning. Nature 515(7528):505–511

Bashan Y, de-Bashan LE, Prabhu S, Hernandez JP (2014) Advances in plant growth-promoting bacterial inoculant technology: formulations and practical perspectives (1998–2013). Plant Soil 378:1–33

Blankinship JC, Fonte SJ, Six J, Schimel JP (2016) Plant versus microbial controls on soil aggregate stability in a seasonally dry ecosystem. Geoderma 272:39–50
Bhattacharjee A, Dubey S, Sharma S (2023) "Next-generation bioformulations" for plant growth promotion and stress mitigation: a promising approach for sustainable agriculture. J Plant Growth Regul 42:1–19
Buragohain P (2019) Role of microbes on carbon sequestration. Int J Microbiol Res. ISSN 0975-5276
Bastida F, Eldridge DJ, García C, Kenny Png G, Bardgett RD, Delgado-Baquerizo M (2021) Soil microbial diversity–biomass relationships are driven by soil carbon content across global biomes. The ISME J 15(7):2081–2091
Canfora L, Costa C, Pallottino F, Mocali S (2021) Trends in soil microbial inoculants research: a science mapping approach to unravel strengths and weaknesses of their application. Agriculture 11(2):158
Chakraborty T, Akhtar N (2021) Biofertilizers: prospects and challenges for future. In: Biofertilizers: study and impact. Wiley, pp 575–590
Chen FYH, Jung HW, Tsuei CY, Liao JC (2020) Converting Escherichia coli to a synthetic methylotroph growing solely on methanol. Cell 182(4):933–946
Cotrufo MF, Soong JL, Horton AJ, Campbell EE, Haddix ML (2015) Formation of soil organic matter via biochemical and physical pathways of litter mass loss. Nat Geosci 8(10):776–779
Diallo A, Donkor E, Owusu V (2020) Climate change adaptation strategies, productivity and sustainable food security in Southern Mali. Clim Chang 159(3):309–327
Dogan E, Hodzic S, Sikic TF (2022) A way forward in reducing carbon emissions in environmentally friendly countries: the role of green growth and environmental taxes. Econ Res Ekon Istraž 35(1):5879–5894
Dong F, Wang Y, Su B, Hua Y, Zhang Y (2019) The process of peak CO_2 emissions in developed economies: a perspective of industrialization and urbanization. Resour Conserv Recycl 141:61–75
Duport C, Jobin M, Schmitt P (2016) Adaptation in *Bacillus cereus*: from stress to disease. Front Microbiol 7:1–18
Ekblad A, Wallander H, Godbold DL, Cruz C, Johnson D, Baldrian P, Björk R, Epron D, Kieliszewska-Rokicka B, Kjøller R (2013) The production and turnover of extra-matrical mycelium of ectomycorrhizal fungi in forest soils: role in carbon cycling. Plant Soil 366:1–27
Elnahal ASM, El-Saadony MT, Saad AM, Desoky EM, El-Tahan AM, Rady MM, AbuQamar SF, El-Tarabily KA (2022) The use of microbial inoculants for biological control, plant growth promotion, and sustainable agriculture: a review. Eur J Plant Pathol 162(4):759–792
Fang H, Cheng S, Wang Y, Yu G, Xu M, Dang X, Li L, Wang L (2014) Changes in soil heterotrophic respiration, carbon availability, and microbial function in seven forests along a climate gradient. Ecol Res 29:1077–1086
Fawzy S, Osman AI, Doran J, Rooney DW (2020) Strategies for mitigation of climate change: a review. Environ Chem Lett 18:2069–2094
Fernandez-Bou AS, Ortiz-Partida JP, Pells C, Classen-Rodriguez LM, Espinoza V, Rodríguez-Flores JM, Medellín-Azuara J (2021) Regional report for the San Joaquin Valley region on impacts of climate change. California Natural Resources Agency. Publication number: SUM-CCCA4-2021-003
Ferrero HEW, Gonzalez MG, Giuffre L, Ciarlo E (2016) Glomalins and their relationship with soil carbon. Int J Appl Sci Technol 6(2):1–5
Glick BR, Nascimento FX (2021) *Pseudomonas* 1-aminocyclopropane-1-carboxylate (ACC) deaminase and its role in beneficial plant-microbe interactions. Microorganisms 9:2467. https://doi.org/10.3390/microorganisms9122467
Gomiero T, Pimentel D, Paoletti M (2011) Is there a need for a more sustainable agriculture? Crit Rev Plant Sci 30(1–2):6–23
Gougoulias C, Joanna MC, Shaw LJ (2014) The role of soil microbes in the global carbon cycle: tracking the below-ground microbial processing of plant-derived carbon for manipulating carbon dynamics in agricultural systems. J Sci Food Agric 94(12):2362–2371

Guduru RK, Gupta AA, Dixit U (2022) Biological processes for CO_2 capture. In: Khalid M, Dharaskar SA, Sillanpa M, Siddiqui H (eds) Emerging carbon capture technologies. Elsevier, pp 371–400

Hartmann M, Six J (2023) Soil structure and microbiome functions in agroecosystems. Nat Rev Earth Environ 4:4–18

Heijden VD (2008) The unseen majority: soil microbes as drivers of plant diversity and productivity in terrestrial ecosystems. Ecol Lett 11:296–310

Huisingh D, Zhang Z, Moore JC, Qiao Q, Li Q (2015) Recent advances in carbon emissions reduction: policies, technologies, monitoring assessment and modeling. J Clean Prod 103:1–12

Janzen HH (2004) Carbon cycling in earth systems- a soil science perspective. Agric Ecosyst Environ 104(3):399–417

Jones WD (2020) Carbon capture and conversion. J Am Chem Soc 11:4955–4957

Kaur T, Devi R, Kour D, Yadav A, Yadav AN, Dikilitas M, Abdel-Azim AM, Ahluwalia AS, Saxina AK (2021) Plant growth promoting soil microbiomes and their potential implications for agricultural and environmental sustainability. Biologia 76(9):2687–2709

Khan KS, Mack R, Castillo X, Kaiser M, Joergensen RG (2016) Microbial biomass, fungal and bacterial residues, and their relationships to the soil organic matter C/N/P/S ratios. Geoderma 271:115–123

Lal R (2008) Soils and sustainable agriculture. A review. Agron Sustain Dev 28:57–64

Lange M, Eisenhauer N, Sierra CAH, Engels C (2015) Plant diversity increases soil microbial activity and soil carbon storage. Nat Commun 6:6707

Li C, Jia Z, Zhang S, Li T, Ma S, Cheng X, Chen M, Nie H, Zhai L, Zhang B (2023) The positive effects of mineral solubilizing microbial inoculants on asymbiotic nitrogen fixation of mine soils are driven by keystone phylotype. Sci Total Environ 882:163663

Liu X, Le Roux X, Salles JF (2022) The legacy of microbial inoculants in agroecosystems and potential for tackling climate change challenges. iScience 25:103821. https://doi.org/10.1016/j.isci.2022.103821

Mawarda PC, Mallon CA, Le Roux X, van Elsas JD, Salles JF (2022) Interactions between bacterial inoculants and native soil bacterial community: the case of spore-forming Bacillus spp. FEMS Microbiol Ecol 98:fiac127

Mellado-Vazquez PG, Lange M, Glexiner G (2019) Soil microbial communities and their carbon assimilation are affected by soil properties and season but not by plants differing in their photosynthetic pathways (C3 *vs.* C4). Biogeochemistry 142:175–187

Mitra D, Mondal R, Koshru B (2022) Actino- bacteria enhanced plant growth, nutrient acquisition, and crop protection: advances in soil, plant and microbial multifactorial interactions. Pedosphere 32(1):149–170

Moitinho MR, Fernandes C, Truber PV, Marcelo AV, Corá JE, Bicalho E (2020) Arbuscular mycorrhizal fungi and soil aggregation in a no-tillage system with crop rotation. J Plant Nutr Soil Sci 183(4):482–491

Moreira ZPM, Chen MY, Ortuno DLY, Haney CH (2023) Engineering plant microbiomes by integrating ecoevolutionary principles into current strategies. Curr Opin Plant Biol 71:102316

Muluneh MG (2021) Impact of climate change on biodiversity and food security: a global perspective- a review article. Agric Food Sec 10(1):1–25

Naik SN, Goud VV, Rout PK, Dalai AK (2010) Production of first- and second-generation biofuels: a comprehensive review. Renew Sust Energ Rev 14:578–597. https://doi.org/10.1016/j.rser.2009.10.003

National Library of Medicine (2022) Microbes. wwwcbinlmnlhgov. Updated April 5, 2022

Nwachukwu OI, Pulford ID (2011) Microbial respiration as an indication of metal toxicity in contaminated organic materials and soil. J Hazard Mater 185:1140–1147. www.elsevier.com/jhazmat

O'Callaghan M, Ballard RA, Wright D (2022) Soil microbial inoculants for sustainable agriculture: limitations and opportunities. Soil Use Manag 38(3):1340–1369

Olson KR (2013) Soil organic carbon sequestration, storage, retention and loss in US croplands: issues paper for protocol development. Geoderma 195:201–206

Ontl TA, Schulte LA (2012) Soil carbon storage. Nat Educ Knowl 3:1–10

Polko JK, Kieber JJ (2019) 1-aminocyclopropane 1-carboxylic acid and its emerging role as an ethylene-independent growth regulator. Front Plant Sci 10:1602. https://doi.org/10.3389/fpls.2019.01602

Panchal P, Preece C, Peñuelas J, Giri J (2022) Soil carbon sequestration by root exudates. Trends Plant Sci 27(8):749–757

Poppeliers SW, Sanchez-Gil JJ, de Jonge R (2023) Microbes to support plant health: understanding bioinoculant success in complex conditions. Curr Opin Microbiol 73:102286

Powlson DS, Whitmore AP, Goulding KW (2011) Soil carbon sequestration to mitigate climate change: a critical re-examination to identify the true and the false. Eur J Soil Sci 62:42–55

Regnier P, Resplandy L, Najjar RG, Ciais P (2022) The land-to-ocean loops of the global carbon cycle. Nature 603(7901):401–410

Sakimoto KK, Wong AB, Yang P (2016) Self-photosensitization of non-photosynthetic bacteria for solar-to-chemical production. Science 351(6268):74–77

Santos MS, Nogueira MA, Hungria M (2019) Microbial inoculants: reviewing the past, discussing the present and previewing an outstanding future for the use of beneficial bacteria in agriculture. AMB Express 9:205

Singh BK, Bardgett RD, Smith P, Reay DS (2010) Microorganisms and climate change: terrestrial feedbacks and mitigation options. Nat Rev Microbiol 8:779–790

Singh BK, Delgado-Baquerizo M, Egidi E, Guirado E, Leach JE, Liu H, Trivedi P (2023) Climate change impacts on plant pathogens, food security and paths forward. Nat Rev Microbiol 21(10):640–656

Six J, Frey S, Thiet R, Batten K (2006) Bacterial and fungal contributions to carbon sequestration in agroecosystems. Soil Sci Soc Am J 70:555–569

Sofo A, Zanella A, Ponge J (2022) Soil quality and fertility in sustainable agriculture, with a contribution to the biological classification of agricultural soils. Soil Use Manag 38(2):1085–1112

Sosa-Hernández MA, Leifheit EF, Ingraffia R, Rillig MC (2019) Subsoil arbuscular mycorrhizal fungi for sustainability and climate-smart agriculture: a solution right under our feet? Front Microbiol 10:744. https://doi.org/10.3389/fmicb.2019.00744

Suyal DC, Soni R, Sai S, Goel R (2016) Microbial inoculants as biofertilizer. In: Microbial inoculants in sustainable agricultural productivity, Research perspectives, vol 1, pp 311–318

Voroney RP, Heck RJ (2015) The soil habitat. In: Paul EA (ed) Soil microbiology, ecology and biochemistry, 4th edn. Academic, Boston, pp 15–39

Wang Y, He X, Meng L, Zhou Y, Wu Q (2023) Extra radical mycorrhizal hyphae promote soil carbon sequestration through difficultly extractable glomalin-related soil protein in response to soil water stress. Microb Ecol 86(2):1023–1034

Wardle DA, Bardgett RD, Klironomos JN, Setala H, Van Der Putten WH, Wall DH (2004) Ecological linkages between aboveground and belowground biota. Science 304:1629–1633

Xiao L, Lian B, Hao J, Liu C, Wang S (2015) Effect of carbonic anhydrase on silicate weathering and carbonate formation at present day CO_2 concentrations compared to primordial values. Sci Rep 5:1–10

Yang PF, Spanier N, Aldredge P, Shahid N, Coleman A, Lyons J, Langley JA (2023) Will free living microbial community composition drive biogeochemical responses to global change? Biogeochemistry 162:285–307

Yunusa M, Ibrahim M, Ibrahim A (2022) Use of biochar for enhance carbon sequestration to mitigate climate change and growth of maize in Sudan savanna zone of Nigeria. Braz J Sci 1(12):63–75. ISSN: 2764-3417

Zhang J, Tang X, Zhong S, Yin G, Gao Y, He X (2017) Recalcitrant carbon components in glomalin-related soil protein facilitate soil organic carbon preservation in tropical forests. Sci Rep 7:2391. https://doi.org/10.1038/s41598-017-02486-6

Zhang Z, Lian B, Hou W, Chen M, Li X, Li Y (2011) Bacillus mucilaginosus can capture atmospheric CO_2 by carbonic anhydrase. Afr J Microbiol Res 5:106–112

Chapter 7
Microbial Biotechnology in Alleviating Heavy Metal Contamination in Foods

Olubukola Blessing Oyetibo, Nwanneka Marie Akinyemi, Oyetayo Olaoluwa Adefiranye, Ganiyu Oladunjoye Oyetibo, and Olukayode Oladipo Amund

Abstract Foods are required substances by humans and animals for the growth and healthy living. Unfortunately, these substances are often contaminated with heavy metals (HMs) at various stages of processing from farm to table of human consumers. While there is no food products that is free from potential HMs contamination, vegetables, cereals, legumes, poultry, and sea foods top the list of foods through which humans are exposed to HMs intoxication. Upon contamination, such food materials are laced with quantities of HMs (notably Hg, Cd, Pb, As, Ni, Co etc.) at threshold, which render the food unwholesome and unsafe for consumption due to the toxicity of the metals thereof. Incidental ingestion of foods laden with HMs distorts human metabolomes with various clinical manifestations that mostly result to morbidity and even fatality. Producers of food, across all stages of supply chain, are saddled with responsibility of making available quality foods, where quality assurance with emphasis on food safety is foremost. Food products contaminated with HMs are subject to rejection leading to huge economic losses that require various remedial process that recovers food from the contaminants. On contrary to the inefficient physical and traditional approaches to free HMs from foods, biotechnological techniques are sought in this study as laudable technique to decontaminate

O. B. Oyetibo
Department of Microbiology, Lagos State University, Ojoo, Lagos State, Nigeria

N. M. Akinyemi · O. O. Adefiranye
Centre for Bioresources and Nanotechnology, University of Lagos, Akoka, Lagos State, Nigeria

G. O. Oyetibo (✉)
Centre for Bioresources and Nanotechnology, University of Lagos, Akoka, Lagos State, Nigeria

Geomicrobiology and Environmental Biotechnology Group, Department of Microbiology, Faculty of Science, University of Lagos, Akoka, Lagos State, Nigeria
e-mail: goyetibo@unilag.edu.ng

O. O. Amund
Department of Microbiology, Faculty of Science, University of Lagos, Akoka, Lagos State, Nigeria

O. O. Babalola et al. (eds.), *Ecofriendly Frontiers*,
https://doi.org/10.1007/978-3-031-98700-7_7

foods of bound HMs before any possible consumption by humans. The biological based techniques proposed in this study for alleviating HMS in contaminated foods include fermentation and probiotics premised on biocatalyses of microorganisms as evidenced with biosorption, bioaccumulation and various redox reactions of the contaminated HMs.

Keywords Bioaccumulation · Biosorption · Fermentation · Food safety · Metal-microbe interaction · Probiotics

7.1 Introduction

Food is referred to substances consumed to provide nutritional support regarding energy for life processes, macromolecules for growth, repair and maintenance for humans to maintain optimal health. Human food is classified into (i) plant-based (e.g., tubers, fruits, vegetables, and grains), (ii) animal-based (e.g., meat, dairy, eggs), (iii) fungi (e.g., mushrooms), and (iv) microorganism-based (e.g., yogurt, fermented foods, single cell proteins). Wholesomeness of food is crucial to human nutrition, food science, and public health. Often, the healthiness and safety of food is compromised upon exposure to toxic chemical substances, of which heavy metals (HMs) at low or elevated concentrations are gaining global attentions because they impact food chain security and human health (Machate 2023). Exposure of food to HMs leads to a type of food contamination that is hazardous to human health because of their toxic effects. HMs are generally earth elements with molecular weights, density and atomic number greater than 53 g, 6 g/cm^3 and 20, respectively (AMAP/UNEP 2013).

A wide range of elements fall into this category but those of interest to food contaminations, which are regarded as “bad metals”, include mercury (Hg) and lead (Pb) because they lack any metabolic and physiological merit to humans (AMAP/UNEP 2013; Oyetibo et al. 2017b, 2023). Equally regarded as “bad metals” but reported to exhibited metabolic functions in nature are cadmium (Cd) and chromium (Cr). Metabolic role of Cd in phytoplankton and protozoan involves functional substitution for cobalt (Co) and zinc (Zn) at the region where metal binds to carbonic anhydrase to facilitate the reversible reaction that forms bicarbonate and protons from water and carbon: $CO_2 + H_2O \leftrightarrow HCO_3^- + H^+$ (Lane and Morel 2000; Lane et al. 2005; Morel et al. 2020; Urbański et al. 2020). Also, Cr is essential to processes regulating insulin activity, metabolic syndrome, and cardiovascular disease Hummel et al. 2007; Gossa Al-Saadde et al. 2023). However, some other HMs (including iron [Fe], copper [Cu], manganese [Mn], molybdenum [Mo], zinc [Zn] etc.) otherwise called “trace elements” or micronutrients are with established metabolic functions. This category of HMs is required at low but critical concentrations to drive redox processes, osmotic pressure regulation, and enzymes regulations (co-factors) that are critical to healthy growth and reproduction in living organisms (AMAP/UNEP 2013). Equally important elements in food contamination are arsenic (As) and selenium (Se) that are classified as metalloids, but often grouped with toxic metals as HMs whenever the need to discuss ecotoxicology of metals arises. Thus, As and Se will also be treated as toxic HMs in this essay on food contamination with HMs. In all, all these metals are regarded toxic when found in such concentrations above the acceptable limits in food, from where they are biomagnified and exacerbate various toxicological effects on animals and humans. An attempt to safe food from HMs contamination and their consequential toxicities on consumers involves concerted strategies of prevention, mitigation, and remediation of which microorganisms and microbial biocatalyses are critical.

7.2 Sources of Heavy Metal Contamination in Food

The sources by which HMs get introduced to foods vary with respect to the type of HMs and particular food product. According to Nerín et al. (2016), the sources include: (i) Environmental contamination of the external part of raw foods; (ii) Raw materials transportation to the factory prior to processing; (iii) Application of conditioning processes such as storage, preheating, disinfection, cleaning and sterilisation steps; (iv) Heating procedures including boiling, cooking, baking, frying or enrichment with other ingredients in an oven at elevated temperature or in a reactor; (v) Packaging methods; (vi) Transport of packaged food; and (vii) Storage and distribution of packaged food. This phenomenon has expanded due to enormous anthropogenic activities as reflected in various industrialisation, urbanisation, robust waste management, and enhanced farming practices (Desoky et al. 2020). The continual reliance on chemical fertilisers, irrigation with wastewater sludge and composting during farming enables addition of elevated concentrations of HMs to agricultural soils, which adversely impact the soil health (Kumar 2013; Alzahrani and Rady 2019; Egbe et al. 2021; Hu and Seyfferth 2021). For example, Pb^{2+} contamination in foods can occur through the use of lead-containing pesticides (Desoky et al. 2019), Pb-based solder in food cans (Huang et al. 2018); contaminated water used for irrigation (Oyetibo et al. 2019a, 2023), and petroleum hydrocarbons residues in soil (Oyetibo et al. 2017b). Cd concentrations were observed to be usually less than 0.5 mg/kg in normal soil, but can up-scale to 3.0 mg/kg (Nazar et al. 2012) or higher (Egbe et al. 2021) depending upon the soil origin-based materials. Pb can accumulate in foods like leafy vegetables, root crops, fruits, grains, and seafood (Li et al. 2024). Hg contamination in foods mainly originates from industrial sources such as power plants that are fired with coal, mining, and mercury-containing products. Arsenic contamination of foods can result from natural sources, industrial activities, and application of As-containing pesticides and herbicides in agricultural practices. Moreover, healthy crops with no known history of HMs contamination could be exposed to HMs during various domestic and/or industrial processing.

7.3 Heavy Metals Toxicity

HMs provokes alteration in forms, structures and physiologies of soil microbiome with consequential soil degradation, negative impacts on several physiologies that hamper crop performances, growth and productivity (Majer et al. 2002; Oyetibo et al. 2019a, 2023; Egbe et al. 2021).

7.4 Consequences of Toxic Metals to the Contaminated Food Crops

Inadvertently, crops cultivated in contaminated soils uptake parts of the HMs from the soil-food crop subsystems and accumulate them intracellularly in their tissues over times. The toxicity of HMs in crop tissues alters the wholesomeness of food crops, posing risks to animal and human consumers. Cd accumulates primarily in grains, leafy vegetables, root crops, and shellfish. Accumulation of Cd^{2+} in plant tissues alters their biochemistry, physiology, and structure, making the crops/vegetables of such plants unwholesome (Rady et al. 2019). Specifically, Cd exposure increases the endogenous abscisic acid (ABA) concentrations that perturb the growth and cellular metabolism of rice (Kim et al. 2014). Crops under Cd^{2+} toxicity, like other non-redox HMs (e.g., Ni, Al, Zn, Hg, and As), display oxidative stress with free radicals' overproduction (Yadav 2010) upon depleting the glutathione (GSH) levels. The mechanisms used by non-redox HMs to exert oxidative stress in food crops include: (i) binding to sulfhydryl (-SH) groups to inhibit oxidative enzymes, (ii) increase in NADPH oxidases levels, (iii) lipid per-oxidation induction, (iv) elevation of malondialdehyde (MDA) levels due to fatty acid de-esterification, and (v) regulation of anti-oxidant enzymes (including superoxide dismutase, SOD; peroxidase, POD; catalase, CAT; and ascorbate glutathione, AsA-GSH) cycles (Bücker-Neto et al. 2017; AbdElgawad et al. 2020; Martínez-Castillo et al. 2022). Cd stress elicits lipoxygenase in seedlings leading to a decreased SOD and CAT activity causing reduced germination, chlorosis, necrosis, impairment of genetic materials, cell death/apoptosis due to ROS generation, lipid peroxidation, disruption of photosynthetic apparatus, diminished chlorophyll and proline, reduced total biomass and crop yields, reduced nutrient use efficiency of phosphorus, decline cell elongation, repressed stomatal opening, induced stomatal closure and altered redox status (Thakur et al. 2022). On the contrary, redox HMs (e.g., Cr, Cu, Mn, and Fe) undergoes redox-cycling reactions where their toxicity spurs potential oxidative stress on food crops via Haber-Weiss and Fenton reactions, for example (Bücker-Neto et al. 2017). In *Citrus aurantium* (sour oranges), Pb exposure causes disruption of photosynthetic apparatus, integrity of membrane, mineral nutrition, and elevated lipid peroxidation upon oxidative damage to the tissues (Giannakoula et al. 2021).

7.5 Consequences of Foods Contaminated with Heavy Metals on Consumers

HMs are non-biodegradable, and once absorbed by certain crops and aquatic animals, they remain stable in their tissues for a longer period to negatively influence both the safety of the food supply and the public health (WHO 2019; Turanlı and Gedik 2021; Ahmed et al. 2023; Sharma et al. 2023, 2024). Elevated concentrations of HMs have been reported in various foods including algal formulations, crustaceans, cereals/grains, cocoa-based products, meat, poultry organ meats, oilseeds, seaweeds, water molluscs, vegetables etc. (Authority 2012; Mahmood et al. 2016; Aziz et al. 2021; Turanlı and Gedik 2021). Food crops like *Cajanus cajan* (pigeon pea), *Zea mays* (maize), *Oryza sativa* (rice), *Triticum aestivum* (wheat), *Hordeum vulgare* (barley), *Brassica napus* (rapeseed), *Vigna radiata* (mung bean), *Lens culinaris/Lens esculenta* (lentil), *Solanum lycopersicum* (tomato) fruit, *Glycine max* (soybean) and *Nigella sativa* (black cumin) are well known as source of HMs intake for humans upon retention of potent HMs in the tissues (Hu et al. 2014, 2016; Zou et al. 2021; Shamshir et al. 2022; Thakur et al. 2022; Khan et al. 2023). In contaminated food crops, rice for example, ionic HMs like Cd^{2+} bind with carboxyls in glutamic acid and aspartic acid, imidazolyls in histidine and sulfydryls in cysteine of the crop protein (Huo et al. 2016; Slope and Peacock 2016). Driven by electron donation, functional groups are completely deprotonated at neutral condition (pH 7) to produce their respective anionic forms, which coordinate binding with various HMs ions. These functional groups include (i) oxygen atom in carboxyl group (pKRa = 4.60), (ii) the three nitrogen atoms of histidine side chain in imidazole group (pKRa = 7.00), (iii) the primary amine group at the *N*-terminal of the main chain of the peptide chain that provides electrons via the nitrogen atom (α-NH_2, pKRa = 8.9), the amino group at the end of the lysine side chain (ε-NH_2, pKRa = 10.79), and the disulfide bond by cysteine side chain sulfhydryl group (pKRa = 8.8). Furthermore, disulfide bridges, hydrophobic reactions, hydrogen bonding that spur protein folding serve as excellent ligands for the complexation of metal ions via multidentate coordination.

The content of HMs in marine and freshwater fish is commonly very high, making fish a vital bioindicator for HMs hazard assessment of water pollution (Turanlı and Gedik 2021). Seafood and fish are the primary sources by which humans get exposed to Hg, particularly via predatory fish species that bioaccumulate Hg through the food chain. Rice, grains, vegetables, and seafood can contain As, Cd (Hu et al. 2014, 2016) and other HMs, with rice being a significant dietary source in regions where it is a staple food. Further, the kidney and liver of mammals are organs of choice for high levels of Cd when mammals are fed with Cd-rich diets and sea foods like crustaceans, scallops, oysters, and mussels. Other food and food products that have been reported to contribute to dietary intake of HMs in humans include milk (Rofrano et al. 2023), All HMs are poisonous irrespective of their metabolic importance, causing dangerous morbidity and mortality in animals and humans once their acceptable concentration is exceeded in food (Ray and Ray 2009). The health implications of HMs exposure are influenced by certain factors including the type, concentration, exposure duration, and susceptibility of individual metal. HMs get access to the body of animals/humans via the alimentary system despite the innate barriers imposed by the gastrointestinal tract anatomy (Bonelli et al. 2019). Chronic exposure to HMs contained in foods leads to various pernicious health effects in form of neurological disorders, kidney damage, cancer, cardiovascular diseases, and developmental delays in children.

Thus, specific effects of some HMs of interest to food safety in humans are enumerated herewith and further highlighted in Table 7.1:

- Pb toxicity is broad spectrum, affecting almost all organs but mostly affect the nervous system. Pb intoxication upon consuming contaminated foodstuff manifests as induced miscarriage, intelligence deficiency, bone pre-maturation, articulation weakness, increased blood pressure, renal dysfunction and allergic; immune-modulatory, inflammatory, oxidative, respiratory and cardiovascular diseases; neurotoxicity, blood-brain disorder, gastrointestinal and cancer in humans (Mani et al. 2019; Steenland and Boffetta 2000). Pb hampers the intellectual performance and cognitive development of children (Flora et al. 2006), and even deaths as children in their hundreds died from mining-associated Pb poisoning in Northern Nigeria in 2010 (Dooyema et al. 2012). Thousands of the children in Zamfara State alone have dangerous concentrations of Pb in their blood due to consumption of water and food processed with the mining-impacted water (CDC 2024). Moreover, chronic exposure to Pb causes anaemia, hypertension and reduced fertility in males. Whilst in pregnant women, high load of Pb in blood triggers premature birth with characteristic low birth weight.
- Cd poisoning causes cancer, poor reproductive capacity, kidney function impairment, hepatic dysfunction, hypertension, and ultimately damage to organs (like liver, brain, and kidney) and tissues (like bone etc.) (Ankit et al. 2021; Zhang et al. 2021; Ihunwo and Ibezim-Ezeani 2022; Wang et al. 2022). The "Itai-Itai disease" firstly reported in Japan in the 1950s was occasioned because of the extended consumption of rice contaminated with Cd (Naksen et al. 2022). In 2015, Cd was rated a group 1 carcinogen because an estimated more than 2000 deaths worldwide were linked to foodborne Cd intoxication (Gibb et al. 2019). Cd in contaminated food enters the food chain (Genchi et al. 2020), distressing the homeostasis of organs, tissues, and the entirety of animals and humans (Rost-Roszkowska et al. 2022). Disorders triggered by Cd toxicity begin at cellular level where physiological functions of cell organelles are impaired, activating either degenerative or repair processes (Lipovšek et al. 2022). Cd induces neurotoxicity and damages to the brain when it uses olfactory pathway or alters the permeability of the blood–brain barrier to enter the central nervous system (CNS) (Mohajeri et al. 2017), thereby affecting the metabolism, release, and reuptake of various neurotransmitters (Gumilar et al. 2019; Lim et al. 2019). Cd released from the terminals of amygdala neurons affects the degree and balance of excitation inhibition in synaptic neurotransmission, causing disordering of the CNS in form of olfactory damage. Also linked to Cd toxicity is dysfunction of renal

Table 7.1 Sources and health implications of heavy metals in contaminated foods

Heavy metal	Source	Food crop	Health implication	References
Hg	Natural phenomena (volcanic eruption, earthquakes, tsunami, flood), human activities (mining, coal-burning power plants, biomass burning, chlor-alkali industry)	Fish, seaweed, and other sea foods	Cardiovascular disease risk factor Reproductive and developmental toxicity Neurotoxicity, nephrotoxicity Impairment of natural defence	Genchi et al. (2017)
As	Natural phenomena (earth crust), man-made (industries dealing with: glass making, pesticides, pharmaceutics)	Portable water, rice grains, vegetables	Inorganic forms of as induced cancer of the lung, skin, and bladder Risk factor for the lesions of the skin, and diseases related to cardiovascular, developmental toxicity, and type II diabetes. Toxic to neurons Increase disruption of glucose metabolism	Cubadda et al. (2017), Silano and Silano (2017), Upadhyay et al. (2019)
Pb	Anthropogenic (urbanization, agricultural practices), natural phenomena (volcanic eruption, weathering of rocks and soils)	Sugar beet, orange, pomegranate, grapes lemon,strawberry,peach esfenugreek, rosemary	Neurotoxic, reduce cognitive abilities, reduce intelligence at formative age Damage the circulatory system and kidneys Reduce fertility Poorly develop foetal during pregnancy	Kumar et al. (2018, 2019)
Cd	Natural phenomena, anthropogenic activities (phosphate containing fertilizers, production of equipment, electroplating of utensils, and plastic manufacturing)	Seafood, meat, cereals, vegetables, grains, tubers, starchy roots	Osteoporosis causative factors, aetiologies of rheumatoid arthritis, and osteoarthritis Disruptor of endocrine Human carcinogenicity	Kumar et al. (2019), Reyes-Hinojosa et al. (2019), Schaefer et al. (2020).

tubular manifesting as proteinuria, diabetes, and aminoaciduria, which can lead to hypercalciuria and the formation of kidney stones since calcium metabolism is affected (Rinaldi et al. 2017). Chronic Cd poisoning up-regulated alanine aminotransferase (ALT), aspartate aminotransferase (AST), and alkaline phosphatase (ALP) that are markers of liver injury in the blood, liver dysfunction and hepatocyte necrosis occur as Cd^{2+} ions traverse the hepatocyte membrane by binding to voltage-gated Ca^{2+} channels and Fe^{2+}/Zn^{2+} transporters (Nordberg et al. 2018; Tang et al. 2022).

- Chromium usually occur as Cr(VI) and Cr(III) in food items. Cr(VI) (classified as a Type A human carcinogen or mutagen) has cytotoxicity that is several thousand times greater than that of Cr(III) because of its high reactivity, mobility, and solubility (Zhang et al. 2024). Cr intoxication often leads to severe cardiovascular, respiratory, renal, gastrointestinal, haematological, hepatic, and neurological effects that may eventually be lethal. Asthma, internal bleeding, atopic dermatitis, nausea, and damage to the kidneys in humans have also been linked to Cr exposure (Mpouras et al. 2021).
- Arsenic (As) is a toxic metalloid with consequential risk factor to public health in its inorganic form. The dietary intake of As-rich food is carcinogenic and induces neuro-cardiovascular, respiratory, gastrointestinal and liver diseases. Other ailment associated with as intoxication is diabetes, anaemia, thrombocytopenia effects and leucopenia, genotoxicity and cytotoxicity effects, numerous enzymes denaturation, decreased ATP production, increased pro-inflammatory agents, stillbirth, spontaneous absorption, preterm birth, male sexual impotence affecting spermatogenesis, lower weight, oxidative stress, lipid peroxidation, and other dysfunctions (Machate 2023).
- Hg in food intake causes ataxia, neurological diseases, cognitive damage, weakness, chewing and deglutition disorder, cellular metabolism unbalance, oxidative stress, kidney dysfunction (proximal tubules deleterious), liver damage, non-alcoholic fatty liver disease, and inflammation diseases.

7.6 Methods of Heavy Metal Detection in Contaminated Food

The major approaches applied to analysing HMs contained in food items include various modifications to spectrophotometry upon quantification of atomised chemicals via graphite furnace atomic absorption spectrometry (GFAAS), flame atomic absorption spectrometry (FAAS), cold vapor atomic absorption spectrometry (CVAAS), inductively coupled plasma mass spectrometry (ICP-MS), and inductively coupled plasma atomic emission spectrometry (ICP-AES) (Zukowska and Biziuk 2008). These techniques revolve round the use of highly expensive equipment, which are not handy and require expertise to operate. Thus, detection of HMs in food available to poor and upcoming economies where HMs toxicities are prevalent becomes mostly difficult and impossible. Cheap, fast, and convenient systems of detecting HMs are essential in attempt to limit the impact of HMs toxicity on poor communities that remains the most affected population. This necessitated the need to develop biotechnological techniques for detecting HMs, named biosensors. Biosensors are sensitive, precise, rapid, non-destructive and cheap biotechnologies (Nevetha et al. 2021). The organic constituents of biosensors that are responsible for the detection of HMs include enzymes, antibodies, aptamers, and whole cells. The transduction impulse is based on electrochemistry (amperimetry, voltammetry, cyclic voltammetry, potentiometry, conductometry), optical (fluorescence, surface plasmon resonance [SPR], chemiluminescence) or colorimetry (colour changes). Chen et al. (2019) developed a multiplex biosensor that simultaneously detects chemical toxicants like Pb^{2+}, kanamycin, and estradiol in food.

The whole-cell biosensors make use of immobilized whole-cells designed for use in the field, where reporter genes are designed to detect chemical contaminants in food crops. Usually, the sensitivity and specific adaptations of biosensors need optimised for practical use in the field. The design and building sensitive bacterial biosensors as shown in Fig. 7.1 involves (i) identifying promoters that are potentially inducible by toxicants of interest in desired operon *in silico* within the bacterial genomes, (ii) using reverse transcription polymerase chain reaction (RT-PCR) to confirm the gene encoding toxicant (HMs) regulation, and (iii) signal transducer. Magnetotactic bacteria have been used as interesting cellular hosts models for bacterial biosensors, relying on their intrinsic magnetism for easy sedimentation and entrapment towards a magnetic field to boost their HMs-response signals. Biosensors based on transcriptional fusion between the bacterial luciferase *luxCDABE* operon

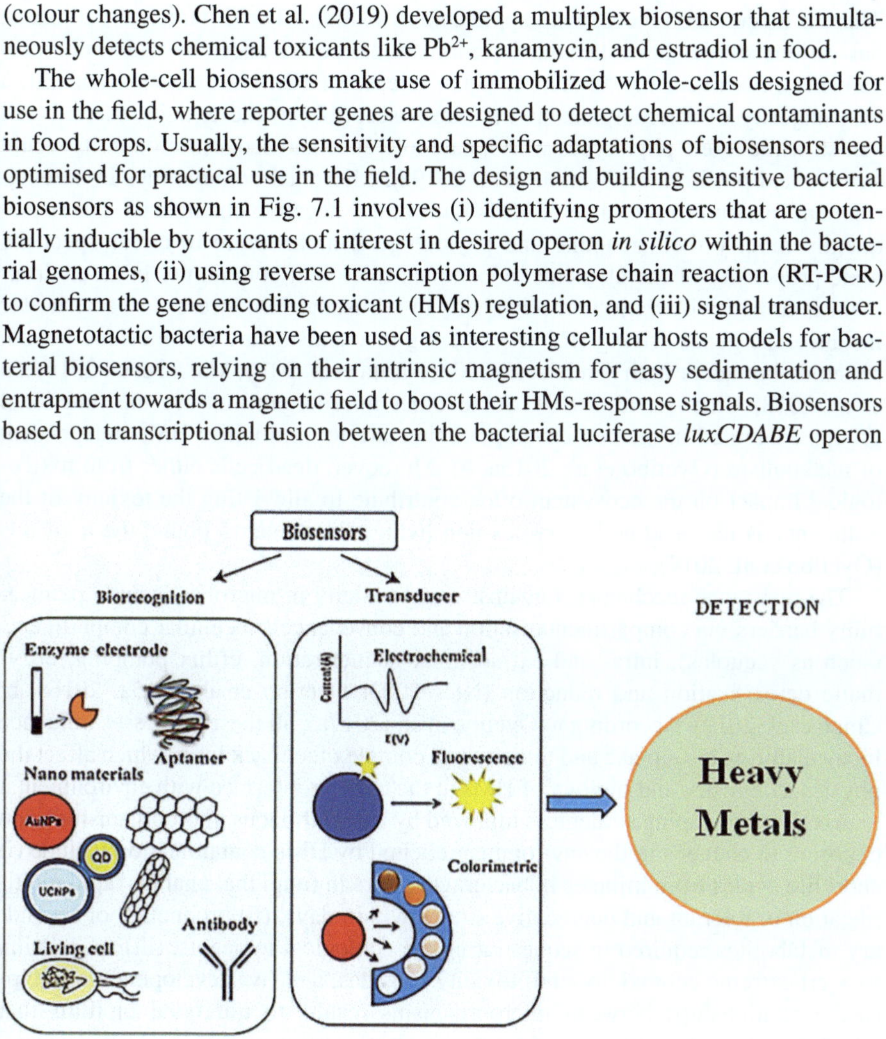

Fig. 7.1 Illustration of various biosensors that is applied to heavy metal detection in food. (Adapted from Dieudonné et al. 2020)

and the arsenic-inducible promoters were reported to produce an element-specific response of 0.5 μM detection limit of an arsenite within 30 min (Dieudonné et al. 2020). Other specific cases where biosensors were deployed to detect HMs in foods are summarised in Table 7.2.

7.7 Microbial Interactions with Heavy Metals that Facilitate Remediation of Contaminated Food

There is no common mechanism applicable to abating all environmental pollutants, but various studies have discussed combinational contrivances by which microorganisms circumvent eco-toxicities. Upon release of HMs into an environment, a vast majority of microorganisms that are susceptible to the toxicity exerted will vanish while a few phylotypes with innate tolerance and sequestration traits would survive and proliferate (Oyetibo et al. 2019a). Microbial communities in an environment respond to presence of HMs with initial negative chemotaxis, but subsequently modify their genomic and physiological architecture to circumspect the toxic propensity of the elements. A complex microbial community in an environment is armed with various mechanisms to withstand HM stress. In bacteria, for instance, the genes driving HM sequestering mechanisms are chromosome-bound, plasmids-bound, or both (Endo et al. 2002; Narita et al. 2010; Matsui et al. 2016; Xu et al. 2018; Chien et al. 2019). Some other microorganisms use the functional groups contained in the cell-envelop architecture to immobilise HMs independent of metabolism (Oyetibo et al. 2015a, b). Moreover, dead cells either from toxicological impact on the ecosystem often contribute to alleviating the toxicity of the pollutants as the dead cells serve as non-living adsorbents to bound the toxicants (Oyetibo et al. 2014).

The resistance mechanisms against metal toxicity in microbes include permeability barriers via compartmentalisation and conveyance subcellular compartments (such as vacuoles), intra- and extracellular sequestration, efflux pumping, enzymatic detoxification and reduction (Nies 2003; Oyetibo et al. 2015a, 2016a, b; Chien et al. 2019). According to Oyetibo et al. (2017a), all these processes influence bioavailability, bio-uptake and toxicity in a complex feedback loop, which affect the physics, chemistry and biology of HMs at the cellular interface with environmental matrix. Further coping strategies employed by autochthonous microorganisms upon exposure to changes in the environment elicited by HMs contamination include (i) short life cycles from minutes in bacteria to hours in fungi that enables rapid multiplication of tolerant and degradative strains within days, (ii) production of secondary metabolites required in sequestrating the ecological toxicants, (iii) sporulation to avert extreme ecotoxicity until toxicity subsides, and (iv) development of reciprocate relationships between microorganisms results to microbial biofilms that wield coping strategies to perturbations via slower diffusion rates and toxicants co-metabolism (Mangwani et al. 2017).

Table 7.2 Various biosensors that have been deployed to detecting heavy metals in foods and water used in food processing

Organic compound	Heavy metal ions	Limits of detection	Electrode material	Method	Food sample	References
Single-stranded DNA	Pb^{2+},	1.0×10^{-10} mol/L,	Stationary mercury-film electrode	Amperometric	Milk, water	Babkina and Ulakhovich (2004)
	Cd^{2+},	1.0×10^{-9} mol/L,				
	Fe^{3+}	1.0×10^{-7} mol/L				
Fluorescence-labelled DNA containing T-T mismatch structure	Hg^{2+}	1.2 nm	Fibre optical sensor	Optical	Water	Long et al. (2013)
Alkaline phosphatase	Cd^{2+}, Zn^{2+}, Co^{2+}, Ni^{2+}, Pb^{2+}	0.5–40 mg/L	Gold electrodes	Conductometric	Water	Berezhetskyy et al. (2008)
Glucose oxidase	Cd^{2+}, Cu^{2+}, Pb^{2+}, Zn^{2+}	1–9 µg/L	Carbon film electrode	Conductometric	Milk	Ghica and Brett (2008)
	Hg^{2+}, Ag^{+}, Cu^{2+}, Cd^{2+}, Fe^{3+}, Co^{2+}, Ni^{2+}, Pb^{2+}, Zn^{2+}, Mn^{2+},Cr^{3+},andCrO_4^{2-}	0.05–48 µm	A Pt-wire used as counter electrode + a saturated calomel electrode (SCE) as reference	Amperometric	–	Guascito et al. (2008)
DNA	Hg^{2+}	0.92 nm	Water soluble graphene oxide (GO) sheets plus functionalized ssDNA aptamer	Fluorescent, colorimetric	–	Li et al. (2013)
Urease	Pb^{2+}	9.66 µm	Nylon membrane with pH indicator	Colorimetric and potentiometric	Milk	Kaur et al. (2014)
	Cd^{2+}, then Cu^{2+}, Pb^{2+}	0.1–10 mm	Screen-printed electrode	Conductometric	Juices	Ilangovan et al. (2006)
	Cu^{2+}, Cd^{2+}, Zn^{2+}, Ni^{2+}, Pb^{2+}, Cr^{3+}	10–300 µm	Glass pad	Optical	Water	Tsai et al. (2003)

(continued)

Table 7.2 (continued)

Organic compound	Heavy metal ions	Limits of detection	Electrode material	Method	Food sample	References
Bacillus sphaericus	Lead ions	2.4×10^{-3} nm	Carbon based electrode	Potentiometric	Milk	Verma et al. (2011)
	Ni^{2+}	0.03–0.68 nm	NH_4^+ selective electrode		Wheat flour	Verma and Singh (2006)
Invertase, mutarotase, glucose oxidase	Ag^+, Hg^{2+}	100 nm, 25 nm	Ceramic pad	Conductometric	–	Soldatkin et al. (2012)
Horseradish peroxidase (HRP)	Cr^{3+}	0.27–1.15 μm,	Carbon film electrodes, carbon film electrodes, + electropolymerized neutral red	Amperometric and voltammetric	–	Attar et al. (2014)
	Cr^{6+}	0.09–1.6 μm				

7.8 Specific Microbial Biocatalyses Leading to Bio-Removal of Heavy Metals

Microbial communities relate with bioavailable HMs via development of tolerance or outright resistance physiologies leading to changing the oxidation state of the metals (Fig. 7.2). It is noteworthy to establish that HMs cannot be degraded but microbes can depend on them for electrons either as electron donor or electron acceptor. In both ways, bioavailable HMs are sequestered into innocuous forms and thereby removed from exerting ecotoxicity in the environments. Processes involving bio-removal of HMs in intrinsic and extrinsic food matrixes include extracellular detoxification, intracellular detoxification, compartmentalisation, altered permeability, and efflux pumping.

7.8.1 *Extracellular Detoxification of Heavy Metals*

Some microorganisms produce biomolecules, which chelate and complex HMs ions or alter their oxidation states to stem their toxicity. For example, *Rhodospirillum rubrum* decreases Cd bioaccumulation as *Yarrowia* species apparently volatilises Hg^{2+} via the production of extracellular ligand (Smiejan et al. 2003; Oyetibo et al.

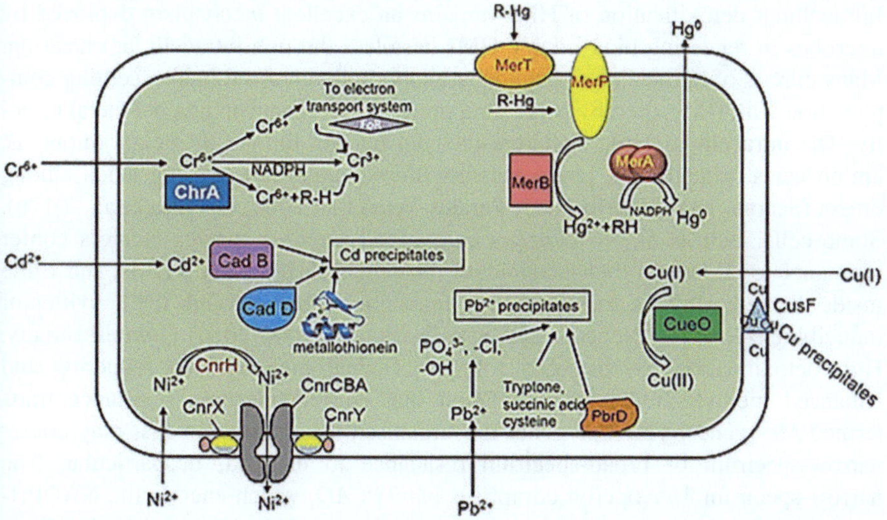

Fig. 7.2 General illustration of the genetic strategies that drive bacterial interaction with heavy metals, leading to bacterial-based removal of contaminating heavy metals from food (Modelled from Oyetibo et al. 2015a, 2016b, 2019b; Kabiraj et al. 2022; Cai et al. 2023). These strategies of heavy metals bio-removal is ascribed to efflux systems, genes encoding metal-resistance and detoxification, precipitation, biopolymer chelation, aromatic ring fission, biosorption, or bioaccumulation

2015a, 2016b, 2019b). The oxalic acid excreted by *Penicillium* has been found to complex with Cu outside the hyphae (Nies 2003; Schmidt et al. 2005). Algae, bacteria and fungi produce heterogeneous extracellular polymeric substances (EPS) capable for circumventing HMs toxicity within cell vicinity (Worms et al. 2006; Oyetibo et al. 2016b). Moreover, *Emiliana huxleyi* secreted thiol-rich exudates and amino acids to form strong complex with HMs in impacted matrix (Dupont and Ahner 2005). Chelators produced by some microbes decrease bio-availability, uptake and toxicity of HMs, and consequently reduced production of metal binding ligand (Worms et al. 2006; Oyetibo et al. 2016b, 2019b;Cai et al. 2023). Most arsenite-resistant microorganisms oxidise As(III) that act as electron acceptors, through the secreted S-adenosylmethionine methyl-transferase (ArsM) to methylate As (Kabiraj et al. 2022). For example, *Thiomonas arsenivorans* used arsenite oxidase (*AoxB*) at its outer membrane to meet its energy need through efficient oxidation of dissolved As(III) and grew autotrophically (Cai et al. 2022). Mn-oxides and Fe-oxides produced during microbial oxidation of Fe-Mn have been reported to form complexes with cells, inhibiting As(III) adsorption (Battaglia-Brunet et al. 2021) and effectively oxidized As(III) to As(V) leading to complete arsenite detoxification (Zhao et al. 2022) in a phenomenon likened to co-metabolism.

7.8.2 Intracellular Detoxification of Heavy Metals

Intracellular detoxification of HMs remains an excellent mechanism deployed by microbes in rendering bioavailable HMs harmless through intracellular chelation. Many classes of intracellular chelators are reportedly involved in intracellular complexation with HMs, thereby conferring protection mechanism against metal toxicity. The intracellular HMs chelators include, but not limited to metallothioneins, amino acids, glutathione, phytochelatins, thioredoxins and organic acids among others (Scarano and Morelli 2003; Perales-Vela et al. 2006; Oyetibo et al. 2017b). Some cells such as *Staphylococcus aureus* and *Bacillus* species etcetera confer resistance to HMs with their plasmid, or the cells directly take up HMs, and while inside the cell, HM are transformed to innocuous forms through the activities of inducible enzymes like reductase and lyase (Narita et al. 2002, 2010). Quintessentially, Hg^{2+} detoxification by microbes follows efficient mercuric ion reduction and enhanced methylmercury demethylation that confers mercury resistance traits termed *Mer* genes. The *Mer* genes are organised into an operon that may confer narrow-spectrum or broad-spectrum resistance to bacteria, in particular. The narrow-spectrum *Mer* operon comprises *merTPCAD*, which encode the NADPH-dependent flavin oxidoreductase, mercuric reductase (*MerA*) that detoxify Hg^{2+}; two inner-membrane proteins (*MerT* and *MerC*), and a periplasmic protein (*MerP*) that jointly drive Hg^{2+} uptake; while regulatory proteins *MerR* and *MerD* are for Hg detection (Barkay et al. 2003; Chen et al. 2008; Matsui et al. 2016; Chien et al. 2019). Unlike multicellular organisms, some bacterial cells ideally use narrow-spectrum *mer* proteins to reduce soluble intracellular Hg^{2+} to volatile Hg^0 that escapes into the atmosphere. However, some Hg-resistant bacteria (classified as broad spectrum) possess organomercuryl lyase (*MerB*) with *MerA* in the *Mer* operon to form a more complex *merRTPCAGBDE* operon. Thus, broad-spectrum mercury-resistance operon use *MerB* to synthesize lyase protein that cleaves Hg^{2+} from organic mercury (e.g., NH_3-Hg), and *MerA* to produce reductase protein that reduce the cleaved Hg^{2+} to Hg^0 (Barkay et al. 2003; Narita et al. 2010; Matsui et al. 2016). It is noteworthy that periplasmic protein *MerG* provides resistance against organomercurial in some Gram-negative bacteria that lack *MerB*, making such bacterial strain appeared broad-spectrum mercury-resistant (Boyd and Barkay 2012). As such, presence of *MerA* along with *MerG* exhibits the physiology of broad-spectrum mercury bio-detoxification. Evidence of Hg^{2+} reduction have also been reported in fungi (Oyetibo et al. 2015b), although the gene responsible for this reduction in fungi have not been elucidated. Similar enzymatic reductions of chromium, arsenic and selenium species also occur in some bacteria.

7.8.3 Compartmentalisation of Heavy Metals

Compartmentalisation of HMs has been deployed by biota to stem toxicity of excess metal ions in the cytosol (Escher and Hermens 2004). This involves HMs being complexed by phytochelatins, polyphosphate bodies (Pan-Hou et al. 2002; Jensen et al. 2003; Kobayashi et al. 2005), peptides (Nies 2003), and organic acids (el-Enany and Issa 2001) to form organometallic complexes that are confined to cellular internal compartments (vacuoles and chloroplasts) by transporter proteins. This mechanism of HMs detoxification via compartmentalisation has variously been observed with *Yarrowia* (Oyetibo et al. 2016a), *Chlamydomonas acidophila* (Dean et al. 2019), *Schizosaccharomyces pombe* (Perales-Vela et al. 2006) and so on, where HMs-chelator complexes are internalised using Mg-ATP energised transport system.

7.8.4 Altered Permeability of Heavy Metals Bio-Removal

This occurs when cells uptake HMs without gaining entrance into the cytoplasm. HMs including Cd and Hg have been reportedly denied entrance into cell cytoplasm upon alteration to cell membrane permeability aided by plasmid bound trait, and thus the HMs were retained within cell wall (Wu et al. 2006a; Kumar and Bharadvaja 2020). Also, *Micrococcus* and *Azotobacter* removed Pb through rapid extracellular immobilisation within the cell wall and prevent transportation across the cell membrane but accumulate the complexed HMs on the cell wall. Removal of HMs from cell and cell environment may also occur through plasmid conferred resistance to Hg upon methylation to form volatile methyl mercury by methyl transferases as found in *Escherichia coli* and *Pseudomonas aeruginosa* (Narita et al. 2002, 2010).

Moreover, dissolution and eventual oxidation of volatile Hg^0 to Hg^{2+} through electron donation from sulphur molecules and forms insoluble HgS nanoparticles that cannot cross into the cytoplasm was reported for *Yarrowia* spp. (Oyetibo et al. 2019b). Although, the transport of organomercuryl compound is poorly understood, bacteria resist phenyl-mercury through altered permeability mechanism that was reportedly actualised by polypeptides coded by *merG* situated amid *merA* and *merB* of the *mer* operon (Boyd and Barkay 2012). The *merG* biomolecule was assumed to bind phenyl mercury, altering its permeability and prevent it from accessing the cytoplasm of the bacteria (Boyd and Barkay 2012).

7.8.5 *Efflux Pump for Heavy Metals Bio-Removal*

Unlike eukaryotes, prokaryotes employ ATP-consuming efflux of HMs to detoxify such metals (Liu et al. 2002; Nies 2003; Schmidt et al. 2005). Three main classes of transporters well-known as P-type ATPases (Groot et al. 2005; Wu et al. 2006a, b), zinc regulated transporter/iron-regulated transporter (*ZRT/IRT*), and macrophage proteins (*Nramp*) proton symporter are linked to natural resistance of HMs. P-type ATPases appeared to facilitate both the internalisation and detoxification of trace metals (Wu et al. 2006a, b). While the *ZRT/IRT1* proteins are involved in the uptake and homeostasis of Zn and Fe (Chung and Bird 2019), and also permit Cd into cells. The *Nramp* family of proteins support the cellular uptake of Cd and Pb (Agranoff et al. 2005). HMs transport P-type ATPases (or CPx-ATPases), located at the sixth transmembrane helix, form putative cation transduction channel through which monovalent and divalent HMs cross the membrane of microbes (Liu et al. 2002). In *Staphylococcus aureus,* for instance, the resistance to Zn^{2+} and Co^{2+} is due to *zntR,* and *zntA* that encodes transmembrane structural protein for driving efflux of Zn^{2+} and Co^{2+} ions, and a regulatory biomolecule that probably auto-regulates expression of itself (Chung and Bird 2019). HMs resistance in the mutants of bacteria could be proportional to overexpression of the HMs efflux pump *CzcCBA* and its cognate two-component regulator genes, *czcR-czcS* (Perron et al. 2004; Chung and Bird 2019). Microbes increase production or affinity of the transporters in adapting to low metal concentration (Worms et al. 2006; Chung and Bird 2019). Some of the efflux pumps systems responsible for HMs sequestrations are presented in Table 7.3.

7.9 Remedial Strategies to Mitigate Heavy Metal Contamination of Food

Regulatory agencies worldwide have constituted stringent limits for the HMs concentrations permissible in food and food products that need followed by food manufacturers (Balali-Mood et al. 2021). Industries dealing with food production face

Table 7.3 Some varieties of efflux pump systems that encode resistance to heavy metal in bacteria

s/n	Efflux pump systems	Metal ions
1	czrCBA	Zn^{2+}, Cd^{2+},
2.	nczCBA	Co^{2+}, Ni^{2+}
3.	CznABC	Zn^{2+}, Cd^{2+}, Ni^{2+}
4.	P-type ATPase	Pb^{2+}, Zn^{2+}, Cd^{2+}
5.	CusCBA	Cu^{2+}, Ag^{2+}
6.	arsK genes	As^{3+}
7.	PbrT and PbrA	Pb^{2+}
8.	P1B-4-Type ATPase (PmtA)	Fe^{2+}
9.	CzcCBA	Zn^{2+}, Cd^{2+}
10.	CusCFBA	Cu^{+}, Ag^{+}
11.	P-type ATPase (CadA), CzcCBA	Zn^{2+}
12	P1B-4-ATPase (CzcP)	Cd^{2+}, Zn^{2+}, Co^{2+}
13.	P1B-1 ATPase (CopA)	Cu^{2+}
14.	P1B-type ATPases (ZntA), CopA1b	Zn^{2+}, Cd^{2+}, Pb^{2+}
15.	P1B-type ATPases (PbrA)	Zn^{2+}, Cd^{2+}, Pb^{2+}

enormous challenges to ensure food safety from HMs intoxication via adherence to strict regulatory standards for HMs levels in foods. Aside safety issues, the incidence of HMs in food triggers consumer apprehensions, loss of confidence on the industry, and reputation damage to food brands, that altogether results in huge financial deficit for the food industry (Handford et al. 2016). To overcome these obstacles, all stakeholders of food business do cooperate to ensure food safety and sustainability. Once the sources at which foods get exposed to HMs are identified, it is possible to consider the best approach(es) to remove the HMs leading to improving food safety. Remediation of HMs contaminated foods and food products focuses on reducing or outright removal of HMs from foods, or the environments where food crops are cultivated or processed. Thus, two broad strategies, targeting at pre-harvest by reducing HMs in the environment where crops are grown or plants where foods are processed and post-harvest via reducing HMs concentrations in foods, are suggested for the mitigation of HMs in contaminated foods.

7.9.1 Pre-harvest Approaches

These approaches of mitigating HMs contamination of food crops are preventive measures of producing HMs-free crops in environmental matrixes decommissioned of HMs. Thus, soil and water remediation, sustainable agricultural practices, monitoring and surveillance programs, and regulatory measures are essential to producing HMs-free food crops.

7.9.2 Soil Remediation

Exposure of food crops to HMs often originate from contaminated soil where such foods are cultivated. Implementing soil remediation techniques can reduce the HMs bioavailability and prevent HMs uptake by crops. Several strategies have been used to remediate HMs in agricultural soils, thereby preventing HMs contamination of food crops. These strategies include chemical-biological remediation, microbial augmentation, electrokinetic microbial phytoremediation, phytobial remediation (using plants to extract contaminants from soil), and remediation mediated with organic bio-stimulants (Huang et al. 2018; Alzahrani and Rady 2019; Selvi et al. 2019).

7.9.3 Water Remediation

This encompasses the usage of microorganisms and/or microbial biomolecules to precipitate, immobilise, bio-sorb or bio-accumulate HMs from environmental matrices onto functional groups of microbes (Oyetibo et al. 2014, 2015a, b; Hoang et al. 2021). This technique is applied to stem toxicity of HMs via reduction of toxicants solubility, mobility and bioavailability in matrixes. The humic substances (including fulvic acids, humic acids, and humin) found in biosolids showed excellent sorption capacity to HMs at alkaline pH via their multivalent functional groups. The humic substances interact with HMs using electrostatic interactions, hydrophobic effects, π–π bonds, and hydrogen bonding to form biosolid-HMs aggregates, allowing microorganisms have direct access to HMs (Chianese et al. 2020). Additionally, dissolved organic matter (DOM) in matrixes increases solubility and mass transfer of HMs into the aqueous phase for microorganisms to directly access the toxicants bound to DOM (Cai et al. 2017). For example, *Pseudomonas putida* G7 reportedly exhibited chemotaxis through bacterial DOM-enhanced mobility that brought toxicants in close proximity for the bacterium to sequester (Jimenez-Sanchez et al. 2015). Bioaccumulation of HMs frequently occurs in hostile, toxic milieus where the required water, appropriate electron acceptors or nutrients for microbial growth is lacking. During biofilm colonisation on a surface or interface, microbes are sheathed with EPS or "slime" that permits microbes to share nutrients and precipitate toxicants while they remain sheltered from environmental hassles (Guibaud et al. 2008; Oyetibo et al. 2016b; Mangwani et al. 2017).

7.9.4 Phytoremediation

Phytoremediation is a plant-mediated purification process that extracts HMs from a contaminated soil (Shah and Daverey 2020; Thakur et al. 2021; Liu et al. 2023). Certain microbes in symbiotic relationships with plants help the plants to absorb

HMs present in the soil. This process can be enhanced by using specific microbial strains that promote metal uptake. Phytoremediation is possible via:

(a) **Phytoextraction** involves the use of plant root system to uptake HMs from the media and translocate the HMs to harvestable parts such as shoots and leaves. With *Pteris vittata*, Huang et al. (2015) confirmed the removal of arsenate from aqueous system, where 80% of the arsenate was borne intracellular of the fern fronds.
(b) **Phytovolatilisation** occurs when HMs are released into the atmosphere after uptake and transpiration.
(c) **Phytostabilisation** occurs when plants disable HMs and prevent them from migrating through accumulation and absorption, thus reducing bioavailability of HMs in the environment.

7.9.5 Agricultural Practices

Intercropping: This is a practice where food crops are cultivated with HMs hyper-accumulators. It is critical to achieving simultaneous safe food crop production, and decommissioning of polluted agricultural soils for an improved soil health and sustainable ecosystem output (Maitra et al. 2021; Mahmoud et al. 2022). Intercropping removes toxic concentrations of HMs from soil and eventually from food crops via variety of HMs transporters (Kumar et al. 2021), and bivalent element transport proteins (Shahid et al. 2017). Previous study revealed remarkable decrease in the HMs concentrations in the soil and wheat when intercropped with *Sedum plumbizincicola* (Zou et al. 2021). Significant improvement of plant biomass vis-à-vis reduction in Cd^{2+} concentrations of wheat shoots and roots was reportedly facilitated by intercropping (Hussein et al. 2019). Mei et al. (2022) also demonstrated an efficient transfer of Pb^{2+} from the roots to the shoot of *Sonchus asper* (L.) Hill as Pb^{2+} content reduction occurred in the roots and shoots of maize when both plants were intercropped together.

Crop selection and management: Certain crops have a higher affinity for accumulating HMs from the soil. By selecting low-accumulating crop varieties and implementing proper agricultural practices such as crop rotation and soil amendment with organic matter, the uptake of HMs by crops can be minimized. Thus, biotechnological approach involving the use of multi-omics comprising integrating genomics, transcriptomics, proteomics, metabolomics, lipidomics and phenomics is apt to the selection plant hyperacculators, and management of food crops free from HMs load (Liu et al. 2023). Crop management such as the use of slow-release nitrogen fertilisers is capable of reducing HMs in plants that have low efficiency of translocation to the edible parts of the plant. Also, fallowing with hyper-accumulating plants provides avenue for producing food crops that are free from HMs contaminations.

Water management: Irrigation water contaminated with heavy metals can contribute to food contamination. Implementing proper water management practices, such as using clean water sources for irrigation or treating contaminated water before use, can help prevent HMs contamination in crops. Several techniques used to remove HMs from water include ion exchange, chemical precipitation, electrochemical, flotation, nanotechnologies, and coagulation-flocculation. Advanced technologies applied to removing HMs from water since 1990 follows processes relying on polysulfone technologies adsorption membranes for Pb^{2+}; adsorption, membrane filtration like osmosis and nanofiltration membrane for Cd; graphene-based microbots for capture, transfer and removal of Pb^{2+}; nanofiltration for Pb^{2+}, Ni^{2+}, As^{3+}, Cr^{6+} and Cd^{2+}; carbon-nanoparticle for Pb^{2+}, and Cr^{6+}; organic polyethersulfone and inorganic Fe–Mn binary oxide to eliminate As(III); the hybrid membrane of nanoparticle of Fe-Ag/carbon nanotube/polyethersulphone for Cr(VI) among others. Biotechnological strategies applied to remove HMs from water primarily include (i) Biosorption, which is a passive adsorption process for eradicating HMs from contaminated media; and (ii) Bioaccumulation that is an active absorption process where HMs are transferred from media into the cellular system and HMs are retained intracellular.

Post-harvest Approaches

The development of HMs-free soil, which remains unrealistic, and HMs-resistant crop varieties appeared to be the most efficient but a laborious strategy for HMs removal from food. Thus, it is of practical importance to make available HMs-free foods for consumers during processing of food products. These approaches include, but not limited to physical removal (soaking, washing, cooking etc.) and chemical desorption (with oxalic acid, for example) (Zayed et al. 2025), hydrolysis, mechanical/chemical disinfection, baking, heating, fermentation, probiotics (Ullah et al. 2025) among others as some are explained below.

7.9.6 Physical Removal of Heavy Metals from Food

This method removes or reduces HMs from food without chemical reaction. It includes improving exactness of food, direct soaking in water and cooking with water. The foods that are readily rinsed include grains (maize, rice, millets, barley etc.) and vegetables involve using several strategies (Table 7.4). The study of Sharafi et al. (2019) reported 27.6% Cd removal when Cd-contaminated rice was soaked in water for 5 h before cooking. Pre-cooking processes (including washing, soaking and other rinsing methods) exhibited efficient removal of toxic metals from rice than old-styled cooking methods (Shariatifar et al. 2020). Merits of reducing HMs in food items through physical method include its simple operation, lack of chemical residue, and stable morphology of food crops. This makes physical removal of HMs from food the easiest technique to apply in real life. However, separation of HMs from food crops (Cd in rice, for example) is difficult due to HMs and protein

Table 7.4 Approaches and risk reduction of heavy metals during washing of whole raw fruits and vegetables (Culled from Sharafi et al. 2019; Shariatifar et al. 2020)

	Method	Risk reductions	Comments
1.	Rinsing or spraying under uninterrupted running and draining water.	Rotating the food ensures thorough washing/rinsing of all surfaces. Water velocity sufficiently frees food surface from soil and particles. Washing separate portions or small sets of food limits chances of recontamination.	Improper submerged of produce in water is liable to cross-contamination and microbial infiltration. Rinsing multiple pieces simultaneously could result in cross-contamination. Avoid splashing of water onto other food or food-contact surfaces. This approach appears not feasible for food items in large volumes.
2.	Rinsing/spraying in a vessel under a constant stream of running water with a unceasing overflow	Reduce potential infiltration by ensuring the water temperature remains warmer than that of produce pulp. Provide enough water velocity that allows loosening of dirt from the surface, where the loosened dirt/particles float off in the overflow The food produce has equal exposure to the water flow when the food is stirred in the container. Produce is washed in small batches.	Possible cross-contamination risks are lessened through continuous flowing and draining of water. Rinsing of several food items at the same time could lead to cross-contamination. Water splashing onto other produce or contact surfaces of foods should be avoided. This technique appears unrealistic for food items in large volumes.
3.	Submerging/spraying/ rinsing under running water amended with registered antimicrobial agents that regulatory agencies permit in the water for food processing.	Manufacturer's instructions must be followed Accurate concentration of the registered antimicrobial substance must be used. Ensure all food items get exposed to the treated water through agitation, allowing the items to release dirt and surface contaminants.	Check the label of registered antimicrobials for ability to control (approx. 10^3 reduction) pathogens (such as *Listeria monocytogenes*, *Salmonella*, and *E. coli* O157:H7) in the wash water. Risks of cross-contamination through pathogen infiltration and water splashing are lessened when pathogens addition into the water by contaminated food is reduced. Addition of antimicrobials in the wash mitigates pathogen contamination menace from wash water through infiltration when reduction of temperature differential between the food and the water is impossible. Decrease soaking time of food items reduces time of water infiltration. Amendment of water with verified concentration of antimicrobials should follow manufacturer's label instructions.

(continued)

Table 7.4 (continued)

	Method	Risk reductions	Comments
4.	Submerging in water amended with an approved food additive, or generally recognized as safe (GRAS), or subject of a food contact notification (FCN) based on Food and Drug Administration (FDA) agency Food Code 7–204.12 but unregistered as an antimicrobial by EPA	Follow manufacturer's guidelines Ensure the temperature of water is warmer than that of food pulp to lessen potential infiltration. Develop a policy for the frequency of changing the water Agitate to release dirt and contaminants on the surfaces of food.	The wash-products may aid the loosening and removal of dirt and other contaminants on food surfaces, and the partial antimicrobial effects on the pathogens that get into the water from infected food produce. Wash-products are unregistered by EPA, and are not lethal to pathogen or have reduction claims. Water infiltration rate is reduced with decreasing soaking time of food items. Prepare the treated water with verified concentration according to the manufacturer label instructions.
5.	Submerging in water that is free of any additives	Potential infiltration is reduced by maintaining warmer water temperature than that of food pulp. Develop a policy for the frequency of changing the water Agitate food item for dirt and contaminants to loosen from surfaces of food.	The technique provides the fewest preventive controls. Water may ease of dirt and some pathogens from the food surface without reducing pathogens in the water, which consequently increases the hazard of water-borne cross-contamination and infiltration of pathogens. Decrease soaking time of food for the reduction of water infiltration rate.

complexation. Thus, the consequence of using the physical method to lessen HMs in food crops is somehow insignificant, and it has certain limits for food crops (rice, for example) that have somewhat higher HMs content.

7.9.7 Chemical Desorption of Heavy Metals from Complexation with Food Molecules

This technique changes the acid-base conditions to eliminate the existing HMs binding to the protein of food crops; or destroying the bonds that coordinate complexation of HMs with molecules of food crops, causing the ions of HMs to migrate into the solution. As earlier discussed, complexation of HMs ions with crop proteins via bonds between metallic ions and the functional groups (carbonyls, sulfydryls, and imidazolyls) of amino acids can be destroyed by competitive actions of chemical agents and metabolic important metals. The competitors such as coordination agents (including sodium acetate, odium citrate, sodium pyrophosphate and disodium edentate dehydrate) and metal ions (calcium, copper, iron and zinc) that are of food grade have been described (Evangelou et al. 2007; Feng et al. 2019). Destruction of the complexation bonds of HMs-bound food crops is achievable through alkali and acid methods. The alkali solution destroys the architecture of food starch and protein, and the inter-protein secondary bond to dissociate the polar groups, separate protein and starch from the HMs-bound food, and reduces the HMs in food molecules. Wang et al. (2020) achieved significant reduction of Cd content in rice to 0.04 mg/kg from 0.38 mg/kg upon extraction of starch from Cd contaminated rice followed by alkali treatment of the extracted starch and carboxymethylation. The acid leaching method involves formation of a stable complex when a part of the acid interacts with Cd, as the pH value reduces to dissolve protein and lessened binding ability of Cd and protein. Thereby, part of Cd dissociates into ions and leach into the solution. Substantial Cd removal from contaminated rice has been achieved via soaking in varied glacial acetic acid concentrations at 30 °C after 15 min. (Al-Naimi and Al-Ghouti 2020). Similarly, incubating contaminated paddy rice in a molar solution of citric acid at ambient temperature for 12 h, followed by incubation in one molar solution of calcium carbonate for another 12 h, and then neutralized the acidic pH value before finally cooking the rice (Pogoson et al. 2021) efficiently reduced Cd. In particular, 94% Cd removal was reported when the treatment condition changed to 0.15 M citric acid at 40 °C for 60 min and 15 mL/g liquid–solid ratio was optimised (Zou et al. 2019).

Surfactant adsorption, deep eutectic solvent washing, and complexation have variously been used to remove HMs from contaminated food. Surfactant adsorption relies on surfactant weakening of the binding forces between HMs ions and protein to release HMs from food through the electrostatic attraction and complexation of functional groups of surfactant and HMs ions. Shen et al. (2019) adopt a two-step scrubbing process that uses rhamnolipid biosurfactant alongside Pluronic F127/

polyacrylic acid (F127/PAA) hydrogel to remove 92% of the Cd bound to rice's protein, leave behind a residual Cd content lower than the safety level (<0.2 mg/kg). Other innovative green technological approaches used to abate HMs contaminations of food involves ultrasonic-assisted extraction (Luo et al. 2021), where natural deep eutectic solvents based on choline chloride and glycerine as detergent are used (Huang et al. 2018); and acidification with HCl or EDTA disodium solution is used to soak rice protein. Combining ultrasound with other methods like ozone, ultraviolet and pulsed electric field appears to be a better strategy to decontaminate foods laden with HMs (Yuan et al. 2021). Sound wave at a frequency 20–100 kHz acts on liquid food to generate positive and negative pressure, which produces periodic compression and expansion on the food molecules (Yu et al. 2020). During the negative pressure expansion, ultrasonic wave produces miniature bubbles or cavities in the liquid food into free radicals that eliminate HMs from the contaminated foods. These cavities continue absorbing acoustic energy until a critical size is achieved and break violently, and release high energy that rapidly eliminates HMs from the complexes formed with the contaminated food molecules (Askari et al. 2017; Chen et al. 2020a). Condón-Abanto et al. (2018) applied ultrasound wave (frequency, 35 kHz; power, 200 W; temperature, 50 °C) to release 22.8% Cd from crab to water. Likewise, mushroom have been decontaminated of its As using ultrasound (Chen et al. 2020b), while low frequency (20 kHz) ultrasound was reportedly used to eliminate HMs from milk to safe level as the sound waves reduced size of fat and protein particles in milk, leading to increasing permeability of reduced-size proteins on polyhepan (Porova et al. 2014). For hydrochloric acid soaking (Feng et al. 2019) lowering pH to 4 have efficiently reduced Cd contents in rice and rice floor by 51–96%. Also, hydrions (hydrogen ions, H^+, in a solution) could dissociate the protein-HMs complexes just as the coordinating agents destabilises the HMs binding to food crop proteins. For example, Ca, Cu and Zn started restricting rice proteins-Cd^{2+} binding at pH 6.5 once the concentration of metal ion reached 500 mg/kg (Feng et al. 2019). Chemical leaching of HMs from the contaminated food has been efficient, but the consequence of chemical solvent on food features, chemical solvent recovery and the chemical residues in the treated food and food products limit its application in reducing HMs in contaminated foods. Moreover, the application of non-thermal technology like ultrasound treatment may adversely affect food characteristics like degradation of some phenolic compounds and vitamins, colour changes, and anthocyanin loss (Yuan et al. 2021).

7.9.8 Biotechnological Techniques

Alleviating heavy metal contamination in foods is critical to ensuring safety of food and public health protection. Biotechnological strategies to decommission HMs from contaminated foods and food products use whole microbial cells and or their metabolites for fermentation and self-adsorption of HMs from food molecules, or stimulate the dissolution of HMs binding biomolecules. These involve:

7.10 Fermentation

Food fermentation involves biocatalysis of microorganisms or their metabolites in the degradation of hemicellulose and cellulose of food substances to form more porous and loose structures and polysaccharides. Recently, precision fermentation that combines traditional fermentation methods with programming microorganisms through recombinant DNA technology to unique proteins has enabled the production of safe food products upon ameliorating toxicant HMs (Rezaei-Tavirani and Farahani 2024; Jin et al. 2025). Lactic acid bacteria (LAB) and yeast are the commonly used microorganisms in food fermentation. Zhang et al. (2017) reported more than 80% Cd removal efficiency from rice after fermentation at optimal time, liquid ratio, temperature, inoculum size, and washing times. Similarly, 87.6% and 93.4% Cd removal efficiency have been reported after fermentation of contaminated rice using strains of *Lactibacillus plantarum* (Zhai et al. 2019) and mixed LAB (Zhu et al. 2019), respectively, to produce high quality rice product. During fermentation of tea, dietary fiber modifies, leading to an improved specific surface area and removal efficiency of contaminating HMs ions like As^{3+}, Pb^{2+}, and Cd^{2+} (Jin et al. 2018; Chen et al. 2020a; Raji et al. 2023). HMs adsorption rate from the contaminated tea during dietary fibre fermentation can be increased with the addition of 5% NaCl to the process to prevent the exposure of more active groups (hydroxyl or carboxyl groups) that are capable to form a stable complex with HMs ions (Jin et al. 2018). It is noteworthy that the percentage of HMs released during fermentation is relative to the rate of fermentation, to the extent that the dietary fibre fermented is directly varied with the percentage of HMs released. For example, millet bran fermentation by *Bacillus natto* breakdown cellulose and hemicellulose structures and boosted the binding capacities for HMs to stem the toxicity of contaminated HMs (Chu et al. 2019).

While fermentation prides in high quality food products through higher hardness and chew-ability, less cooking loss, shorter rehydration time, and crushing rate; a few limitations have been reported. According to Zhai et al. (2019) and Vanderschueren et al. (2020), possible limitations of fermentation include: (i) alterations in the chemical and physical properties of food, like the reduction of fat, protein, and ash content; (ii) improvement of gelatinization properties; (iii) the change of crystal structure of rice starch granules; (iv) the appearance of porous micromorphology; and loss of flavour (in cocoa products) due to acidity. However, there have been instances where these limitations are circumvented by fixing *Lactobacillus plantarum* to prepare diatomite co-immobilized microbial pellets for the removal of more than 90% of Cd from contaminated rice with attendant little impact on the rice quality. Generally, biotechnological method for decommissioning HMs in contaminated food and food products is safer than the chemical method because the treatment conditions are relatively mild, the damage to the structure of food molecules is small, and the product quality can be improved further.

7.11 Probiotics

Consumption of HMs-contaminated foods often exerts a negative impact on the human health in terms of intestinal disorders due to perturbation of the forms, structure and function of the gut microbiome (Nogacka et al. 2019; He et al. 2020; Ansari et al. 2024), leading to gut dysbiosis (Andrade et al. 2020; Zloch et al. 2018; Fateminasab et al. 2025). Presently, an innovative strategy to mitigate the risk of foodborne HMs poisoning is bio-detoxification via biocatalysis of probiotic strains and their enzymes (Fateminasab et al. 2025). The effectiveness and feasibility of probiotics in preventing HMs-induced HMs dysbiosis and alleviating their toxicity are well reported (Reddivari et al. 2017; Ju et al. 2019; Zloch et al. 2018; Ansari et al. 2024). Probiotic microorganisms have competence to remove HMs from contaminated foods via cellular binding (immobilization) onto cell surface and/or enzymatic sequestration of HMs to innocuous forms after altering their oxidation states (Wang et al. 2016; Chiocchetti et al. 2018; Muhialdin et al. 2020). The most common probiotics, otherwise called traditional probiotics, used for HMs sequestration during food fermentation are strains of LABs, *Bifidobacterium* sp., and *Saccharomyces* sp. (Armando et al. 2012). General characteristics of probiotics that become handy for HMs sequestration include strong binding capability, HMs tolerance or detoxification via redox activities; elevated tolerance to acid and bile; tough attachment to gut mucosa; and strong antioxidant or immuno-regulatory abilities needed for coping with the altered gut system (Muhialdin et al. 2020; Abdel-Megeed 2021; Ansari et al. 2024). The mechanisms used by probiotics to sequester HMs toxicity is summarily illustrated in Fig. 7.3. HMs adsorption onto the LABs cell envelop emanates from the net negative charges of the functional groups like carbonyl and phosphoryl groups provided for by mainly peptidoglycan, and other minor polymers including lipoteichoic acids and teichoic acids (Kanamarlapudi et al. 2018; Lili et al. 2018; Rogowska et al. 2019). For example, strains of *Lactobacillus* with capability of immobilizing HMs have been demonstrated. Oral probiotic supplementation was reportedly alleviated the adverse health effects of HMs contaminated food via modulation of the gut microflora's composition (Abdel-Megeed 2021; Ansari et al. 2024), reduction of the oxidative stress (Chattopadhyay et al. 2019), improvement of the intestinal barrier function (Duan et al. 2020), modulation of the host gene expression with consequential capability of the host to sequester HMs (Wang et al. 2018). *Bifidobacterium longum* 46 and *Lactobacillus fermentum* ME3 were reportedly bound and removed HMs using the capsular polysaccharides and teichoic acids of the cell surface (Teemu et al. 2008). Similarly, *Lactobacillus plantarum* MF04218 was reported to have efficiently bound Pb^{2+} and Cd^{2+} at highly acidic pH (Ameen et al. 2020). Furthermore, attenuation of probiotics with heat or acid partially breakdown peptidoglycan scaffold, denature protein, and increase the surface area for the exposure of HMs binding sites (Lili et al. 2018). On the contrary, the *Saccharomyces* yeast cell wall mainly consists β-1,3-glycan-chitin architecture in the inner layer, the highly glycosylated mannoproteins that covalently bound to the network occupy the outer layer, and altogether provide several

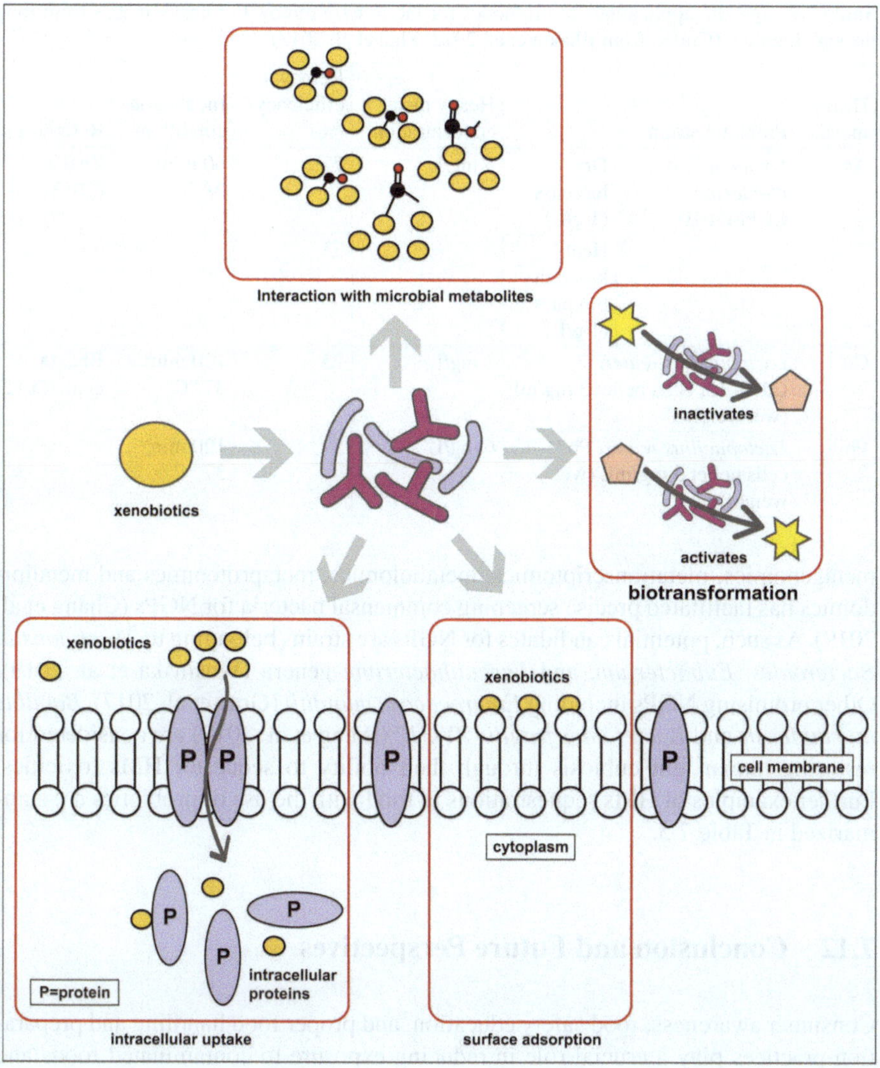

Fig. 7.3 The mechanisms used by probiotic microorganisms to detoxify food xenobiotics (HMs). (Culled from Średnicka et al. 2021)

phosphodiester bridges in the carbohydrate side chains of the surface proteins useful for HMs uptake. Moreover, probiotics can adsorb HMs associate foods in the alimentary canal, bind to the cell wall via its functional groups, thus lessening the bioavailability and toxicity of HMs (Bisanz et al. 2014; Adjei-Fremah et al. 2018). Lately, there are initiatives to shift focus from traditional probiotics to exploiting gut microflora and its health significance in human for ostensible Next Generation Probiotics (NGPs) (Langella et al. 2019). The advanced knowledge in

Table 7.5 Specific applications of probiotics for the *in vitro* removal of heavy metals from food through binding. (Culled from Bhakta et al. 2012; Zhai et al. 2013)

Heavy metals	Probiotic strain		Heavy metal concentration	Removal efficiency (%)	Incubation conditions	References
As	*Lactobacillus plantarum* CCFM8610	Dry biomass (1 g/L)	5 mg/L	77	60 min; 28 °C	Zhai et al. (2013)
		Heat-killed dry biomass (1 g/L)		73		
Cd	*Lactobacillus reuteri* Cd70–13, cells pellet 3 mg/ml (wet weight)		1 mg/L	25	120 min; 37 °C	Bhakta et al. (2012)
Pb	*Lactobacillus reuteri* Pb71–1, cells pellet 3 mg/mL (wet weight)		6 mg/L	59	120 min; 37 °C	

metagenomics, metatranscriptomics, metabolomics, metaproteomics and metalipidomics has facilitated precise screening commensal bacteria for NGPs (Chang et al. 2019). As such, potential candidates for NGPs are strains belonging to *Akkermansia, Bacteroides, Eubacterium,* and *Faecalibacterium* genera (Wosinska et al. 2019). Other promising NGPs including *Enterococcus munditii* (Grau et al. 2017), *Bacillus thuringiensis* and *Bacteroides fragilis ZY-312* (Wang et al. 2017) are considered for restoring human gut eubiosis through their ability to sequester HMs toxicities. Further examples of HMs sequestrations in food with the use of probiotics are summarized in Table 7.5.

7.12 Conclusion and Future Perspectives

Consumer awareness, food safety education, and proper food handling and preparation practices play a crucial role in reducing exposure to contaminated foods and protecting public health. While microbial biotechnology offers promising solutions for mitigating heavy metal contamination in foods, it's essential to consider the safety and regulatory aspects of using microorganisms in food production and remediation processes. Additionally, implementing robust quality control measures and regular monitoring of heavy metal levels in foods can help identify contamination issues early and prevent contaminated foods from reaching consumers. Further research is needed to optimize the techniques of metals toxicities in contaminated food for practical applications and ensure their effectiveness in real-world settings should be upcoming focus. Such researches should centre round exploring eco-physiologies of microorganisms, particularly those with probiotic properties, targeted at transforming toxic metals to innocuous forms in food crops and food products using cutting-edge technics.

References

AbdElgawad H, Zinta G, Hamed BA, Selim S, Beemster G, Hozzein WN, Wadaan MAM, Asard H, Abuelsoud W (2020) Maize roots and shoots show distinct profiles of oxidative stress and antioxidant defense under heavy metal toxicity. Environ Pollut 258:113705. https://doi.org/10.1016/j.envpol.2019.113705

Abdel-Megeed RM (2021) Probiotics: a promising generation of heavy metal detoxification. Biol Trace Elem Res 199:2406–2413. https://doi.org/10.1007/s12011-020-02350-1

Adjei-Fremah S, Ekwemalor K, Asiamah EK, Ismail H, Ibrahim S, Worku M (2018) Effect of probiotic supplementation on growth and global gene expression in dairy cows. J Appl Anim Res 46:257–263. https://doi.org/10.1080/09712119.2017.1292913

Agranoff D, Collins L, Kehres D, Harrison T, Maguire M, Krishna S (2005) The Nramp orthologue of *Cryptococcus neoformans* is a pH-dependent transporter of manganese, iron, cobalt and nickel. Biochem J 385:225–232

Ahmed M, Shafqat SS, Javed A, Sanaullah M, Shakoor A, Shafiq MI, Shahzadi SK, Wani TA, Zargar S (2023) Exposure assessment of essential and potentially toxic metals in wheat-based sweets for human consumption: multivariate analysis and risk evaluation studies. Molecules 28(21):7365. https://doi.org/10.3390/molecules28217365

Al-Naimi M, Al-Ghouti MA (2020) Effects of soaking, acidity and temperature on cadmium and lead removal from rice. Food Chem 310(2020):125591. https://doi.org/10.1016/j.foodchem.2019.125591

Alzahrani Y, Rady MM (2019) Compared to antioxidants and polyamines, the role of maize grain-derived organic biostimulants in improving cadmium tolerance in wheat plants. Ecotoxicol Environ Saf 182:109378. https://doi.org/10.1016/j.ecoenv.2019.109378

AMAP/UNEP (2013) Global releases of mercury to aquatic environments. In: Technical background report for global mercury assessment, p 263

Ameen FA, Hamdan AM, El-Naggar MY (2020) Assessment of the heavy metal bioremediation efficiency of the novel marine lactic acid bacterium, *Lactobacillus plantarum* MF042018. Sci Rep 10:314. https://doi.org/10.1038/s41598-019-57210-3

Andrade JC, Almeida D, Domingos M, Seabra CL, Machado D, Freitas AC, Gomes AM (2020) Commensal obligate anaerobic bacteria and health: production, storage, and delivery strategies. Front Bioeng Biotechnol 8(550). https://doi.org/10.3389/fbioe.2020.00550

Ankit, Saha L, Kumar V, Tiwari J, Sweta, Rawat S, Singh J, Bauddh K (2021) Electronic waste and their leachates impact on human health and environment: global ecological threat and management. Environ Technol Innov 24:102049. https://doi.org/10.1016/j.eti.2021.102049

Ansari F, Lee CC, Rashidimehr A, Eskandari S, Ashaolu TJ, Mirzakhani E, Pourjafar H, Jafari SM (2024) The role of probiotics in improving food safety; detoxification of heavy metals and chemicals. Toxin Rev 43(1):63–91

Armando MR, Pizzolitto RP, Dogi CA, Cristofolini A, Merkis C, Poloni V, Dalcero AM, Cavaglieri LR (2012) Adsorption of ochratoxin A and zearalenone by potential probiotic *Saccharomyces cerevisiae* strains and its relation with cell wall thickness. J Appl Microbiol 113:256–264. https://doi.org/10.1111/j.1365-2672.2012.05331.x

Askari P, Faraji A, Khayatian G, Mohebbi S (2017) Effective ultrasound-assisted removal of heavy metal ions As (III), Hg (II), and Pb (II) from aqueous solution by new MgO/CuO and MgO/MnO 2 nanocomposites. J Iran Chem Soc 14:613–621. https://doi.org/10.1007/s13738-016-1011-y

Attar A, Ghica ME, Amine A, Brett CMA (2014) Poly (neutral red) based hydrogen peroxide biosensor for chromium determination by inhibition measurements. J Hazard Mater 279:348–355

Authority, E.F.S (2012) Cadmium dietary exposure in the European population. EFSA J 10(1):2551

Aziz B, Zubair M, Irshad N, Ahmad KS, Mahmood M, Tahir MM, Shah KH, Shaheen A (2021) Biomonitoring of toxic metals in feathers of birds from North-Eastern Pakistan. Bull Environ Contam Toxicol 106(5):805–811. https://doi.org/10.1007/s00128-021-03184-w

Babkina SS, Ulakhovich NA (2004) Amperometric biosensor based on denatured DNA for the study of heavy metals complexing with DNA and their determination in biological, water and food samples. Bioelectrochemistry 63:261–265. https://doi.org/10.1016/j.bioelechem.2003.11.006

Balali-Mood M, Naseri K, Tahergorabi Z, Khazdair MR, Sadeghi M (2021) Toxic mechanisms of five heavy metals: mercury, lead, chromium, cadmium, and arsenic. Front Pharmacol 12:643972. https://doi.org/10.3389/fphar.2021.643972

Barkay T, Miller SM, Summers AO (2003) Bacterial mercury resistance from atoms to ecosystems. FEMS Microbiol Rev 27:355–384

Battaglia-Brunet F, Casiot C, Fernandez-Rojo L, Hery M, Le Pape P, Tris H et al (2021) Laboratory-scale bio-treatment of real arsenic-rich acid mine drainage. Water Air Soil Poll 232:330

Berezhetskyy AL, Sosovska OF, Durrieu C, Chovelon J-M, Dzyadevych SV, Tran-Minh C (2008) Alkaline phosphatase conductometric biosensor for heavy-metal ions determination. IRBM 29(2–3):136–140. https://doi.org/10.1016/j.rbmret.2007.12.007

Bhakta JN, Ohnishi K, Munekage Y, Iwasaki K, Wei MQ (2012) Characterization of lactic acid bacteria-based probiotics as potential heavy metal sorbents. J Appl Microbiol 112:1193–1206. https://doi.org/10.1111/j.1365-2672.2012.05284.x

Bisanz JE, Enos MK, Mwanga JR, Changalucha J, Burton JP, Gloor GB, Reid G (2014) Randomized open-label pilot study of the influence of probiotics and the gut microbiome on toxic metal levels in Tanzanian pregnant women and school children. mBio 5:e01580–e01514. https://doi.org/10.1128/mBio.01580-14

Bonelli M, Bruno D, Caccia S, Sgambetterra G, Cappellozza S, Jucker C, Tettamanti G, Casartelli M (2019) Structural and functional characterization of *Hermetia illucens* larval midgut. Front Physiol 10:204. https://doi.org/10.3389/fphys.2019.00204

Boyd ES, Barkay T (2012) The mercury resistance operon: from an origin in a geothermal environment to an efficient detoxification machine. Front Microbiol 3:349. https://doi.org/10.3389/fmicb.2012.00349

Bücker-Neto L, Paiva ALS, Machado RD, Arenhart RA, Margis-Pinheiro M (2017) Interactions between plant hormones and heavy metals responses. Genet Mol Biol 40(1 suppl 1):373–386. https://doi.org/10.1590/1678-4685-GMB-2016-0087

Cai D, Yang X, Wang S, Chao Y, Morel J-L, Qiu R (2017) Effects of dissolved organic matter derived from forest leaf litter on biodegradation of phenanthrene in aqueous phase. J Hazard Mater 324:516–525

Cai X, Xue C, Owens G, Chen Z (2023) Removal of As(III) using a microorganism sustained secrete laccase-straw oxidation system. J Hazard Mater 448:130967. https://doi.org/10.1016/j.jhazmat.2023.130967

Cai XL, Zhang ZN, Yin NY, Lu WY, Du HL, Yang M et al (2022) Controlling microbial arsenite oxidation and mobilization in arsenite-adsorbed iron minerals: the Influence of pH conditions and mineralogical composition. J Hazard Mater 433:128778

CDC Report (2024) Lead poisoning investigation in Northern Nigeria. https://www.cdc.gov/one-health/php/stories/lead-poisoning.html#:~:text=Investigations%20of%20other%20villages%20in,of%20lead%20poisoning%20in%20history. Assessed at 10:44 am WAT, on August 15, 2024

Chang C-J, Lin T-L, Tsai Y-L, Wu T-R, Lai W-F, Lu C-C, Lai H-C (2019) Next-generation probiotics in disease amelioration. J Food Drug Anal 27:615–622. https://doi.org/10.1016/j.jfda.2018.12.011

Chattopadhyay S, Khatun S, Maity M, Jana S, Perveen H, Dash M, Dey A, Jana LR, Maity PP (2019) Association of vitamin B12, lactate dehydrogenase, and regulation of NF-κB in the mitigation of sodium arsenite-induced ROS generation in uterine tissue by commercially available probiotics. Probiotics Antimicrob Proteins 11:30–42. https://doi.org/10.1007/s12602-017-9333-3

Chen C-Y, Hsieh J-L, Silver S, Endo G, Huang C-C (2008) Interaction between two *MerR* regulator/promoter region in the mercury resistance module of *Bacillus megaterium*. Biosci Biotechnol Biochem 72:2403–2410

Chen S, Kimatu BM, Fang D, Chen X, Chen G, Hu Q, Zhao L (2020b) Effect of ultrasonic treatment on transformations of arsenic species in edible mushrooms. Anal Lett 53(1):102–121. https://doi.org/10.1080/00032719.2019.1639056

Chen Z, Zhang X, Ma Y, Chen Song Y, Li Y, Geng G, Huang Y (2020a) Anammox biofilm system under the stress of Hg(II): nitrogen removal performance, microbial community dynamic and resistance genes expression. J Hazard Mater 395:122665. https://doi.org/10.1016/j.jhazmat.2020.122665

Chen X, Wang J, Shen HY, Su X, Cao Y, Li T, Gan N (2019) Microfluidic chip for multiplex detection of trace chemical contaminants based on magnetic encoded aptamer probes and multibranched DNA nanostructures as signal tags. ACS Sensors 4:2131–2139 https://doi.org/10.1021/acssensors.9b00963

Chianese S, Fenti A, Iovino P, Musmarra D, Salvestrini S (2020) Sorption of organic pollutants by humic acids: a review. Molecules 25(4):918. https://doi.org/10.3390/molecules25040918

Chien MF, Ho YN, Yang HE, Narita M, Miyauchi K, Endo G, Huang CC (2019) Identification of a novel arsenic resistance transposon nested in a mercury resistance transposon of Bacillus sp. MB24. Microorganisms 7(11):566. https://doi.org/10.3390/microorganisms7110566

Chiocchetti GM, Jadán-Piedra C, Monedero V, Zúñiga M, Vélez D, Devesa V (2018) Use of lactic acid bacteria and yeasts to reduce exposure to chemical food contaminants and toxicity. Crit Rev Food Sci Nutr 59:1534–1545. https://doi.org/10.1080/10408398.2017.1421521

Chu J, Zhao H, Lu Z, Lu F, Bie X, Zhang C (2019) Improved physicochemical and functional properties of dietary fiber from millet bran fermented by Bacillus natto. Food Chem 294:79–86. https://doi.org/10.1016/j.foodchem.2019.05.035

Chung H, Bird AJ (2019) Zinc signals in biology. In: Fukada T, Kambe T (eds) Zinc signaling. Springer, Singapore. https://doi.org/10.1007/978-981-15-0557-7_19

Condón-Abanto S, Raso J, Arroyo C, Lyng JG, Condón S, Álvarez I (2018) Evaluation of the potential of ultrasound technology combined with mild temperatures to reduce cadmium content of edible crab (Cancer pagurus). Ultrason Sonochem 48:550–554. https://doi.org/10.1016/j.ultsonch.2018.07.019

Cubadda F, Jackson BP, Cottingham KL, Van Horne YO, Kurzius-Spencer M (2017) Human exposure to dietary inorganic arsenic and other arsenic species: state of knowledge, gaps and uncertainties. Sci Total Environ 579:1228–1239. https://doi.org/10.1016/j.scitotenv.2016.11.108

Dean AP, Hartley A, McIntosh OA, Smith A, Feord HK, Holmberg NH, King T, Yardley E, White KN, Pittman JK (2019) Metabolic adaptation of a *Chlamydomonas acidophila* strain isolated from acid mine drainage ponds with low eukaryotic diversity. Sci Total Environ 647:75–87. https://doi.org/10.1016/j.scitotenv.2018.07.445

Desoky E-SM, Merwad A-RM, Semida WM, Ibrahim SA, El-Saadony ET, Rady MM (2020) Heavy metals-resistant bacteria (HM-RB): potential bioremediators of heavy metals-stressed *Spinacia oleracea* plant. Ecotoxicol Environ Saf 198:110685. https://doi.org/10.1016/j.ecoenv.2020.110685

Desoky EM, Elrys AS, Rady MM (2019) Integrative moringa and licorice extracts application improves *Capsicum annuum* fruit yield and declines its contaminant contents on a heavy metal contaminated saline soil. Ecotoxicol Environ Saf 169:50–60. https://doi.org/10.1016/j.ecoenv.2018.10.117

Dieudonné A, Prévéral S, Pignol D (2020) A sensitive magnetic arsenite-specific biosensor hosted in magnetotactic bacteria. Appl Environ Microbiol 86(14):e00803–e00820. https://doi.org/10.1128/AEM.00803-20

Dooyema CA, Neri A, Lo YC, Durant J, Dargan PI, Swarthout T, Biya O, Gidado SO, Haladu S, Sani-Gwarzo N, Nguku PM (2012) Outbreak of fatal childhood lead poisoning related to artisanal gold mining in northwestern Nigeria, 2010. Environ Health Perspect 120(4):601–607. https://doi.org/10.1289/ehp.1103965

Duan H, Yu L, Tian F, Zhai Q, Fan L, Chen W (2020) Gut microbiota: a target for heavy metal toxicity and a probiotic protective strategy. Sci Total Environ 742(140429):140429. https://doi.org/10.1016/j.scitotenv.2020.140429

Dupont C, Ahner BA (2005) Effects of copper, cadmium and zinc on the production and exudation of thiols by Emiliania huxleyi. Limnol Oceanogr 50:508–515

Egbe CC, Oyetibo GO, Ilori MO (2021) Ecological impact of organochlorine pesticides consortium on autochthonous microbial community in agricultural soil. Ecotoxicol Environ Saf 207:111319. https://doi.org/10.1016/j.ecoenv.2020.111319

el-Enany AE, Issa A (2001) Proline alleviates heavy metal stress in Scenedesmus armatus. Folia Microbiol (Praha) 46:227–230

Endo G, Narita M, Huang C-C, Silver S (2002) Heavy metal resistance transposons and plasmids: potential use for environmental biotechnology. J Environ Biotechnol 2:71–82

Escher BI, Hermens JL (2004) Internal bioavailability to effects. Environ Sci Technol 38:455A–462A

Evangelou MW, Ebel M, Schaeffer A (2007) Chelate assisted phytoextraction of heavy metals from soil. Effect, mechanism, toxicity, and fate of chelating agents. Chemosphere 68(6):989–1003. https://doi.org/10.1016/j.chemosphere.2007.01.062

Fateminasab F, Koushki M, Jafarian F, Mansouri V, Ahmadzadeh A, Amiri-Dashatan N, Farahani M (2025) Assessing efficacy of the microbial interventions in heavy metal decontamination of the environment and food production systems. Appl Food Biotechnol 11(2):e8–e8. https://doi.org/10.22037/afb.v11i2.46832

Feng L, Chen S, Zhang L, Qu W, Chen Z (2019) Bisphenol A increases intestinal permeability through disrupting intestinal barrier function in mice. Environ Pollut 254(2019):112960. https://doi.org/10.1016/j.envpol.2019.112960

Flora SJS, Flora G, Saxena G (2006) Environmental occurrence, health effects and management of lead poisoning. Elsevier Science BV, Amsterdam, pp 158–228. https://doi.org/10.1016/B978-044452945-9/50004-X

Genchi G, Sinicropi MS, Lauria G, Carocci A, Catalano A (2020) The effects of cadmium toxicity. Int J Environ Res Public Health 17:3782. https://doi.org/10.3390/ijerph17113782

Genchi G, Sinicropi M, Carocci A, Lauria G, Catalano A (2017) Mercury exposure and heart diseases. Int J Environ Res Publ Health 14(74). https://doi.org/10.3390/ijerph14010074

Ghica ME, Brett CMA (2008) Glucose oxidase inhibition in poly(neutral red) mediated enzyme biosensors for heavy metal determination. Microchim Acta 163:185–193. https://doi.org/10.1007/s00604-008-0018-1

Giannakoula A, Therios I, Chatzissavvidis C (2021) Effect of lead and copper on photosynthetic apparatus in citrus (Citrus aurantium L.) plants. The role of antioxidants in oxidative damage as a response to heavy metal stress. Plants 10(1):155. https://doi.org/10.3390/plants10010155

Gibb HJ, Barchowsky A, Bellinger D, Bolger PM, Carrington C, Havelaar AH et al (2019) Estimates of the 2015 global and regional disease burden from four foodborne metals—arsenic, cadmium, lead and methylmercury. Environ Res 174:188–194. https://doi.org/10.1016/j.envres.2018.12.062

Gossa Al-Saadde DL, Haider AM, Ali A, Abdu Musad Saleh E, Turki Jalil A, Abdulelah FM, Romero-Parra RM, Tayyib NA, Ramírez-Coronel AA, Alkhayyat AS (2023) The role of chromium supplementation in cardiovascular risk factors: a comprehensive reviews of putative molecular mechanisms. Heliyon 9(9):e19826. https://doi.org/10.1016/j.heliyon.2023.e19826

Grau T, Vilcinskas A, Joop G (2017) Probiotic *Enterococcus mundtii* isolate protects the model insect *Tribolium castaneum* against *Bacillus thuringiensis*. Front Microbiol 8(1261). https://doi.org/10.3389/fmicb.2017.01261

Groot MN, Klaassens E, de Vos WM, Delcour J, Hols P, Kleerebezen M (2005) Genome-based in silico detection of putative manganese transport systems in *Lactobacillus plantarum* and their genetic analysis. Microbiology 151:1229–1238

Guascito MR, Malitesta C, Mazzotta E, Turco A (2008) Inhibitive determination of metal ions by an amperometric glucose oxidase biosensor. Study of the effect of hydrogen peroxide decomposition. Sensors Actuators B Chem 131:394–402. https://doi.org/10.1016/j.snb.2007.11.049

Guibaud G, Bordas F, Saaid A, D'abzac P, van Hullebursch E (2008) Effect of pH on cadmium and lead binding by extracellular polymeric substances (EPS) extracted from environmental bacterial strains. Colloid Surf B Biointerf 63:48–54

Gumilar F, Bras C, Aggio P, Domínguez S, Bartos M, Gallegos C, Monaco N, Minetti A (2019) Prenatal exposure to cadmium during organogenesis impairs memory in young rats. Int J Toxicol 38(4):312–318. https://doi.org/10.1177/1091581819850579

Handford CE, Campbell K, Elliott CT (2016) Impacts of milk fraud on food safety and nutrition with special emphasis on developing countries. Compr Rev Food Sci Food Saf 15(1):130–142. https://doi.org/10.1111/1541-4337.12181

He X, Qi Z, Hou H, Qian L, Gao J, Zhang X-X (2020) Structural and functional alterations of gut microbiome in mice induced by chronic cadmium exposure. Chemosphere 246(125747):125747. https://doi.org/10.1016/j.chemosphere.2019.125747

Hoang SA, Sarkar B, Seshadri B, Lamb D, Wijesekara H, Vithanage M, Liyanage C, Kolivabandara PA, Rinklebe J, Lam SS, Vinu A, Wang H, Kirkham MB, Bolan NS (2021) Mitigation of petroleum-hydrocarbon-contaminated hazardous soils using organic amendments: a review. J Hazard Mater 416:125702. https://doi.org/10.1016/j.jhazmat.2021.125702

Hu R, Seyfferth AL (2021) Rice Cd levels in Cambodia ranged 3 orders of magnitude due to season and soil Cd levels. ACS Omega 6(30):19876–19882. https://doi.org/10.1021/acsomega.1c02741

Hu T, Zhu S, Tan L, Qi W, He S, Wang G (2016) Overexpression of OsLEA4 enhances drought, high salt and heavy metal stress tolerance in transgenic rice (Oryza sativa L.). Environ Exp Bot 123:68–77. https://doi.org/10.1016/j.envexpbot.2015.10.002

Hu Y, Wang D, Wei L, Zhang X, Song B (2014) Bioaccumulation of heavy metals in plant leaves from Yan′ an city of the Loess Plateau, China. Ecotoxicol Environ Saf 110:82–88. https://doi.org/10.1016/j.ecoenv.2014.08.021

Huang D, Deng R, Wan J, Zeng G, Xue W, Wen X, Zhou C, Hu L, Liu X, Xu P, Guo X, Ren X (2018) Remediation of lead-contaminated sediment by biochar-supported nanochlorapatite: accompanied with the change of available phosphorus and organic matters. J Hazard Mater 348:109–116. https://doi.org/10.1016/j.jhazmat.2018.01.024

Huang Y, Miyauchi K, Inoue C, Endo G (2015) Development of suitable hydroponics system for phytoremediation of arsenic contaminated water using an arsenic hyperaccumulator plant Pteris vittata. Biosci Biotechnol Biochem 80:714–718

Hummel M, Standl E, Schnell O (2007) Chromium in metabolic and cardiovascular disease. Horm Metab Res 39(10):743–751. https://doi.org/10.1055/s-2007-985847

Huo Y, Du H, Xue B, Niu M, Zhao S (2016) Cadmium removal from rice by separating and washing protein isolate. J Food Sci 81(6):T1576–T1584. https://doi.org/10.1111/1750-3841.13323

Hussein Y, Almin G, Askora A, Gahin H (2019) Phytotoxicity remediation in wheat (Triticum aestivum L.) cultivated in Cadmium-contaminated soil by intercropping design. Biosci Res 16:2159–2181

Ihunwo OC, Ibezim-Ezeani MU (2022) Metal accumulation in muscle and oxidative stress response in the liver of juvenile *Oreochromis niloticus* from contaminated sediment under a simulation of increasing temperature. Environ Res Commun 4(7):075008. https://doi.org/10.1088/2515-7620/ac803d

Ilangovan R, Daniel D, Krastanov A, Zachariah C, Elizabeth R (2006) Enzyme based biosensor for heavy metal ions determination. Biotechnol Biotechnol Equip 20:184–189. https://doi.org/10.1080/13102818.2006.10817330

Jensen LT, Ajua-Alemanji M, Culotta VC (2003) The Saccharomyces cerevisiae high affinity phosphate transporter encoded by PHO84 also functions in manganese homeostasis. J Biol Chem 278:42036–42040

Jimenez-Sanchez C, Wick LY, Cantos M, Ortega-Calvo J-J (2015) Impact of dissolved organic matter on bacterial tactic motility, attachment, and transport. Environ Sci Technol 49:4498–4505

Jin J, den Besten HM, Rietjens IM, Widjaja F (2025) Chemical and microbiological hazards arising from new plant-based foods, including precision fermentation–produced food ingredients. Annu Rev Food Sci Technol 16. https://doi.org/10.1146/annurev-food-111523-122059

Jin Y, Luan Y, Ning YI, Wang L (2018) Effects and mechanisms of microbial remediation of heavy metals in soil: a critical review. Appl Sci 8(1336):2–17

Ju J, Shen L, Xie Y, Yu H, Guo Y, Cheng Y, Qian H, Yao W (2019) Degradation potential of bisphenol A by *Lactobacillus reuteri*. LWT (Lebensm-Wiss & Technol) 106:7–14. https://doi.org/10.1016/j.lwt.2019.02.022

Kabiraj A, Biswas R, Halder U, Bandopadhyay R (2022) Bacterial arsenic metabolism and its role in arsenic bioremediation. Curr Microbiol 79(5):131

Kanamarlapudi SLRK, Chintalpudi VK, Muddada S (2018) Application of biosorption for removal of heavy metals from wastewater. In: Biosorption. InTech. https://doi.org/10.5772/intechopen.77315

Kaur H, Kumar S, Verma N (2014) Enzyme-based colorimetric and potentiometric biosensor for detecting Pb (II) ions in milk. Braz Arch Biol Technol 57:613–619

Khan ZI, Ahmad K, Ahmad T, Zafar A, Alrefaei AF, Ashfaq A, Akhtar S, Mahpara S, Mehmood N, Ugulu I (2023) Evaluation of nickel toxicity and potential health implications of agriculturally diversely irrigated wheat crop varieties. Arab J Chem 16(8):104934. https://doi.org/10.1016/j.arabjc.2023.104934

Kim YH, Khan AL, Kim DH, Lee SY, Kim KM, Waqas M, Jung HY, Shin JH, Kim JG, Lee IJ (2014) Silicon mitigates heavy metal stress by regulating P-type heavy metal ATPases, Oryza sativa low silicon genes, and endogenous phytohormones. BMC Plant Biol 14:1–13. https://doi.org/10.1186/1471-2229-14-13

Kobayashi I, Fujiwara S, Shimogawara K, Sakuma C, Shida Y, Kaise T, Usuda H, Tsuzuki M (2005) High intracellular phosphorus contents exhibit a correlation with arsenate resistance in Chlamydomonas mutants. Plant Cell Physiol 46:489–496

Kumar L, Bharadvaja N (2020) Microbial remediation of heavy metals. In: Shah M (ed) Microbial bioremediation & biodegradation. Springer, Singapore. https://doi.org/10.1007/978-981-15-1812-6_2

Kumar M (2013) Crop plants and abiotic stresses. J Biomol Res Ther 3(1). https://doi.org/10.4172/2167-7956.1000e125

Kumar N, Kumari V, Ram C, Thakur K, Tomar SK (2018) Bio-prospectus of cadmium bioadsorption by lactic acid bacteria to mitigate health and environmental impacts. Appl Microbiol Biotechnol 102:1599–1615. https://doi.org/10.1007/s00253-018-8743-9

Kumar S, Prasad S, Yadav KK, Shrivastava M, Gupta N, Nagar S, Bach Q-V, Kamyab H, Khan SA, Yadav S, Malav LC (2019) Hazardous heavy metals contamination of vegetables and food chain: role of sustainable remediation approaches – a review. Environ Res 179:108792. https://doi.org/10.1016/j.envres.2019.108792

Kumar V, Pandita S, Sidhu GPS, Sharma A, Khanna K, Kaur P, Bali AS, Setia R (2021) Copper bioavailability, uptake, toxicity and tolerance in plants: a comprehensive review. Chemosphere 262:127810. https://doi.org/10.1016/j.chemosphere.2020.127810

Lane TW, Morel FM (2000) A biological function for cadmium in marine diatoms. Proc Natl Acad Sci 97(9):4627–4631. https://doi.org/10.1073/pnas.090091397

Lane TW, Saito MA, George GN, Pickering IJ, Prince RC, Morel FM (2005) A cadmium enzyme from a marine diatom. Nature 435(7038):42–42. https://doi.org/10.1038/435042a

Langella P, Guarner F, Martín R (2019) Next-generation probiotics: from commensal bacteria to novel drugs and food supplements. Front Microbiol 10:1973. https://doi.org/10.3389/fmicb.2019.01973

Li A, Li A, Luo C, Liu B (2024) Assessing heavy metal contamination in *Amomum villosum* Lour. fruits from plantations in Southern China: soil-fungi-plant interactions. Ecotoxicol Environ Saf 269:115789. https://doi.org/10.1016/j.ecoenv.2023.115789

Li M, Zhou X, Ding W, Guo S, Wu N (2013) Fluorescent aptamer-functionalized graphene oxide biosensor for label-free detection of mercury(II). Biosens Bioelectron 41:889–893. https://doi.org/10.1016/j.bios.2012.09.060

Lili Z, Junyan W, Hongfei Z, Baoqing Z, Bolin Z (2018) Detoxification of cancerogenic compounds by lactic acid bacteria strains. Crit Rev Food Sci Nutr 58:2727–2742. https://doi.org/10.1080/10408398.2017.1339665

Lim HJ, Park JH, Jo C, Yoon K, Koh YH (2019) Cigarette smoke extracts and cadmium induce COX-2 expression through γ-secretase-mediated p38 MAPK activation in C6 astroglia cells. PLoS One 14(2):e0212749. https://doi.org/10.1371/journal.pone.0212749

Lipovšek S, Novak T, Dariš B, Hofer F, Leitinger G, Letofsky-Papst I (2022) Ultrastructure of spherites in the midgut diverticula and Malpighian tubules of the harvestman *Amilenus aurantiacus* during the winter diapause. Histochem Cell Biol 157(1):107–118. https://doi.org/10.1007/s00418-021-02046-0

Liu T, Nakashima S, Shibasaka M, Katsuhara M, Kasamo K (2002) A novel histidine-rich CPx-ATPase from the filamentous Cyanobacterium *Oscillatoria brevis* related to multiple-heavy-metal cotolerance. J Bacteriol 184:5027–5035

Liu Y, Huang L, Wen Z, Fu Y, Liu Q, Xu S, Li Z, Liu C, Yu C, Feng Y (2023) Effects of intercropping on safe agricultural production and phytoremediation of heavy metal-contaminated soils. Sci Total Environ 875:162700. https://doi.org/10.1016/j.scitotenv.2023.162700

Long F, Zhu A, Shi H, Wang H, Liu J (2013) Rapid on-site/in-situ detection of heavy metal ions in environmental water using a structure-switching DNA optical biosensor. Sci Rep 3:1–7. https://doi.org/10.1038/srep02308

Luo Z, Duan H, Yang Y, Zhang W, Ramaswamy HS, Wang C (2021) Ultrasound assisted extraction of cadmium for decontamination of rice and its influence on structure/texture of cooked rice. J Cereal Sci 97:103142. https://doi.org/10.1016/j.jcs.2020.103142

Machate DJ (2023) Anthropogenic hyperactivity for natural resources increases heavy metals concentrations in the environment: toxicity of healthy food and cancer risks estimated. J Trace Elem Min 4:100057. https://doi.org/10.1016/j.jtemin.2023.100057

Mahmood T, Qadosi IQ, Fatima H, Akrim F, Rais M (2016) Metal concentrations in common skittering frog (Euphlyctis cyanophlyctis) inhabiting Korang River, Islamabad, Pakistan. Basic Appl Herpetol 30:25–38

Mahmoud R, Casadebaig P, Hilgert N, Alletto L, Freschet GT, de Mazancourt C, Gaudio N (2022) Species choice and N fertilization influence yield gains through complementarity and selection effects in cereal-legume intercrops. Agron Sustain Dev 42(2):12. https://doi.org/10.1007/s13593-022-00754-y

Maitra S, Hossain A, Brestic M, Skalicky M, Ondrisik P, Gitari H, Brahmachari K, Shankar T, Bhadra P, Palai JB, Jena J (2021) Intercropping—A low input agricultural strategy for food and environmental security. Agronomy 11(2):343. https://doi.org/10.3390/agronomy11020343

Majer BJ, Tscherko D, Paschke A (2002) Effects of heavy metal contamination of soils on micronucleus induction in Tradescantia and on microbial enzyme activities: a comparative investigation. Mut Res 515(2002):111–124. https://doi.org/10.1016/S1383-5718(02)00004-9

Mangwani N, Kumari S, Das S (2017) Marine bacterial biofilms in bioremediation of polycyclic aromatic hydrocarbons (PAHs) under terrestrial condition in a soil microcosm. Pedosphere 27(3):548–558. https://doi.org/10.1016/s1002-0160(17)60350-3

Mani MS, Kabekkodu SP, Joshi MB, Dsouza HS (2019) Ecogenetics of lead toxicity and its influence on risk assessment. Hum Exp Toxicol 38(9):1031–1059. https://doi.org/10.1177/0960327119851253

Martínez-Castillo JI, Saldaña-Robles A, Ozuna C (2022) Arsenic stress in plants: a metabolomic perspective. Plant Stress 3:100055. https://doi.org/10.1016/j.stress.2022.100055

Matsui K, Yoshinami S, Narita M, Chien M-F, Phung LT, Silver S, Endo G (2016) Mercury resistance transposons in Bacilli strains from different geographical regions. FEMS Microbiol Lett 363. https://doi.org/10.1093/femsls/fnw013

Mei X, Wang Y, Li Z, Larousse M, Pere A, da Rocha M, Zhan F, He Y, Pu L, Panabières F, Zu Y (2022) Root-associated microbiota drive phytoremediation strategies to lead of Sonchus Asper (L.) Hill as revealed by intercropping-induced modifications of the rhizosphere microbiome. Environ Sci Pollut Res 29:1–15. https://doi.org/10.1007/s11356-021-17353-1

Mohajeri M, Rezaee M, Sahebkar A (2017) Cadmium-induced toxicity is rescued by curcumin: a review. Biofactors 43(5):645–661. https://doi.org/10.1002/biof.1376

Morel FM, Lam PJ, Saito MA (2020) Trace metal substitution in marine phytoplankton. Annu Rev Earth Planet Sci 48(1):491–517. https://doi.org/10.1146/annurev-earth-053018-060108

Mpouras T, Polydera A, Dermatas D, Verdone N, Vilardi G (2021) Multi wall carbon nanotubes application for treatment of Cr (VI)-contaminated groundwater; Modeling of batch & column experiments. Chemosphere 269:128749. https://doi.org/10.1016/j.chemosphere.2020.128749

Muhialdin BJ, Saari N, Meor Hussin AS (2020) Review on the biological detoxification of mycotoxins using lactic acid bacteria to enhance the sustainability of foods supply. Molecules 25(11):2655. https://doi.org/10.3390/molecules25112655

Naksen P, Boonruang S, Yuenyong N, Lee HL, Ramachandran P, Anutrasakda W, Amatatongchai M, Pencharee S, Jarujamrus P (2022) Sensitive detection of trace level Cd (II) triggered by chelation enhanced fluorescence (CHEF) “turn on”: nitrogen-doped graphene quantum dots (N-GQDs) as fluorometric paper-based sensor. Talanta 242:123305. https://doi.org/10.1016/j.talanta.2022.123305

Narita M, Lin K-H, Matsui K, Huang C-C, Endo G (2010) Organomercurials removal by heterogeneous *merB* genes harboring bacterial strains. J Biosci Bioeng 110:94–98

Narita M, Yamagata T, Ishii H, Huang C-C, Endo G (2002) Simultaneous detection and removal of organomercurial compounds by using the genetic expression system of an organomercury lyase from the transposon *TnMERI1*. Appl Microbiol Biotechnol 59:86–90

Nazar R, Iqbal N, Masood A, Khan M, Syeed S, Khan N (2012) Cadmium toxicity in plants and role of mineral nutrients in its alleviation. Am J Plant Sci 3(2012):1476–1489. https://doi.org/10.4236/ajps.2012.310178

Nerín C, Aznar M, Carrizo D (2016) Food contamination during food process. Trends Food Sci Technol 48:63–68. https://doi.org/10.1016/j.tifs.2015.12.004

Nevetha R, Kumar S, Yashini M, Rajeshwari S, Mamathi CA, Nirmal Thirunavookarasu S, Sunil CK (2021) Recent advances in surface plasmon resonance (SPR) biosensors for food analysis: a review. Crit Rev Food Sci Nutr 63:1055–1077. https://doi.org/10.1080/10408398.2021.1958745

Nies DH (2003) Efflux-mediated heavy metal resistance in prokaryotes. FEMS Microbiol Rev 27:313–339

Nogacka AM, Gómez-Martín M, Suárez A, González-Bernardo O, de los Reyes-Gavilán, C.G., González, S. (2019) Xenobiotics formed during food processing: their relation with the intestinal microbiota and colorectal cancer. Int J Mol Sci 20:2051. https://doi.org/10.3390/ijms20082051

Nordberg GF, Bernard A, Diamond GL, Duffus JH, Illing P, Nordberg M, Bergdahl IA, Jin T, Skerfving S (2018) Risk assessment of effects of cadmium on human health (IUPAC Technical Report). Pure Appl Chem 90(4):755–808. https://doi.org/10.1515/pac-2016-0910

Oyetibo GO, Adebusoye SA, Ilori MO, Amund OO (2023) Heavy metals assessment of ecosystem polluted with wastewaters and taxonomic profiling of multi-resistant bacteria with potential for petroleum hydrocarbon catabolism in nitrogen-limiting medium. World J Microbiol Biotechnol 39:84. https://doi.org/10.1007/s11274-023-03524-4

Oyetibo GO, Chien M-F, Otsubo-Ikeda W, Obayori OS, Adebusoye SA, Ilori MO, Endo G (2017a) Biodegradation of crude oil and phenanthrene by heavy metal resistant *Bacillus subtilis* isolated from multi-polluted environments. Int Biodeterior Biodegradation 120:143–151. https://doi.org/10.1016/j.ibiod.2017.02.021

Oyetibo GO, Miyauchi K, Ilori MO, Amund OO, Endo G (2017b) Biotechnological remedies for polluted estuarine environment where heavy metals and persistent organic pollutants co-exist. Int Biodeterior Biodegradation 119:614–625. https://doi.org/10.1016/j.ibiod.2016.10.005

Oyetibo GO, Ilori MO, Obayori OS, Amund OO (2014) Equilibrium studies of cadmium biosorption by presumed non-viable bacterial strains isolated from polluted sites. Int Biodeterior Biodegradation 91:37–44

Oyetibo GO, Ilori MO, Obayori OS, Amund OO (2015a) Metal biouptake by actively growing cells of metal-tolerant bacterial strains. Environ Monit Assess 187:525. https://doi.org/10.1007/s10661-015-4731-z

Oyetibo GO, Ishola ST, Ikeda-Ohtsubo W, Miyauchi K, Ilori MO, Endo G (2015b) Mercury bioremoval by *Yarrowia* strains isolated from sediments of mercury-polluted estuarine water. Appl Microbiol Biotechnol 99:3651–3657. https://doi.org/10.1007/s00253-014-6279-1

Oyetibo GO, Miyauchi K, Huang Y, Ikeda-Ohtsubo W, Chien M-F, Ilori MO, Amund OO, Endo G (2019a) Comparative geochemical evaluation of toxic metals pollution and bacterial communities of industrial effluent tributary and a receiving estuary in Nigeria. Chemosphere 227:638–646. https://doi.org/10.1016/j.chemosphere.2019.04.048

Oyetibo GO, Miyauchi K, Suzuki H, Endo G (2016a) Mercury removal during growth of mercury resistant and self-aggregating *Yarrowia* spp. AMB Express 6:99–110. https://doi.org/10.1186/s13568-016-0271-3

Oyetibo GO, Miyauchi K, Suzuki H, Endo G (2019b) Bio-oxidation of elemental mercury during growth of mercury resistant yeasts in simulated hydrosphere. J Hazard Mater 373:243–249. https://doi.org/10.1016/j.jhazmat.2019.02.075

Oyetibo GO, Miyauchi K, Suzuki H, Ishikawa S, Endo G (2016b) Extracellular mercury sequestration by exopolymeric substances produced by *Yarrowia* spp. during growth: thermodynamics, mass action equilibrium, and kinetics studies. J Biosci Bioeng 122:701–707. https://doi.org/10.1016/j.jbiosc.2016.05.009

Pan-Hou H, Kiyono M, Omura H, Omura T, Endo G (2002) Polyphosphate production in recombinant *Escherichia coli* confers mercury resistance. FEMS Microbiol Lett 10325:159–164

Perales-Vela HV, Pena-Castro JM, Canazares-Villanueva RO (2006) Heavy metal detoxification in eukaryotic microalgae. Chemosphere 64:1–10

Perron MJ, Stremlau M, Song B, Ulm W, Mulligan RC, Sodroski J (2004) *CzcR-CzcS,* a two-component system involved in heavy metal and carbapenem resistance in *Pseudomonas aeruginosa**. J Biol Chem 279:8761–8768

Pogoson E, Carey M, Meharg C, Meharg AA (2021) Reducing the cadmium, inorganic arsenic and dimethylarsinic acid content of rice through food-safe chemical cooking pre-treatment. Food Chem 338:127842. https://doi.org/10.1016/j.foodchem.2020.127842

Porova N, Botvinnikova V, Krasulya O, Cherepanov P, Potoroko I (2014) Effect of ultrasonic treatment on heavy metal decontamination in milk. Ultrason Sonochem 21(6):2107–2111. https://doi.org/10.1016/j.ultsonch.2014.03.029

Rady MM, Elrys AS, Abo El-Maati MF, Desoky EM (2019) Interplaying roles of silicon and proline effectively improve salt and cadmium stress tolerance in *Phaseolus vulgaris* plant. Plant Physiol Biochem 139:558–568. https://doi.org/10.1016/j.plaphy.2019.04.025

Raji Z, Karim A, Karam A, Khalloufi S (2023) A review on the heavy metal adsorption capacity of dietary fibers derived from agro-based wastes: opportunities and challenges for practical applications in the food industry. Trends Food Sci Technol 137:74–91. https://doi.org/10.1016/j.tifs.2023.05.004

Ray S, Ray MK (2009) Bioremediation of heavy metal toxicity-with special reference to chromium. Al Ameen J Med Sci 2(2):57–63

Reddivari L, Veeramachaneni DNR, Walters WA, Lozupone C, Palmer J, Hewage MKK, Bhatnagar R, Amir A, Kennett MJ, Knight R, Vanamala JKP (2017) Perinatal bisphenol A exposure induces chronic inflammation in rabbit offspring via modulation of gut bacteria and their metabolites. mSystems 2:1–16. https://doi.org/10.1128/msystems.00093-17

Reyes-Hinojosa D, Lozada-Pérez CA, Zamudio Cuevas Y, López-Reyes A, Martínez-Nava G, Fernández-Torres J, Olivos-Meza A, Landa-Solis C, Gutiérrez-Ruiz MC, Rojas del Castillo E, Martínez-Flores K (2019) Toxicity of cadmium in musculoskeletal diseases. Environ Toxicol Pharmacol 72:103219. https://doi.org/10.1016/j.etap.2019.103219

Rezaei-Tavirani M, Farahani M (2024) Synthetic and systems biotechnology in food sciences. Appl Food Biotechnol 11(2):e1–e1. https://doi.org/10.22037/afb.v11i2.47151

Rinaldi M, Micali A, Marini H, Adamo EB, Puzzolo D, Pisani A, Trichilo V, Altavilla D, Squadrito F, Minutoli L (2017) Cadmium, organ toxicity and therapeutic approaches: a review on brain, kidney and testis damage. Curr Med Chem 24(35):3879–3893. https://doi.org/10.2174/0929867324666170801101448

Rofrano G, Esposito M, Pizzolante A, Miniero R, Danese A, Signorelli D, D'Auria LJ, Gallo A, Di Stasio A, Gallo P, Cerino P (2023) Inorganic elements profiling in milk from buffalo farms of the Campania Region (Italy). J Trace Elem Miner 3:100046. https://doi.org/10.1016/j.jtemin.2023.100046

Rogowska A, Pomastowski P, Rafińska K, Railean-Plugaru V, Zloch M, Walczak J, Buszewski B (2019) A study of zearalenone biosorption and metabolisation by prokaryotic and eukaryotic cells. Toxicon 169:81–90. https://doi.org/10.1016/j.toxicon.2019.09.008

Rost-Roszkowska M, Poprawa I, Chajec L, Chachulska-Żymełka A, Wilczek G, Skowronek M, Student S, Leśniewska M (2022) Hazards related to the presence of cadmium in food—studies on the European soil centipede, *Lithobius forficatus*. Sci Total Environ 845:157298. https://doi.org/10.1016/j.scitotenv.2022.157298

Scarano G, Morelli E (2003) Properties of phytochelatin-coated CdS nanocrystallites formed in a marine phytoplanktonic alga (*Phaeodactylum tricornutum*, Bohlin) in response to Cd. Plant Sci 165:803–810

Schaefer HR, Dennis S, Fitzpatrick S (2020) Cadmium: mitigation strategies to reduce dietary exposure. J Food Sci 85:260–267. https://doi.org/10.1111/1750-3841.14997

Schmidt A, Haferburg G, Sineriz M, Merten D, Buchel G, Kothe E (2005) Heavy metal resistance mechanisms in actinobacteria for survival in AMD contaminated soils. Chem Erde 65:131–144

Selvi A, Rajasekar A, Theerthagiri J, Ananthaselvam A, Sathishkumar K, Madhavan J, Rahman PKSM (2019) Integrated remediation processes toward heavy metal removal/recovery from various environments—a review. Front Environ Sci 7:66. https://doi.org/10.3389/fenvs.2019.00066

Shah V, Daverey A (2020) Phytoremediation: a multidisciplinary approach to clean up heavy metal contaminated soil. Environ Technol Innov 18:100774. https://doi.org/10.1016/j.eti.2020.100774

Shahid M, Dumat C, Khalid S, Niazi NK, Antunes PM (2017) Cadmium bioavailability, uptake, toxicity and detoxification in soil-plant system. Rev Environ Contam Toxicol 241:73–137. https://doi.org/10.1007/398_2016_8

Shamshir F, Abbas G, Amjad M, Rizwan M, Akram M, Ahmad S, Tahir M, Ali S, Farooq ABU (2022) Physiological and biochemical characterization of Kalongi (Nigella sativa) against arsenic stress: implications for human health risk assessment. Environ Pollut 298:118829. https://doi.org/10.1016/j.envpol.2022.118829

Sharafi K, Yunesian M, Mahvi AH, Pirsaheb M, Nazmara S, Nodehi RN (2019) Advantages and disadvantages of different pre-cooking and cooking methods in removal of essential and toxic metals from various rice types-human health risk assessment in Tehran households. Iran Ecotoxicol Environ Saf 175:128–137. https://doi.org/10.1016/j.ecoenv.2019.03.056

Shariatifar N, Rezaei M, Alizadeh Sani M, Alimohammadi M, Arabameri M (2020) Assessment of rice marketed in Iran with emphasis on toxic and essential elements; effect of different cooking methods. Biol Trace Element Res 198:721–731. https://doi.org/10.1007/s12011-020-02110-1

Sharma AK, Sharma M, Sharma AK, Sharma M (2023) Mapping the impact of environmental pollutants on human health and environment: a systematic review and meta-analysis. J Geochem Explor 255:107325. https://doi.org/10.1016/j.gexplo.2023.107325

Sharma AK, Sharma M, Sharma S, Malik DS, Sharma M, Sharma AK (2024) A systematic review on assessment of heavy metals toxicity in freshwater fish species: current scenario and remedial approaches. J Geochem Explor 262:107472. https://doi.org/10.1016/j.gexplo.2024.107472

Shen C, Tang S, Meng Q (2019) Cadmium removal from rice protein via synergistic treatment of rhamnolipids and F127/PAA hydrogels. Colloids Surf B: Biointerfaces 181:734–739. https://doi.org/10.1016/j.colsurfb.2019.06.019

Silano M, Silano V (2017) Food and feed chemical contaminants in the European Union: regulatory, scientific, and technical issues concerning chemical contaminants occurrence, risk assessment, and risk management in the European Union. Crit Rev Food Sci Nutr 57:2162–2217. https://doi.org/10.1080/10408398.2015.1059313

Slope LN, Peacock AF (2016) *De Novo* design of Xeno-Metallo coiled coils. Chem Asian J 11(5):660–666. https://doi.org/10.1002/asia.201501173

Smiejan A, Wilkinson KJ, Rossier C (2003) Cd bioaccumulation by a freshwater bacterium, *Rhodospirillum rubrum*. Environ Sci Technol 37:701–706

Soldatkin OO, Kucherenko IS, Pyeshkova VM, Kukla AL, Jaffrezic-Renault N, El'skaya AV, Dzyadevych SV, Soldatkin AP (2012) Novel conductometric biosensor based on three-enzyme system for selective determination of heavy metal ions. Bioelectrochemistry 83:25–30. https://doi.org/10.1016/j.bioelechem.2011.08.001

Średnicka P, Juszczuk-Kubiak E, Wójcicki M, Akimowicz M, Roszko ML (2021) Probiotics as a biological detoxification tool of food chemical contamination: a review. Food Chem Toxicol 153:112306. https://doi.org/10.1016/j.fct.2021.112306

Steenland K, Boffetta P (2000) Lead and cancer in humans: where are we now? Am J Ind Med 38(3):295–299. https://doi.org/10.1002/1097-0274(200009)38:3<295::AID-AJIM8>3.0.CO;2-L

Tang J, Zhu Q, Xu Y, Zhou Y, Zhu L, Jin L, Wang W, Gao L, Chen G, Zhao H (2022) Total arsenic, dimethylarsinic acid, lead, cadmium, total mercury, methylmercury and hypertension among Asian populations in the United States: NHANES 2011–2018. Ecotoxicol Environ Saf 241:113776. https://doi.org/10.1016/j.ecoenv.2022.113776

Teemu H, Seppo S, Jussi M, Raija T, Kalle L (2008) Reversible surface binding of cadmium and lead by lactic acid and Bifidobacteria. Int J Food Microbiol 125:170–175. https://doi.org/10.1016/j.ijfoodmicro.2008.03.041

Thakur M, Praveen S, Divte PR, Mitra R, Kumar M, Gupta CK, Kalidindi U, Bansal R, Roy S, Anand A, Singh B (2022) Metal tolerance in plants: molecular and physicochemical interface determines the "not so heavy effect" of heavy metals. Chemosphere 287:131957. https://doi.org/10.1016/j.chemosphere.2021.131957

Thakur N, Sharma M, Sharma M (2021) Phytoremediation-a green technology adapted to eradication of harmful heavy toxic metals from contaminated soil. J Innov Agric 8(1):26–31. https://doi.org/10.37446/jinagri/ra/8.1.2021.26-31

Tsai H-C, Doong R-A, Chiang H-C, Chen K-T (2003) Sol–gel derived urease-based optical biosensor for the rapid determination of heavy metals. Anal Chim Acta 481:75–84

Turanlı N, Gedik K (2021) Spatial trace element bioaccumulation along with consumer risk simulations of Mediterranean mussels in coastal waters of Turkey. Environ Sci Pollut Res 28:41746–41759. https://doi.org/10.1007/s11356-021-13506-4

Ullah H, Khan AM, Ullah S, Asif M, Imtiaz M, Shazia B (2025) Perspective chapter: Heavy metals-mediated chemical contamination in foods, associated health risks, and remediation techniques. In: Heavy metals in the environment-contamination, risk, and remediation. IntechOpen. https://doi.org/10.5772/intechopen.1007943

Upadhyay MK, Shukla A, Yadav P, Srivastava S (2019) A review of arsenic in crops, vegetables, animals and food products. Food Chem 276:608–618. https://doi.org/10.1016/j.foodchem.2018.10.069

Urbański LJ, Di Fiore A, Azizi L, Hytönen VP, Kuuslahti M, Buonanno M, Monti SM, Angeli A, Zolfaghari Emameh R, Supuran CT, De Simone G (2020) Biochemical and structural characterisation of a protozoan beta-carbonic anhydrase from Trichomonas vaginalis. J Enzyme Inhib Med Chem 35(1):1292–1299. https://doi.org/10.1080/14756366.2020.1774572

Vanderschueren R, De Mesmaeker V, Mounicou S, Isaure MP, Doelsch E, Montalvo D, Delcour JA, Chavez E, Smolders E (2020) The impact of fermentation on the distribution of cadmium in cacao beans. Food Res Int 127:108743. https://doi.org/10.1016/j.foodres.2019.108743

Verma N, Kaur H, Kumar S (2011) Whole cell based electrochemical biosensor for monitoring lead ions in milk. Biotechnology 10:259–266. https://doi.org/10.3923/biotech.2011.259.266

Verma N, Singh M (2006) A Bacillus sphaericus based biosensor for monitoring nickel ions in industrial effluents and foods. J Autom Methods Manag Chem 2006:1–4. https://doi.org/10.1155/JAMMC/2006/83427

Wang H, Wang A, Wang X, Zeng X, Xing H (2022) AMPK/PPAR-γ/NF-κB axis participates in ROS-mediated apoptosis and autophagy caused by cadmium in pig liver. Environ Pollut 294:118659. https://doi.org/10.1016/j.envpol.2021.118659

Wang J, Ji H, Wang S, Liu H, Zhang W, Zhang D, Wang Y (2018) Probiotic *Lactobacillus plantarum* promotes intestinal barrier function by strengthening the epithelium and modulating gut microbiota. Front Microbiol 9:1953. https://doi.org/10.3389/fmicb.2018.01953

Wang Y, Xie Z, Wu Q, Song W, Liu L, Wu Y, Gong Z (2020) Preparation and characterization of carboxymethyl starch from cadmium-contaminated rice. Food Chem 308:125674. https://doi.org/10.1016/j.foodchem.2019.125674

Wang Y, Wu Y, Wang Y, Xu H, Mei X, Yu D, Wang Y, Li W (2017) Antioxidant properties of probiotic bacteria. Nutrients 9(521). https://doi.org/10.3390/nu9050521

Wang YS, Wu T-H, Yang Y, Zhu C-L, Ding C-L, Dai C-C (2016) Binding and detoxification of chlorpyrifos by lactic acid bacteria on rice straw silage fermentation. J Environ Sci Heal Part B 51:316–325. https://doi.org/10.1080/03601234.2015.1128744

WHO, World Health Organization (2019) Exposure to cadmium: A major public health concern: 2019. Internet: http://www.who.int/ipcs/features/cadmium.pdf

Worms I, Simon DF, Hassler CS, Wilkinson KJ (2006) Bioavailability of trace metals to aquatic microorganisms: importance of chemical, biological and physical processes on biouptake. Biochimie 88:1721–1731

Wosinska L, Cotter PD, O'Sullivan O, Guinane C (2019) The potential impact of probiotics on the gut microbiome of athletes. Nutrients 11(2270). https://doi.org/10.3390/nu11102270

Wu CC, Gardarin A, Catty P, Guillain F, Mintz E (2006a) CadA, the Cd^{2+}-ATPase from *Listeria monocytogenes*, can use Cd^{2+} as co-substrate. Biochimie 88:1687–1692

Wu CH, Wood TK, Mulchandani A, Chen W (2006b) Engineering plant-microbe symbiosis for rhizoremediation of heavy metals. Appl Environ Microbiol 72:1129–1134

Xu X, Liu W, Tian S, Wang W, Qi Q, Jiang P, Gao X, Li F, Li H, Yu H (2018) Petroleum hydrocarbon-degrading bacteria for the remediation of oil pollution under aerobic conditions: a perspective analysis. Front Microbiol 9:2885. https://doi.org/10.3389/fmicb.2018.02885

Yadav SK (2010) Heavy metals toxicity in plants: an overview on the role of glutathione and phytochelatins in heavy metal stress tolerance of plants. South Afr J Bot 76:167–179. https://doi.org/10.1016/j.sajb.2009.10.007

Yu H, Liu Y, Li L, Guo Y, Xie Y, Cheng Y, Yao W (2020) Ultrasound-involved emerging strategies for controlling foodborne microbial biofilms. Trends Food Sci Technol 96:91–101. https://doi.org/10.1016/j.tifs.2019.12.010

Yuan S, Li C, Zhang Y, Yu H, Xie Y, Guo Y, Yao W (2021) Ultrasound as an emerging technology for the elimination of chemical contaminants in food: a review. Trends Food Sci Technol 109:374–385. https://doi.org/10.1016/j.tifs.2021.01.048

Zayed A, Adly GM, Farag MA (2025) Management strategies for the anti-nutrient oxalic acid in foods: a comprehensive overview of its dietary sources, roles, metabolism, and processing. Food Bioprocess Technol 18:1–21. https://doi.org/10.1007/s11947-024-03726-0

Zhai Q, Liu Y, Wang C, Qu D, Zhao J, Zhang H, Tian F, Chen W (2019) *Lactobacillus plantarum* CCFM8661 modulates bile acid enterohepatic circulation and increases lead excretion in mice. Food Funct 10:1455–1464. https://doi.org/10.1039/C8FO02554A

Zhai Q, Wang G, Zhao J, Liu X, Tian F, Zhang H, Chen W (2013) Protective effects of *Lactobacillus plantarum* CCFM8610 against acute cadmium toxicity in mice. Appl Environ Microbiol 79:1508–1515. https://doi.org/10.1128/AEM.03417-12

Zhang L, Lei Q, Cheng Y, Xie Y, Qian H, Guo Y, Chen Y, Yao W (2017) Study on the removal of cadmium in rice using microbial fermentation method. J Food Sci 82(6):1467–1474. https://doi.org/10.1111/1750-3841.13734

Zhang T, Feng L, Cui J, Tong W, Zhao H, Wu T, Zhang P, Wang X, Gao Y, Su J, Fu X (2024) Hexavalent chromium induces neurotoxicity by triggering mitochondrial dysfunction and ROS-mediated signals. Neurochem Res 49(3):660–669. https://doi.org/10.1007/s11064-023-04063-y

Zhang Y, Liu Z, He Q, Wu F, Xiao Y, Chen W, Jin Y, Yu D, Wang Q (2021) Construction of mode of action for cadmium-induced renal tubular dysfunction based on a toxicity pathway-oriented approach. Front Genet 12:696892. https://doi.org/10.3389/fgene.2021.696892

Zhao XX, Xie ZM, Li YW, Liu TK, Liu DW (2022) Synergy of Fe and biogenic Mn oxide components mediated by a newly isolated indigenous bacterium to enhance As(III/V) immobilization in groundwater. Appl Geochem 140:105266

Zhu Y, Zhang T, Xu D, Wang S, Yuan Y, He S, Cao Y (2019) The removal of pesticide residues from pakchoi (*Brassica rape* L. ssp. chinensis) by ultrasonic treatment. Food Control 95:176–180. https://doi.org/10.1016/j.foodcont.2018.07.039

Zloch Z, Sedlacek P, Langmajerová J, Mullerova D (2018) Intake and profile of plant polyphenols in the diet of the Czech population. Pol J Food Nutr Sci 68:57–62

Zou M, Zhou S, Zhou Y, Jia Z, Guo T, Wang J (2021) Cadmium pollution of soil-rice ecosystems in rice cultivation dominated regions in China: a review. Environ Pollut 280:116965. https://doi.org/10.1016/j.envpol.2021.116965

Zou Y, Zhang C, Ju X, Wang Z, Wu Y, Yuan J, Chen W, He R (2019) Effect of removing cadmium with citric acid on the physicochemical and microstructure properties of rice bran. Food Control 98:290–296. https://doi.org/10.1016/j.foodcont.2018.11.044

Zukowska J, Biziuk M (2008) Methodological evaluation of method for dietary heavy metal intake. J Food Sci 73(2):R21–R29. https://doi.org/10.1111/j.1750-3841.2007.00648.x

Chapter 8
Symbiotic Strategies for Plant Stress Tolerance: The Role of Beneficial Microbes and Omics in Sustainable Agriculture

Akinlolu Olalekan Akanmu, Modupe Stella Ayilara, and Olubukola Oluranti Babalola

Abstract Abiotic and biotic stresses pose significant challenges to agricultural productivity, often resulting in substantial yield losses. Reliance on chemical pesticides has not satisfactorily mitigated these challenges. Hence, there is a need to explore more sustainable management measure that leverages the beneficial microbe-plant symbiotic relationship, an association that is fundamental to achieving crop resilience, nutrient acquisition, and plant tolerance to stress and diseases. The beneficial microbes consisting of the plant growth-promoting rhizobacteria (PGPR), nitrogen-fixing bacteria, and mycorrhizal fungi enhance plant productivity and health through their modulation of phytohormone levels, nutrient solubilization, production of bioactive compounds, and ability to induce systemic resistance mechanism against pathogens, drought, or saline conditions. Hence, this review explores the functional roles of microbe-plant interaction that confers induced tolerance to plants. The current advances in biotechnology offer new insight into engineering the microbial consortia that is tailored for specific crop and stress conditions. It further explores the new frontiers of research in the microbe-plant symbiosis, highlighting its application in sustainable agriculture and the development of climate-resilient cropping systems. Understanding and harnessing the benefits of this interaction will result in the practice of sustainable agriculture and stress management while improving climate resilience and crop productivity without dependency on chemical use.

A. O. Akanmu · M. S. Ayilara
Food Security and Safety Focus Area, Faculty of Natural and Agricultural Sciences, North-West University, Mmabatho, South Africa

O. O. Babalola (✉)
Department of Life Sciences, Imperial College London, Ascot, Berkshire, UK

Food Security and Safety Focus Area, Faculty of Natural and Agricultural Sciences, North-West University, Mmabatho, South Africa
e-mail: olubukola.babalola@nwu.ac.za

O. O. Babalola et al. (eds.), *Ecofriendly Frontiers*,
https://doi.org/10.1007/978-3-031-98700-7_8

Keywords Abiotic stress · Biotic stress · Cropping system · Genes · Induced systemic resistance · Next-Generation sequencing

8.1 Introduction

Plants encounter multiple environmental stress factors that endanger their yield, normal growth processes, and development. Plants face two main environmental stresses which are classified as abiotic and biotic stresses (Adeleke et al. 2024b). The abiotic stress includes heavy metal contamination, drought, and salinity, while the biotic stress includes invading pests and diseases (Fig. 8.1). To alleviate the effects of biotic and abiotic stresses, plants use complex defence mechanisms through their relationships with beneficial microbes. The interactions between plants and microbes are essential for enhancing plant health conditions and increasing plant resilience for improved yield (Ojuederie et al. 2021; Chamkhi et al. 2022). These relationships are controlled by complex signaling pathways and regulatory mechanisms. The identification of these mechanisms plays an essential role in using microbial symbiosis as an operational tool in sustainable agricultural systems and the management of the ecosystem. The interaction between plants and rhizobia is a notable and likely the most famous symbiotic relationship resulting in plant atmospheric nitrogen acquisition (Ayilara et al. 2022). The release of flavonoids by leguminous plant roots attracts rhizobia bacteria, which manufacture Nod factors that activate root nodule development processes (Shumilina et al. 2023). The rhizobia bacteria produce bacteroid inside the root nodules to transform atmospheric nitrogen fixation into ammonia. The leghemoglobin plant protein generates an oxygen-regulated atmosphere that allows nitrogenase enzyme function to proceed. The

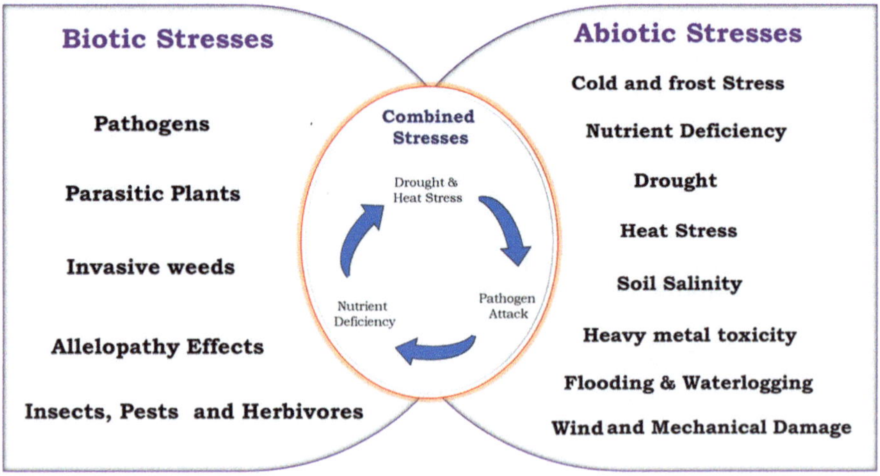

Fig. 8.1 Sources of biotic and abiotic stresses in crops

exchange of resources between plants and bacteria becomes mutually beneficial as the plant supplies carbon-enriched compounds for bacterial benefit. Through this mechanism, the biological nitrogen fixation process decreases plant demands for synthetic nitrogen fertilizers, resulting in valuable outcomes in terms of ecology and economic benefits.

Similarly, plants effectively acquire phosphorus and micronutrients through symbiotic relationships with mycorrhizal fungi (Das et al. 2022). Through its relationship with plants, Arbuscular mycorrhizal fungi penetrate cortical cells and then form arbuscular structures that help transfer nutrients (Das et al. 2022). Specifically, extending the fungal hyphae in the soil helps plants obtain more nutrients by expanding plant root absorptive areas. Mycorrhizal fungi encase plant roots externally to improve wood plants and tree access to nutrients. The symbiotic relationship between plants and mycorrhizal fungi transforms root architecture to create better soil structure and boosts disease resistance, nutrient acquisition abilities, and drought tolerance (Das et al. 2022). Endophytic microbes consisting of fungi and bacteria establish inside plant tissues without causing damage (Adeleke et al. 2024c). Instead, they are advantageous to plants as they boost their development through defence system stimulation, essential nutrient acquisition and phytohormone production. A significant number of endophytic microbes activate systemic resistance through mechanisms that induce plants to fight against pathogens and environmental stress agents (induce systemic resistance) (Akanmu et al. 2021; Ayilara et al. 2024a, b). The broad species host range and flexible nature of endophytes make them strong potential candidates as agricultural biostimulants compared to rhizobia and mycorrhizal fungi (Adeleke et al. 2022). Sophisticated molecular networks control the establishment of plant-microbe symbioses by enabling proper communication and recognition between symbiotic partners (Adeleke et al. 2024a). Thus, the recent advances in the Next-Generation Sequencing have therefore unveiled the in-depth of the microbial to microbe and microbe to plant communications, while also elucidating their functional roles. This tool has therefore been leveraged on to enhance crop resilience to both biotic and abiotic stresses.

Omics technologies have experienced rapid development which revolutionize the knowledge about plant-microbe interactions and plant disease systems. The combination of genomics with transcriptomics, proteomics and metabolomics has enabled researchers to better understand complex biological relationships which aids in disease recognition along with improving crop resistance (Jain et al. 2024). This section enumerates the omics approaches in analyzing the plant-microbe interactions while discussing their application for sustainable agriculture and disease prevention (Kumar et al. 2023a). Genomic research benefits from Next-Generation Sequencing (NGS) development through which scientists can sequence various microorganisms. Pathogen evolution together with adaptation and their phenotypic traits become accessible through the study of comparative genomics and phylogenomics as well as population genomics, while the identification of resistant genes and understanding of the host-pathogen relationship gets a deeper understanding with these technical approaches (Hardison 2003). Transcriptomics serves as the primary tool to evaluate changes in gene expression patterns occurring in crops while helping to study plant reactions against different biotic stressors, while proteomics serves as a fundamental tool to identify and characterize the functional proteins during interactions between plants and their pathogens. This allows scientists to analyze the cell wall-degrading enzymes, effector proteins, and virulence factors present in fungal and bacterial pathogens (Zhuang et al. 2014). Research now leverages proteome analysis for enhancing crop resilience to stress while advancing biopesticide development. Furthermore, metabolomics reveals how lower-molecular-weight metabolites contribute to plant-microbe interaction. The field of metabolomics drives scientific understanding of plant resistance mechanisms and helps identify biomarkers related to disease immunity and uncovers the biochemical roles of symbiotic relationships (Moshood et al. 2025). Hence, the advancement of sustainable agricultural practices is deeply reliant on omics technologies, which enhance our knowledge of plant-microbe dynamics and drive innovative solutions for crop protection and resilience.

8.2 Microbial Contributions to Abiotic and Biotic Stress Tolerance

Plant resilience against stresses is enhanced by plant endophytic bacteria, mycorrhizal fungi, and growth-promoting rhizobacteria, which use their molecular and physiological regulatory functions. The bacteria enable plants to survive under abiotic stresses by generating osmolytes, antioxidants, and phytohormone regulation. In addition, they produce antimicrobial compounds and trigger systemic resistance to defend against pathogens (Ayilara et al. 2023). Research into microbial contributions to crop productivity remains essential as it helps develop agricultural systems that decrease synthetic chemical usage while improving crop resistance against climate change effects and environmental damage.

Plants use receptor proteins as sensing elements that identify signaling molecules from beneficial microbes, including Nod factors produced by rhizobia (Ghantasala and Roy Choudhury 2022) and Myc factors produced by mycorrhizal fungi (Nasir et al. 2021). The microbial signals activate signaling pathways in the cells that involve calcium oscillations and mitogen-activated protein kinase activation, which results in the control of symbiotic gene expression. The root development and microbial colonization process are also affected by gibberellins, auxins and cytokinin, which work jointly to transmit hormonal signals.

The regulation of genes remains indispensable for sustaining the equilibrium between mutualistic relationships and blocking unwanted microbial infections. The symbiosis genes of plants operate as common genetic elements by coordinating proper microbial accommodations (Goyal and Habtewold 2023). Symbiotic microorganisms adjust plant immune defence signals through these effectors to establish their colonies without activating extensive protection mechanisms. The symbiotic signaling mechanisms with immune regulation networks function to develop healthy, beneficial microbial associations that do not harm plant systems. Farmers can significantly boost crop sustainability and resilience by including symbiotic microbes in their agricultural strategies. Plant-microbe interactions allow farmers to minimize their usage of chemical pesticides while also improving soil quality, enabling higher productivity, especially in harsh environmental circumstances (Babalola et al. 2020; Dlamini et al. 2022). Future studies need to optimize microbial consortia matching crop types and soil conditions to develop potential biofertilizer and biostimulant solutions while exploring microbial-based crop management strategies. The knowledge and practice of native symbiotic relationships will be essential in constructing resilient food production systems that address worldwide agricultural challenges.

8.2.1 Microbial Role in Abiotic Stress Tolerance

Drought stress poses a major challenge to global agriculture because it hinders plant water absorption abilities while disrupting photosynthesis and metabolic processes. However, plants exhibit improved drought tolerance through the presence of beneficial microbes, which help with root growth and function enhancement, increased moisture retention and modified biochemical mechanisms that protect against water shortage (Ojuederie et al. 2019). Under drought conditions, microbes make osmolytes such as glycine trehalose betaine, salicylic acid and proline available, enhancing plants' resilience to stress (Agunbiade and Babalola 2024). These organic molecules prevent the dehydration damage of plants' macromolecules and maintain cellular balance in drought situations. Plant Growth Promoting Rhizobacteria such as *Pseudomonas* sp. and *Bacillus* sp. have been reported to successfully increase proline production in plants, which helps them survive under drought conditions (Gamalero and Glick 2022). Furthermore, plants' microbial partners help them organize their antioxidant defence abilities under drought conditions. When plants experience drought, they produce reactive oxygen species, which include superoxide radicals, superoxide anion and hydrogen peroxide, that cause oxidative stress in plants, consequently destroying plant DNA, proteins and lipids (Oyedoh et al. 2023). On the other hand, beneficial plant microbes trigger plants to synthesize antioxidants such as enzymes peroxidase, superoxide dismutase and catalase, which capture reactive oxygen species to stop cellular destruction (Khan et al. 2023b). Mycorrhizal fungi, especially arbuscular mycorrhizal fungi, help plants take up more water by growing their hyphae deep into the soil to access moisture and prevent drought damage. Moreover, bacteria help plants defend against water loss through the synthesis of abscisic acid, which makes stomata close to prevent water loss when plants are exposed to drought conditions (Bharath et al. 2021). Many specific plant-beneficial microbes synthesize metabolites such as auxin, gibberellins and indole-3-acetic acid that promote root development and enhance soil water absorption (Grover et al. 2021). The microbes adjust plant hormone availability, which enables plants to metabolize efficiently in drought situations.

High salt levels in the soil damage plant growth by changing how ions behave in the system while diminishing water intake and producing dangerous free radicals (Rahim et al. 2024). High salt exposure makes plants retain too much sodium and chloride ions, which destroy their tissues (Alharbi et al. 2022). Beneficial microbes help plants fight salinity stress by stabilizing ions, initiating stress defence genes and synthesizing exopolysaccharides. In support, microbes regulate sodium and potassium levels in plants by guiding against the buildup of toxic ions and modifying ion transporter expression (Sardans and Peñuelas 2021). For example, *Exiguobacterium* sp., *Bacillus* sp., and *Halomonas* sp. improve sodium and potassium transportation to maintain proper levels of these ions within the plants (Kumar et al. 2023b). When microbes produce exopolysaccharides, they create an environment to reduce salinity near plant roots and keep the soil moist. In the same way microbes protect plants from extreme dryness, they strengthen defence mechanisms against plant salt damage (Sofy et al. 2021). Beneficial microbes protect plant cells against salinity effects by producing antioxidants and compounds that dissolve excess salt in plants. In saline environments, phytohormones, specifically gibberellins and cytokinin, help plants grow better by reducing the rate of leaf senescence and enhancing plant cell division.

The presence of heavy metals in soils, owing to chemical fertilizers and pesticide overuse, mining and industrial activities, detrimentally affects plant growth and food security and safety (Alengebawy et al. 2021). Heavy metals such as arsenic, cadmium and lead, when present in agricultural lands induce oxidative stress in plants and alter their cellular functions and enzyme activities, which the plants are able to cope with due to the presence of beneficial microorganisms (Ghorbani et al. 2024). Under heavy metal stress, beneficial microbes produce siderophores (metal-chelating compounds), which makes toxic metals unavailable for uptake by plant roots (Pal et al. 2022). Similarly. exopolysaccharides produced by microbes are also capable of binding heavily in soil, preventing metal toxicity in plants. Furthermore, microorganisms reduce heavy metal toxicity by synthesizing metal-reducing enzymes. Examples of these enzymes include phytochelatins and metallothionein (Srinath et al. 2024). Also, endophytic bacteria help transform toxic metal compounds into less toxic forms for plants (Adeleke et al. 2022).

8.2.2 Microbial Role in Biotic Stress Tolerance

Aside from helping plants cope with abiotic stress, beneficial microbes protect plants from pathogens by triggering plant immunity and antagonism and competing with pathogens for nutrients (Olanrewaju et al. 2021). Beneficial bacteria suppress plant diseases by producing antibiotics and enzymes that destroy pathogens. Examples of such pathogens include hydrolytic enzymes, antibiotics and lipopeptides (Dimkić et al. 2022). In a study by Le et al. (2022) revealed that *Streptomyces* sp. is capable of producing a metabolite named Streptomycin to control plant pathogens, while Gao et al. (2023) reported the ability of the same species to produce

Actinomycin, which also helps to control plant diseases. Other examples of microbes that protect plants against plant pathogens are detailed in Table 8.1 below.

Beneficial microorganisms within the rhizosphere area occupy the spaces that should be inhabited by pathogenic organisms and compete for nutrients with them to prevent them from developing and reproducing (Ayangbenro et al. 2022). In the same vein, arbuscular mycorrhizal fungi create root barriers that prevent the penetration of pathogens into plant roots. Microbes also induce Systemic Response in plants to defend them from pathogens. This approach is different from the pathogen triggered immunity as it offers a broad-spectrum protection, and microbes initiate it (Ayilara et al. 2022). The defence systems of plants become active through jasmonic acid and ethylene signals, which quickly protect the plant after contact with harmful

Table 8.1 Microbes that prevent plant pathogens and their mechanisms of actions

Species	Metabolite produced	Mode of action	References
Streptomyces sp.	Fungichromin	Disruption of pathogen cell membranes	Wan et al. (2023)
	Phenazines	Generation of reactive oxygen species	Dong et al. (2023)
Bacillus sp.	Iturins	Disruption of pathogen cell membranes	Zhang et al. (2022)
	Surfactins	Induced Systemic Resistance	Dong et al. (2022)
	Bacillomycin	Disruption of pathogen cell membranes	Ajuna et al. (2024)
Pseudomonas sp.	Pyrrolnitrin	Inhibition of pathogen's respiration	Dimkić et al. (2022)
	Pyoluteorin	Inhibition of spore germination	Chen et al. (2023)
	Rhamnolipids	Destruction of biofilms	Mehmood et al. (2023)
Trichoderma sp.	Gliotoxin	Inhibition of cellular respiration	Jayalakshmi et al. (2021)
	Viridin	Disruption of pathogen cell membranes	Suyanto and Astar (2024)
	6-Pentyl-α-pyrone (6-PP)	Inhibition of fungal pathogen sporulation and growth	Degani et al. (2021)
Lactobacillus sp.	Lactic Acid	Acidification of the environment	Jaffar et al. (2023)
	Bacteriocins	Pathogen cell destruction	Chakchouk-Mtibaa et al. (2024)
Aspergillus sp.	Aspergillomarasmine A	Inhibition of cell wall cell synthesis	Zhu et al. (2024)
	Fumigaclavines	Pathogen growth inhibition	Wang et al. (2024)
	Tryptoquivaline	Pathogen growth inhibition	Zhu et al. (2023)
Burkholderia sp.	Malleilactone	Suppression of pathogen growth	Deng et al. (2023)
	Cepacins	Inhibition of fungal pathogen spore formation	Webster et al. (2023)
	Pyochelin	Inhibition of pathogen growth	Ho et al. (2021)

pathogens. This approach helps to confer protection in plants and reduces the application of chemical pesticides on crops.

8.3 Impact of Microbial Symbiosis on Soil Health and Nutrient Cycling

Sustainable agriculture depends on healthy and fertile soils, which determine crop yield and ecosystem resilience (Adeleke et al. 2023; Akanmu et al. 2023). Soil biogeochemical cycles depend mainly on soil microbes, as this process is essential in maintaining soil nutrients and preserving its long-term productivity. Plant-available nitrogen remains available through the nitrogen cycle, which depends on nitrogen-fixing bacteria and ammonifying decomposers alongside nitrifying bacteria and denitrifiers to prevent buildup. Microorganisms, known as phosphate-solubilizing bacteria together with mycorrhizal fungi, help transfer phosphorus to plants while controlling environmental erosion (Tian et al. 2021). Soil organic matter maintenance and improved moisture retention happen through the carbon cycle which is managed by fungal carbon sequestration combined with microbial decomposition. The combined microbial actions strengthen soil fertility while maintaining ecological stability, which enables long-term agricultural sustainability. These functions and processes are discussed in detail below.

Plants need nitrogen as a major growing nutrient, as it is the major block of nucleic acids, amino acids and proteins, and since plants cannot obtain usable nitrogen from atmospheric nitrogen, the intervention of different microbial species such as the *Frankia* sp., *Azospirillum* sp., *Azotobacter* sp., *Rhizobium* sp. and *Bradyrhizobium* sp. through nitrogen fixation, ensures that nitrogen is available for plant growth and development (Saranraj et al. 2021; Bhattacharyya et al. 2024). Soil fertility through these organisms enables improved access of plants to nitrogen and reduces reliance on synthetic nitrogen fertilizers. Equally, the intervention of these organisms helps reduce farmers' expenses on synthetic nitrogen fertilizers. It also promotes a sustainable environment by preventing the leaching of nitrate and the emission of greenhouse gases associated with chemical fertilizer applications (Olowe et al. 2022).

In addition, plants need phosphorus as an essential nutrient because it works as the catalyst for root development, photosynthesis and energy transference (Khan et al. 2023a). A large part of phosphorus stored in soil is insoluble, preventing plants from accessing it for growth. Hence, the bioconversion of insoluble phosphates into accessible forms happens through biochemical processes performed by plants' symbiotic microbes, specifically the phosphate-solubilizing bacteria and fungi. The process of microbial phosphate solubilization is performed through organic acid secretion, where organic compounds such as the malic, gluconic and citric acids chelate and make bound phosphorus available for plant use (Pan and Cai 2023). On the other hand, certain microbes, through phosphatase production, solubilize bound organic phosphorus compounds into inorganic phosphate, increasing phosphorus availability in the soil (Tian et al. 2021). Examples of these microorganisms include

Penicillium sp., *Pseudomonas sp.* and *Bacillus* sp. which have been developed for commercial usage (Aberathna et al. 2023).

The availability of phosphorus in the soil, owing to the presence of phosphate-solubilizing microbes, enhances plant root systems and improves flowering and fruit production (Ramos Cabrera et al. 2024). These microbes used for biofertilization help reduce cultivation dependence on phosphate fertilizers extracted from non-renewable rock phosphate, thus minimizing environmental damage (Ayilara et al. 2024a). Soil ecosystems rely on organic matter decomposition by soil microbes as an essential ecological process that helps maintain its structure and function (Ayilara et al. 2020). Soil receives major nutrients through the breakdown of materials such as animal waste products, crop residues and plant remain through decomposing actinomycetes, bacteria and fungi.

Different studies have been carried out to unravel specific microbial species that carry out these decompositions Xiao et al. (2022) reported the ability of *Syntrophomonadaceae* sp., *Ruminiclostridium* sp., *Ruminococcaceae* sp., *Mobilitalea* sp., *Papillibacte*r sp., *Eubacterium* sp., *Anaerobacteriu* sp., *Xylanophilum* sp. and Sporomusa sp. in enhancing the decomposition of organic matter in paddy soil. Similarly, Zhou et al. (2023) reported the ability of *Bacillus* sp. and *Geobacillus* sp. in decompose cow dung. Equally fungal species such as *Trichoderma* sp., *Aspergillus sp. and* actinomycetes are capable of breaking down recalcitrant plant biomass, which contains cellulose and lignin through the release of metabolites that can degrade these compounds, consequently resulting in the release of soil nutrient and enhancing soil aggregation and porosity (Shinde et al. 2022; Li et al. 2023b; Lima et al. 2024; Akanmu and Babalola 2024).

A healthy soil structure in agriculture helps prevent water loss from the soil and facilitates root growth and soil aeration. The microorganisms create soil aggregates through the production of extracellular polymeric substances that promote the binding of soil particles (Ali et al. 2024). The aggregated soil structure proves beneficial to plants because it enhances aeration and reduces erosion while improving water retention and nutrients, creating optimal conditions for plant growth (Cárceles Rodríguez et al. 2022).

8.4 Multi-Omics Approaches for Deciphering Microbial Role in Plant Resilience

Genomics, metagenomics, transcriptomics, proteomics and metabolomics collectively enable researchers to study gene functional attributes, analyzing microbial communities, sequence transcripts while mapping protein interactions and building metabolic network models. The integration of these approaches provides deeper insights into plant-microbial relationships which support the advancement of sustainable farming systems (Roychowdhury et al. 2023). The research evaluates how each omics technique enhances our understanding of beneficial plant-microbe relationships while promoting sustainable agricultural methods (Table 8.2).

Table 8.2 Applications of Omics in Plant-Microbe Interactions towards stress resilience

Omics Approach	Description	Contributions to Plant Resilience	Applications in Plant-Microbe Interactions	References
Genomics	Investigates the complete genetic material (DNA) of both microorganisms and plants.	Identification of genes responsible for pathogen resistance, drought resistance and nutrient absorption processes. Discovery of genes which regulate microbial symbiosis, e.g. nitrogen fixation genes.	Identification of genes associated with beneficial traits, such as growth-promotion, pathogen-inhibition mechanisms and nitrogen-fixing capacities in microorganisms. Agricultural scientists use genetic engineering to improve crop resistance by incorporating particular genes into crop genomes.	Ku et al. (2021), Singh et al. (2019), Dlamini et al. (2023)
Transcriptomics	Analysis of gene expression through RNA sequencing, revealing which genes are active under certain conditions.	Biochemical models for studying how defense genes switch on when plants face biotic attacks and abiotic stressors from drought and salt. Identification of microbial genes that contribute to stress alleviation in plants.	Mapping gene expression profiles of beneficial microbes in response to environmental stresses Identifying plant genes which regulate both symbiotic interaction and stress response pathways.	Meena et al. (2017), Ji et al. (2023)
Proteomics	The large-scale study of proteins, including their functions, structures, and interactions.	Detection of stress-related proteins Identification of plant proteins involved in immune signalling pathways, oxidative stress management and manage nutrient cycling.	Identifying the microbial proteins that synthesize phytohormones such as IAA and cytokinins. Identifying microbial enzymes which degrade plant pathogens or releases antimicrobial compounds	Mustafa and Komatsu (2021), Akhtar et al. (2020)

(continued)

Table 8.2 (continued)

Omics Approach	Description	Contributions to Plant Resilience	Applications in Plant-Microbe Interactions	References
Metabolomics	Comprehensive study of metabolites and small biochemical molecules which function during cellular processes.	Detection of plant stress metabolites which include osmolytes together with antioxidants and signaling molecules. Identification of defense-related plant secondary metabolites which defend plants against attacks from pathogens or herbivores.	Discovery of microbial metabolites e.g antibiotic compounds, growth-promoting compounds and organic acid) Investigation of chemical communication mechanisms between plants and microorganisms for regulating growth and stress resilience.	Rodziewicz et al. (2014), Olanrewaju et al. (2024)

8.4.1 Microbial Genomics and Biotechnological Advancements

Genomic study of complete organism genetic composition has had significant advancements in the last 20 years. DNA sequencing technologies and computational biology represent the main forces behind the significant growth experienced in the field. The practice uses recombinant DNA methods together with sequencing analysis and bioinformatics to conduct genomic sequence, assemble and analysis that reveals deep structural and functional insights on evolutionary processes in diverse organisms including microorganisms (Dweh et al. 2023). The microbial genomics enables the understanding of genomes of bacterial, archaeal, fungal, viral and other microorganism genomes (Diaz 2021). This encompasses genome mapping, sequencing, editing and functional analysis, scientists have explored the microbial diversity and the microbial evolution as well as the environmental interactions. The advent of Next-generation sequencing (NGS) platforms such as Illumina HiSeq, PacBio RS and Oxford Nanopore Technologies has facilitated rapid and high-throughput sequencing at a cost-efficient rate, while enhancing the accessibility to whole genomes data (Marudamuthu et al. 2023). Whole Genome Sequencing (WGS) represents an important technology for biological investigations. WGS techniques have been employed to investigate the pathogen varieties and locate virulence elements as well as monitor microbial evolutionary patterns (Aminu et al. 2024). An instance is the study of Phytophthora genomes which have uncovered distinct patterns of simple sequence repeats among different isolates, revealing pathogenesis-related genes in repeat-rich regions (Mandal et al. 2022). Furthermore, studies based on comparative genomic assessments of *Phytophthora cactorum* isolates have identified the biological mechanisms behind host specialization as well as possible determinants of host specificity (Mandal et al. 2022). The field of microbial genomics are categorized into structural genomics and functional genomics. Structural genomics studies the arrangement of genes and chromosomal organizations, while functional genomics integrates genome data with transcriptomic and proteomic techniques to study gene functions and interactions. Comparative genomics enables the identification of conserved genetic elements across species, shedding light on genome architecture dynamics in fungal pathogens (Martin and van Der Heijden 2024; Merényi et al. 2023).

8.4.2 Metabolomics Roles and Pathways in Stress Mitigation

Metabolomics is the comprehensive study of endogenous, low-molecular-weight compounds (metabolites) present in cells, tissues or organisms. Several essential biochemical processes depend on the activity of these metabolites to aid the understanding of cellular stress reactions to natural and biotic environmental factors (Nandy et al. 2022). Metabolites within the plant kingdom number between 0.2 and 1.0 million distinct metabolites, with varying concentrations across the species. These metabolites possess unique physiochemical properties, chemical structures and polarity which allows them to function in multiple plant activities including growth development, defense mechanisms and adaptive processes (Alseekh et al. 2021).

The domain of metabolites divides into primary as well as secondary metabolites. The normal growth as well as developmental processes depend on essential primary metabolites including sugars and amino acids and Krebs cycle intermediates. Secondary metabolites function as defense compounds that help plants adapt to biotic and abiotic stresses. The stress response of these secondary metabolites includes alkaloids, flavonoids, terpenes, polyphenols and phenolic acids (Ruiz-Villafán et al. 2022). Plants employ highly developed metabolic mechanisms to adapt when faced by environmental threats. Under drought condition, plants experience osmoprotectant accumulation through proline and glycine betaine and trehalose compounds for sustaining cellular function. Plants through the antioxidant enzymes including superoxide dismutase, peroxidase, catalase and glutathione reductase defend themselves from oxidative stress (Hossain et al. 2021).

Earlier studies have shown that in response to environmental stress, plant metabolites provide insight into the plant resistance mechanisms. Polyamines including spermidine, spermine and putrescine have been found to promote both normal plant development and enhance stress tolerance (González-Hernández et al. 2022). Cotton use the *Put* gene which is regulated by *Arginine decarboxylase2* (*ADC2*) gene to develop salinity tolerance (Ren et al. 2022). Similarly, drought stress has been reported as triggered the accumulation of phenolic compounds in species such as *Hypericum polyanthemum*, *Salvia officinalis*, rice, and barley (Moroldo et al. 2024). More so, Trehalose as a non-reducing disaccharide has demonstrated its use in improving drought resistance through its protective actions against damage to macromolecules and PSII photo-oxidation (Chaves and Oliveira 2004). In another report, the response of plants to heat stress leads was found to result in elevated flavonoid and phenylpropanoid and phenolic metabolite production which strengthens their defense system, while the heat stress factor HsfB1 present in tomatoes triggers the production of flavonoids alongside phenylpropanoid pathway products which leads to increased thermotolerance in the plants (Rehman et al. 2024).

Plant adaptation to environmental stress occurs through essential involvement of metabolic pathways. The ascorbate–glutathione cycle functions as a detoxifying reactive oxygen species (ROS) against oxidative stress. Cellular macromolecules benefit from anti-oxidative protection through Glutathione's capability to reduce ROS levels and protect them from oxidative damage. Other metabolites that reinforce plant defense techniques include proline, polyphenols, ascorbic acid, carotenoids and tocopherols that act as non-enzymatic antioxidants (Das and Roychoudhury 2014). The hormonal changes resulting from salt stress modify plant metabolism. As a proton pump inhibitor omeprazole improves tomato salt tolerance by controlling abscisic acid expression while it decreases both auxins along with cytokinin and gibberellic acid concentrations (Van Oosten et al. 2017). Studies show that salt stress conditions lead to observed complex metabolic interactions between polyamines and alkaloids and sesquiterpenes through conjugation reactions. Malic acid stands out as a vital drought tolerance compound for tropical grasses along with cotton and other plant species among organic metabolites. The expression of galacturonic acid reductase enzyme in potatoes leads to increased ascorbic acid production and enhanced water stress tolerance (Owusu et al. 2023).

8.4.3 Transcriptomics and Metatranscriptomics in Plant Stress Mitigation

The next-generation sequencing (NGS) technologies in transcriptomics enable the detection and quantification of RNA molecules in biological specimens to elucidate gene functions across various conditions (Kulski 2016). These technologies, including microarray analysis, SOLiD-SAGE and RNA sequencing (RNA-seq) facilitate the identification of regulatory alterations, sequence variants, differential gene expression, and alternative splicing (Jain et al. 2024). Transcriptome analysis plays an important role in quantifying gene expression and detecting sequence variations in order to study plant-microbe relationships particularly as they shift between mutualistic and pathogenic states (Jain et al. 2024).

Transcriptomic analysis has been employed in identifying key genes involved in plant defense. An instance was the using transcriptomic methods to unveil plant defense-related genes activated during plant's interactions with microbe *Pseudomonas fluorescens* which induces systemic resistance functions in *Arabidopsis*. Similarly, transcriptomics has been used to evaluate the pathogen responses triggered by *Penicillium expansum* in blueberries both in terms of infections that lead to fruit spoilage and mycotoxin development. The mechanisms of plant growth-promoting bacterial interactions with plant roots become more comprehensible through RNA-seq transcriptomic analyses which study bacterial colonization dynamics in crops such as canola and soybean (Bhimani et al. 2024).

Real-time monitoring of microbial functional profiles and physiological activity becomes possible through metatranscriptomics since it analyzes gene expression patterns across entire microbial communities. The implementation of this approach has boosted research into plant-associated microorganisms by revealing dynamic interactions that shape plant health and disease resistance (Kimotho and Maina 2024). The study of complete mRNA through metatranscriptomics aids in discovering active biological functions of microorganisms across different ecosystems, enabling the functional characterization of plant-microbe interactions. Comparative metatranscriptomic analyses have proven valuable in evaluating microbial responses under varying environmental conditions, such as contaminated versus uncontaminated soils, thereby informing strategies for phytoremediation. More so, metatranscriptomics has been applied to study transcriptome-wide changes in microbial gene expression during pathogenic infections such as in *Xylella fastidiosa* causing Pierce's disease in grapevines (Jain et al. 2024).

8.4.4 Proteomics in Plant Stress Mitigation

Unlike the genome of organisms that remains relatively stable, the proteome is a dynamic entity that responds to developmental stages and stress conditions and external stimuli. Proteomics presents a comprehensive analysis of the of the complete set of proteins expression within particular cells and tissues or developmental periods. The examination of protein expression patterns through proteomic studies provides important data about plant growth processes, stress responses and pathogen resistance mechanisms (Aebersold and Mann 2016). Proteomic analysis requires the identification and quantification of amino acid sequences, and estimating protein abundance, characterizing post-translational modifications, and elucidating protein-protein interactions (PPIs). Functional and structural proteomics together with PPI analysis has enable further insights into the roles of protein in plant defense mechanisms (Wetie et al. 2014). Researchers have identified protein expression modifications using proteomic techniques that monitor the reaction of organisms to environmental stresses including pathogen attacks. Also, plants exhibit physiological and biochemical alterations upon exposure to pathogens by causing reductions in chlorophyll levels and photosynthesis efficiency.

Gains from proteomic research reveal the importance of fungal pathogenicity factors through the identification of pectinases along with esterases, xylanases, pectate lyases, and galacturonases which degrade fungal cell walls (de Souza and Kawaguti 2021). Research shows that *Lasiodiplodia theobromae, Magnaporthe oryzae* and *Fusarium graminearum* pathogens elevate levels of these enzymes which assists in breaking down host tissues and establishing infections (Yan et al.

2018). In another investigation, it was demonstrated that the virulence mechanism of *Colletotrichum higginsianum* and *Alternaria alternata* involves the hydrolase glucan-β-glucosidase. Pathogenic fungi release effector proteins that contain avirulence (Avr) factors which block pathogen-associated molecular pattern (PAMP)-triggered immunity through immune response suppression. R genes in plants contain receptors which detect the avirulence factors to activate robust defense mechanisms while triggering effector-triggered immunity (Hiruma et al. 2010).

Proteomic studies of plant-pathogen interactions enable the discovery of new genes and proteins which improve crop stress tolerance potential (Ashwin et al. 2017). While combining proteomic research with genomic and transcriptomic analysis enables researchers to create specific approaches that strengthen plant resistance against diseases. Also, proteomics contributes to the understanding of pesticide functionality and degradation, thereby facilitating the development of environmentally sustainable agricultural practices. The progression of proteomics technology will continue to extend its agricultural biotechnology applications, offering innovative solutions for crop protection and sustainable food production (Sarethy and Saharan 2021).

8.5 Biotechnological and Agricultural Applications of Symbiotic Microbes

The research of microbial roles in plant resilience has undergone revolutionary changes due to modern multi-omics technologies The application of genomics, transcriptomics, proteomics and metabolomics research in the identification of complex molecular structures which enhance the plants survival and adaptation to both abiotic and biotic stressors (Jain et al. 2024). Rhizobial species particularly *Rhizobium* and *Bradyrhizobium* use their *NodA* and *FixA* genes to convert atmospheric nitrogen into a usable plant form thus boosting plant resilience. The soil-fertility improvement through microbial nitrogen fixation remains essential for nurturing plants in nitrogen-poor soils (Khan et al. 2022).

Toward promoting plant resilience PGPR produce phytohormones including indole-3-acetic acid (IAA) that increases root development while improving water uptake and nutrient acquisition (Ranjan et al. 2024). Plant growth under stressful conditions is dependent on genome-regulated modifications of the hormonal balance through genes such as *iaaM* and *iaaH*. Resistance to plant stress can be achieved in *Pseudomonas* spp. The enzyme, ACC deaminase breaks down ACC into ethylene precursor reducing stress levels while improving root development for drought tolerance (Buragohain et al. 2024). Plant resilience benefits from microbial phosphate solubilization as a basic functional mechanism. Through *PhoA* enzyme production *Pseudomonas* and *Bacillus* solubilize inorganic phosphorus nutrients which become accessible for plant growth particularly in locations with reduced phosphorus content (Li et al. 2023a). In biocontrol, microbes such as *Trichoderma* and *Bacillus* species produce chitinases and bioactive compounds that protect plants from pathogenic fungi and pests. The microbes secrete siderophores that sequester iron and prevent pathogen growth (Adeleke and Babalola 2022). Furthermore, the induced systemic resistance (ISR) in plant by beneficial microbes activates plant defense genes such as *PR1*, *PR2*, and *PR5*, thereby enhancing plant resistance to pathogens (Yu et al. 2022; Akanmu et al. 2021). More so, the production of exopolysaccharides (EPS) by microbes enhances plants' defense mechanism, while enhancing resilience to drought and pathogenic attack (Santoyo et al. 2024). Hence, application of the multi-omics technology is providing valuable insights into the microbial functions, especially in the plant-microbe interactions, and development of sustainable agricultural strategies to ensure food security in a changing climate (Table 8.3).

8.6 Future Applications and Conclusion

Omics technologies have facilitated an-in-depth exploration of plant—microbe relationship by unveiling the mechanisms underlying microbial assisted plant defense against stress and disease. The integration of genomics, transcriptomics, proteomics and metabolomics has advanced our understanding of the microbial communities and their beneficial roles in plant health. The implementation of this integrated method revolutionized plant disease ecology leading to targeted solutions which strengthen crop resistance.

Research in omics currently relies on in-silico computational analyses to predict crops responses to engineered modifications across different environmental conditions in the effort to optimize disease-resistant breeding programs. The future development of multi-omics technologies will address the fundamental issues by identifying rare proteins and mapping their distribution their distribution within cellular compartments. These advancements will lead to improved efficiency for plant disease control and sustainable agricultural practices.

Analysis of multi-omics data requires a robust bioinformatic tools and well curated reference databases to handle their complexity. However, data integration currently constitutes challenges because of the diverse formats and sources of multi-omics dataset. Hence, the need to develop standardized protocol for managing genetic traits towards ensuring accuracy and reliability of results. Therefore, establishment of comprehensive microbial species and metabolite databases will further improve the reliability of omics research and its applications in sustainable agriculture. Future developments in harnessing the benefit of symbiotic plant-microbe interactions will benefit from such analytic tools to unveil the plant-microbe molecular processes, towards the formulation of tailored specific microbial inoculants as biofertilizers and biopesticides to meet specific crop or soil requirements.

Table 8.3 The contribution of beneficial microorganism to plant resilience

Key Contribution	Gene(s)	Microbial Source	Function/Role in Plant Resilience	References
Nitrogen Fixation and Symbiosis Genes	NodA, NodB, FixA, FixB, FixC	*Rhizobium*, *Bradyrhizobium*	Legume root nodule formation occurs when rhizobia infect roots to fix nitrogen which improves fertility and drives plant growth.	Wekesa et al. (2021)
Plant Growth Promotion (PGPR) Genes	IAA, ACC deaminase	*Azospirillum*, *Pseudomonas*	The bacteria promote plant development by synthesizing IAA hormone which extends roots and breaks down ACC for decreased ethylene stress factors that create drought tolerance in plants.	Yuan et al. (2022)
Abiotic Stress Tolerance Genes	TPS (Trehalose-6-phosphate synthase), P5CS (Pyrroline-5-carboxylate synthase)	*Bacillus*, *Rhizobium*	The use of osmoprotectants such as proline and trehalose enable both plants and microbes to endure droughts and salt stresses as well as various other environmental pressures.	Chattaraj et al. (2025)
Phosphate Solubilization Genes	PhoA (Alkaline Phosphatase), PstS (Phosphate transporter)	*Pseudomonas*, *Bacillus*	Plant growth and nutrient availability get enhanced by facilitating organic phosphate breakdown to available forms in phosphorus-deficient soils.	Lu et al. (2023)
Antagonistic Gene Products in Biocontrol	ChiA, ChiB (Chitinase), Cry genes (Bacillus thuringiensis)	*Trichoderma*, *Bacillus*	The synthesis of chitinases and Cry proteins by the plants facilitates pathogen cell wall breakdown and insect pest defense which enhances overall plant health and pathogen resistance.	Kumar et al. (2022)
Siderophore Production Genes	EntB (Enterobactin synthetase)	*Pseudomonas*	The bacteria produce siderophores which acquire environmental iron thus providing plant access to essential nutrients while blocking pathogens from acquiring these nutrients.	Klebba et al. (2021)

(continued)

Table 8.3 (continued)

Key Contribution	Gene(s)	Microbial Source	Function/Role in Plant Resilience	References
Induced Systemic Resistance (ISR) Genes	PR1, PR2 (Pathogenesis-Related genes), Chi1 (Chitinase)	*Pseudomonas fluorescens*, *Bacillus subtilis*	Plants activate their defense system when exposed to certain signals which then develops systemic immunity and results in improved resistance to stress.	Kashyap et al. (2021)
Exopolysaccharide (EPS) Production Genes	eps genes	*Rhizobium*, *Pseudomonas*	Exopolysaccharides produced by the bacteria create biofilm layers around roots which provide protection against diseases and maintain root stability under stressful conditions.	Bhagat et al. (2021)

Acknowledgments OOB recognizes National Research Foundation (NRF South Africa) for grants (UID123634 and UID132595) that support this research.

Authors' Contributions The authors contributed equally to this work.

References

Aberathna AAU, Satharasinghe DA, Jayasooriya AP, Jinadasa HRN, Manopriya S, Jayaweera BA, Fernando CAN, Weerathilake WDV, Prathapasinghe G, Liyanage JA (2023) Managing soil and plant nutrients: role of microbial phosphate solubilisation. In: Phosphorus in soils and plants. MDPI AG, Basel, Switzerland

Adeleke BS, Babalola OO (2022) Roles of plant endosphere microbes in agriculture-a review. J Plant Growth Regul 41(4):1411–1428

Adeleke BS, Fadiji AE, Ayilara MS, Igiehon ON, Nwachukwu BC, Babalola OO (2022) Strategies to enhance the use of endophytes as bioinoculants in agriculture. Horticulturae 8(6):498

Adeleke BS, Ayilara MS, Sharma D, Sultana R, Mathur M, Bhandari G, Chaudhary A (2023) Nanotextured materials for soil health. In: Advances in nanotechnology for smart agriculture. CRC Press, London, United Kingdom, pp 305–322

Adeleke BS, Ayilara MS, Adeyemi UT, Uthman QO, Chaudhary A, Abou Fayssal S (2024a) Role of microbes in plant biotic stress management. In: Microbial inoculants: applications for sustainable agriculture. Springer Nature Singapore Pte Ltd, Singapore, pp 145–162

Adeleke BS, Chaudhary P, Ayilara MS, Ojo FM, Erinoso SM, Upadhayay VK, Adeyemo AI, Akinola SA (2024b) Rhizosphere microbiomes mediating abiotic stress mitigation for improved plant nutrition. Ecologies 5(3):375–401

Adeleke BS, Olowe OM, Ayilara MS, Fasusi OA, Omotayo OP, Fadiji AE, Onwudiwe DC, Babalola OO (2024c) Biosynthesis of nanoparticles using microorganisms: a focus on endophytic fungi. Heliyon 10(21):e39636

Aebersold R, Mann M (2016) Mass-spectrometric exploration of proteome structure and function. Nature 537(7620):347–355
Agunbiade VF, Babalola OO (2024) Drought stress amelioration attributes of plant-associated microbiome on agricultural plants. Bioinform Biol Insights 18:11779322241233442
Ajuna HB, Lim H-I, Moon J-H, Won S-J, Choub V, Choi S-I, Yun J-Y, Ahn YS (2024) The prospect of antimicrobial peptides from Bacillus species with biological control potential against insect pests and diseases of economic importance in agriculture, forestry and fruit tree production. Biotechnol Biotechnol Equip 38(1):2312115
Akanmu AO, Babalola OO (2024) Characterization and biocontrol potential of some rhizobacteria against fungal pathogens causing foliar diseases in maize. Studia Universitatis Babeş-Bolyai. Biologia 69:183–203
Akanmu AO, Babalola OO, Venturi V, Ayilara MS, Adeleke BS, Amoo AE, Sobowale AA, Fadiji AE, Glick BR (2021) Plant disease management: leveraging on the plant-microbe-soil interface in the biorational use of organic amendments. Front Plant Sci 12:700507
Akanmu AO, Akol AM, Ndolo DO, Kutu FR, Babalola OO (2023) Agroecological techniques: adoption of safe and sustainable agricultural practices among the smallholder farmers in Africa. Front Sustain Food Syst 7:1143061
Akhtar SS, Mekureyaw MF, Pandey C, Roitsch T (2020) Role of cytokinins for interactions of plants with microbial pathogens and pest insects. Front Plant Sci 10:1777
Alengebawy A, Abdelkhalek ST, Qureshi SR, Wang M-Q (2021) Heavy metals and pesticides toxicity in agricultural soil and plants: ecological risks and human health implications. Toxics 9(3):42
Alharbi K, Al-Osaimi AA, Alghamdi BA (2022) Sodium chloride (NaCl)-induced physiological alteration and oxidative stress generation in Pisum sativum (L.): a toxicity assessment. ACS Omega 7(24):20819–20832
Ali N, Abbas SAAA, Sharif L, Shafiq M, Kamran Z, Haseeb M, Shahid MA (2024) Microbial extracellular polymeric substance and impacts on soil aggregation. In: Bacterial secondary metabolites. Elsevier Inc., UK, pp 221–237
Alseekh S, Aharoni A, Brotman Y, Contrepois K, D'Auria J, Ewald J, Ewald C, Fraser PD, Giavalisco P, Hall RD (2021) Mass spectrometry-based metabolomics: a guide for annotation, quantification and best reporting practices. Nat Methods 18(7):747–756
Aminu S, Ascandari A, Laamarti M, Safdi NEH, El Allali A, Daoud R (2024) Exploring microbial worlds: a review of whole genome sequencing and its application in characterizing the microbial communities. Crit Rev Microbiol 50(5):805–829
Ashwin N, Barnabas L, Ramesh Sundar A, Malathi P, Viswanathan R, Masi A, Agrawal GK, Rakwal R (2017) Advances in proteomic technologies and their scope of application in understanding plant–pathogen interactions. J Plant Biochem Biotechnol 26:371–386
Ayangbenro AS, Chukwuneme CF, Ayilara MS, Kutu FR, Khantsi M, Adeleke BS, Glick BR, Babalola OO (2022) Harnessing the rhizosphere soil microbiome of organically amended soil for plant productivity. Agronomy 12(12):3179
Ayilara MS, Olanrewaju OS, Babalola OO, Odeyemi O (2020) Waste management through composting: challenges and potentials. Sustainability 12(11):4456
Ayilara MS, Adeleke BS, Babalola OO (2022) Bioprospecting and challenges of plant microbiome research for sustainable agriculture, a review on soybean Endophytic bacteria. Microb Ecol 86(2):1454
Ayilara MS, Adeleke BS, Akinola SA, Fayose CA, Adeyemi UT, Gbadegesin LA, Omole RK, Johnson RM, Uthman QO, Babalola OO (2023) Biopesticides as a promising alternative to synthetic pesticides: a case for microbial pesticides, phytopesticides, and nanobiopesticides. Front Microbiol 14:1040901
Ayilara MS, Akinola SA, Adebajo MT (2024a) Benefits and drawbacks of microbial inoculant in terms of human health and the environment. In: Microbial inoculants: applications for sustainable agriculture. Springer Nature Singapore Pte Ltd, Singapore, pp 411–435

Ayilara MS, Akinola SA, Adeleke BS, Gbadegesin LA, Adejumo GD, Glick BR, Babalola OO (2024b) Biopesticides versus synthetic pesticides usage in Africa. In: Microbiome-based decontamination of environmental pollutants. Elsevier Ltd. (Academic Press), Amsterdam, Netherlands, pp 417–450

Babalola OO, Fadiji AE, Enagbonma BJ, Alori ET, Ayilara MS, Ayangbenro AS (2020) The nexus between plant and plant microbiome: revelation of the networking strategies. Front Microbiol 11:548037

Bhagat N, Raghav M, Dubey S, Bedi N (2021) Bacterial exopolysaccharides: insight into their role in plant abiotic stress tolerance. J Microbiol Biotechnol 31(8):1045–1059

Bharath P, Gahir S, Raghavendra AS (2021) Abscisic acid-induced stomatal closure: an important component of plant defense against abiotic and biotic stress. Front Plant Sci 12:615114

Bhattacharyya PN, Islam NF, Sarma B, Nath BC, Al-Ani LKT, Lesueur D (2024) Frankia-actinorhizal symbiosis: a non-chemical biological assemblage for enhanced plant growth, nodulation and reclamation of degraded soils. Symbiosis 92(1):1–26

Bhimani P, Mahavar P, Rajguru B, Bhatt VD, Nathani NM, Shri M (2024) Unveiling the green dialogue: advancements in omics technologies for deciphering plant–microbe interactions in soil. Discov Plants 1(1):4

Buragohain K, Tamuly D, Sonowal S, Nath R (2024) Impact of drought stress on plant growth and its management using plant growth promoting rhizobacteria. Indian J Microbiol 64(2):287–303

Cárceles Rodríguez B, Durán-Zuazo VH, Soriano Rodríguez M, García-Tejero IF, Gálvez Ruiz B, Cuadros Tavira S (2022) Conservation agriculture as a sustainable system for soil health: a review. Soil Syst 6(4):87

Chakchouk-Mtibaa A, Mechri S, Azabou MC, Triki MA, Smaoui S, Mellouli L (2024) The novel bacteriocin BacYB1 produced by Leuconostoc mesenteroides YB1: from recent analytical characterization to biocontrol Verticillium dahliae and Agrobacterium tumefaciens. Microb Pathog 192:106680

Chamkhi I, Cheto S, Geistlinger J, Zeroual Y, Kouisni L, Bargaz A, Ghoulam C (2022) Legume-based intercropping systems promote beneficial rhizobacterial community and crop yield under stressing conditions. Ind Crop Prod 183:114958

Chattaraj S, Samantaray A, Ganguly A, Thatoi H (2025) Employing plant growth-promoting rhizobacteria for abiotic stress mitigation in plants: with a focus on drought stress. Discov Appl Sci 7(1):68

Chaves MM, Oliveira MM (2004) Mechanisms underlying plant resilience to water deficits: prospects for water-saving agriculture. J Exp Bot 55(407):2365–2384

Chen S, Haga M, Imai I, Sakai R, Fujita MJ (2023) Function of the algicidal bacterium Pseudomonas sp. Go58 isolated from the biofilm on a water plant, and its active compounds, pyoluteorins. Sci Total Environ 872:162088

Das K, Roychoudhury A (2014) Reactive oxygen species (ROS) and response of antioxidants as ROS-scavengers during environmental stress in plants. Front Environ Sci 2:53

Das PP, Singh KR, Nagpure G, Mansoori A, Singh RP, Ghazi IA, Kumar A, Singh J (2022) Plant-soil-microbes: a tripartite interaction for nutrient acquisition and better plant growth for sustainable agricultural practices. Environ Res 214:113821

de Souza TS, Kawaguti HY (2021) Cellulases, hemicellulases, and pectinases: applications in the food and beverage industry. Food Bioprocess Technol 14(8):1446–1477

Degani O, Khatib S, Becher P, Gordani A, Harris R (2021) Trichoderma asperellum secreted 6-pentyl-α-pyrone to control Magnaporthiopsis maydis, the maize late wilt disease agent. Biology 10(9):897

Deng P, Jia J, Foxfire A, Baird SM, Smith LJ, Lu S-E (2023) A polyketide synthetase gene cluster is responsible for antibacterial activity of Burkholderia contaminans MS14. Phytopathology 113(1):11–20

Diaz PI (2021) Subgingival fungi, Archaea, and viruses under the omics loupe. Periodontol 85(1):82–89

Dimkić I, Janakiev T, Petrović M, Degrassi G, Fira D (2022) Plant-associated Bacillus and Pseudomonas antimicrobial activities in plant disease suppression via biological control mechanisms-a review. Physiol Mol Plant Pathol 117:101754

Dlamini SP, Akanmu AO, Babalola OO (2022) Rhizospheric microorganisms: the gateway to a sustainable plant health. Front Sustain Food Syst 6:925802.s

Dlamini SP, Akanmu AO, Babalola OO (2023) Variations in the functional diversity of rhizosphere microbiome of healthy and Northern Corn leaf blight infected maize (Zea mays L.). *Spanish.* J Soil Sci 13:10964

Dong L, Wang P, Zhao W, Su Z, Zhang X, Lu X, Li S, Ma P, Guo Q (2022) Surfactin and fengycin contribute differentially to the biological activity of Bacillus subtilis NCD-2 against cotton verticillium wilt. Biol Control 174:104999

Dong D, Li M, Zhang T, Niu Z, Xue G, Bai H, Zhao W, Yu J, Jiang W, Wu H (2023) Antagonistic activity of Streptomyces Alfalfae 11F against Fusarium Wilt of watermelon and transcriptome analysis provides insights into the synthesis of Phenazine-1-Carboxamide. Plants 12(22):3796

Dweh TJ, Pattnaik S, Sahoo JP (2023) Assessing the impact of meta-genomic tools on current cutting-edge genome engineering and technology. Int J Biochem Mol Biol 14(4):62–75

Gamalero E, Glick BR (2022) Recent advances in bacterial amelioration of plant drought and salt stress. Biology 11(3):437

Gao L, Kumaravel K, Xiong Q, Liang Y, Ju Z, Jiang Y, Zhang J (2023) Actinomycins produced by endophyte Streptomyces sp. GLL-9 from navel orange plant exhibit high antimicrobial effect against Xanthomonas citri susp. citri and Penicillium italicum. Pest Manag Sci 79(11):4679–4693

Ghantasala S, Roy Choudhury S (2022) Nod factor perception: an integrative view of molecular communication during legume symbiosis. Plant Mol Biol 110(6):485–509

Ghorbani A, Emamverdian A, Pehlivan N, Zargar M, Razavi SM, Chen M (2024) Nano-enabled agrochemicals: mitigating heavy metal toxicity and enhancing crop adaptability for sustainable crop production. J Nanobiotechnol 22(1):91

González-Hernández AI, Scalschi L, Vicedo B, Marcos-Barbero EL, Morcuende R, Camañes G (2022) Putrescine: a key metabolite involved in plant development, tolerance and resistance responses to stress. Int J Mol Sci 23(6):2971

Goyal RK, Habtewold JZ (2023) Evaluation of legume–rhizobial symbiotic interactions beyond nitrogen fixation that help the host survival and diversification in hostile environments. Microorganisms 11(6):1454

Grover M, Bodhankar S, Sharma A, Sharma P, Singh J, Nain L (2021) PGPR mediated alterations in root traits: way toward sustainable crop production. Front Sustain Food Syst 4:618230

Hardison RC (2003) Comparative genomics. PLoS Biol 1(2):e58

Hiruma K, Onozawa-Komori M, Takahashi F, Asakura M, Bednarek P, Okuno T, Schulze-Lefert P, Takano Y (2010) Entry mode–dependent function of an indole glucosinolate pathway in Arabidopsis for nonhost resistance against anthracnose pathogens. Plant Cell 22(7):2429–2443

Ho Y-N, Hoo SY, Wang B-W, Hsieh C-T, Lin C-C, Sun C-H, Peng C-C, Lin C, Yang Y-L (2021) Specific inactivation of an antifungal bacterial siderophore by a fungal plant pathogen. ISME J 15(6):1858–1861

Hossain A, Pramanick B, Bhutia KL, Ahmad Z, Moulick D, Maitra S, Ahmad A, Aftab T (2021) Emerging roles of osmoprotectant glycine betaine against salt-induced oxidative stress in plants: a major outlook of maize (Zea mays L.). In: Frontiers in plant-soil interaction. Academic Press (Elsevier Inc.), London, UK, pp 567–587

Jaffar NS, Jawan R, Chong KP (2023) The potential of lactic acid bacteria in mediating the control of plant diseases and plant growth stimulation in crop production-a mini review. Front Plant Sci 13:1047945

Jain A, Sarsaiya S, Singh R, Gong Q, Wu Q, Shi J (2024) Omics approaches in understanding the benefits of plant-microbe interactions. Front Microbiol 15:1391059

Jayalakshmi R, Sobanbabu G, Oviya R, Mehetre S, Kannan R, Paramasivam M, Santhanakrishnan V, Kumar K, Theradimani M, Ramamoorthy V (2021) Evaluation of gliotoxin phytotoxicity and gliotoxin producing Trichoderma virens for the suppression of damping off of tomato. J Biol Control 35:187–195

Ji C, Liang Z, Cao H, Chen Z, Kong X, Xin Z, He M, Wang J, Wei Z, Xing J (2023) Transcriptome-based analysis of the effects of compound microbial agents on gene expression in wheat roots and leaves under salt stress. Front Plant Sci 14:1109077

Kashyap PL, Kumar S, Aggarwal SK, Kaul N, Jasrotia P, Gupta A, Singh GP (2021) Resistance inducers and their role in reinforcing wheat defense system against fungal pathogens. J Cereal Res 112810, 13(3):229–254

Khan NF, Rasool A, Mansoor S, Saleem S, Baba TR, Haq SM, Rehman SA, Adetunji CO (2022) Potential applications of Rhizobacteria as eco-friendly biological control, plant growth. In: Sustainable crop production: recent advances. IntechOpen, London, United Kingdom, p 151

Khan F, Siddique AB, Shabala S, Zhou M, Zhao C (2023a) Phosphorus plays key roles in regulating plants' physiological responses to abiotic stresses. Plants 12(15):2861

Khan M, Ali S, Al Azzawi TNI, Saqib S, Ullah F, Ayaz A, Zaman W (2023b) The key roles of ROS and RNS as a signaling molecule in plant–microbe interactions. Antioxidants 12(2):268

Kimotho RN, Maina S (2024) Unraveling plant–microbe interactions: can integrated omics approaches offer concrete answers? J Exp Bot 75(5):1289–1313

Klebba PE, Newton SM, Six DA, Kumar A, Yang T, Nairn BL, Munger C, Chakravorty S (2021) Iron acquisition systems of gram-negative bacterial pathogens define TonB-dependent pathways to novel antibiotics. Chem Rev 121(9):5193–5239

Ku Y-S, Wang Z, Duan S, Lam H-M (2021) Rhizospheric communication through mobile genetic element transfers for the regulation of microbe–plant interactions. Biology 10(6):477

Kulski J (2016) Next-generation sequencing—an overview of the history, tools, and "Omic" applications. In: Next generation sequencing-advances, applications and challenges. InTechOpen, Rijeka, Croatia, pp 3–60

Kumar M, Chakdar H, Pandiyan K, Thapa S, Shahid M, Singh A, Srivastava AK, Saxena AK (2022) Bacterial chitinases: genetics, engineering and applications. World J Microbiol Biotechnol 38(12):252

Kumar U, Raj S, Sreenikethanam A, Maddheshiya R, Kumari S, Han S, Kapoor KK, Bhaskar R, Bajhaiya AK, Gahlot DK (2023a) Multi-omics approaches in plant–microbe interactions hold enormous promise for sustainable agriculture. Agronomy 13(7):1804

Kumar V, Raghuvanshi N, Pandey AK, Kumar A, Thoday-Kennedy E, Kant S (2023b) Role of halotolerant plant growth-promoting rhizobacteria in mitigating salinity stress: recent advances and possibilities. Agriculture 13(1):168

Le KD, Yu NH, Park AR, Park D-J, Kim C-J, Kim J-C (2022) *Streptomyces* sp. AN090126 as a biocontrol agent against bacterial and fungal plant diseases. Microorganisms 10(4):791

Li H-P, Han Q-Q, Liu Q-M, Gan Y-N, Rensing C, Rivera WL, Zhao Q, Zhang J-L (2023a) Roles of phosphate-solubilizing bacteria in mediating soil legacy phosphorus availability. Microbiol Res 272:127375

Li J, Wiebenga A, Lipzen A, Ng V, Tejomurthula S, Zhang Y, Grigoriev IV, Peng M, de Vries RP (2023b) Comparative genomics and transcriptomics analyses reveal divergent plant biomass-degrading strategies in fungi. J Fungi 9(8):860

Lima PC, Karimian P, Johnston E, Hartley CJ (2024) The use of Trichoderma spp. for the bioconversion of agro-industrial waste biomass via fermentation: a review. Fermentation 10(9):442

Lu Z, He S, Kashif M, Zhang Z, Mo S, Su G, Du L, Jiang C (2023) Effect of ammonium stress on phosphorus solubilization of a novel marine mangrove microorganism Bacillus aryabhattai NM1-A2 as revealed by integrated omics analysis. BMC Genomics 24(1):550

Mandal K, Dutta S, Upadhyay A, Panda A, Tripathy S (2022) Comparative genome analysis across 128 Phytophthora isolates reveal species-specific microsatellite distribution and localized evolution of compartmentalized genomes. Front Microbiol 13:806398

Martin FM, van Der Heijden MG (2024) The mycorrhizal symbiosis: research frontiers in genomics, ecology, and agricultural application. New Phytol 242(4):1486–1506

Marudamuthu B, Sharma T, Purru S, Soam S, Rao CS (2023) Next-generation sequencing technology: a boon to agriculture. Genet Resour Crop Evol 70(2):353–372
Meena KK, Sorty AM, Bitla UM, Choudhary K, Gupta P, Pareek A, Singh DP, Prabha R, Sahu PK, Gupta VK (2017) Abiotic stress responses and microbe-mediated mitigation in plants: the omics strategies. Front Plant Sci 8:172
Mehmood N, Saeed M, Zafarullah S, Hyder S, Rizvi ZF, Gondal AS, Jamil N, Iqbal R, Ali B, Ercisli S (2023) Multifaceted impacts of plant-beneficial pseudomonas spp. in managing various plant diseases and crop yield improvement. ACS Omega 8(25):22296–22315
Merényi Z, Krizsán K, Sahu N, Liu X-B, Bálint B, Stajich JE, Spatafora JW, Nagy LG (2023) Genomes of fungi and relatives reveal delayed loss of ancestral gene families and evolution of key fungal traits. Nat Ecol Evol 7(8):1221–1231
Moroldo M, Blanchet N, Duruflé H, Bernillon S, Berton T, Fernandez O, Gibon Y, Moing A, Langlade NB (2024) Genetic control of abiotic stress-related specialized metabolites in sunflower. BMC Genomics 25(1):199
Moshood AY, Abdulraheem MI, Li L, Zhang Y, Raghavan V, Hu J (2025) Deciphering nutrient stress in plants: integrative insight from metabolomics and proteomics. Funct Integr Genomics 25(1):1–26
Mustafa G, Komatsu S (2021) Plant proteomic research for improvement of food crops under stresses: a review. Mol Omics 17(6):860–880
Nandy S, Das T, Tudu CK, Mishra T, Ghorai M, Gadekar VS, Anand U, Kumar M, Behl T, Shaikh NK (2022) Unravelling the multi-faceted regulatory role of polyamines in plant biotechnology, transgenics and secondary metabolomics. Appl Microbiol Biotechnol 106(3):905–929
Nasir F, Bahadur A, Lin X, Gao Y, Tian C (2021) Novel insights into host receptors and receptor-mediated signaling that regulate arbuscular mycorrhizal symbiosis. J Exp Bot 72(5):1546–1557
Ojuederie OB, Olanrewaju OS, Babalola OO (2019) Plant growth promoting rhizobacterial mitigation of drought stress in crop plants: implications for sustainable agriculture. Agronomy 9:712
Ojuederie OB, Chukwuneme CF, Samuel O, Olanrewaju MA, Adegboyega TT, Babalola OO (2021) Contribution of microbial inoculants in sustainable maintenance of human health, including test methods and evaluation of safety of microbial pesticide microorganisms. In: Biopesticides: botanicals and microorganisms for improving agriculture and human health. Logos Verlag Berlin GmbH, Berlin, Germany, pp 207–240
Olanrewaju OS, Ayilara MS, Ayangbenro AS, Babalola OO (2021) Genome mining of three plant growth-promoting Bacillus species from maize rhizosphere. Appl Biochem Biotechnol 193(12):3949–3969
Olanrewaju OS, Glick BR, Babalola OO (2024) Metabolomics-guided utilization of beneficial microbes for climate-resilient crops. Curr Opin Chem Biol 79:102427
Olowe OM, Nicola L, Asemoloye MD, Akanmu AO, Babalola OO (2022) Trichoderma: potential bio-resource for the management of tomato root rot diseases in Africa. Microbiol Res 257:126978
Owusu AG, Lv Y-P, Liu M, Wu Y, Li C-L, Guo N, Li D-H, Gao J-S (2023) Transcriptomic and metabolomic analyses reveal the potential mechanism of waterlogging resistance in cotton (Gossypium hirsutum L.). Front Plant Sci 14:1088537
Oyedoh OP, Yang W, Dhanasekaran D, Santoyo G, Glick BR, Babalola OO (2023) Rare rhizo-Actinomycetes: a new source of agroactive metabolites. Biotechnol Adv 67:108205
Pal A, Bhattacharjee S, Saha J, Sarkar M, Mandal P (2022) Bacterial survival strategies and responses under heavy metal stress: a comprehensive overview. Crit Rev Microbiol 48(3):327–355
Pan L, Cai B (2023) Phosphate-solubilizing bacteria: advances in their physiology, molecular mechanisms and microbial community effects. Microorganisms 11(12):2904
Rahim HU, Ali W, Uddin M, Ahmad S, Khan K, Bibi H, Alatalo JM (2024) Abiotic stresses in soils, their effects on plants, and mitigation strategies: a literature review. Chem Ecol 41:1–34
Ramos Cabrera EV, Delgado Espinosa ZY, Solis Pino AF (2024) Use of phosphorus-solubilizing microorganisms as a biotechnological alternative: a review. Microorganisms 12(8):1591

Ranjan A, Rajput VD, Prazdnova EV, Gurnani M, Sharma S, Bhardwaj P, Shende SS, Mandzhieva SS, Sushkova S, Minkina T (2024) Augmenting abiotic stress tolerance and root architecture: the function of phytohormone-producing PGPR and their interaction with nanoparticles. S Afr J Bot 167:612–629

Rehman A, Khan I, Farooq M (2024) Secondary metabolites mediated reproductive tolerance under heat stress in plants. J Plant Growth Regul 43(9):2993–3011

Ren G, Mo H, Xu R (2022) Arginine decarboxylase gene ADC2 regulates fiber elongation in cotton. Genes 13(5):784

Rodziewicz P, Swarcewicz B, Chmielewska K, Wojakowska A, Stobiecki M (2014) Influence of abiotic stresses on plant proteome and metabolome changes. Acta Physiol Plant 36:1–19

Roychowdhury R, Das SP, Gupta A, Parihar P, Chandrasekhar K, Sarker U, Kumar A, Ramrao DP, Sudhakar C (2023) Multi-omics pipeline and omics-integration approach to decipher plant's abiotic stress tolerance responses. Genes 14(6):1281

Ruiz-Villafán B, Cruz-Bautista R, Manzo-Ruiz M, Passari AK, Villarreal-Gómez K, Rodríguez-Sanoja R, Sanchez S (2022) Carbon catabolite regulation of secondary metabolite formation, an old but not well-established regulatory system. Microb Biotechnol 15(4):1058–1072

Santoyo G, Kumar A, Babalola OO (2024) The role of microbial seed endophytes in agriculture: mechanisms and applications. Cereal Res Commun 53:1–13

Saranraj P, Sivasakthivelan P, Al-Tawaha A, Sudha A, Al-Tawaha A, Sirajuddin S (2021) Diversity and evolution of Bradyrhizobium communities relating to soybean cultivation: a review. In: IOP conference series: earth and environmental science. IOP Publishing, Bristol, England, United Kingdom, 788(1):012208

Sardans J, Peñuelas J (2021) Potassium control of plant functions: ecological and agricultural implications. Plants 10(2):419

Sarethy IP, Saharan A (2021) Genomics, proteomics and transcriptomics in the biological control of plant pathogens: a review. Indian Phytopathol 74(1):3–12

Shinde R, Shahi DK, Mahapatra P, Naik SK, Thombare N, Singh AK (2022) Potential of lignocellulose degrading microorganisms for agricultural residue decomposition in soil: a review. J Environ Manag 320:115843

Shumilina J, Soboleva A, Abakumov E, Shtark OY, Zhukov VA, Frolov A (2023) Signaling in legume–Rhizobia symbiosis. Int J Mol Sci 24(24):17397

Singh D, Raina TK, Kumar A, Singh J, Prasad R (2019) Plant microbiome: a reservoir of novel genes and metabolites. Plant Gene 18:100177

Sofy MR, Aboseidah AA, Heneidak SA, Ahmed HR (2021) ACC deaminase containing endophytic bacteria ameliorate salt stress in Pisum sativum through reduced oxidative damage and induction of antioxidative defense systems. Environ Sci Pollut Res 28:40971–40991

Srinath N, Vivek P, Rajakumari K, Meenambiga S, Yuvaraj D (2024) Biotechnological innovations in radioactive waste management. Radioact Pollut 7:137

Suyanto A, Astar I (2024) Effect of using Tithonia diversifolia green manure and Trichoderma sp. secondary metabolites on the growth and yield of tomato plants (Solanum lycopersicum L.) on alluvial soil. J Smart Sci Technol 4(2):54–63

Tian J, Ge F, Zhang D, Deng S, Liu X (2021) Roles of phosphate solubilizing microorganisms from managing soil phosphorus deficiency to mediating biogeochemical P cycle. Biology 10(2):158

Van Oosten MJ, Silletti S, Guida G, Cirillo V, Di Stasio E, Carillo P, Woodrow P, Maggio A, Raimondi G (2017) A benzimidazole proton pump inhibitor increases growth and tolerance to salt stress in tomato. Front Plant Sci 8:1220

Wan M, Gan L, Li Z, Wang M, Chen J, Chen S, Hu J, Li J (2023) Enhancement of fungichromin production of *Streptomyces* sp. WP-1 by genetic engineering. Appl Microbiol Biotechnol 107(17):5415–5425

Wang W-H, Li C-R, Qin X-J, Yang X-Q, Xie S-D, Jiang Q, Zou L-H, Zhang Y-J, Zhu G-L, Zhao P (2024) Novel alkaloids from Aspergillus fumigatus VDL36, an endophytic fungus associated with Vaccinium dunalianum. J Agric Food Chem 72(19):10970–10980

Webster G, Mullins AJ, Petrova YD, Mahenthiralingam E (2023) Polyyne-producing Burkholderia suppress Globisporangium ultimum damping-off disease of Pisum sativum (pea). Front Microbiol 14:1240206

Wekesa CS, Furch AC, Oelmüller R (2021) Isolation and characterization of high-efficiency rhizobia from Western Kenya nodulating with Common bean. Front Microbiol 12:697567

Wetie AGN, Woods AG, Darie CC (2014) Mass spectrometric analysis of post-translational modifications (PTMs) and protein–protein interactions (PPIs). In: Advancements of mass spectrometry in biomedical research. Springer International Publishing, Cham, Switzerland, pp 205–235

Xiao M, Luo Y, Zhang H, Yu Y, Yao H, Zhu Z, Chadwick DR, Jones D, Chen J, Ge T (2022) Microplastics shape microbial communities affecting soil organic matter decomposition in paddy soil. J Hazard Mater 431:128589

Yan JY, Zhao WS, Chen Z, Xing QK, Zhang W, Chethana KT, Xue MF, Xu JP, Phillips AJ, Wang Y (2018) Comparative genome and transcriptome analyses reveal adaptations to opportunistic infections in woody plant degrading pathogens of Botryosphaeriaceae. DNA Res 25(1):87–102

Yu Y, Gui Y, Li Z, Jiang C, Guo J, Niu D (2022) Induced systemic resistance for improving plant immunity by beneficial microbes. Plants 11(3):386

Yuan Y, Zu M, Sun L, Zuo J, Tao J (2022) Isolation and Screening of 1-aminocyclopropane-1-carboxylic acid (ACC) deaminase producing PGPR from Paeonia lactiflora rhizosphere and enhancement of plant growth. Sci Hortic 297:110956

Zhang R-S, Wang F-G, Qi Z-Q, Qiao J-Q, Du Y, Yu J-J, Yu M-N, Liang D, Song T-Q, Yan P-X (2022) Iturins produced by Bacillus velezensis Jt84 play a key role in the biocontrol of rice blast disease. Biol Control 174:105001

Zhou L, Yang X, Wang X, Feng L, Wang Z, Dai J, Zhang H, Xie Y (2023) Effects of bacterial inoculation on lignocellulose degradation and microbial properties during cow dung composting. Bioengineered 14(1):2185945–2185228

Zhu J, Song L, Shen S, Fu W, Zhu Y, Liu L (2023) Bioactive alkaloids as secondary metabolites from plant endophytic Aspergillus genus. Molecules 28(23):7789

Zhu Z, Ma Y, Xie X, Wang Z, Han L, Song J, Liang Y, Poelarends GJ, Chen Y, Lu M (2024) Investigation of AMA synthase from Pyrenophora teres f. teres 0–1 for the synthesis of Aspergillomarasmine A analogues and non-canonical amino acids. Adv Synth Catal 366(24):5160–5170

Zhuang J, Zhang J, Hou X-L, Wang F, Xiong A-S (2014) Transcriptomic, proteomic, metabolomic and functional genomic approaches for the study of abiotic stress in vegetable crops. Crit Rev Plant Sci 33(2–3):225–237

Part III
Microbial Biocontrol and Crop Protection

Chapter 9
Microbial Biopesticides: Ecofriendly Alternatives for Crop Protection

Lateef Babatunde Salam

Abstract The application of microbial biopesticides as an environmentally friendly substitute for chemical pesticides in crop protection is covered in length in this chapter. Microbial pesticides consist of bacteria, fungi, viruses, nematodes, and protozoa that are used to control several diseases in food crops across the globe. The use of biopesticides as an eco-benign approach in pest management is gaining traction due to their multiple modes of action, including regulating gastrointestinal disturbance, disrupting pest growth and metabolism, activating target-poisoning mechanisms, demonstrating multisite inhibitory effects, and releasing neuromuscular toxins and bioactive substances. The chapter starts by stating that food security and environmental safety are not mutually exclusive, thus a need to devise ecofriendly ways to ensure food security via crop protection without jeopardizing environmental safety is imperative. The chapter then comprehensively discusses microbial biopesticides (bacteria, fungi, viruses, nematodes, and protozoa), their mode of action, infectivity range, and their strengths and limitations as biocontrol agents. The last part of the chapter examines issues associated with the field application of microbial biopesticides and offers some recommendations that could enhance its acceptability, usage, and performance.

Keywords Biopesticides · Microbial biopesticides · Biocontrol agents · Entomopathogens

9.1 Introduction

Food loss is linked to a reduction in the amount and quality of food produced or stored (FAO 2019). These losses frequently serve as the primary source of hunger and malnutrition and influence food security and health globally. Additionally,

L. B. Salam (✉)
Department of Biological Sciences, Microbiology Unit, Elizade University, Ilara-Mokin, Ondo State, Nigeria

O. O. Babalola et al. (eds.), *Ecofriendly Frontiers*,
https://doi.org/10.1007/978-3-031-98700-7_9

natural resource constraints and the potential for environmental degradation affect preventive measures intended to reduce food losses (Peng and Berry 2018). Depending on the commodity and the local socio-economic milieu, fruits, vegetables, and cereals can lose up to 40% of their supply and market value during production, postharvest storage, shipping, and marketing (Beausang et al. 2017; FAO 2019). Agricultural product production and commercialization always need substantial human labor input and use precious and frequently finite resources (Liu et al. 2022). An important cause of food loss is biotic damage, brought on by microorganisms (Beausang et al. 2017; Magalhães et al. 2021). Diverse synthetic biocontrol agents have been developed and commercialized by agrochemical industries to control and avoid these losses. However, improper application of these agents can result in food contamination, bioaccumulation of toxic residues, insect pest resistance, deleterious impacts on beneficial arthropods, ecological balance disruption, and general environmental and public health problems (Jepson et al. 2020; Shi et al. 2021).

The US Environmental Protection Agency defines biopesticides as substances manufactured by genetically modified plants, naturally occurring benign biochemical pesticides, and microorganisms and their metabolic products (microbial pesticides) (Sabbahi et al. 2022). Multiple mechanisms of action, such as regulating gastrointestinal disturbance and disrupting pest growth and metabolism, contribute to the effectiveness of biopesticides in pest management. Protein denaturation, metabolic disruption and paralysis, target-poisoning mechanisms elicitation, antagonistic effects at multiple sites, and the release of neuromuscular toxins and bioactive chemicals are the ways that biopesticides function (Sparks and Nauen 2015; Dar et al. 2021; Fenibo et al. 2021). Unlike chemical pesticides, these diverse activities give biopesticides the ability to alter the path of pest resistance. According to studies, biopesticides are environmentally benign, have low toxicity, biodegrade, and work selectively, having little to no negative effects on organisms that are not their intended targets (Kalpana and Anil 2021).

Microbial pesticides consist of viruses, and prokaryotic and eukaryotic microorganisms that are deployed to control diseases in food crops across the globe (Thakur et al. 2020). Diverse techniques are used to apply microbial pesticides to the environment (De Oliveira et al. 2021). In augmentative biocontrol techniques, microbial biopesticides are frequently employed as a prophylactic approach to control infections. Since the required population of microbial biocontrol agents that serve as the basis for those biopesticide products can only be maintained for a brief period, the products must be applied multiple times to maintain their efficacy (Glare and O'Callaghan 2019). This chapter comprehensively discusses the various types of microbial biopesticides (bacteria, fungi, viruses, nematodes, and protozoa), their mode of action, infectivity range, and their strengths and limitations as biocontrol agents.

9.2 Viruses as Microbial Biopesticides

9.2.1 Bacteriophages

Bacteriophages are viruses, either virulent or temperate, that infect and multiply in live prokaryotes. Several reports have indicated virulent phages' ability to effectively control bacterial phytopathogens (Kazi and Annapure 2016; Korniienko et al. 2022). Bacteriophages are harmless and benign to humans, plants, and animals due to their exclusive affinity for prokaryotic cells (Vero et al. 2023). However, some phages have narrow specificity, making their use as biopesticides problematic. A cocktail of phages has been used to enhance their effectiveness, such as inhibiting *Pectobacterium*-induced soft rot in potatoes (Zaczek-Moczydłowska et al. 2020). Contrastively, broad host range phages can infect numerous species within a genus and bacteria from other genera, such as a Myoviridae phage with broad specificity for *Erwinia, Cronobacter*, and *Pectobacterium* bacterial phytopathogens (Buttimer et al. 2017).

Bacteriophages have biocontrol ability against various bacterial phytopathogens, including those that cause postharvest losses in potatoes (*Pectobacterium* spp., *Dickeya* spp.) (Zaczek-Moczydłowska et al. 2020), *Pseudomonas syringae* pathovars phytopathogenic activities on tomato seedlings, sweet cherry plantlets, kiwifruit leaves, and cherry leaves (Hernandez et al. 2020; Pinheiro et al. 2020; Rabiey et al. 2020; Akbaba and Ozaktan 2021), and activities of Gram-positive phytopathogens like *Streptomyces* and *Clavibacter michiganensis* on potatoes and maize seeds (Kimmelshue et al. 2019). There are numerous phage-based biocontrol agents on the market. These include four commercial products called AgriPhage™ (Omnilytics, Sandy, UT, USA) that control bacterial spots and specks (*Xanthomonas campestris* pv. vesicatoria and *Pseudomonas syringae* pv. Tomato), citrus canker (*Xanthomonas citri* subsp. Citri), apple tree fire blight (*Erwinia amylovora*), and tomato bacterial canker (*Clavibacter michiganensis* subsp. Michiganensis) (Vero et al. 2023). An additional noteworthy commercial product is Erwiphage Plus™ (Enviroinvest, Pécs, Hungary) made from a cocktail of phages to suppress fire blight in apple trees. Also worth mentioning is Biolyse-PB™ from APS biocontrol, which is designed to prevent bacterial soft rot in potatoes by combining a variety of phages (Vero et al. 2023).

Phage treatment is a promising method for controlling bacterial phytopathogens with certain inherent limitations. These include the potential for phages to come into touch with and infect vulnerable hosts, their susceptibility to UV radiation and other environmental conditions that could diminish their efficacy, and their possible detrimental effects on beneficial bacteria. When designing stable and efficient bacteriophage formulations, these constraints should be carefully taken into account (Vero et al. 2023).

9.2.2 *Baculoviruses*

Baculoviruses are characterized by their circular, dsDNA genome, rod-shaped virions, and 37 main genes (Garavaglia et al. 2012). Baculoviruses are found only in arthropods, particularly insects, making them safe alternatives for controlling pest insects (Rohrmann 2013). Within the host cell, they form crystalline occlusion bodies that contain infectious particles, which are linked to their infectivity. Baculoviruses are infectious orally, with occlusion-derived virions (ODV) infecting epithelial cells in the midgut (Hubbard et al. 2014). Baculovirus infections typically involve two enveloped virion phenotypes: ODV and budded virions (BV). ODV, which causes horizontal insect transmission, are implanted in occlusion bodies, while BV spreads the infection throughout the host's tissues (Clem and Passarelli 2013; Hubbard et al. 2014). Only baculoviruses are known to encode a DNA-directed RNA polymerase, which is utilized in the baculovirus expression vector system to transcribe late and very late viral genes and express foreign genes (Clem and Passarelli 2013).

According to the structure of their occlusion bodies, baculoviruses are separated into two types: granuloviruses (GV) and nucleopolyhedroviruses (NPV) (Szewczyk et al. 2011). GVs are larger and more complex viruses with different protective protein coats. GVs uncoat inside the nuclear pore complex, while NPVs uncoat inside the nucleus (Kalawate 2014). Before the virus replicates in the midgut cells, NPVs can penetrate the intestinal epithelium and cause systemic hemocoel infection after ingestion. GVs, on the other hand, do not pass through midgut cells as quickly and have a longer developmental cycle (Kalawate 2014). Baculoviruses are considered attractive biocontrol agents against insect crop pests for a variety of reasons. These include their extensive research history; their established safety and environmental acceptability profile; their high pathogenicity against some of the most significant crop pests in the world, including *Spodoptera* spp., *Plutella xylostella*, and various *Heliothis*/*Helicoverpa* species; and, lastly, the fact that many baculoviruses have highly developed commercially viable mass production systems, making their use as biological pesticides feasible (Lacey et al. 2015; Grzywacz and Moore 2017; Wilson et al. 2020).

9.2.2.1 Mode of Action

Two varieties of nucleocapsids and encapsulated virions are produced in the nuclei of infected cells. A single nucleocapsid exits the nucleus and buds out of the cell, getting an envelope from the plasma membrane to generate the first virus, known as the budded virus (BV) (Hubbard et al. 2014). Peplomers are structures that concentrate the BV attachment and fusion protein, either GP64 or F, at one end of the virion (Clem and Passarelli 2013; Kalawate 2014). When nucleocapsids take up an envelope from the inner nuclear membrane, they produce the second kind, known as an occlusion-derived virus (ODV). ODV embeds itself in proteinaceous crystals called occlusion bodies, which are primarily composed of the viral protein polyhedrin. The occlusion bodies are trapped inside the nucleus until they are released by cell lysis (Fig. 9.1). The baculovirus replication cycle begins when a vulnerable insect larva ingests viral occlusion bodies that contaminate their food source. In the larva's alkaline midgut, the occlusion bodies disintegrate, releasing ODV, which sticks to the midgut epithelial cells' microvillar membranes. The basal side of infected midgut epithelial cells is where the BV that infects tracheal epithelial cells sprouts. Tracheal cell infection is one way that BV can penetrate the midgut basal lamina and infect the entire insect. When the infected insects die, they liquefy, which makes it possible for the obstructed corpses to spread and facilitate further infections (Clem and Passarelli 2013; Kalawate 2014; Williams et al. 2017).

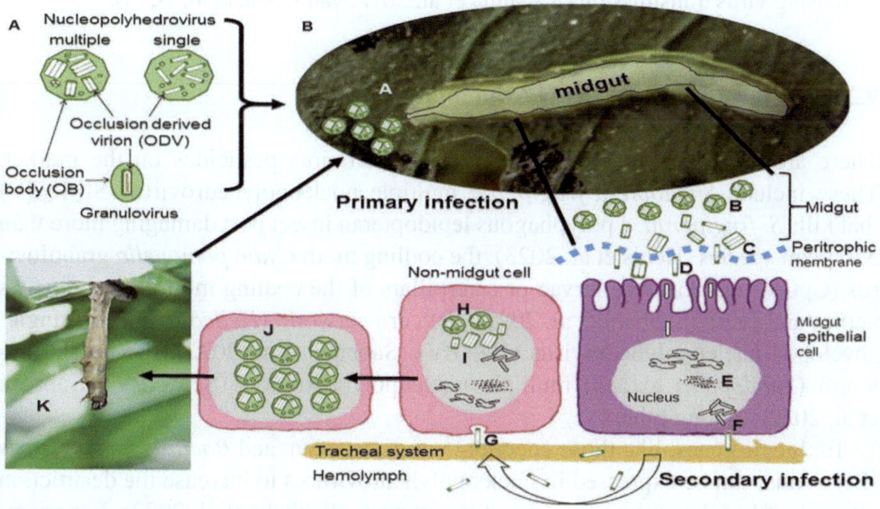

Fig. 9.1 Schematics of the baculovirus structure and the infection cycle. (A) Occlusion-derived virions (ODVs) are surrounded by nucleopolyhedrovirus occlusion bodies (OBs), which are primarily made of crystalline polyhedrin. One nucleocapsid or many nucleocapsids can be found within each ODV. In the granule-shaped outer membrane (OB) of granuloviruses, a single ODV with a single nucleocapsid is encased in granulin, a crystalline protein. In all cases, each nucleocapsid contains a single viral genome. (B) The stages involved in the transmission and reproduction of nucleopolyhedroviruses. When feeding on contaminated vegetation during the initial infection, (A) OBs are consumed. (B) ODVs are released and solubilized by OBs in the midgut of insects, where they merge with the microvilli of the midgut epithelial cells (D) after passing through the peritrophic membrane (C). Nucleocapsids release the viral DNA to initiate replication once they reach the nucleus. (E) Viruses multiply in virogenic stroma. The GP64 or F fusion protein, which is found in the virus-modified cell membrane, is present in the envelope that is produced when nucleocapsid offspring form and bud across the basal membrane (F). During the secondary phase of infection, these budded virions (BVs) spread throughout the hemolymph or along the cells of the insect tracheal system to infect cells in other insect organs. (G) After entering by endocytosis, BVs proliferate in the cell nucleus. Recently constructed nucleocapsids (H) can either sprout out of the cell or produce ODVs that are enveloped and occluded into OBs (I). At the end of the infectious cycle, OBs accumulate in the nucleus (J). The larvae typically dangle from the upper leaves of the host plant (K), and when the larvae die, their tegument rupture, releasing OBs that contaminate foliage and start new cycles of horizontal transmission (Williams et al. 2017)

Baculovirus-infected larvae exhibit pathogen-modified physiology and behavior, such as inhibiting larval growth and releasing viral occlusion bodies. By deactivating ecdysone, the main hormone that causes insect molting, the viral enzyme ecdysteroid UDP-glucosyltransferase (EGT) stops larvae from molting (O'Reilly and Miller 1989). Infected larvae enlarge beyond normal before liquefying, releasing millions of occlusion bodies per milligram of larval tissue. Another instance of pathogen-modified behavior is the larval climbing behavior of baculovirus-infected gypsy moths induced by the expression of the *egt* gene by the virus (Hoover et al. 2011), which ascend to the top of trees to die and release the occluded virus onto the foliage below. Additionally, the expression of protein tyrosine phosphatase (PTP) elicits increased larval locomotory activity, which promotes increased wandering behavior, thereby inadvertently increasing virus transmission (Katsuma et al. 2012; van Houte et al. 2012).

9.2.2.2 Baculovirus Pesticide Availability

There are several commercially available baculovirus pesticides on the market. These include *Spodoptera frugiperda* multiple nucleopolyhedrovirus (SfMNPV) that kills *S. frugiperda*, a polyphagous lepidopteran insect pest damaging more than 350 plant species (Kenis et al. 2023); the codling moth *Cydia pomonella* granulovirus (CpGV) that kills the larvae or caterpillars of the codling moth, which attacks apples and pears (Sabbahi et al. 2022); the corn earworm (*Helicoverpa zea*) single-enveloped nucleopolyhedrovirus (HzNPV) (Sabbahi et al. 2022); the beet armyworm (*Spodoptera exigua*) multiple nucleopolyhedrovirus (SeMNPV) (Sabbahi et al. 2022); among others.

Toxigenic genes, like those encoding wasp, scorpion, and *Bacillus thuringiensis* (Bt) toxins, can be expressed in nucleopolyhedroviruses to increase the destruction efficacy of baculovirus against target insect pests (Sabbahi et al. 2022). For example, the recombinant baculovirus ColorBtrus is created when the baculoviral OB and the Bt toxin are combined. Consequently, OBs harboring the Bt Cry1Ac toxin insecticidal protein are more pathogenic and have a faster rate of action than their wild-type counterparts (Sabbahi et al. 2022).

The short persistence of baculoviruses on crops due to UV radiation degradation is a significant barrier to their widespread use as biocontrol agents (Grzywacz and Moore 2017). This degradation reduces their half-life in temperate cropping systems to 2–10 days and 8 hours or less in the tropics or unshaded crop surfaces (Shapiro et al. 2012; Wilson et al. 2020). Farmers are demotivated by this vulnerability, as it requires more frequent weekly applications than competitive chemical pesticides, increasing expenses (Lacey et al. 2015). To overcome this constraint, a recent study developed a unique formulation that micro-encapsulates the baculovirus using a UV absorbent (titanium dioxide) and ENTOSTAT wax. This capsule dissolves in the insect's stomach and releases the virus, but it stops the viral DNA from degrading in the sun. Compared to current virus formulations, the new formulation extends the biopesticide's effectiveness beyond a few hours, potentially necessitating lower application rates or longer spray intervals. It has a minimum 6-month shelf life at 30 °C and is not phytotoxic to host plants (Wilson et al. 2020).

9.2.3 Mycoviruses

Mycoviruses are viruses that reproduce exclusively within the cells of the fungus they infect, without an extracellular phase. They can spread intracellularly through hyphal fusion, cell division, or sporogenesis within the same thallus (Son et al. 2015). They are found in various fungi taxonomic groups and have over 300 genomes sequenced. Most mycoviruses have linear double-stranded RNA (dsRNA) and some positive linear single-stranded RNA, with only a small number of circular single-strand DNA or negative linear single-stranded RNA (Vero et al. 2023).

It has been discovered that plant pathogenic fungi, including Magnaphorte oryzae, Alternaria alternata, Rosellinia necatrix, Penicillium digitatum, Fusarium graminearum, and Botryosphaeria dothidea, contain mycoviruses linked to hypovirulence (Vero et al. 2023). Mycovirus FGHV2, which lives on *F. graminearum*, decreases mycelial growth and deoxynivalenol production, a mycotoxin that is commonly present in infected wheat grains (Li et al. 2019). *Sclerotinia sclerotiorum* strains harbor hypoviruses including SsHV1 and SsHV2, infect over 400 species of plant, and host several mycoviruses (Wang et al. 2022). Fungal hosts of the SsBRD2 virus, found in the Botybirnaviridae family, are excellent prospective biocontrol agents as they can cause hypovirulence in the host fungus, inhibit fungal growth, and prevent sclerotia development (Ran et al. 2016).

Mycoviruses are used to transfer viruses from hypovirulent to virulent fungal strains, enabling biocontrol of fungal plant diseases. This restriction has been addressed by the use of artificial transfection techniques (Vero et al. 2023). For example, *Botrytis cinerea* protoplasts were transfected with partitivirus virions from a hypovirulent strain of *S. sclerotiorum*, reducing its virulence and germination in its host plant (Xiao et al. 2014). Introduction of MyRV3, a mycoreovirus from *Rosellinia necathrix* into protoplasts of other phytopathogenic fungi results in hypovirulence, altered pigmentation, and decreased growth rates in all recipient strains (Kanematsu et al. 2014).

9.3 Fungi as Microbial Biopesticides

9.3.1 Entomopathogenic Fungi

9.3.1.1 *Beauveria bassiana*

Beauveria bassiana is a widely known entomopathogenic fungus that can infect over two hundred species of insects (Nakahara et al. 2009). It produces several toxins that cause infections exogenously due to its vigorous multiplication (Naqqash et al. 2016).

The first stage of entomopathogenic fungal infection is "insect cuticle penetration," that commences with the use of mechanical forces, followed by activities of cuticle-degrading enzymes, and specific infection structures (appressoria) produced by hyphae that enter the host cell and multiply. The chitin and protein cuticle of insects is broken down by extracellular fungal proteinases, allowing hyphae to enter the host hemolymphoid. Hydrocarbons and lipid cuticular layers of the insects are rapidly assimilated by primary hydroxylating enzymes (Wang et al. 2021a, 2021, c). According to Lewis et al. (2009a, 2009b), given the right conditions, fungal conidia, which sprout into hyphae, produce a variety of insecticidal toxins that are non-specifically absorbed through the insect's epidermis. Most of the toxins produced by *B. bassiana* are secondary metabolites and small compounds, including beauvericin, oosporein, oxalic acid, bassianin, bassianolide, tenellin, calcium oxalate crystals, beauverolides, and several beauvericin analogs (Wang et al. 2021a, b, c). Among the most significant toxins is mycelial-secreted beauvericin, which also has nematicidal activity (Safavi 2013). it is noteworthy that new beauvericin compounds have insecticidal and cytotoxic properties (Xu et al. 2007). for instance, aside from the fungicidal and bactericidal properties of oosporein, it also prevents the growth of tumor cells (Feng et al. 2015).

The fungal insecticidal effect of *B. bassiana* toxins is believed to be a result of multiple mechanisms, including increasing virulence factors, preventing the host immune system from activating, destroying the insect host's epidermis, interfering with nerve conduction pathways, absorbing water and nutrients from the host body, and clogging spiracles (Ortiz-Urquiza and Keyhani 2013; Pedrini et al. 2013). The host insect provides the high energy needed for these processes through its hydrocarbons, which the fungi use as carbon and energy sources (Luo et al. 2015). According to Chu et al. (2017), toxins cause a host insect to exhibit a variety of symptoms, including severe dehydration, odd behavior, disorientation, convulsions, feeding difficulty, and metabolic abnormalities, all of which eventually result in the insect's death. Mycelia from the dead host generates many conidia, which are dispersed via the air to infect new hosts, continuing the infection cycle (Feng et al. 2015; Pedrini 2018). Other vectors in close contact with insects, such as leaves, roots, water, and soil also participate in the spread of infection, but they also run the risk of non-targeted infection, potentially harming beneficial insects (Wang et al. 2021a, b, c).

Recent discoveries of novel *B. bassiana* toxins and enzymes have led to their industrialization for pharmaceutical and agricultural applications (Pedrini et al. 2015). Adequate knowledge of toxin diversity can help develop eco-friendly and efficient bioinsecticides. The *B. bassiana* 70 strains yielded the insect-toxic protein Bb70P, which is most active in weakly acidic conditions. It causes *Galleria mellonella* (Greater wax moth) to undergo melanism, which is an immunoreaction that changes the color of the insect's body (Khan et al. 2016). A member of aminoacyl-tRNA synthetases, lysyl-tRNA synthetases (Krs), are used to develop *B. bassiana* bioinsecticides (Wang et al. 2021a, b, c). Chitosanase-like protein (Bclp), a hydrophilic 28-kDa toxin, was found in the *B. bassiana* 618 strains, killing *Galleria mellonella* through melanization and damaging epithelial cells in the insect's epidermis and trachea resulting in hemocyte infiltration and injury (Fuguet et al. 2004). Vip3A exhibited specific insecticidal action in Lepidopteran insects, demonstrating the potent insecticidal activity of vegetative insecticidal proteins (Vips), which were initially derived from *B. thuringiensis* (Wang et al. 2021a, b, c).

According to Montalva et al. (2016), *B. bassiana* is a biological control agent that has broad-spectrum insecticidal action against both medically significant arthropods and agricultural pests. Taxonomic intricacy, unreported phenotypic traits, and insect resistance have raised doubts about its broader adoption despite decades of commercial use (Wang et al. 2021a, b, c). Effective virulence strategies have been developed by *B. bassiana*, including mechanical penetration, immune substance release suppression, phagocyte avoidance, and secretory epidermis-degradative enzymes (Pedrini et al. 2013). Its employment may be restricted, nevertheless, by drawbacks such as a lengthy time to kill the target insect, impact on non-target invertebrates, human allergies, reliance on ambient temperature and humidity, increasing insect resistance, and an unclear taxonomic status of the fungi (Wang et al. 2021a, b, c). Some of the tradenames of commercially available *B. bassiana* formulations include Bauveril, BotaniGard ES, BotaniGard 22 WP, Boverol, Boverin, Mycotrol ES Mycotrol-O, Naturalis, Ostrinil, Proecol, among others (Bamisile et al. 2021).

9.3.1.2 *Metarhizium anisopliae*

Metarhizium anisopliae is a widespread insect pathogen that can be discovered as a saprobe in arthropod carcasses or in the soil of plant rhizospheres (Rehner and Kepler 2017). It can infect human disease vectors and agricultural pests. This fungus is asexual and uses conidia as its infection form (Lovett et al. 2019). According to Gao et al. (2011), the infection process consists of adhesion, spore germination, appressorium formation, host penetration, in vivo development, and host death. *M. anisopliae* kills more slowly than chemical pesticides, taking around five days to kill under optimal conditions (Fang et al. 2007). Additionally, environmental conditions like temperature, humidity, and UV radiation can easily impact how effective it is at controlling pests (Gašić and Tanović 2013). Pathogenicity of *M. anisopliae.*

Conidia Adhesion

Conidia of *M. anisopliae* stick to the host cuticle during the early stages of infection, and the hydrophobic epidermis and conidia are encouraged to cling by hydrophobic proteins in the outer layer (Schrank and Vainstein 2010; San and Mun 2017). Exogenous lectins found in *M. anisopliae* conidia's cell walls mix with the host's epidermal glycoprotein to create a specific bond that identifies the host (Samson et al. 1988). Hydrophobic interactions between spore surface proteins and the lipid layers covering arthropod cutin are necessary for adhesion (Schrank and Vainstein 2010). The significance of lipid enzymes is reinforced by the fact that lipids in the cuticular layer serve as the initial line of defense against pathogenic microbes. *M. anisopliae*'s prepenetration growth on the host epidermis is linked to lipid degradation. This growth is essential for detecting vulnerable hosts and generating the initial nutrient molecules that facilitate conidia germination (Pedrini et al. 2007; Schrank and Vainstein 2010). When the conidia cell wall changes, the adhesin protein Mad1 may replace hydrophobins that have been degraded, which is essential for the attachment of swollen conidia to the host surface (Peng et al. 2022).

Conidia Germination and Development

Conidia of *M. anisopliae* produce varying lengths of germ tubes by adhering to the host cuticle during germination (Ment et al. 2010). Trehalose, a regular source of glucose in the host hemolymph, is used by the early-stage enzyme trehalase (San and Mun 2017). Following germination, spores enlarge and develop germ tubes that develop into appressorium, which adhere via mucus by the hypha apex to the upper epidermis surface (Lovett and St Leger 2015). To create infection pegs, the germ tube lengthens (Hui et al. 2004). A variety of substances, including proteins, carbohydrates, lipids, and amino acids, can help conidia germinate (Peng et al. 2022). Mad1 and Mad2, two genes that encode adhesin, are implicated in plant root colonization and insect pathogenesis (Butt et al. 2013). Since the protein is also involved in cell division and appropriate cytoskeleton organization, it has been observed that deletion of the Mad1 gene delays conidia germination (Barelli et al. 2011).

Appressorium Formation

Before entering the epidermis, the germ tube of *M. anisopliae* develops into the appressorium, which is home to a large number of mitochondria, Golgi bodies, endoplasmic reticulum, and ribosomes (Senthil-Nathan 2015; Peng et al. 2022). The appressorium has a thick layer of mucilage covering it, and its shape ranges from elliptic to irregular, with mucous buildup giving it a mulberry shape (Hui et al. 2004). The thallus, host, and nutritional conditions all affect the quantity and composition of appressoria. Appressoria formation is caused by the genes for ornithine decarboxylase (ODC1) and *Metarhizium* perilipin-like protein (MPL1). The MPL1 gene controls lipid homeostasis and appressorium differentiation, while the ODC1 gene is upregulated during conidia germination and germ tube differentiation (San and Mun 2017). Appressorium development and the expression of cuticle-degrading enzymes can be delayed by protein kinase A inhibitors (Fang et al. 2009).

Cuticle Penetration

M. anisopliae, an insect pathogen, penetrates the host's cuticle through physical pressure and cuticle degradation. According to Fang et al. (2009), protease plays a critical role in the host's epidermal breakdown, which determines the ability to infect. *M. anisopliae* produces several enzymes that break down the protein-rich plasma membrane of arthropods, including chymotrypsin, carboxypeptidase, acid protease (PR3), subtilisin-like enzyme (PR1), and trypsin-like enzyme (PR2) (Moraes et al. 2003). M. anisopliae can infect a variety of hosts because the kinds and quantities of protein it produces vary depending on the host (San and Mun 2017). Chitin degradation products control chitinase activity and PR1 pretreatment can increase subsequent activity (Lin et al. 2011). P450 monooxygenase, which has ideal pH values of about 8, is essential for cuticle penetration. According to, the diseased epidermis's alkalinity is a physiological indicator of toxicity factors. Carbon competition between pathogens and hosts may affect the rate and breadth of infection, and chitinase is the primary predictor of pathogenicity (Seyoum et al. 2002; Hui et al. 2004). Conidia and the host can interact more hydrophobically when lipase on their surfaces releases free fatty acids (San and Mun 2017). The cuticle is broken down by protease, lipase, and chitinase together, allowing for efficient penetration and nutrition uptake for a successful infection (Santi et al. 2010).

Colonization of Hemolymph

M. anisopliae invades and proliferates in infected insects' hemolymph, potentially leading to their death (Seyoum et al. 2002). In vivo and in infected insect cultures, it generates cyclic peptide toxins known as destruxins (DTXs) (Hsiao and Ko 2001). Trehalases, which may be an adaptation to the host environment, turn trehalose from the host hemolymph into glucose for energy production (Santi et al. 2010). To shield conidia from UV light and reactive oxygen species, the fungus additionally generates catalase and peroxidases (San and Mun 2017). The utilization of sugar as a host nutrient is decreased by the fungus's ability to ingest trehalose and use sugar as a nutrient. Hyphae can quickly multiply and differentiate in the host's hemolymph because the Mad1 protein starts gene expression in the cell cycle (San and Mun 2017).

Extrusion and Sporulation

The fungus produces hyphae and conidia on the surface of the dead host as it exhausts nutrients during host colonization (Gabarty et al. 2014). The infection is thereby terminated when *M. anisopliae* develops a dense network of green spores on the cadavers of the infected hosts (San and Mun 2017).

M. anisopliae Toxins

DTXs are among the fungal secondary metabolites produced by *M. anisopliae* that cause host insects to become paralyzed, depolarize their membranes, and eventually die (Hsiao and Ko 2001). These toxins are cyclic peptide insect-borne fungal toxins that are insecticidal, phytotoxic, and antiviral (Schrank and Vainstein 2010).

M. anisopliae is the primary producer of the 38 DTXs. There are five classes of DTXs, and destruxins A, E, and B have the strongest insecticidal effects (Pedras et al. 2002; Sree et al. 2008). In human hepatoma Hep3B cells, destruxins B (DB) and desmethyl-DB can inhibit the expression of the hepatitis B virus surface antigen and are phytotoxic to Brassica plants (Chen et al. 1997; Hu et al. 2006). By destroying the muscular system and malpighian tubules, DTXs also impair the host's immunological system, which results in problems with nutrition, excretion, and mobility (Schrank and Vainstein 2010).

Commercial Formulations of *M. anisopliae*

There are several commercially available formulations of *M. anisopliae* in the market. These include formulations against *Tetranynchus urticae* (Achieve; Mozambique), Isoptera (*Kalotermitidae, Rhinotermitidae, Termopsidae*) (BioBlast; USA), Sugarcane pest; grayback canegrub (Bio-Cane; Australia), Hemiptera (*Aleyrodidae, Aphididae*) (Bio-Catch-M; India), Coleoptera (*Scarabaeidae*) (Bio-Green Granules; Australia), Coleoptera (*Curculionidae, Scarabaeidae*), Hemiptera (*Cercopidae*) (Bio-Magic; India), Mealybugs (Campaign; Ghana), Locust and grasshopper (Green Muscle; UK/France), and Mealybugs, thrips, leafminers (Real metarhizium 69; South Africa). Others include the Sugarcane spittle bug; *Mahanarva postica* (Metaquino; Brazil), Aphids (Mazao supreme; South Africa), Blattodea (*Blattellidae, Blattidae*) (BioPath; USA), and so on (Bamisile et al. 2021).

9.3.2 *Mycoparasitism*

Mycoparasitism is a phenomenon that describes a parasitic lifestyle in which one fungus (the mycoparasite) directly attacks another fungus (the prey) with the sole aim of exploiting the prey as a nutrient source (Leveau and Preston 2008). The parasitic interaction can be biotrophic or necrotrophic.

9.3.2.1 Biotrophic Mycoparasitism

Biotrophic mycoparasitism is a method where a parasite relies on a living host to obtain nutrients without killing them. *Ampelomyces* spp. are well-known biotrophic mycoparasites that can control powdery mildew on crops caused by members of the *Erysiphaceae* family (Kiss et al. 2004). They are a safe biocontrol agent with no antagonistic effects on beneficial microbiota because of their distinct method of action (Manjunatha et al. 2020). Unlike most biopesticides, this mycoparasite is safer since it parasitizes plant tissues only when there is an existing infection. Biotrophic mycoparasites also lessen the possibility of pathogenic strains becoming resistant. Several commercially marketed biofungicide products, including AQ10®

and Q-fect®, are derived from strains of *Ampelomyces* spp. and are available in different nations (Palmieri et al. 2022).

9.3.2.2 Necrotrophic Mycoparasitism

Numerous fungal plant pathogens are examples of necrotrophic mycoparasites, which penetrate host spores or hyphal cells after killing them. *Trichoderma* and *Clonostachys* are two well-researched mycoparasitic fungi that are used as active components in commercial formulations for the biocontrol of different fungal plant pathogens. Lalstop G46® WG contains *Clonostachys* spp. as the active ingredient, while commercially formulated products containing *Trichoderma* spp. include Promot WP® and Trichosoil®, among others (Vero et al. 2023). Plant pathogens like *Rhizoctonia solani*, ascomycetes like *Botrytis cinerea, Alternaria alternata, Sclerotinia sclerotiorum, Fusarium* spp., and oomycetes like *Pythium ultimum* are among the many host ranges of *Trichoderma* mycoparasites (Druzhinina et al. 2011). They may colonize and break down fungal structures other than hyphae, including *Giberella zeae* perithecia in wheat straw, and they are comparatively resistant to abiotic stress. This reduces the main source of inoculum for *Fusarium* head blight (Cabrera et al. 2020). Furthermore, *Trichoderma* species can parasitize and degrade phytopathogens sclerotia, such as *Sclerotium rolfsii* and *Sclerotinia sclerotiorum* (Silva et al. 2022). Also, an obligate mycoparasite *Coniothyrium minitans* parasitizes plant pathogens that produce sclerotia. Contans® WG (Bayer), a commercial formulation of this mycoparasite, has been developed (Palmieri et al. 2022).

Although their mycotrophic characteristics are similar, *Trichoderma* and *Clonostachys* species have evolved distinct tactics. According to Vero et al. (2023), *Trichoderma* species are invasive mycoparasites that cling to their host, coil around their hyphae, and create structures resembling appressoria to pierce them. Although *C. rosea* employs a similar tactic, it has not been confirmed that it forms appressoria (Karlsson et al. 2015). *Trichoderma* spp. are drawn to the prey by chemotropism in the intricate, multi-stage process of mycoparasitism. Certain lectins from the cell surface promote parasite adhesion and coiling around the prey hyphae once they come in contact with the prey. To permeabilize and break down the prey's fungal cell wall, the parasite produces hydrolytic enzymes like chitinases, β-1,3-glucanases, and proteases (Karlsson et al. 2017). After entering the prey's cells, the parasite uses the intracellular material as nourishment (Vero et al. 2023).

Because *Trichoderma* and *Clonostachys* species have a significant capacity to create and release a variety of antifungal secondary metabolites that can promote plant growth and induce resistance, their mycoparasitism is highly effective (Vinale et al. 2008). In comparison to other filamentous ascomycetes, genome analysis has shown that mycoparasitic *Trichoderma* species and *C. rosea* are particularly concentrated in genes associated with secondary metabolism (Karlsson et al. 2017). The fungal prey typically reacts to the attack during the mycoparasitic encounter by releasing enzymes, reactive oxygen species, and secondary metabolites. This counterattack must be able to be handled by a successful mycoparasite (Druzhinina et al. 2011). For instance, the potent antifungal compound zearalenone (ZEA) is produced by *Fusarium graminearum*. Through the production of an enzyme that detoxifies ZEA and ABC transporters that release ZEA and its breakdown products from its fungal cell, *C. rosea* has demonstrated the ability to tolerate ZEA while interacting with *F. graminearum* (Utermark and Karlovsky 2007).

9.3.2.3 Volatile Organic Compounds (VOCs)-Producing Yeast

According to Werner et al. (2016), volatile organic compounds (VOCs) are organic molecules with low molecular weight, low polarity, and high vapor pressure. VOCs include a wide range of molecular types, including phenols, benzene derivatives, alcohols, aldehydes, thioalcohols, ketones, alkanes, thioesters, heterocyclic compounds, and hydrocarbons (Morath et al. 2012).

The antifungal volatile organic compounds (VOCs) produced by yeasts have been the subject of numerous research, with a focus on the biocontrol of fungal infections during postharvest storage. Two Antarctic strains of *Candida sake* produced antifungal volatile organic compounds (VOCs) that prevented five apple pathogens from growing in vitro (Arrarte et al. 2017). *P. expansum* development in "Red Delicious" apple wounds was equally inhibited by the volatile organic compounds (VOCs) generated by the *C. sake* strains. According to Ararte et al. (2017), the two yeast strains generated different volatile organic compounds (VOCs), suggesting the strain-dependent nature of the volatilome. Sepúlveda et al. (2022) evaluated the antifungal effect of volatile organic compounds (VOCs) generated by two isolates of *Vishniacozyma victoriae* in dual cultures on different media against *Phlyctema vagabunda*, the agent that causes apple bull's eye rot. They found that the type of culture media affected the chemical composition of the volatilome as well as the pathogen-inhibiting action of volatile organic compounds (VOCs), suggesting that the growth conditions of microorganisms influence their volatilome (Sepúlveda et al. 2022).

Among the low molecular weight lipophilic compounds that come from primary and secondary metabolism, 2-phenylethanol has been repeatedly detected in the volatilome of biocontrol yeasts (Gotor-Vila et al. 2017; Farbo et al. 2018). To determine how exactly 2-phenylethanol inhibits sporulation, development, and the production of ochratoxin, the effect of the whole volatilome produced by the biocontrol yeast *Candida intermedia* 253 on the proteome of the target pathogen (*Aspergillus carbonarius*) was assessed in relation to the proteome obtained with 2-phenylethanol alone (Tilocca et al. 2019). The yeast volatile organic compounds (VOCs) drastically decreased mitochondrial metabolism, protein production, proliferative activity, and the detoxification of toxic substances.

Since VOCs can impact pathogens without being in close contact with them or the matrix that supports their growth, they are thought to be potentially safe biofumigants for foods (Arrarte et al. 2017). However, as certain VOCs may be extremely harmful and carcinogenic, they should be thoroughly profiled before using them as biocontrol agents, especially in the postharvest stage (Li et al. 2021a, 2021b).

9.4 Bacteria as Microbial Biopesticides

9.4.1 Bacillus Species

The genus *Bacillus* is a diverse collection of motile, spore-forming, facultatively anaerobic, aerobic, Gram- and catalase-positive bacteria. Morphologically, they are large, rod-shaped, and straight, with a width of 0.5–2.5 μm and a length of 1.2–10 μm (Zeigler and Perkins 2008). They exhibit high metabolic diversity and can develop heterotrophically or autotrophically using a variety of carbon sources. Because of the dormant endospores they produce, *Bacillus* spp. can thrive in unfavorable conditions (Saxena et al. 2020). The cellular arrangement can be single, in pairs, or chains, with the endospores centrally or terminally located. *Bacillus* spp. are widely distributed in diverse natural habitats, owing in part to the physiological diversity that allows them to grow and sporulate under broad environmental conditions (Piggot and Hilbert 2004). Moreover, their ability to produce biofilms accentuates their colonization, survival, and adaptation (Haque et al. 2020).

Only a small number of species in this genus are harmful, but most of them have a positive impact on plant growth and yield (Miljaković et al. 2020). Beneficial Bacillus species are extensively utilized in agriculture due to their high production of a range of bioactive substances and appropriate culturing characteristics (Sansinenea 2019). Because of its insecticidal properties, which are crucial against phytophagous insects, Bacillus thuringiensis is the most commonly used Bacillus in agriculture. Members of the *Bacillus subtilis* complex are primarily utilized against phytopathogens (Su et al. 2020) and are generally regarded as safe (GRAS). Bacillus species use a variety of biocontrol mechanisms to protect plants from phytopathogens, including the synthesis of antibiotics, lytic enzymes, siderophores, competition for nutrients and space, production of volatile compounds, and induced systemic plant resistance (ISR) (Zhang et al. 2023).

9.4.1.1 Bacillus Thuringiensis (Bt) Toxins

The ability of the Gram-positive, spore-forming bacterium Bacillus thuringiensis (Bt) to produce crystal proteins (Cry) is widely recognized. Since many insects are thought to be poisoned by cry protein, Bt is utilized as a microbial insecticide to modify plants to increase their resistance genetically (Salehi Jouzani et al. 2008). Bt produces secreted insecticidal protein (Sip), vegetative insecticidal protein (Vip), and crystal proteins. Both of these biodegradable substances impact particular targets, particularly Lepidoptera and Coleoptera species (Chakroun et al. 2016). The digestive enzymes in the gut of the insect larvae activate the toxin when they consume the proteinaceous crystal, causing pore creation in the gut's cell membrane, gut paralysis, and ultimately the larvae's death (Vachon et al. 2012). Cry proteins primarily target three types of insects: Diptera (flies and mosquitoes), Coleoptera (beetles and weevils), and Lepidoptera (butterflies and moths). However, according to Domínguez-Arrizabalaga et al. (2020), cry proteins are also harmful to nematodes and hymenopterans, which include wasps and bees.

Bacillus thuringiensis strains produce delta endotoxins, often known as cry and cyt proteins (Bravo et al. 2007). During its stationary phase, the bacterium manufactures them during sporulation. They are formed as crystalline inclusions of paraspora that are triggered by the proteolytic action in the insect's gut following ingestion. Invertebrates are not known to be harmed by these proteins, which are referred to as parasporins (PS) (Ohba et al. 2009). Furthermore, some toxins called secretory insecticidal proteins (Sip) and vegetative insecticidal proteins (Vip) are generated during the vegetative phase. *Bacillus thuringiensis* strain EG2158's culture medium contained Sip (Sip1Aa1), the only protein that is harmful to coleopteran larvae (Palma et al. 2014). With a wide variety of insecticidal activity, the intestinal cry protein (ICP) toxin-encoding crystal (Cry) genes are the most common types. There are around 800 cry genes organized into 75 families (Cry1 to Cry 75), 40 cyt genes organized into 3 families (Cyt 1, Cyt 2, Cyt 3), and 146 Vip genes organized into 4 families (Vip 1 to Vip 4), according to Crickmore et al. (2020).

9.4.1.2 Mode of Action

The principal mode of action of the Bt toxin, which is the most generally accepted and approved, is the lysis of the insect's midgut epithelial cells. The toxin enters the plasma membrane and functions from the outside of the cells; it does not enter the cytoplasm. Columnar cells exhibit swelling and blabbing as well as subsequent lysis in the initial impact. The effect on the goblet cell is slower when contrasted (Liu et al. 2018). In the case of lepidopterans, when a lepidopteran consumes the toxin within the first five minutes, the gut cells absorb more glucose and experience histopathology (Knowles 1994). Five minutes later, the midgut becomes paralyzed. The columnar microvilli bubble and the columnar cells swell in ten minutes. The goblet cells are the site of the histopathology. As a result, the blood's pH rises while the lumen's pH falls. The blood potassium ion level then drops within 30 minutes due to a rise in potassium ion activity in the gastrointestinal cells. As the levels of leucine and glucose drop, the gut cells begin to degrade metabolically. Following lysis, the cells begin to separate from the basement membrane after 30 minutes. The entire body becomes paralyzed in one to seven hours, and death ensues due to starvation after one to three days (Liu et al. 2018). Bt produces a variety of insecticidal protein kinds. About 950 distinct toxin types have been cloned from 74 Cry groups, 3 Vip groups, and 3 Cyt protein groups (Crickmore et al. 2020).

Bt Genetically Modified Crops

Many crops use Bt toxin genes to improve the quality of their insect resistance (Jouzani et al. 2017). About 198 types of Bt crops, which line up eight plants—soybean, potato, aubergine, corn, maize, cotton, poplar, rice, maize, and tomato—have been generated since the introduction of Bt crops. Cry toxins are aggregated crystals of Bt, an insecticidal protein with a parasporal body and crystalline structure, that develop during sporulation. Certain Nematoda, Lepidoptera, Coleoptera, Diptera, and Hymenoptera species are effectively targeted by these toxins (Abbas 2018). Crops that have undergone genetic modification (GM) have one or more *cry* or *vip* genes. The most widely approved Bt-GM crops are potatoes, maize, and cotton, which have 42, 115, and 30 variants, respectively (ISAAA 2016). Cry1A.105, cry1Ab, cry1F, cry1Ac, cry2Ae, cry2Ab, and vip3A are the seven anti-lepidopteran vip and cry genes utilized to increase resistance to lepidopterans. There are currently 111 approved Bt types of anti-coleopteran genes, some of which are anti-lepidopteran genes. According to Jouzani et al. (2017), two genes, cry34Ab1–cry35Ab1 and cry3A, are utilized in around 34 and 60 genetically modified crops, respectively, to create variants resistant to coleopteran pests.

Furthermore, Muddanuru et al. (2019) created a transgenic castor (*Ricinus communis* L.) that expresses the Bt cry1Aa gene to fight lepidopteran insect pests. Both lepidopteran and coleopteran insects are specifically targeted by the Cry1I class of Bt toxins. Berretta et al. (2020) evaluated the toxicity of a Cry1Ia protein from an Argentinian Bt strain against agricultural pests from the Noctuidae, Curculionidae, Tortricidae, and Tenebrionidae families. *Anthonomus grandis, Spodoptera frugiperda*, and *Alphitobius diaperinus* were not affected by the protein, however, *Cydia pomonella* and *Rachiplusia nu* were susceptible. Furthermore, Singh and colleagues (2021a, 2021b) have demonstrated that the Bt protein known as Cry can inhibit the growth of *S. litura, B. cucurbitae*, *C. Pipens*, and *G. malonella,* larvae. This suggests that crops that are resistant to these pests might be developed by plant breeders.

Despite the relative global acceptance of Bt genetically modified crops and their impacts in controlling insect pests and enhancing productivity, there are inherent concerns relating to the insect pests' development of resistance to the Cry toxins. Mutations in the genes encoding for receptor molecules lead to the development of the resistant insect-pest population by interfering with the insect's ability to engage with the toxin (Wang et al. 2016). Chimeric toxins produced via toxin domain swapping and combinatorial ICPs are being used to prevent resistance and offer long-term efficacy against insect pests (Wang et al. 2016). Alternatively, several strategies have been developed and assessed for application in crop improvement initiatives, including RNA interference (RNAi), which seeks to inhibit particular genes implicated in insect development and feeding (Tyagi et al. 2020). A further strategy is to utilize genome editing technologies to modify insect pest receptors connected to *cry* toxin resistance genetically. For instance, the knockdown of cadherin receptors that are genetically linked to the insect's resistance to the Cry1Ac toxin is evidence of successful genome editing in Helicoverpa armigera (Wang et al. 2016). The target areas in midgut receptors responsible for the development of resistance to BtICPs or insecticidal proteins can also be changed using this strategy, according to Tyagi et al. (2020).

9.4.1.3 Mechanisms of Biological Control against Phytopathogens by Other Bacillus Species

Production of Antimicrobial Compounds

Members of the genus *Bacillus* devote 5–8% of their genome to producing several secondary metabolites that aid in antibiosis (Fira et al. 2018). Antimicrobial compounds are classified into three groups according to their biosynthetic pathways: ribosomal peptides (RPs), polyketides (PKs), and non-ribosomal lipopeptides and peptides (NRPs) (Caulier et al. 2019).

Bacteriocins

Bacteriocins are ribosomal peptide (RP) toxins with diverse structural and biochemical groups that elicit bactericidal effects via attachment to certain receptors on the surface of vulnerable bacterial pathogens with consequent alteration to their metabolic, biological, and morphological functions (Daw and Falkiner 1996; Grinter et al. 2012). Lantibiotics are the main biocidal bacteriocin category in *Bacillus* species, despite the discovery of additional bacteriocins such as nisin, subtilisin, subtilin, lichenicidin, ericin, cerein, and thuricin (Abriouel et al. 2011). It has been demonstrated that bacteriocins from B. amyloliquefaciens FZB42, including amylocyclicin (6.4 kDa) toxin (Scholz et al. 2014) and plantazolicin (Microcin, 1.3 kDa) protein toxin (Scholz et al. 2011), has antibacterial activity against closely related bacteria with phytopathogenic potential. Gebhardt et al. (2002) reported bacitracin production by a *Bacillus* sp. strain lU 2184, which exhibited antifungal activity against *B. cinerea*, the causative agent of grey mold in various crops. The study also reported a *Bacillus* endosymbiont *Bacillus* sp. lU 7314 that produces amicoumarin B, a bacteriocin that showed a herbicidal effect against *Lemma minor* (duckweed) in freshwater (Gebhardt et al. 2002). Another study found that two strains of *B. pumilus*, ZeD17 and DFAR8, produced a bacteriocin that inhibited seed germination in certain herbs while simultaneously having antifungal activity against *Rhizoctonia solani* (Dehghanifar et al. 2019).

Thuricin 17, a bacteriocin generated by an endophytic *B. thuringiensis* strain NeB17, inhibited a wide variety of bacterial species, including some closely related strains (Grey et al. 2006). However, a 21 kDa bacteriocin from *B. subtilis* strain 14B reduced the percentage of crown gall infections caused by the *Agrobacterium tumefaciens* pathogen in weeping fig trees (Hammami et al. 2009). Bacteriocins may be more important in encouraging niche colonization by the biocontrol agent since they eliminate or reduce other related bacteria, including certain important phytopathogens (Ajuna et al. 2024).

Cyclic Lipopeptides

Cyclic lipopeptides (LPs) are a diverse class of NRP antibiotics that combat a range of bacterial and fungal pathogens. According to Schneider et al. (2014), LPs work against bacteria by attaching themselves to their cell membrane, causing ion leakage and perforations, depolarising quickly, inhibiting the synthesis of DNA, RNA, and proteins, and finally causing cell death. The primary targets of antifungal LPs are chitin and (1–3)-β-D-glucan synthases, which lead to the disruption of pathogenic fungi cell wall formation, osmotic pressure regulation, and whole cell shape (Helmy and Parang 2023). Aside from the cell membrane, they also impact the synthesis and normal functioning of internal cellular components such as endoplasmic reticulum, proteins, mitochondrial membranes, and nucleotides, (Helmy and Parang 2023). Three major LP families have been identified in *Bacillus* species: fengycin (fengycin, plipastatin, and maltacin), iturin, and surfactin (e.g., surfactin, lichenysin, pumilacidin, and halobacillin) (Köhl et al. 2019). While iturins and fengycins have narrow specificity against only fungal pathogens, surfactins exhibit antibacterial and antifungal properties (Meena and Kanwar 2015).

Surfactins

Non-ribosomal peptide synthetases (NRPSs) produce surfactins, which are amphiphilic cyclic lipopeptides composed of heptapeptides of various amino acids joined by a β-hydroxy fatty acid to form a range of congeners (Fig. 9.2; Fira et al. 2018; Geissler et al. 2019; Théatre et al. 2021). They can obstruct normal phytopathogenic growth by causing pore formation, disrupting cellular organelles, and solubilizing the lipid bilayer in cell membranes (via the cation-carrier effect) because of their amphiphilic properties, which allow them to act on the surface of vulnerable host cells. This causes the cells to expand, become rough, develop apical outgrowths,

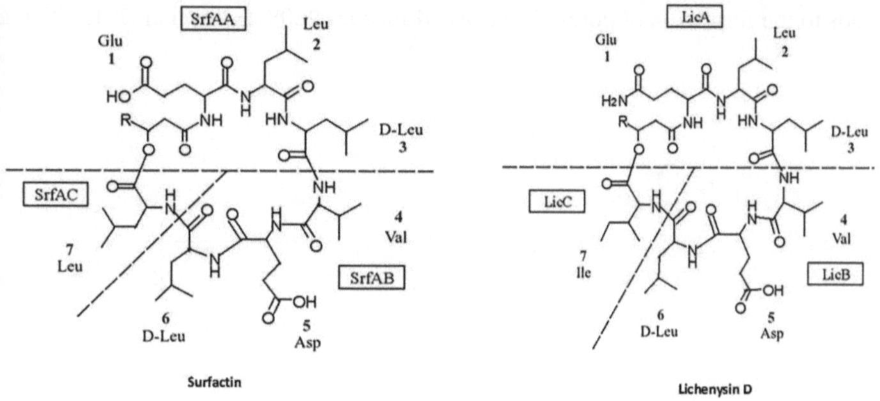

Fig. 9.2 Structures of Non-ribosomal Peptides of the Surfactin Family

deform, and protrude (Zeriouh et al. 2014; Park et al. 2019). Because *Bacillus* sp. surfactins do more than only directly elicit antibiosis, they are crucial for the biocontrol of bacterial phytopathogens and inhibition of their growth through the production of biofilms and the colonization of phyllospheres and rhizospheres (Ajuna et al. 2024). It is commonly acknowledged that surfactins are essential to the biocontrol ability of several Bacillus strains against phytopathogens.

The fungal pathogen *Colletotrichum gloeosporioides*, which causes anthracnose disease in a range of economic crops, was effectively inhibited by surfactins with a C15-lipid chain from *Bacillus velezensis* strain GH1-13, according to a study by Park et al. (2019). The scientists noted that the fungal cell wall's structural integrity was lost as a result of the surfactins' disruption of the chitin, glucans, and glycoproteins (Park et al. 2019). Another study discovered that surfactins from *Bacillus pumilus* strain PtB180 efficiently controlled grey mold disease in tomatoes by preventing the mycelial growth and conidial germination of phytopathogenic fungi, particularly *Botrytis cinerea* (Bouchard-Rochette et al. 2022). Furthermore, Surfactin C, which was obtained from *B. velezensis* strain 1B-23, showed antagonistic activity against a variety of phytopathogenic fungi, such as *Fusarium* sp., which causes Fusarium wilt, *Monilinia fructicola*, which causes brown rot in stone fruits, *Cochliobolus carbonum*, which causes northern leaf spot in maize, and *Rhizoctonia solani*, which causes stem and root rot (and dumping off) in a variety of plants (Li et al. 2021a, b).

Iturins

Numerous *Bacillus* species strains emit iturins, a well-known class of antimicrobial lipopeptides that are heptapeptides linked to a 14–17 carbon β-amino fatty acid chain (Fig. 9.3). They might be useful in the biocontrol of several phytopathogens (Khan et al. 2017; Ongena and Jacques 2008). Iturins have strong fungistatic activity by causing severe morphological abnormalities and conglobation in the hyphae and conidia, mainly because of their capacity to permeabilize membranes, which leads to the formation of pores (Ongena and Jacques 2008; Gong et al. 2015; Zhang

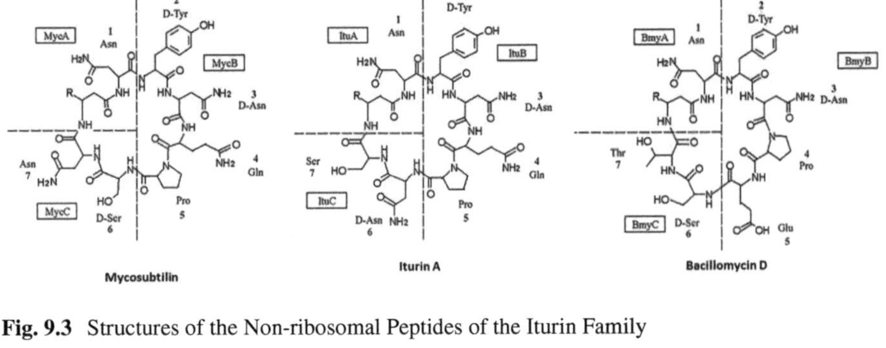

Fig. 9.3 Structures of the Non-ribosomal Peptides of the Iturin Family

et al. 2022). The iturin family also includes mycosubtilins, bacillomycins, and, more recently, mojavensin A due to structural similarities (Dunlap et al. 2019; Geissler et al. 2019). For instance, it has been observed that purified iturin A from B. subtilis strain Z-14 causes the hyphae to fragment, the cell wall to vanish, the cell membrane to degenerate, and the organelles to shrink. In the end, these actions inhibited growth and decreased the infection of the wheat take-all disease-causing Gaeumannomyces graminis var. tritici (ggt) (Xiao et al. 2021). Iturin A from *B. amyloliquefaciens* strain S76-3 significantly altered the morphology of the conidial spores, distorted the hyphae, and caused cellular contents to leak out and become inactive. These actions had a potent fungicidal effect on *Fusarium graminearum*, a pathogen that is well-known for causing problems in a range of plants (Gong et al. 2015).

Some *Bacillus* strains including *B. velezensis* c16, produce iturins that showed efficient suppression of *Alternaria solani* by cytoplasmic component leakage, potentially signifying the development of ion-conducting cell wall pores (Zhang et al. 2021). This resulted in protoplast shrinkage, altered cell shape, hyphae growth inhibition, and suppression of *A. solani* conidial germination, all of which decreased the incidence of potato leaves early blight disease (Zhang et al. 2021). Also, Iturin A5 from *B. amyloliquefaciens* strain S185 recoved from banana rhizosphere was reported to cause severe morphological changes, which in turn prevented *F. oxysporum* f. sp. cubense, the cause of Panama disease/Fusarium wilt disease of bananas, from germination of its spores and proliferation of its hyphae (Singh et al. 2021a, b).

Fengycins

Among the numerous subtypes of cyclic lipopeptides that make up the fengycin family are plipastatin, fengycin A, fengycin B, and fengycin S. With a side-chain length of 16–19 carbon atoms for the β-hydroxy fatty acid, these lipopeptides are partially cyclic decapeptides (Fig. 9.4; Steller and Vater 2000; Geissler et al. 2019). Several strains of *Bacillus* species release these important antimicrobial lipopeptides, which have strong antagonistic activity against a variety of fungal phytopathogens (Ongena and Jacques 2008). It has been demonstrated that fengycins break down the cytoplasm and cellular organelles, causing the hyphae to contract and distort. This destroys the internal cell structure of vulnerable pathogens (Xiao et al. 2021). The fengycin family member plipastatin A, which is produced by the *B. amyloliquefaciens* strain S76-3, induces vacuolation, which causes conglobation in immature hyphae and branch tips. It also causes severe damage to the plasma membranes and separates them from the cell walls (Gong et al. 2015).

Additionally, it was observed that the main antifungal compounds from *B. subtilis* strain NcD-2 that work against *R. solani*, the culprit behind cotton damping-off, are fengycin-type lipopeptides (Guo et al. 2014). It was also shown that the primary lipopeptides causing the antibacterial action against *Ralstonia solanacearum*, the causative agent of tomato bacterial wilt, were fengycins from *B. velezensis* strain FJAt (Chen et al. 2020a, 2020b). *B. velezensis* strain BA-26 is another example; it

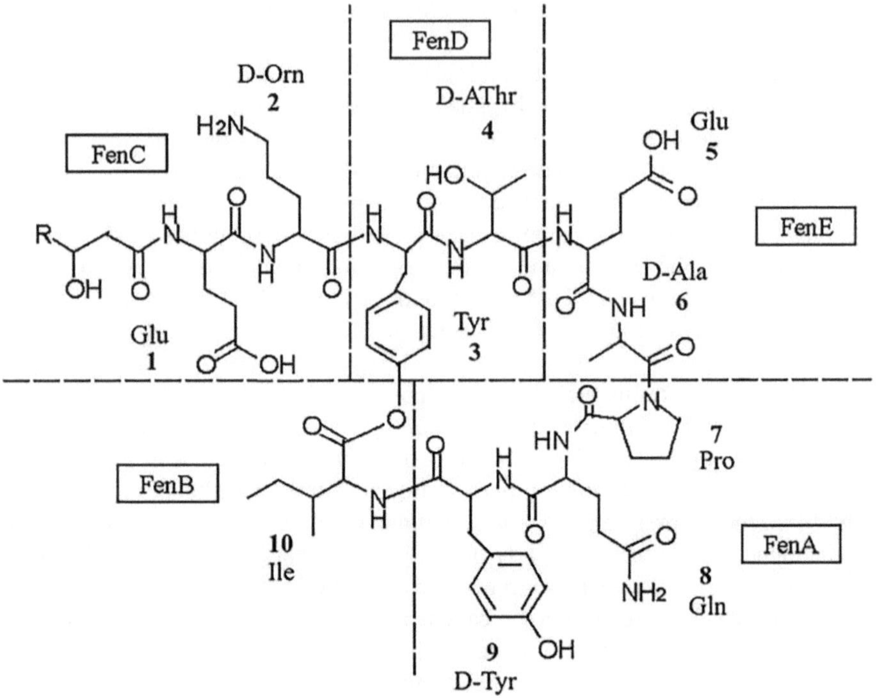

Fig. 9.4 Structure of the Non-ribosomal Peptide Fengycin

secretes fengycin, which prevents the *B. cinerea* pathogen from causing strawberry fruit rot (Wang et al. 2021a, b, c). Additionally, fengycins A and B from the *B. velezensis* strain Pw192 showed emulsifying capacity, decreased surface tension, and inhibited the growth of *Colletotrichum* spp., the pathogen that causes anthracnose disease (Jumpathong et al. 2022).

Bacillibactin

Bacillibactin, a microbial catecholate siderophore NRP is biosynthesized via the condensation of a cyclic amino acid core to three 2,3-dihydroxybenzoate-glycine-threonine units. Acquisition of bacillibactin by Bacillus strains confers significant ecological adaptations to them (Nithyapriya et al. 2021). Through iron scavenging and antibiosis, bacillibactin contributes significantly to Bacillus spp.'s antimicrobial efficacy against phytopathogens (Dimopoulou et al. 2021). The binding of bacillibactin to iron (III) results in bacillibactin-iron complexes, which are carried into bacterial cells by siderophore-binding proteins in the cell membrane. Because iron is crucial for microbial growth, this effectively deprives the phytopathogens of the micronutrient (iron) they need to grow.

Iron (III) is reduced to iron (II) once it enters the bacterial cells, where it is used, particularly in situations where iron is limited, to promote the growth of the biocontrol agent (Fukushima et al. 2013; Rabbee et al. 2019). Since this scavenging mechanism is independent of direct antibiosis mechanisms, it suppresses the proliferation of phytopathogens, including non-susceptible microorganisms. For example, the bacteriostatic activity of the bacillibactin from *B. amyloliquefaciens* strain MBi600 was seen against *P. syringae* pv. tomato, a phytopathogen that is not susceptible, and it was only produced in environments where iron starvation is prevalent (Dimopoulou et al. 2021). Similar to this, under iron deprivation, the growth of phytopathogens such as *V. dahliae, F. oxysporum* f. sp. radicis-lycopersici, *R. solani,* and *A. flavus* that are not susceptible or that have low susceptibility was inhibited in a dose-dependent manner by bacillibactin from *B. amyloliquefaciens* strain MBi600 (Dimopoulou et al. 2021).

In some studies, the presence of genes that synthesize bacillibactin-siderophore in some *Bacillus* species has been used to predict the role of siderophore in the biocontrol of phytopathogens and to estimate the biocontrol potential of the siderophore-producing *Bacillus* species (Dimopoulou et al. 2021; Dunyashev et al. 2021; Shin et al. 2021).

Antimicrobial Polyketides from *Bacillus* spp.

Bacillaene

Trans-acyltransferase polyketide synthetase is an enzyme that produces Bacillaene, a non-ribosomal linear polyketide (PK/NRP hybrid) that comes from *Bacillus* species, primarily *B. subtilis* (Miao et al. 2023). Bacillaene is a powerful antibiotic that may be used to control pathogenic microorganisms (Miao et al. 2023).

Bacillaene from *Bacillus* species has been shown to possess antifungal features against phytopathogens that cause huge financial losses in fruit production before and after harvest. For example, the phytopathogen *Penicillium digitatum*, which causes post-harvest rot disease in citrus fruit, was somewhat reduced by the bacillaene from the *Bacillus amyloliquefaciens* strain DH-4 (Chen et al. 2018). By interfering with the mitochondria and generating vacuolation and cell leakage, bacillaene inhibited mycelial development and decreased the rate of spore germination (Chen et al. 2018). A bacillaene biosynthesis gene cluster that inhibits protein synthesis in plant pathogens that infect pepper seeds and maize, such as *Pythium ultimum,* and *F. graminearum*, was discovered in a different study using genomic analysis of the TSA32-1 strain of *B. velezensis* (Kim et al. 2022). Furthermore, it was suggested that bacillaene moderates the antibacterial activity of *Bacillus amyloliquefaciens* FZB42 against the phytopathogen *Erwinia amylovora*, which causes apple and pear flower fire blight (Chen et al. 2009).

Macrolactins

Macrolactins are highly unsaturated macrolides comprising six 24-carbon ring lactones secreted by deep-sea bacteria that are employed as biocontrol agents and therapeutic antibiotics against bacterial and fungal phytopathogens (Li et al. 2016). fMarine sediments inhabiting *B. subtilis* B5 contain the macrolactins 7-O-2′e-butenoyl macrolactin A and macrolactin A, which target *C. Gloeosporioides* and *pestalotiopsis theae* that cause grey tea blight and anthracnose in tea leaves (Li et al. 2016). Furthermore, *Bacillus amyloliquefaciens* strain Eli149's 7-O-succinyl macrolactin A and macrolactin A demonstrate antifungal activity against soil-borne phytopathogens such as *Moniliophthora roreri* (cacao frosty pod rot), *Rhizoctonia* sp. (damping off disease), and *F. oxysporum* (wilt disease) (Salazar et al. 2020).

Additionally, *B. siamensis* strain NJ08-3's macrolactin R has been shown to limit germ tube elongation (ic50–1.93 mg/l), spore germination, and mycelial development (ic50–2.88 mg/l) in *B. cinerea*, the bacterium that causes grey mold in strawberries. According to Ni et al. (2023), this is because macrolactin R changes membrane permeability, which results in protein and nucleic acid (organelles) leakage from cells and eventually in cell death.

Bacilysin

Bacillus spp. secretes polyketides called bacilysin, a non-ribosomal simple peptide (NRP) antibiotic that contains both the L-Anticapsin and an L-alanine residue (Tojo et al. 2014). With a fairly basic structure, bacilysin is extremely active, depending on the environment, against a wide variety of significant phytopathogens (Parker and Walsh 2013; Rabbee and Baek 2020). Bacilysin is a deadly weapon against plant pathogens because of its potent antimicrobial activity against bacterial and fungal pathogens of plants, which is linked to its irreversible interference with glucosamine synthetase, with consequent disruption of mannoprotein biosynthesis in fungal pathogens or peptidoglycan synthesis in bacterial pathogens (Wu et al. 2018).

Bacilysin has diverse uses in the realm of crop protection, much like pharmaceutical applications. Bacilysin D from *Bacillus velezensis* strain FZB42 is thought to exhibit anti-oomycete activity against *Phytophthora sojae*, the causative agent of soybeans root and stem rot disease, by severely damaging hyphal structures and resulting in the loss of intracellular contents. Additionally, it can counteract other *Phytophthora* sp. phytopathogenic strains (Han et al. 2021a, 2021b).

According to Afsharmanesh et al. (2018), the main antagonistic antimicrobial peptide against *A. flavus* is bacilysin from *B. subtilis* UtB1. This peptide stops dangerous aflatoxins from accumulating in human food and animal feed and aids in the control of cereal grains, nuts, and legumes ear rot diseases. The oxidoreductase enzyme BacC which decimates aflatoxins in pistachio nuts, is encoded by the *bacC* gene, which is also in charge of bacilysin production. Therefore, bacilysin production by *B. subtilis* strain UtB1 directly breaks down aflatoxins and stops the growth of *A. flavus*.

Bacilysin's antimicrobial efficacy against dangerous algal bloom microorganisms was also emphasized in another study (Rabbee and Baek 2020). Here, bacilysin interferes with the glucosamine synthase biosynthesis enzyme, which results in alterations to the peptidoglycans found in the bacterial cell walls.

Production of Hydrolytic Extracellular Enzymes

By dissolving the glycosidic bonds in the fungal cell wall, *Bacillus* species employ hydrolytic enzymes to suppress pathogens, especially phytopathogenic fungi (Garcia-Rubio et al. 2020). Plant defense, growth, and development depend on these enzymes, which include glucanases, cellulases, proteases, chitinases, and chitosanases (Khalil et al. 2022). *Aspergillus, Fusarium, Trichoderma, Bacillus, Azospirillum, Aeromonas, Bradyrhizobium, Vibrio, Streptomyces,* and *Serratia* are among the microorganisms that produce them (Karačić et al. 2024).

Chitinolytic enzymes are essential for controlling fungal phytopathogens, with genes encoding chitinases (Dobrzyński et al. 2023). B. pumilus MCB-7 has shown effectiveness against fungi, with chitinase activity at high saline concentrations and temperatures up to 60 °C (Rishad et al. 2017). Important agricultural phytopathogens like *Aspergillus flavus, Fusarium oxysporum, Ceratorhiza hydrophila, Aspergillus niger*, and *Aspergillus fumigatus* were all susceptible to the fungistatic effects of the *B. pumilus* strain's crude and refined chitinase (Rishad et al. 2017).

Tomato postharvest deterioration caused by *B. cinerea* reportedly decreased when chitosanase-producing *B. subtilis* V26 was used (Kilani-Feki et al. 2016). With its cellulase, chitinase, amylase, lipase, protease, and phytase activities, *B. velezensis* KS04AU showed *in vitro* antagonism against various fungal phytopathogens (Diabankana et al. 2022).

9.4.1.4 Bacillus Species and Its Bacterial Volatile Compounds

Bacterial volatile compounds (BVCs) are species-specific metabolites of bacteria that are greatly impacted by their surroundings and have a high vapor pressure and low water solubility (Almeida et al. 2023). They serve various functions in plant-microbial interactions and have shown antibacterial efficacy against plant diseases (Rani et al. 2023). BVCs induce structural damage and deformations such as lysis of the cell wall, plasmolysis, leakage of cellular contents, organelle enlargement, and retraction of the protoplasm, which impose toxicity against phytopathogenic fungal mycelia (Gao et al. 2017). These structural abnormalities lower disease incidence and severity by changing how the phytopathogen's cells function (Ajuna et al. 2024). Although the effects appear transient, when toxic volatile concentration decreases, phytopathogenic cells usually recover and regain pathogenicity (Gao et al. 2017).

Bacillus species frequently produce microbiostatic BVCs in addition to non-bioactive or unknown bioactive BVCs (Calvo et al. 2020). Both 5-nonylamine and

3-methylbutanoic acid, two BVCs from B. velezensis strain Ce 100, inhibited the spore germination and mycelial development of C. gloeosporioides, which causes fruit tree anthracnose (Choub et al. 2022). According to Gao et al. (2017), BVCs from *B. Velezensis* strain ZSY-1 also prevented *B. cinerea* and *A. solani* from growing and sporulating, suppressing incidences of early blight disease and grey mold in tomatoes. 2,4-di-tert-butylphenol, another BVC generated by *Bacillus subtilis* strain CF-3, is an antifungal substance that prevents the growth of *C. gloeosporioides*, the phytopathogen responsible for anthracnose disease in harvested litchi fruits (Zhao et al. 2019). In summary, BVCs can potentially be used in managing plant diseases and pests. However, their environmental application is still problematic because of their species-specific nature and the impacts of environmental perturbations.

9.4.1.5 Competition for Nutrients and Space by Bacillus Species

Oftentimes, biocontrol agents compete for nutrients and space with phytopathogens, which limit their availability to pathogens and impede their propensity to grow and proliferate. Micronutrients such as iron, copper, manganese, and zinc are fiercely competed for between phytopathogens and biocontrol agents because they occupy similar niches and require the same resources (Dubinkina et al. 2019; Lahlali et al. 2022). Members of the genus *Bacillus* are adept at solubilizing and absorbing nutrients and can deplete resources and reduce pathogen growth conditions (Köhl et al. 2019). They are also well-suited to colonize, endure, adapt to, and withstand various stressors, aiding in their establishment as biocontrol agents.

Recent research indicates the deployment of this mechanism by *Bacillus* spp. to inhibit tomato pathogens. *B. velezensis* strain GF267 demonstrated greater competitive ability than pathogens and the maximum decrease of tomato bacterial spots (Mates et al. 2019). Also, *B. amyloliquefaciens* strains CM-2 and T-5 promoted plant growth and biocontrol of *R. solanacearum* in greenhouse conditions, leading to high colonization and decreased pathogen density in the tomato rhizosphere (Tan et al. 2013). Similarly, *B. amyloliquefaciens* SQYUV 162 effectively suppressed *R. solanacearum* as the pathogen and antagonist competed for root exudates (Wu et al. 2017).

Competition for space can be used to control pathogen development by restricting resource allocation and preventing pathogens from accessing germination cues. Biocontrol agents can inhibit pathogen establishment and infection by fully colonizing a resource-available region (Vero et al. 2023). For example, Bacillus amyloliquefaciens FZB42 produced an antimicrobial peptide upregulated when forming a biofilm, facilitating effective microbial colonization and persistence (Kröber et al. 2016).

9.4.1.6 *Bacillus* Species and Induced Systemic Resistance in Plant Hosts

Bacillus spp. produced elicitor compounds that help host plants develop induced systemic resistance (ISR) to diseases and pests. For instance, *B. cereus* strain AR156 elicited ISR in *Arabidopsis* ecotype Col-0 plants against *P. syringae* pv. tomato strain DC3000, resulting in decreased severity of the disease, reduced pathogen density in leaves, and increased biomass output (Niu et al. 2011). Defense-related genes for signaling pathways were also expressed by the primed plants. In *Arabidopsis* plants infected with *Agrobacterium tumefaciens* strain C58, *B. velezensis* strain CLA178 produced ISR, promoting plant defense genes, reactive oxygen species, and antioxidants (Chen et al. 2020a, b).

According to Li et al. (2015), *B. amyloliquefaciens* strain LJ02 induces ISR in cucumbers that are resistant to *Sphaerotheca fuliginea*'s powdery mildew disease. The strain expresses the PR-1 gene, produces plant defense enzymes, and decreases disease incidence when inoculated into plants. It also produces salicylic acid (SA) in cucumber leaves, which inhibits fungal germination in the rhizosphere. According to Kim et al. (2015), *Bacillus* sp. strain JS inhibits the growth of *P. nicotianae* and *R. solani* and lessens the symptoms of tobacco disease. Tobacco leaves treated with *Bacillus* sp. strain JS's BVCs also exhibit increased expression of genes linked to antifungal enzymes.

Several defense mechanisms, such as competition, antibiosis, and ISR elicitation, are employed by *Bacillus* species to combat phytopathogens. For example, by directly opposing the phytopathogen by cyclic lipopeptide production and inducing ISR through root application, *B. velezensis* strain RHF4.1-25 prevented rice blast disease caused by *Pyricularia oryzae* (Lam et al. 2021). These antibiosis molecules (fengycin and iturin) impede mycelial growth and spore germination, suppressing the TTSS pathogenicity pathway and priming for plant defense proteins.

Research shows that certain *Bacillus* species can cause plant defense reactions against herbivorous phytophagous insects, like aphids. *Schizaphis graminum* Rond, also known as the green bug/wheat aphid, can be effectively suppressed through direct and indirect aphidicidal activity (Veselova et al. 2019). This was achieved when *B. thuringiensis* Berliner strains V-6066 and V-5689 and *B. subtilis* Cohn strain 26 D were applied to wheat plants. This resulted in hydrogen peroxide regeneration, elevated expression of NADPH oxidases, expression of genes linked to PR-6 and PR-9 defense, and peroxidase activity.

9.4.2 Serratia *Species as Microbial Biopesticides*

Members of the genus *Serratia* are members of the *Enterobacteriaceae* family that are physiologically and ecologically diverse, ranging from pathogenic to beneficial for economically valuable plants. They are ideal candidates as biocontrol agents due to their ability to produce a wide variety of naturally occurring bioactive compounds (Matilla et al. 2015; Domik et al. 2016a).

9.4.2.1 Induced Systemic Resistance (ISR)

Serratia marcescens CDP-13 is one strain of *Serratia* that can upregulate ISR in plants. Wheat plants seeded with *S. marcescens* CDP-13 showed a markedly decreased sensitivity to diseases caused by *Fusarium graminearum* in water agar tests (Singh and Jha 2016). Although the precise mechanism underlying plant disease resistance is still unknown, previous research has demonstrated that *Serratia* can increase ISR.

9.4.2.2 Production of Antimicrobial Compounds

Prodigiosin, a bioactive member of the prodiginine family, is produced by *Serratia* and released into the growth medium as a secondary metabolite (Han et al. 2021a, b). S. *plymuthica, S. marcescens, S. rubidaea*, and *S. nematodiphila* are the four species that produce it. Prodigiosin damages and depletes vital intracellular components by targeting the bacterial plasma membrane (Suryawanshi et al. 2017). Methicillin-resistant *Staphylococcus aureus, Bacillus cereus, Enterobacter faecalis, Staphylococcus epidermidis, Acinetobacter anitratus, Staphylococcus saprophyticus, Agrobacterium tumefaciens, Streptococcus pyogenes*, and *Bacillus licheniformis* are all effectively combated by prodigiosin broad-spectrum action (Yip et al. 2021; Nguyen et al. 2022). Additionally, it demonstrates activity against phytopathogenic fungi, insects, and nematodes, among other crop-damaging pathogens (Choi et al. 2021; Nguyen et al. 2021).

Oocydin A, a chlorinated macrolide present in plant epiphytic strain *S. marcescens* MSU97, is one of the naturally occurring antifungal compounds reported in *Serratia* (Strobel et al. 1999). Serratia spp. such as *S. plymuthica* strains A153, 4Rx5, 4Rx13, and *Serratia* sp. B1_6 produce oocydin A (Berg 2000; Matilla et al. 2015), which possess antagonistic activity against *Verticillium dahliae* plant disease. Several strains of *S. plymuthica* have shown antifungal efficacy against fungal pathogens *in vitro* (Berg 2000).

9.4.2.3 Production of Volatile Organic Compounds (VOCs)

S. plymuthica isolates produce a unique compound called sodorifen, a rare and special volatile hydrocarbon. The gene cluster responsible for producing sodorifen has been identified, but its biological function remains unknown. When exposed to *Fusarium culmorum*-generated VOCs, *S. plymuthica* PRI-2C generates sodorifen (Schmidt et al. 2017). Serratia also secretes dimethyl disulfide, which can be synthesized *in vitro* with *S. bockelmannii* and *Serratia ureilytica* (Abreo et al. 2021). This impedes *Pythium cryptoirregulare* growth, suppressing the damping-off of tomato seedlings. *S. marcescens* NBRI 1213 can biologically control *Phytophthora nicotianae*, the causative agent of Piper betle root and foot rot (Lavania et al. 2006).

Rhizoctonia solani in lettuce, and *V. dahliae* in olive, strawberries, and oilseed rapeseed can all be controlled by *S. plymuthica* HRO-C48.

9.4.3 Streptomyces *Species as Microbial Biopesticide*

Members of the phylum Actinomycetota's genus Streptomyces possess a wide range of functional characteristics that enable their application as efficient phytopathogen biocontrol agents. They produce diverse antimicrobials, extracellular hydrolytic enzymes, volatile organic compounds (VOCs, and antihelmintic compounds, which could be deployed in the biocontrol of plant pathogens.

9.4.3.1 *Streptomyces* spp. and Induced Systemic Resistance of Host Plants

Streptomyces can indirectly inhibit plant pathogens by activating host resistance systems and competitively excluding them (Ebrahimi-Zarandi et al. 2022). Certain changes such as localized cell death, accumulation of defense-related molecules, and reinforcements of the cell wall were promoted by ISR, which resulted in a better and robust response to the impending pathogen attack (Lugtenberg and Kamilova 2009; Kurth et al. 2014; Viaene et al. 2016). Oak trees inoculation with *Streptomyces* sp. AcH505 increases its elicitation of pathogenesis-related proteins (Kurth et al. 2014).

9.4.3.2 Antibiosis by *Streptomyces* spp.

Magnaporthe oryzae, Rhizoctonia solani, Gaeumannomyces graminis var. tritici, and *Fusarium* species can all be suppressed *in vitro* by *Streptomyces* (Law et al. 2017). Application of *Streptomyces* BN1 spore preparation to seeds reduces the Fusarium-induced seedling length reduction (Jung et al. 2013). Members of *Streptomyces* are potential biocontrol agents for take-all wheat disease due to their saprotrophic lifestyle and spore-formation, which prop them to thrive under unfavorable environmental conditions (Coombs et al. 2004; van der Meij et al. 2017).

Many secreted proteins having broad extracellular functions are encoded by *Streptomyces* species. Because they may inhibit phytopathogenic fungi and oomycetes, chitinases that break down chitin are being used to control phytopathogens (Chater et al. 2010; Wang et al. 2023). For instance, purified chitinase from *Streptomyces lydicus* WYEC108 can lyse the cell walls of *Pythium*, a phytopathogenic fungus that causes root rot in cereal crops (Mahadevan and Crawford 1997). *Streptomyces* produces volatile organic compounds (VOCs), which have antibacterial activity against several phytopathogens, including *R. solani* (Mendes et al. 2013; Chapelle et al. 2016). Streptomyces VOCs used as biofumigants can inhibit the growth of pathogenic organisms. However, more research is needed to verify their production in vivo in plant roots and their functionality in their native environments (Newitt et al. 2019). Streptomyces' production of potent anthelmintic substances, such as avermectin, can also be used to control nematode populations (Huang et al. 2014), indicating its potential to control nematode populations (Zhang et al. 2020).

9.4.4 Pseudomonas *Species as Microbial Biopesticide*

Pseudomonas belonging to the class *Gammaproteobacteria* and phylum Pseudomonadota are imbued with diverse functional attributes such as the synthesis of bioactive compounds and secondary metabolites that could find application in the control of phytopathogens.

9.4.4.1 Induced Systemic Resistance (ISR)

Hyaloperonospora parasitica, *Pseudomonas syringae* pv. tomato, and *Botrytis cinerea* are mildew pathogens that induce defense responses in *Arabidopsis* through jasmonate- and ethylene-mediated responses. These responses are mediated by DAPG (2,4-Diacetylphloroglucinol), a bacterial secretion found in certain strains of *Pseudomonas* (Weller et al. 2012; Chae et al. 2020). Bacterial secretions including siderophores can cause ISR in plants. Pyoverdine and pyochelin produced by *Pseudomonas aeruginosa* 7NSK2 have been shown to shield plants from phytopathogen diseases including *Botrytis cinerea* and *Pythium splendens* (Aznar and Dellagi 2015). Pyocyanin, a phenazine molecule, and pyochelin can also generate ISR, enhancing ROS accumulation and protecting tomatoes from *B. cinerea*-caused diseases (Audenaert et al. 2002).

9.4.4.2 Disease Suppression by *Pseudomonas* Species

Soil is one of the most common natural environments where strains of *Pseudomonas* are commonly found. Alfalfa, sugarcane, clover, potatoes, tomatoes, radish, and beets are among the crops whose rhizosphere and soil have been found to contain isolates of *P. chlororaphis* (Biessy et al. 2019). *P. syringae* isolates MA-4 and NSA-6 have been reported to successfully inhibit *Rhizopus* and brown rot diseases on peaches (Yang and Hong 2020). According to de Freitas and Taylor (2023), *Agrobacterium rhizogen's* hairy root disease on kalanchoe, soybean, and tomato can be minimized by up to 95% using *P. brassicacearum* 93G8, *P. protegens* 1B1, and *P. chlororaphis* 48G9, respectively.

9.4.4.3 Antibiosis by *Pseudomonas* Species

Pseudomonas secondary metabolites have insecticidal and antimicrobial properties, with phenazines being the most conserved metabolite. Phenazines are redox-active, tricyclic, nitrogen-containing chemicals (Mavrodi et al. 2006). It is a broad-spectrum secondary metabolite with efficacy against fungal, bacterial, and oomycete diseases. Several reports indicated its efficiency against phytopathogens such as *Rhizoctonia solani, Phytophthora infestan,* and *Streptomyces scabies* (Arseneault et al. 2015; Jaaffar et al. 2017; Morrison et al. 2017).

2,4-diacetylphloroglucinol (DAPG) is another metabolite majorly produced by *P. protegens* and *P. corrugata* (Almario et al. 2017). Diseases in roots and seedlings are controlled by DAPG and is effective against fungi, nematodes, bacteria, and oomycetes. *Clavibacter michiganensis* subsp. michiganensis is antagonistically affected by DAPG, which is generated by *Pseudomonas* sp. LBUM300, both *in planta* and *in vitro* (Lanteigne et al. 2012).

Some *Pseudomonas* strains produced pyoluteorin (Ramette et al. 2011). Pyoluteorin from *P. protegens* Pf-5 induces cytoplasmic extravasations and cell hollowing, substantially inhibiting *Pantoea ananatis* DZ-12 growth (Gu et al. 2022). Because of its biological function, *Pseudomonas*'s hydrogen cyanide (HCN) exerts a variety of effects on pathogens. Certain *P. fluorescens* subsets and *P. aeruginosa* are known to manufacture HCN. The synthesis of HCN and DAPG is commonly associated in *Pseudomonas*. *Thielaviopsis basicola*-induced tobacco black root rot is inhibited by HCN-producing *P. protegens* CHA0 (Voisard et al. 1989). Furthermore, termites, aphids, root-knot nematodes, and other insects are suppressed by HCN generated by *Pseudomonas* (Devi and Kothamasi 2009; Flury et al. 2017; Kang et al. 2019).

Limited strains of *Pseudomonas* produce distinct bioactive compounds, such as rhizoxins, sessilin, promysalin, and l-furanomycin. These compounds have antifungal, oomycetic, and antibacterial properties (Trippe et al. 2013; D'Aes et al. 2014, Kaduskar et al. 2017).

9.4.5 Predatory Bacteria as Biocontrol Agents

One of the most prevalent types of microbe-to-microbe attacks in nature is predation (Nair and Velicer 2021). Certain bacteria exhibit microbial predatory behavior, eliminating other bacteria and utilizing their dead remains as food. Predatory bacteria have been recovered from diverse habitats and are members of several phyla, primarily *Proteobacteria, Bacteroidetes,* and *Cyanobacteria* (Madigan et al. 2021). For these bacteria, three basic predatory tactics have been identified: group attack, endobiotic, and epibiotic.

Epibiotic predatory bacteria attach themselves to their prey's cell walls and feed on nutrients in the cytoplasm or periplasm. One member of this type of predatory bacteria is the non-photosynthetic Cyanobacterium *Vampirovibrio chlorellavorus*. It attaches itself to the microalga *Chlorella vulgaris*, its prey, and feeds on it until binary division takes place (Soo et al. 2015).

Endobiotic predators, such as *Daptobacter* and *Bdellovibrio*, enter prey cells, take up residence in their cytoplasm, eat their contents, and reproduce. After the prey cell has been lysed, the predator releases its offspring into the surrounding area. These predators can invade phytopathogenic and other Gram-negative bacteria. Each predatory bacterium's range of prey is determined by its specific species and strains. In vitro studies have shown that eight strains of *B. bacteriovorus* showed varying prey ranges when cocultured with rice pathogenic bacteria (Song 2004). In situ assays have also been used to evaluate the biocontrol effectiveness of these predatory bacterial species. Odooli et al. (2020) found that *Bdellovibrio bacteriovorus* strain SOIR-1 effectively controlled rots induced by *Pantoea* sp. and *Xanthomonas campestris* on onion bulbs and potato slices by engaging in predatory activity against them. Youdkes et al. (2020) reported that two strains of *B. bacteriovorus* successfully prevent potato slice rot caused by *Pectobacterium carotovorum* subsp. brasilense by feeding on the bacterium.

The third category of predatory bacteria consists of bacteria that use a group approach to assault prey cells. Predatory bacteria, including *Myxobacteria* and *Lysobacter* spp., use gliding motility to attack prey cells in swarming groups. They release secondary metabolites and hydrolytic enzymes, which destroy the prey cells and serve as nutrition for nearby microbes and predatory bacteria. The effectiveness of this tactic depends on close contact between the predator and prey, and the population density of predatory bacteria (Pérez et al. 2016). Myxobacteria, Gram-negative bacteria with multicellular morphogenesis, are commonly found in soil and are being researched for their phytopathogens' biocontrol potential based on their remarkable tolerance to environmental challenges and ability to operate as predators of numerous bacteria and fungi plant pathogens. For example, the predatory myxobacterium *Myxococcus xanthus* R31 has shown significant antagonistic action against *Ralstonia solanacearum*, controlling tomato wilt induced by the pathogen (Dong et al. 2022).

9.5 Nematodes as Microbial Biopesticides

Entomopathogenic nematodes (EPNs), which are insect-parasitic soil nematodes, have a considerable biocontrol potential for crop pest population management. *S. scapterisci* (mole cricket nematode), *Steinernema carpocapsae*, *S. riobravis*, *S. feltiae*, and *S. thermophilum* are the most often used EPN species in insecticidal formulations; however, *Heterorhabditis megidis* and *H. bacteriophora* are also frequently utilized (Sabbahi et al. 2022). Infectious juveniles (IJs) of EPN find their hosts in the soil and enter through natural openings such the intersegmental membrane, spiracles, mouth, or anus. The EPNs release symbiotic bacteria into the hemocoel once they are within the host's body (infective phase), which causes bacterial septicaemia and ultimately the host's death. In the necrotrophic phase, the nematode reproduces by feeding on the dead insect remains and symbiotic bacteria. In the event of food scarcity, the reassociation of infectious L3 nematode larval stage with the bacterial symbiont produce IJs that depart from the carcass, and proceed to scavenge for other insects (free phase). While *Heterorhabditis* spp. carry bacteria belonging to the *Photorhabdus* spp. species, *Steinernema* spp. carry bacteria belonging to the *Xenorhabdus* spp. (Koppenhöfer and Gaugler 2009).

Vertebrates, vegetation, non-target invertebrates, and the environment are not harmed by EPNs. The EPNs work better on controlling soil pests or those with life cycles that involve soil since they are soil-dwelling. EPNs biocontrol over diverse insect pests is long-term, are simple to grow, and are pesticide-compatible. Significant advancements have been made over the last decade in the research and use of EPNs. They are now more commercially viable as biocontrol agents due to their exemption from registration, simplicity of mass manufacturing, and their use in *in vitro* or *in vivo* procedures (Sabbahi et al. 2022).

9.6 Protozoa as Microbial Biopesticides

Microsporidia are single-celled eukaryotic organisms with the greatest prospect for microbial biocontrol of insects among the protozoa (Solter et al. 2017). They are related to fungi and are obligatory intracellular pathogenic protists. To infect an insect, the majority of microsporidia must be ingested; nonetheless, natural transmission by parasitoids and predators within a pest population is possible (Sarwar et al. 2021). Three phases can be distinguished in the usual life cycle of microsporidia: the proliferative, sporogony, and the infective or environmental phase (Cali and Takvorian 1999). After consuming *Nosema* spp. spores, the infection cycle starts with sporoplasm being injected into the midgut epithelial cells cytoplasm through the polar tube and germination taking place in the host's midgut lumen. After that, the infection extends to several tissues and organs, sometimes resulting in septicemia and tissue disintegration, before invading the subsequent generation via vertical transmission (Solter et al. 2017). One benefit of this kind of infection is that the weakened insects are hyper-receptive to other biotic variables and unfavorable weather conditions that culminates in their death (Gonzalez et al. 2016; Kononchuk et al. 2021).

Nosema locustae (Canning) is a microsporidian that is reported to infect 121 species of orthopterans and has an exceptionally wide host range (Lange 2005). Numerous companies have been producing this microbial biocontrol agent commercially in the USA since the 1980s, and it has proven effective in managing grasshopper populations (Solter et al. 2012). Because *N. locustae*-based products are harmless for vertebrates and have shown great promise for production and spread at exceedingly low volume, it is a very promising locust control agent in the Asia arid regions and the Sahelian regions of Africa (Yadav et al. 2021). Conversely, *Nosema pyrausta* (Paillot) is a useful agent in controlling *Ostrinia nubilalis* (Hübner), the European maize borer, but commercial usage of it is still in its infancy (Lewis et al. 2009a, b).

9.7 Microbial Biopesticide Usage in the Field: Issues, Recommendations and Conclusions

A microbial biocontrol agent is chosen according to how pathogenic it is to the hosts. To determine which microbial agent has the most potential to be entomopathogenic and which is susceptible to insect pests, screening of microbial agents is conducted in huge collections. The effect of the chosen microbes on beneficial insects, which could have a big influence on the dynamics of pest populations, must also be taken into account when choosing a microorganism. Relevant mortality variables, including host range, dose rate, mechanism of action, speed of kill, and persistence, are taken into consideration after screening. Other important criteria include ease of application by farmers, formulation, mass production capacity with significant yields, as well as human and environmental safety and costs.

One of the most significant aspects of employing biocontrol techniques in pest management is accurately identifying microbial agents. Concerns about the release of bacterial, viral, and fungal strains into the field are often raised due to a lack of clarity on the inoculum's environmental fate and its impact on non-intended insects. Molecular methods that can detect the entomopathogen in infected hosts and concurrently screen a large number of samples and distinguish between isolates are needed to meet this challenge. Patenting economically viable strains would likewise greatly benefit from the use of highly specific markers.

To verify the effectiveness of certain microbial biopesticides, field tests are required. After testing in greenhouses, the strains with the best insecticidal features are put through outdoor trials. The formulation and field efficacy studies, toxicity assessments, and quality control necessary for the eventual commercialisation of biopesticides are conducted by accredited laboratories and private sector partners. In contrast, microbial biopesticides are typically used as wettable powders or pellets, liquid base formulations, and water-dispersible granules. These formulations are chosen according to the kind of available dispensing equipment, end-user preferences, and ease of use. It is frequently possible to generate a single microbial biopesticide in all four formulations; the specific choice will depend on the needs of the field.

Research on entomopathogens as formulated biopesticides has advanced greatly the field of microbial biocontrol agents. Microencapsulation technologies and nano-formulations hold promise in enhancing dependability and residual activity of plant protection agents of microbial origin, hence escalating their field application. Standards for field testing and efficacy evaluation should be updated often to demonstrate performance and facilitate the commercialization of microbial biopesticides.

Microbial biopesticides formulations are typically applied through diverse means, depending on the target pest and the mode of action. By spraying microbial biopesticides directly onto the leaves, foliar treatment suppresses apparent pests. To sustain pest numbers below the economic threshold level, the plant is then treated throughout the growing season. Microbial biopesticide sector owes its growth to foliar sprays' simplicity of use. Industry and academia collaboration is essential to assist the large-scale commercial development of microbial biopesticides and expedite the practical application of laboratory discoveries. More research is needed to determine how well microbial biocontrol treatments work against certain insect pests in different farming systems.

The development and use of entomopathogens as traditional, conservative, and augmentative microbial biocontrol agents have increased significantly in the last several years. Globally, there are about 209 microbial isolates from 94 entomopathogen species that are useful for controlling pests. The US market leads the market for microbial biopesticides, whereas for invertebrate biological control agents, the EU is the largest commercial market. Currently, 220 companies produce microbial biocontrol agents, with small and medium-sized suppliers accounting for the majority of these companies (Sabbahi et al. 2022).

Poor technology transfer particularly in less developed countries, seem to be the major factor contributing to the small number of approved solutions for microbial biocontrol of insect pests. Furthermore, a lack of thorough knowledge of organism-specific research methodologies for bulk production and applications makes the mass manufacture of microbial biocontrol agents in *in vivo* environments time-consuming and costly. Additionally, one of the most challenging parts of employing microbial biocontrol agents is getting them to the correct location at the right time with a sufficient density to be effective, and maintaining them there over time.

By the early 2050s, the use of microbial biopesticides was expected to be comparable to that of conventional insecticides; however, most of the flexibility in these projections comes from the unpredictability of acceptance rates, especially in Southeast Asia and Africa. To encourage farmers to utilize biocontrol agents, insect management technologies should be easy to evaluate and user-friendly, offer advantages in terms of demand, cost, and supply efficiency, and provide technical assistance and knowledge. Moreover, the best pest management strategy should balance the benefits of increased output and decreased risk against outside expenses, taking into consideration non-financial aspects that affect farmers' welfare and decision-making. The availability and commercialization of microbial biopesticides could be accelerated by government pest management regulations, incentives for low-risk materials registration, national policies to minimize the use of synthetic insecticides, and the promotion of IPM techniques through tax incentives for organic farming.

References

Abbas MST (2018) Genetically engineered (modified) crops (*Bacillus thuringiensis* crops) and the world controversy on their safety. Egypt J Biol Pest Control 28(1):1–12

Abreo E, Valle D, González A, Altier N (2021) Control of damping-off in tomato seedlings exerted by *Serratia* spp. strains and identification of inhibitory bacterial volatiles in vitro. Syst Appl Microbiol 44:126177

Abriouel H, Franz CM, Ben Omar N, Gálvez A (2011) Diversity and applications of *Bacillus bacteriocins*. FEMS Microbiol Rev 35(1):201–232

Afsharmanesh H, Perez-Garcia A, Zeriouh H et al (2018) Aflatoxin degradation by *Bacillus subtilis* UTB1 is based on production of an oxidoreductase involved in bacilysin biosynthesis. Food Control 94:48–55

Ajuna HB, Lim H-I, Moon J-H, Won S-J, Choub V, Choi S-I, Yun J-Y, Ahn YS (2024) The prospect of antimicrobial peptides from *Bacillus* species with biological control potential against insect pests and diseases of economic importance in agriculture, forestry and fruit tree production. Biotechnol Biotechnol Equip 38(1):2312115

Akbaba M, Ozaktan H (2021) Evaluation of bacteriophages in the biocontrol of *Pseudomonas syringae* pv. *syringae* isolated from cankers on sweet cherry (*Prunus avium* L.) in Turkey. Egypt J Biol Pest Control 31:35

Almario J, Bruto M, Vacheron J, Prigent-Combaret C, Moënne-Loccoz Y, Muller D (2017) Distribution of 2,4-Diacetylphloroglucinol biosynthetic genes among the *Pseudomonas* spp. reveals unexpected polyphyletism. Front Microbiol 8:1218

Almeida OAC, de Araujo NO, Mulato ATN, Persinoti GF, Sforça ML, Calderan-Rodrigues MJ, Oliveira JVC (2023) Bacterial volatile organic compounds (VOCs) promote growth and induce metabolic changes in rice. Front Plant Sci 13:1056082

Arrarte E, Garmendia G, Rossini C, Wisniewski M, Vero S (2017) Volatile organic compounds produced by Antarctic strains of *Candida sake* play a role in the control of postharvest pathogens of apples. Biol Control 109:14–20

Arseneault T, Goyer C, Filion M (2015) *Pseudomonas fluorescens* LBUM223 increases potato yield and reduces common scab symptoms in the field. Phytopathology 105:1311–1317

Audenaert K, Pattery T, Cornelis P, Höfte M (2002) Induction of systemic resistance to *Botrytis cinerea* in tomato by *Pseudomonas aeruginosa* 7NSK2: role of salicylic acid, pyochelin, and pyocyanin. Mol Plant-Microbe Interact 15:1147–1156

Aznar A, Dellagi A (2015) New insights into the role of siderophores as triggers of plant immunity: what can we learn from animals? J Exp Bot 66:3001–3010

Bamisile BS, Akutse KS, Siddiqui JA, Xu Y (2021) Model application of entomopathogenic fungi as alternatives to chemical pesticides: prospects, challenges, and insights for next-generation sustainable agriculture. Front Plant Sci 12:741804

Barelli L, Padilla-Guerrero IE, Bidochka MJ (2011) Differential expression of insect and plant-specific adhesin genes, Mad1 and Mad2, in *Metarhizium robertsii*. Fungal Biol 115(11):1174–1185

Beausang C, Hall C, Toma L (2017) Food waste and losses in primary production: qualitative insights from horticulture. Resour Conserv Recycl 126:177–185

Berg G (2000) Diversity of antifungal and plant-associated *Serratia plymuthica* strains. J Appl Microbiol 88:952–960

Berretta MF, Pedarros AS, Sauka DH, Pérez MP, Onco MI, Benintende GB (2020) Susceptibility of agricultural pests of regional importance in South America to a *Bacillus thuringiensis* Cry1Ia protein. J Invertebr Pathol 172:107354

Biessy A, Novinscak A, Blom J, Léger G, Thomashow LS, Cazorla FM, Josic D, Filion M (2019) Diversity of phytobeneficial traits revealed by whole-genome analysis of worldwide-isolated phenazine-producing *Pseudomonas* spp. Environ Microbiol 21(1):437–455

Bouchard-Rochette M, Machrafi Y, Cossus L, Nguyen TTA, Antoun H, Droit A, Tweddell RJ (2022) *Bacillus pumilus* PTB180 and *Bacillus subtilis* PTB185: production of lipopeptides, antifungal activity, and biocontrol ability against *Botrytis cinerea*. Biol Control 170:104925

Bravo A, Gill SS, Soberon M (2007) Mode of action of *Bacillus thuringiensis* Cry and Cyt toxins and their potential for insect control. Toxicon 49(4):423–435

Butt TM, Greenfield BPJ, Greig C, Maffeis TGG, Taylor JWD, Piasecka J et al (2013) *Metarhizium anisopliae* pathogenesis of mosquito larvae: a verdict of accidental death. PLoS One 8(12):e81686

Buttimer C, McAuliffe O, Ross RP, Hill C, O'Mahony J, Coffey A (2017) Bacteriophages and bacterial plant diseases. Front Microbiol 8:34

Cabrera M, Garmendia G, Rufo C, Pereyra S, Vero S (2020) *Trichoderma atroviride* as a biocontrol agent of *Fusarium* head blight by reducing the inoculum of the pathogen in wheat straw. Terra Latinoamericana 38:629–651

Cali A, Takvorian PM (1999) Developmental morphology and life cycle of the microsporidia. In: Wittner M, Weiss LM (eds) The microsporidia and microsporidiosis. Wiley, pp 85–128

Calvo H, Mendiara I, Arias E, Gracia AP, Blanco D, Venturini ME (2020) Antifungal activity of the volatile organic compounds produced by *Bacillus velezensis* strains against postharvest fungal pathogens. Postharvest Biol Technol 166:111208

Caulier S, Nannan C, Gillis A, Licciardi F, Bragard C, Mahillon J (2019) Overview of the antimicrobial compounds produced by members of the *Bacillus subtilis* group. Front Microbiol 10:302

Chae DH, Kim DR, Cheong MS, Lee YB, Kwak YS (2020) Investigating the induced systemic resistance mechanism of 2,4-diacetylphloroglucinol (DAPG) using DAPG hydrolase-transgenic Arabidopsis. Plant Pathol J 36:255–266

Chakroun M, Banyuls N, Bel Y, Escriche B, Ferré J (2016) Bacterial vegetative insecticidal proteins (Vip) from entomopathogenic bacteria. Microbiol Mol Biol Rev 80(2):329–350

Chapelle E, Mendes R, Bakker PA, Raaijmakers JM (2016) Fungal invasion of the rhizosphere microbiome. ISME J 10:265–268

Chater KF, Biro S, Lee KJ, Palmer T, Schrempf H (2010) The complex extracellular biology of Streptomyces. FEMS Microbiol Rev 34:171–198

Chen HC, Chou CK, Sun CM, Yeh SF (1997) Suppressive effects of destruxin B on hepatitis B virus surface antigen gene expression in human hepatoma cells. Antivir Res 34(3):137–144

Chen XH, Koumoutsi A, Scholz R, Schneider K, Vater J, Süssmuth R, Piel J, Borriss R (2009) Genome analysis of *Bacillus amyloliquefaciens* FZB42 reveals its potential for biocontrol of plant pathogens. J Biotechnol 140(1–2):27–37

Chen K, Tian Z, Luo Y, Cheng Y, Long CA (2018) Antagonistic activity and the mechanism of *Bacillus amyloliquefaciens* DH-4 against citrus green mold. Phytopathology 108(11):1253–1262

Chen L, Wang X, Ma Q, Bian L, Liu X, Xu Y, Zhang H, Shao J, Liu Y (2020a) *Bacillus velezensis* CLA178-induced systemic resistance of *Rosa multiflora* against crown gall disease. Front Microbiol 11:587667

Chen M, Wang J, Liu B, Zhu Y, Xiao R, Yang W, Ge C, Chen Z (2020b) Biocontrol of tomato bacterial wilt by the new strain *Bacillus velezensis* FJAT-46737 and its lipopeptides. BMC Microbiol 20(1):160

Choi SY, Lim S, Yoon K-H, Lee JI, Mitchell RJ (2021) Biotechnological activities and applications of bacterial pigments violacein and prodigiosin. J Biol Eng 15:1–16

Choub V, Won S-J, Ajuna HB, Moon J-H, Choi S-I, Lim H-I, Ahn YS (2022) Antifungal activity of volatile organic compounds from *Bacillus velezensis* CE 100 against *Colletotrichum gloeosporioides*. Horticulturae 8(6):557

Chu ZJ, Sun HH, Zhu XG, Ying SH, Feng MG (2017) Discovery of a new intravacuolar protein required for the autophagy, development and virulence of *Beauveria bassiana*. Environ Microbiol 19(7):2806–2818

Clem RJ, Passarelli AL (2013) Baculoviruses: sophisticated pathogens of insects. PLoS Pathog 9(11):e1003729

Coombs JT, Michelsen PP, Franco CM (2004) Evaluation of endophytic actinobacteria as antagonists of *Gaeumannomyces graminis* var. tritici in wheat. Biol Control 29:359–366

Crickmore N, Berry C, Panneerselvam S, Mishra R, Connor TR, Bonning BC (2020) A structure-based nomenclature for *Bacillus thuringiensis* and other bacteria-derived pesticidal proteins. J Invertebr Pathol 186:107438

D'aes J, Kieu NP, Léclère V, Tokarski C, Olorunleke FE, De Maeyer K et al (2014) To settle or to move? The interplay between two classes of cyclic lipopeptides in the biocontrol strain Pseudomonas CMR12a. Environ Microbiol 16(7):2282–2300

Dar SA, Wani SH, Mir M, Showkat A, Dolkar T, Dawa T (2021) Biopesticides: mode of action, efficacy and scope in pest management. J Adv Res Biochem Pharmacol 4:1–8

Daw MA, Falkiner FR (1996) Bacteriocins: nature, function and structure. Micron 27(6):467–479

de Freitas CC, Taylor CG (2023) Biological control of hairy root disease using beneficial *Pseudomonas* strains. Biol Control 177:105098

de Oliveira JL, Fraceto LF, Bravo A, Polanczyk RA (2021) Encapsulation strategies for *Bacillus thuringiensis*: from now to the future. J Agric Food Chem 69(16):4564–4577

Dehghanifar S, Keyhanfar M, Emtiazi G (2019) Production and partial purification of thermostable bacteriocins from *Bacillus pumilus* ZED17 and DFAR8 strains with antifungal activity. Mol Biol Res Commun 8:41

Devi KK, Kothamasi D (2009) *Pseudomonas fluorescens* CHA0 can kill subterranean termite *Odontotermes obesus* by inhibiting cytochrome c oxidase of the termite respiratory chain. FEMS Microbiol Lett 300:195–200

Diabankana RGC, Shulga EU, Validov SZ, Afordoanyi DM (2022) Genetic characteristics and enzymatic activities of *Bacillus velezensis* KS04AU as a stable biocontrol agent against phytopathogens. Int J Plant Biol 13:201–222

Dimopoulou A, Theologidis I, Benaki D, Koukounia M, Zervakou A, Tzima A et al (2021) Direct antibiotic activity of bacillibactin broadens the biocontrol range of *Bacillus amyloliquefaciens* MBI600. mSphere 6(4):e0037621

Dobrzyński J, Jakubowska Z, Kulkova I, Kowalczyk P, Kramkowski K (2023) Biocontrol of fungal phytopathogens by *Bacillus pumilus*. Front Microbiol 14:1194606

Domik D et al (2016a) A terpene synthase is involved in the synthesis of the volatile organic compound sodorifen of *Serratia plymuthica* 4Rx13. Front Microbiol 7:737

Domínguez-Arrizabalaga M, Villanueva M, Escriche B, Ancín-Azpilicueta C, Caballero P (2020) Insecticidal activity of *Bacillus thuringiensis* proteins against Coleopteran pests. Toxins 12(7):430

Dong H, Xu X, Gao R, Li Y, Li A, Yao Q, Zhu H (2022) *Myxococcus xanthus* R31 suppresses tomato bacterial wilt by inhibiting the pathogen *Ralstonia solanacearum* with secreted proteins. Front Microbiol 12:801091

Druzhinina IS, Seidl-Seiboth V, Herrera-Estrella A, Horwitz BA, Kenerley CM, Monte E et al (2011) Trichoderma: the genomics of opportunistic success. Nat Rev Microbiol 9(10):749–759

Dubinkina V, Fridman Y, Pandey PP, Maslov S (2019) Multistability and regime shifts in microbial communities explained by competition for essential nutrients. elife 8:e49720

Dunlap CA, Bowman MJ, Rooney AP (2019) Iturinic lipopeptide diversity in the *Bacillus subtilis* species group–important antifungals for plant disease biocontrol applications. Front Microbiol 10:1794

Dunyashev T, Laptev GY, Yildirim E et al (2021) Identification of genes associated with the synthesis of siderophores by the *Bacillus subtilis*. J Livest Sci 12(4):287–291

Ebrahimi-Zarandi M, Saberi Riseh R, Tarkka MT (2022) Actinobacteria as effective biocontrol agents against plant pathogens, an overview on their role in eliciting plant defense. Microorganisms 10:1739

Fang W, Pei Y, Bidochka MJ (2007) A regulator of a G protein signalling (RGS) gene, cag8, from the insect-pathogenic fungus *Metarhizium anisopliae* is involved in conidiation, virulence, and hydrophobin synthesis. Microbiology 153(4):1017–1025

Fang W, Pava-Ripoll M, Wang S, Leger RS (2009) Protein kinase A regulates production of virulence determinants by the entomopathogenic fungus, *Metarhizium anisopliae*. Fungal Genet Biol 46(3):277–285

FAO (2019) The state of food and agriculture. Moving forward on food loss and waste reduction. Food and Agriculture Organization of the United Nations (FAO), Rome

Farbo MG, Urgeghe PP, Fiori S, Marcello A, Oggiano S, Balmas V (2018) Effect of yeast volatile organic compounds on ochratoxin A-producing *Aspergillus carbonarius* and *A. ochraceus*. Int J Food Microbiol 284:1–10

Feng P, Shang Y, Cen K, Wang C (2015) Fungal biosynthesis of the bibenzoquinone oosporein to evade insect immunity. Proc Natl Acad Sci USA 112(36):11365–11370

Fenibo EO, Ijoma GN, Matambo T (2021) Biopesticides in sustainable agriculture: a critical sustainable development driver governed by green chemistry principles. Front Sus Food Syst 5:619058

Fira D, Dimkić I, Berić T, Lozo J, Stanković S (2018) Biological control of plant pathogens by *Bacillus* species. J Biotechnol 285:44–55

Flury P, Vesga P, Péchy-Tarr M, Aellen N, Dennert F, Hofer N et al (2017) Antimicrobial and insecticidal: cyclic Lipopeptides and hydrogen cyanide produced by plant-beneficial *Pseudomonas* strains CHA0, CMR12a, and PCL1391 contribute to insect killing. Front Microbiol 8:100

Fuguet R, Théraud M, Vey A (2004) Production in vitro of toxic macromolecules by strains of *Beauveria bassiana*, and purification of a chitosanase-like protein secreted by a melanizing isolate. Comp Biochem Physiol C Toxicol Pharmacol 138(2):149–161

Fukushima T, Allred BE, Sia AK, Nichiporuk R, Andersen UN, Raymond KN (2013) Gram-positive siderophore-shuttle with iron-exchange from Fe-siderophore to apo-siderophore by *Bacillus cereus* YxeB. Proc Natl Acad Sci USA 110(34):13821–13826

Gabarty A, Salem HM, Fouda MA, Abas AA, Ibrahim AA (2014) Pathogencity induced by the entomopathogenic fungi *Beauveria bassiana* and *Metarhizium anisopliae* in *Agrotis ipsilon* (hufn.). J Radiat Res Appl Sci 7(1):95–100

Gao Q, Jin K, Ying S-H, Zhang Y, Xiao G, Shang Y et al (2011) Genome sequencing and comparative transcriptomics of the model entomopathogenic fungi *Metarhizium anisopliae* and *M. acridum*. PLoS Genet 7(1):e1001264

Gao Z, Zhang B, Liu H, Han J, Zhang Y (2017) Identification of endophytic *Bacillus velezensis* ZSY-1 strain and antifungal activity of its volatile compounds against *Alternaria solani* and *Botrytis cinerea*. Biol Control 105:27–39

Garavaglia MJ, Miele SA, Iserte JA, Belaich MN, Ghiringhelli PD (2012) The ac53, ac78, ac101, and ac103 genes are newly discovered core genes in the family Baculoviridae. J Virol 86(22):12069–12079

Garcia-Rubio R, de Oliveira HC, Rivera J, Trevijano-Contador N (2020) The fungal cell wall: *Candida*, *Cryptococcus*, and *Aspergillus* species. Front Microbiol 10:2993

Gašić S, Tanović B (2013) Biopesticide formulations, possibility of application and future trends. Pestic Fitomed 28(2):97–102

Gebhardt K, Schimana J, Müller J, Fiedler HP, Kallenborn HG, Holzenkämpfer M et al (2002) Screening for biologically active metabolites with endosymbiotic bacilli isolated from arthropods. FEMS Microbiol Lett 217(2):199–205

Geissler M, Heravi KM, Henkel M et al (2019) Lipopeptide biosurfactants from *Bacillus* species. In: Douglas GH, Daniel KYS, Ashby RD (eds) Biobased surfactants, 2nd edn. AOCS Press, Urbana, pp 205–240

Glare TR, O'Callaghan M (2019) Microbial biopesticides for control of invertebrates: Progress from New Zealand. J Invertebr Pathol 165:82–88

Gong AD, Li HP, Yuan QS, Song XS, Yao W, He WJ, Zhang JB, Liao YC (2015) Antagonistic mechanism of iturin A and plipastatin A from *Bacillus amyloliquefaciens* S76-3 from wheat spikes against *Fusarium graminearum*. PLoS One 10(2):e0116871

Gonzalez F, Tkaczuk C, Dinu MM, Fiedler Ż, Vidal S, Zchori-Fein E, Messelink GJ (2016) New opportunities for the integration of microorganisms into biological pest control systems in greenhouse crops. J Pest Sci (2004) 89:295–311

Gotor-Vila A, Teixido N, Di Francesco A, Usall J, Ugolini L, Torres R, Mari M (2017) Antifungal effect of volatile organic compounds produced by *Bacillus amyloliquefaciens* CPA-8 against fruit pathogen decays of cherry. Food Microbiol 64:219–225

Gray EJ, Lee KD, Souleimanov AM, Di Falco MR, Zhou X, Ly A et al (2006) A novel bacteriocin, thuricin 17, produced by plant growth promoting rhizobacteria strain *Bacillus thuringiensis* NEB17: isolation and classification. J Appl Microbiol 100(3):545–554

Grinter R, Milner J, Walker D (2012) Bacteriocins active against plant pathogenic bacteria. Biochem Soc Trans 40(6):1498–1502

Grzywacz D, Moore S (2017) Production, formulation, and bioassay of Baculoviruses for pest control. In: Microbial control of insect and mite pests: from theory to practice. Elsevier, pp 109–124

Gu Q, Qiao J, Wang R, Lu J, Wang Z, Li P et al (2022) The role of Pyoluteorin from *Pseudomonas protegens* Pf-5 in suppressing the growth and pathogenicity of *Pantoea ananatis* on maize. Int J Mol Sci 23(12):6431

Guo Q, Dong W, Li S, Lu X, Wang P, Zhang X, Wang Y, Ma P (2014) Fengycin produced by *Bacillus subtilis* NCD-2 plays a major role in biocontrol of cotton seedling damping-off disease. Microbiol Res 169(7–8):533–540

Hammami I, Rhouma A, Jaouadi B, Rebai A, Nesme X (2009) Optimization and biochemical characterization of a bacteriocin from a newly isolated *Bacillus subtilis* strain 14B for biocontrol of *Agrobacterium* spp. strains. Lett Appl Microbiol 48(2):253–260

Han R, Xiang R, Li J, Wang F, Wang C (2021a) High-level production of microbial prodigiosin: a review. J Basic Microbiol 61:506–523

Han X, Shen D, Xiong Q, Bao B, Zhang W, Dai T et al (2021b) The plant-beneficial rhizobacterium *Bacillus velezensis* FZB42 controls the soybean pathogen *Phytophthora sojae* due to bacilysin production. Appl Environ Microbiol 87(23):e0160121

Haque MA, Wang Y, Shen Z, Li X, Saleemi MK, He C (2020) Mycotoxin contamination and control strategy in human, domestic animal and poultry: a review. Microb Pathog 142:104095

Helmy NM, Parang K (2023) Cyclic peptides with antifungal properties derived from bacteria, fungi, plants, and synthetic sources. Pharmaceuticals (Basel) 16(6):892

Hernandez CA, Salazar AJ, Koskella B (2020) Bacteriophage-mediated reduction of bacterial speck on tomato seedlings. Phage 1:205–212

Hoover K, Grove M, Gardner M, Hughes DP, McNeil J, Slavicek J (2011) A gene for an extended phenotype. Science 333(6048):1401

Hsiao YM, Ko JL (2001) Determination of destruxins, cyclic peptide toxins, produced by different strains of *Metarhizium anisopliae* and their mutants induced by ethyl methane sulfonate and ultraviolet using HPLC method. Toxicon 39:837–841

Hu QB, Ren SX, Wu JH, Chang JM, Musa PD (2006) Investigation of destruxin A and B from 80 *Metarhizium* strains in China, and the optimization of cultural conditions for the strain MaQ10. Toxicon 48(5):491–498

Huang WK, Sun JH, Cui JK, Wang GF, Kong LA, Peng H, Chen SL, Peng DL (2014) Efficacy evaluation of fungus *Syncephalastrum racemosum* and nematicide avermectin against the root-knot nematode *Meloidogyne incognita* on cucumber. PLoS One 9(2):e89717

Hubbard M, Hynes RK, Erlandson M, Bailey KL (2014) The biochemistry behind biopesticide efficacy. Sustain Chem Proc 2:18

Hui Q, Zhen-Qiang WU, Shi-Zhong L (2004) *Metarhizium anisopliae* and its mechanism for killing insects. Pesticides 08:4

ISAAA (2016) International service for the acquisition of agribiotech applications. GM Approval Database https://www.isaaa.org/gmapprovaldatabase/

Jaaffar AKM, Parejko JA, Paulitz TC, Weller DM, Thomashow LS (2017) Sensitivity of rhizoctonia isolates to phenazine-1-carboxylic acid and biological control by phenazine-producing Pseudomonas spp. Phytopathology 107:692–703

Jepson PC, Murray K, Bach O, Bonilla MA, Neumeister L (2020) Selection of pesticides to reduce human and environmental health risks: a global guideline and minimum pesticides list. Lancet Planet Health 4(2):e56–e63

Jouzani GS, Valijanian E, Sharafi R (2017) *Bacillus thuringiensis*: a successful insecticide with new environmental features and tidings. Appl Microbiol Biotechnol 101(7):2691–2711

Jumpathong W, Intra B, Euanorasetr J, Wanapaisan P (2022) Biosurfactant-producing *Bacillus velezensis* PW192 as an anti-fungal biocontrol agent against *Colletotrichum gloeosporioides* and *Colletotrichum musae*. Microorganisms 10(5):1017

Jung B, Park SY, Lee YW, Lee J (2013) Biological efficacy of Streptomyces sp. strain BN1 against the cereal head blight pathogen *Fusarium graminearum*. Plant Pathol J 29:52–58

Kaduskar RD, Scala GD, Al Jabri ZJH, Arioli S, Musso L, Oggioni MR, Dallavalle S, Mora D (2017) Promysalin is a salicylate-containing antimicrobial with a cell-membrane-disrupting mechanism of action on Gram-positive bacteria. Sci Rep 7(1):8861

Kalawate AS (2014) Microbial viral insecticides. In: Sahayaraj K (ed) Basic and applied aspects of biopesticides. Springer India, pp 47–68

Kalpana T, Anil K (2021) A review of biopesticides and their plant phytochemicals information. Ann Rom Soc Cell Biol 25:3576–3588

Kanematsu S, Shimizu T, Salaipeth L, Yaegashi H, Sasaki A, Ito T, Suzuki N (2014) Genome rearrangement of a mycovirus Rosellinia necatrix megabirnavirus 1 affecting its ability to attenuate virulence of the host fungus. Virol 450–451:308–15

Kang BR, Anderson AJ, Kim YC (2019) Hydrogen cyanide produced by *Pseudomonas chlororaphis* O6 is a key aphicidal metabolite. Can J Microbiol 65:185–190

Karačić V, Miljaković D, Marinković J, Ignjatov M, Milošević D, Tamindžić G, Ivanović M (2024) *Bacillus* species: excellent biocontrol agents against tomato diseases. Microorganisms 12(3):457

Karlsson M, Durling MB, Choi J, Kosawang C, Lackner G, Tzelepis GD et al (2015) Insights on the evolution of mycoparasitism from the genome of *Clonostachys rosea*. Genome Biol Evol 7(2):465–480

Karlsson M, Atanasova L, Jensen DF, Zeilinger S (2017) Necrotrophic mycoparasites and their genomes. Microbiol Spectr 5(2)

Katsuma S, Koyano Y, Kang W, Kokusho R, Kamita SG, Shimada T (2012) The baculovirus uses a captured host phosphatase to induce enhanced locomotory activity in host caterpillars. PLoS Pathog 8(4):e1002644

Kazi M, Annapure US (2016) Bacteriophage biocontrol of foodborne pathogens. J Food Sci Technol 53:1355–1362

Kenis M, Benelli G, Biondi A, Calatayud P-A, Day R, Desneux N et al (2023) Invasiveness, biology, ecology, and management of the fall armyworm, *Spodoptera Frugiperda*. Entomol Gen 43(2):187–241

Khalil MSM, Hassan MHAR, Mahmoud AF, Morsy KMM (2022) Involvement of secondary metabolites and extracellular lytic enzymes produced by plant growth promoting rhizobacteria in inhibiting the soilborne pathogens in faba bean plants. J Trop Plant Pests Dis 22:100–108

Khan S, Nadir S, Lihua G, Xu J, Holmes KA, Dewen Q (2016) Identification and characterization of an insect toxin protein, Bb70p, from the entomopathogenic fungus, *Beauveria bassiana*, using *Galleria mellonella* as a model system. J Invertebr Pathol 133:87–94

Khan N, Maymon M, Hirsch AM (2017) Combating *Fusarium* infection using *Bacillus*-based antimicrobials. Microorganisms 5(4):75

Kilani-Feki O, Khedher SB, Dammak M, Kamoun A, Jabnoun-Khiareddine H, Daami-Remadi M, Tounsi S (2016) Improvement of antifungal metabolites production by *Bacillus subtilis* V26 for biocontrol of tomato postharvest disease. Biol Control 95:73–82

Kim JS, Lee J, Lee CH, Woo SY, Kang H, Seo SG, Kim SH (2015) Activation of pathogenesis-related genes by the rhizobacterium, *Bacillus* sp. JS, which induces systemic resistance in tobacco plants. Plant Pathol J 31(2):195–201

Kim JA, Song JS, Kim PI, Kim DH, Kim Y (2022) *Bacillus velezensis* TSA32-1 as a promising agent for biocontrol of plant pathogenic fungi. J Fungi (Basel) 8(10):1053

Kimmelshue C, Goggi AS, Cademartiri R (2019) The use of biological seed coatings based on bacteriophages and polymers against *Clavibacter michiganensis* subsp. *nebraskensis* in maize seeds. Sci Rep 9(1):17950

Kiss L, Russell JC, Szentivanyi O, Xu X, Jeffries P (2004) Biology and biocontrol potential of *Ampelomyces* mycoparasites, natural antagonists of powdery mildew fungi. Biocontrol Sci Tech 14:635–651

Knowles BH (1994) Mechanism of action of *Bacillus thuringiensis* insecticidal δ-endotoxins. Adv In Insect Phys 24:275–308

Köhl J, Kolnaar R, Ravensberg WJ (2019) Mode of action of microbial biological control agents against plant diseases: relevance beyond efficacy. Front Plant Sci 10:845

Kononchuk AG, Martemyanov VV, Ignatieva AN, Belousova IA, Inoue MN, Tokarev YS (2021) Susceptibility of the gypsy moth *Lymantria dispar* (Lepidoptera: Erebidae) to *Nosema pyrausta* (Microsporidia: Nosematidae). Insects 12(5):447

Koppenhöfer HS, Gaugler R (2009) Entomopathogenic nematode and bacteria mutualism. In: Defensive mutual. Microb. Symbiosis. CRC Press

Korniienko N, Kharina A, Budzanivska I, Burketová L, Kalachova T (2022) Phages of phytopathogenic bacteria: high potential, but challenging application. Plant Prot Sci 582:81–91

Kröber M, Verwaaijen B, Wibberg D, Winkler A, Pühler A, Schlüter A (2016) Comparative transcriptome analysis of the biocontrol strain *Bacillus amyloliquefaciens* FZB42 as response to biofilm formation analyzed by RNA sequencing. J Biotechnol 231:212–223

Kurth F, Mailänder S, Bönn M, Feldhahn L, Herrmann S, Große I et al (2014) *Streptomyces*-induced resistance against oak powdery mildew involves host plant responses in defense, photosynthesis, and secondary metabolism pathways. Mol Plant-Microbe Interact 27(9):891–900

Lacey LA, Grzywacz D, Shapiro-Ilan DI, Frutos R, Brownbridge M, Goettel MS (2015) Insect pathogens as biological control agents: Back to the future. J Invertebr Pathol 132:1–41

Lahlali R, Ezrari S, Radouane N, Kenfaoui J, Esmaeel Q, El Hamss H, Belabess Z, Barka EA (2022) Biological control of plant pathogens: a global perspective. Microorganisms 10(3):596

Lam VB, Meyer T, Arias AA, Ongena M, Oni FE, Höfte M (2021) *Bacillus* cyclic lipopeptides iturin and fengycin control rice blast caused by *Pyricularia oryzae* in potting and acid sulfate soils by direct antagonism and induced systemic resistance. Microorganisms 9(7):1441

Lange CE (2005) The host and geographical range of the grasshopper pathogen *Paranosema* (Nosema) *locustae* revisited. J Orthoptera Res 14:137–141

Lanteigne C, Gadkar VJ, Wallon T, Novinscak A, Filion M (2012) Production of DAPG and HCN by *Pseudomonas* sp. LBUM300 contributes to the biological control of bacterial canker of tomato. Phytopathology 102:967–973

Lavania M, Chauhan PS, Chauhan SV, Singh HB, Nautiyal CS (2006) Induction of plant defense enzymes and phenolics by treatment with plant growth-promoting rhizobacteria Serratia marcescens NBRI1213. Curr Microbiol 52(5):363–8

Law JW, Ser HL, Khan TM, Chuah LH, Pusparajah P, Chan KG, Goh BH, Lee LH (2017) The potential of *Streptomyces* as biocontrol agents against the rice blast fungus, *Magnaporthe oryzae* (*Pyricularia oryzae*). Front Microbiol 8:3

Leveau JHJ, Preston GM (2008) Bacterial mycophagy: definition and diagnosis of a unique bacterial-fungal interaction. New Phytol 177(4):859–876

Lewis LC, Bruck DJ, Prasifka JR, Raun ES (2009a) Nosema pyrausta: its biology, history, and potential role in a landscape of transgenic insecticidal crops. Biol Control 48:223–231

Lewis MW, Robalino IV, Keyhani NO (2009b) Uptake of the fluorescent probe FM4-64 by hyphae and haemolymph-derived in vivo hyphal bodies of the entomopathogenic fungus *Beauveria bassiana*. Microbiology (Reading) 155(Pt 9):3110–3120

Li Y, Gu Y, Li J, Xu M, Wei Q, Wang Y (2015) Biocontrol agent *Bacillus amyloliquefaciens* LJ02 induces systemic resistance against cucurbits powdery mildew. Front Microbiol 6:883

Li W, Tang XX, Yan X, Wu Z, Yi ZW, Fang MJ, Su X, Qiu YK (2016) A new macrolactin antibiotic from deep sea-derived bacteria *Bacillus subtilis* B5. Nat Prod Res 30(24):2777–2782

Li P, Bhattacharjee P, Wang S, Zhang L, Ahmed I, Guo L (2019) Mycoviruses in *Fusarium* species: an update. Front Cell Infect Microbiol 9:257

Li AJ, Pal KV, Kennan KK (2021a) A review of environmental occurrence, toxicity, biotransformation and biomonitoring of volatile organic compounds. Environ Chem Ecotoxicol 3:91–116

Li MSM, Piccoli DA, McDowell T, MacDonald J, Renaud J, Yuan ZC (2021b) Evaluating the biocontrol potential of Canadian strain *Bacillus velezensis* 1B-23 via its surfactin production at various pHs and temperatures. BMC Biotechnol 21(1):31

Lin L, Fang W, Liao X, Wang F, Wei D, St Leger RJ (2011) The MrCYP52 cytochrome P450 monooxygenase gene of *Metarhizium robertsii* is important for utilizing insect epicuticular hydrocarbons. PLoS One 6(12):e28984

Liu L, Chen Z, Yang Y, Xiao Y, Liu C, Ma Y et al (2018) A single amino acid polymorphism in ABCC2 loop 1 is responsible for differential toxicity of *Bacillus thuringiensis* Cry1Ac toxin in different Spodoptera (Noctuidae) species. Insect Biochem Mol Biol 100:59–65

Liu W, Liu X, Yang H, Ciais P, Wada Y (2022) Global water scarcity assessment incorporating green water in crop production. Water Resour Res 58:e2020WR028570

Lovett B, St Leger RJ (2015) Stress is the rule rather than the exception for *Metarhizium*. Curr Genet 61:253–261

Lovett B, Bilgo E, Millogo SA, Ouattarra AK, Sare I, Gnambani EJ et al (2019) Transgenic *Metarhizium* rapidly kills mosquitoes in a malaria-endemic region of Burkina Faso. Science 364(6443):894–897

Lugtenberg B, Kamilova F (2009) Plant-growth-promoting rhizobacteria. Ann Rev Microbiol 63:541–556

Luo F, Wang Q, Yin C, Ge Y, Hu F, Huang B et al (2015) Differential metabolic responses of *Beauveria bassiana* cultured in pupae extracts, root exudates and its interactions with insect and plant. J Invertebr Pathol 130:154–164

Madigan MY, Bender KS, Buckley DH, Sattley WM, Stahl DA (2021) Brock biology of microorganisms, 16th edn. Pearson, New York

Magalhães VSM, Ferreira LMDF, Silva C (2021) Using a methodological approach to model causes of food loss and waste in fruit and vegetable supply chains. J Clean Prod 283:124574

Mahadevan B, Crawford DL (1997) Properties of the chitinase of the antifungal biocontrol agent *Streptomyces lydicus* WYEC108. Enzym Microb Technol 20:489–493

Manjunatha L, Singh S, Ravikumara BM, Reddy GN, Senthilkumar M (2020) Ampelomyces. In: Amaresan N, Senthil Kumar M, Annapurna K, Krishna Kumar A (eds) Beneficial microbes in agro-ecology. Academic, Amsterdam, pp 833–860

Mates APK, Pontes NC, Halfeld-Vieira BA (2019) *Bacillus velezensis* GF267 as a multi-site antagonist for the control of tomato bacterial spot. Biol Control 137:104013

Matilla MA, Leeper FJ, Salmond GP (2015) Biosynthesis of the antifungal haterumalide, oocydin A, in *Serratia*, and its regulation by quorum sensing, RpoS and Hfq. Environ Microbiol 17:2993–3008

Mavrodi DV, Blankenfeldt W, Thomashow LS (2006) Phenazine compounds in fluorescent *Pseudomonas* spp. biosynthesis and regulation. Annu Rev Phytopathol 44:417–445

Meena KR, Kanwar SS (2015) Lipopeptides as the antifungal and antibacterial agents: applications in food safety and therapeutics. Biomed Res Int 2015:473050

Mendes R, Garbeva P, Raaijmakers JM (2013) The rhizosphere microbiome: significance of plant beneficial, plant pathogenic, and human pathogenic microorganisms. FEMS Microbiol Rev 37:634–663

Ment D, Gindin G, Rot A, Soroker V, Glazer I, Barel S, Samish M (2010) Novel technique for quantifying adhesion of *Metarhizium anisopliae* conidia to the tick cuticle. Appl Environ Microbiol 76:3521–3528

Miao S, Liang J, Xu Y, Yu G, Shao M (2023) Bacillaene, sharp objects consist in the arsenal of antibiotics produced by *Bacillus*. J Cell Physiol 239:e30974. https://doi.org/10.1002/jcp.30974
Miljaković D, Marinković J, Balešević-Tubić S (2020) The significance of *Bacillus* spp. in disease suppression and growth promotion of field and vegetable crops. Microorganisms 8(7):1037
Montalva C, Rocha LFN, Fernandes ÉKK, Luz C, Humber RA (2016) *Conidiobolus macrosporus* (Entomophthorales), a mosquito pathogen in Central Brazil. J Invertebr Pathol 139:102–108
Moraes CKD, Schrank A, Vainstein MH (2003) Regulation of extracellular chitinases and proteases in the entomopathogen and acaricide *Metarhizium anisopliae*. Curr Microbiol 46(3):205–210
Morath SU, Hung R, Bennett JW (2012) Fungal volatile organic compounds: a review with emphasis on their biotechnological potential. Fungal Biol Rev 26:73–83
Morrison CK, Arseneault T, Novinscak A, Filion M (2017) Phenazine-1-carboxylic acid production by *Pseudomonas fluorescens* LBUM636 Alters *Phytophthora infestans* growth and late blight development. Phytopathology 107:273–279
Muddanuru T, Polumetla AK, Maddukuri L, Mulpuri S (2019) Development and evaluation of transgenic castor (Ricinus communis L.) expressing the insecticidal protein Cry1Aa of Bacillus thuringiensis against lepidopteran insect pests. Crop Protect 119:113–125
Nair RR, Velicer GJ (2021) Predatory bacteria select for sustained prey diversity. Microorganisms 9:2079
Nakahara Y, Shimura S, Ueno C, Kanamori Y, Mita K, Kiuchi M, Kamimura M (2009) Purification and characterization of silkworm hemocytes by flow cytometry. Dev Comp Immunol 33(4):439–448
Naqqash MN, Gökçe A, Bakhsh A, Salim M (2016) Insecticide resistance and its molecular basis in urban insect pests. Parasitol Res 115(4):1363–1373
Newitt JT, Prudence SM, Hutchings MI, Worsley SF (2019) Biocontrol of cereal crop diseases using streptomycetes. Pathogens 8:78
Nguyen TH et al (2021) Utilization of by-product of groundnut oil processing for production of prodigiosin by microbial fermentation and its novel potent anti-nematodes effect. Agronomy 12:41
Nguyen TH, Wang S-L, Nguyen VB (2022) Recent advances in eco-friendly and scaling-up bioproduction of prodigiosin and its potential applications in agriculture. Agronomy 12:3099
Ni J, Yu L, Li F, Li Y, Zhang M, Deng Y, Liu X (2023) Macrolactin R from *Bacillus siamensis* and its antifungal activity against *Botrytis cinerea*. World J Microbiol Biotechnol 39(5):117
Nithyapriya S, Lalitha S, Sayyed RZ, Reddy MS, Dailin DJ, El Enshasy HA, Luh Suriani N, Herlambang S (2021) Production, purification, and characterization of bacillibactin siderophore of *Bacillus subtilis* and its application for improvement in plant growth and oil content in sesame. Sustainability 13(10):5394
Niu DD, Liu HX, Jiang CH, Wang YP, Wang QY, Jin HL, Guo JH (2011) The plant growth-promoting rhizobacterium *Bacillus cereus* AR156 induces systemic resistance in *Arabidopsis thaliana* by simultaneously activating salicylate- and jasmonate/ethylene-dependent signaling pathways. Mol Plant-Microbe Interact 24(5):533–542
O'Reilly DR, Miller LK (1989) A baculovirus blocks insect molting by producing ecdysteriod UDP-glucosyl transferase. Science 245:1110–1112
Odooli S, Roghanian R, Emtiazi G, Mohkam M, Ghasemi Y (2020) Characterization of the first highly predatory *Bdellovibrio bacteriovorus* from Iran and its potential lytic activity against principal pathogenic *Enterobacteriaceae*. Iran J Basic Med Sci 23:1275–1285
Ohba M, Mizuki E, Uemori A (2009) Parasporin, a new anticancer protein group from *Bacillus thuringiensis*. Anticancer Res 29(1):427–433
Ongena M, Jacques P (2008) *Bacillus* lipopeptides: versatile weapons for plant disease biocontrol. Trends Microbiol 16(3):115–125
Ortiz-Urquiza A, Keyhani NO (2013) Action on the surface: Entomopathogenic fungi versus the insect cuticle. Insects 4(3):357–374
Palma L, Muñoz D, Berry C, Murillo J, Caballero P (2014) *Bacillus thuringiensis* toxins: an overview of their biocidal activity. Toxins 6(12):3296–3325

Palmieri D, Ianiri G, Del Grosso C, Barone G, De Curtis F, Castoria R, Lima G (2022) Advances and perspectives in the use of biocontrol agents against fungal plant diseases. Horticulturae 8:577
Park G, Nam J, Kim J, Song J, Kim PII, Min HJ, Lee CW (2019) Structure and mechanism of surfactin peptide from *Bacillus velezensis* antagonistic to fungi plant pathogens. Bull Korean Chem Soc 40(7):704–709
Parker JB, Walsh CT (2013) Action and timing of *BacC* and *BacD* in the late stages of biosynthesis of the dipeptide antibiotic bacilysin. Biochemistry 52(5):889–901
Pedras MS, Irina Zaharia L, Ward DE (2002) The destruxins: synthesis, biosynthesis, biotransformation, and biological activity. Phytochemistry 59(6):579–596
Pedrini N (2018) Molecular interactions between entomopathogenic fungi (Hypocreales) and their insect host: perspectives from stressful cuticle and hemolymph battlefields and the potential of dual RNA sequencing for future studies. Fungal Biol 122(6):538–545
Pedrini N, Crespo R, Juárez MP (2007) Biochemistry of insect epicuticle degradation by entomopathogenic fungi. Comp Biochem Physiol C Toxicol Pharmacol 146(1–2):124–137
Pedrini N, Ortiz-Urquiza A, Huarte-Bonnet C, Zhang S, Keyhani NO (2013) Targeting of insect epicuticular lipids by the entomopathogenic fungus *Beauveria bassiana*: hydrocarbon oxidation within the context of a host-pathogen interaction. Front Microbiol 4:24
Pedrini N, Ortiz-Urquiza A, Huarte-Bonnet C, Fan Y, Juárez MP, Keyhani NO (2015) Tenebrionid secretions and a fungal benzoquinone oxidoreductase form competing components of an arms race between a host and pathogen. Proc Natl Acad Sci USA 112(28):E3651–E3660
Peng W, Berry EM (2018) Global nutrition 1990-2015: a shrinking hungry, and expanding fat world. PLoS One 13(3):e019482
Peng Z-Y, Huang S-T, Chen J-T, Li N, Wei Y, Nawaz A, Deng S-Q (2022) An update of a green pesticide: *Metarhizium anisopliae*. All Life 15(1):1141–1159
Pérez J, Moraleda-Muñoz A, Marcos-Torres FJ, Muñoz-Dorado J (2016) Bacterial predation. Environ Microbiol 18:766–779
Piggot PJ, Hilbert DW (2004) Sporulation of *Bacillus subtilis*. Curr Opin Microbiol 7(6):579–586
Pinheiro LAM, Pereira C, Barreal ME, Gallego PP, Balcão VM, Almeida A (2020) Use of phage ϕ6 to inactivate *Pseudomonas syringae* pv. *actinidiae* in kiwifruit plants: in vitro and ex vivo experiments. Appl Microbiol Biotechnol 104(3):1319–1330
Rabbee FM, Baek KH (2020) Antimicrobial activities of lipopeptides and polyketides of *Bacillus velezensis* for agricultural applications. Molecules 25(21):4973
Rabbee MF, Ali MS, Choi J, Hwang BS, Jeong SC, Baek KH (2019) *Bacillus velezensis:* a valuable member of bioactive molecules within plant microbiomes. Molecules 24(6):1046
Rabiey M, Roy SR, Holtappels D, Franceschetti L, Quilty BJ, Creeth R et al (2020) Phage biocontrol to combat *Pseudomonas syringae* pathogens causing disease in cherry. Microb Biotechnol 13(5):1428–1445
Ramette A, Frapolli M, Fischer-Le Saux M, Gruffaz C, Meyer JM, Défago G, Sutra L, Moënne-Loccoz Y (2011) *Pseudomonas protegens* sp. nov., widespread plant-protecting bacteria producing the biocontrol compounds 2,4-diacetylphloroglucinol and pyoluteorin. Syst Appl Microbiol 34(3):180–188
Ran H, Liu L, Li B, Cheng J, Fu Y, Jiang D, Xie J (2016) Co-infection of a hypovirulent isolate of *Sclerotinia sclerotiorum* with a new botybirnavirus and a strain of a mitovirus. Virol J 6:92
Rani A, Rana A, Dhaka RK, Singh AP, Chahar M, Singh S et al (2023) Bacterial volatile organic compounds as biopesticides, growth promoters and plant-defense elicitors: current understanding and future scope. Biotechnol Adv 63:108078
Rehner SA, Kepler RM (2017) Species limits, phylogeography and reproductive mode in the *Metarhizium anisopliae* complex. J Invertebr Pathol 148:60–66
Rishad KS, Rebello S, Shabanamol PS, Jisha MS (2017) Biocontrol potential of Halotolerant bacterial chitinase from high yielding novel *Bacillus Pumilus* MCB-7 autochthonous to mangrove ecosystem. Pestic Biochem Physiol 137:36–41
Rohrmann GF (2013) Baculovirus Molecular Biology, 3rd edn. National Center for Biotechnology Information, Bethesda, p 154

Sabbahi R, Hock V, Azzaoui K, Saoiabi S, Hammouti B (2022) A global perspective of entomopathogens as microbial biocontrol agents of insect pests. J Agric Food Res 10:100376
Safavi SA (2013) In vitro and in vivo induction, and characterization of Beauvericin isolated from *Beauveria bassiana* and its bioassay on *Galleria mellonella* larvae. J Agric Sci Technol 15:1–10
Salazar F, Ortiz A, Sansinenea E (2020) A strong antifungal activity of 7-O-succinyl macrolactin a vs macrolactin a from *Bacillus amyloliquefaciens* ELI149. Curr Microbiol 77(11):3409–3413
Salehi Jouzani G, Seifinejad A, Saeedizadeh A, Nazarian A, Yousefloo M, Soheilivand S et al (2008) *Bacillus thuringiensis* strains of Iran and evaluation of their toxicity on free-living and plant-parasitic nematodes. Can J Microbiol 54(10):812–822
Samson RA, Evans HC, Latgé J-P (1988) Fungal pathogenesis. In: Samson RA, Evans HC, Latgé J-P (eds) Atlas of entomopathogenic fungi. Springer Berlin Heidelberg, Berlin/Heidelberg, pp 128–139
San AKM, Mun HS (2017) Mode of infection of *metarhizium spp.* fungus and their potential as biological control agents. J Fungi 3(2):30
Sansinenea E (2019) *Bacillus* spp. as plant growth-promoting bacteria. In: Singh H, Keswani C, Reddy M, Sansinenea E, García-Estrada C (eds) Secondary metabolites of plant growth promoting Rhizomicroorganisms. Springer, Singapore, pp 225–237
Scholz R, Vater J, Budiharjo A, Wang Z, He Y, Dietel K, et al (2014) Amylocyclicin, a novel circular bacteriocin produced by Bacillus amyloliquefaciens FZB42. J Bacteriol 196(10):1842–52
Santi L, Beys da Silva WO, Berger M, Guimarães JA, Schrank A, Vainstein MH (2010) Conidial surface proteins of *Metarhizium anisopliae*: source of activities related with toxic effects, host penetration and pathogenesis. Toxicon 55(4):874–880
Sarwar M, Shad NA, Batool R (2021) Entomopathogenic protozoa roles in the management of insects pest populations. In: Biopesticides in organic farming. CRC Press
Saxena AK, Kumar M, Chakdar H, Anuroopa N, Bagyaraj DJ (2020) *Bacillus* species in soil as a natural resource for plant health and nutrition. J Appl Microbiol 128(6):1583–1594
Schmidt R, Jager V, Zühlke D, Wolff C, Bernhardt J, Cankar K et al (2017) Fungal volatile compounds induce production of the secondary metabolite Sodorifen in *Serratia plymuthica* PRI-2C. Sci Rep 7(1):862
Schneider T, Müller A, Miess H, Gross H (2014) Cyclic lipopeptides as antibacterial agents - potent antibiotic activity mediated by intriguing mode of actions. Int J Med Microbiol 304(1):37–43
Scholz R, Molohon KJ, Nachtigall J, Vater J, Markley AL, Süssmuth RD, Mitchell DA, Borriss R (2011) Plantazolicin, a novel microcin B17/streptolysin S-like natural product from *Bacillus amyloliquefaciens* FZB42. J Bacteriol 193(1):215–224
Schrank A, Vainstein MH (2010) *Metarhizium anisopliae* enzymes and toxins. Toxicon 56(7):1267–1274
Senthil-Nathan S (2015) A review of biopesticides and their mode of action against insect pests. In: Thangavel P, Sridevi G (eds) Environmental sustainability: role of green technologies. Springer India, New Delhi, pp 49–63
Sepúlveda X, Silva D, Ceballos R, Vero S, López MD, Vargas M (2022) Endophytic yeasts for the biocontrol of *Phlyctema vagabunda* in apples. Horticulturae 8:535
Seyoum E, Bateman RP, Charnley AK (2002) The effect of *Metarhizium anisopliae* var acridum on haemolymph energy reserves and flight capability in the desert locust, *Schistocerca gregaria*. J Appl Entomol 126(2–3):119–124
Shapiro M, El Salamouny S, Jackson DM, Shepard BM (2012) Field evaluation of a kudzu/cottonseed oil formulation on the persistence of the beet armyworm nucleopolyhedrovirus. J Entomol Sci 47(3):197–207
Shi L, Dong M, Lian L, Zhang J, Zhu Y, Kong W et al (2021) Genome-wide association study reveals a new quantitative trait locus in rice related to resistance to brown planthopper *Nilaparvata lugens* (Stål). Insects 12(9):836
Shin JH, Park BS, Kim HY, Lee KH, Kim KS (2021) Antagonistic and plant growth-promoting effects of *Bacillus velezensis* BS1 isolated from rhizosphere soil in a pepper field. Plant Pathol J 37(3):307–314

Silva LG, Camargo RC, Mascarin GM, Nunes PSO, Dunlap C, Bettiol W (2022) Dual functionality of *Trichoderma*: biocontrol of *Sclerotinia sclerotiorum* and biostimulant of cotton plants. Front Plant Sci 13:983127

Singh RP, Jha PN (2016) The multifarious PGPR *Serratia marcescens* CDP-13 augments induced systemic resistance and enhanced salinity tolerance of wheat (*Triticum aestivum* L.). PLoS One 11:e0155026

Singh D, Samiksha TSM, Sohal SK, Kesavan AK (2021a) Exploration of insecticidal potential of Cry protein purified from *Bacillus thuringiensis* VIID1. Int J Biol Macromol 174:362–369

Solter LF, Becnel JJ, Oi DH (2012) Microsporidian entomopathogens. Insect Pathol (2nd Ed), 221–263

Singh P, Xie J, Qi Y, Qin Q, Jin C, Wang B, Fang W (2021b) A thermotolerant marine *Bacillus amyloliquefaciens* S185 producing iturin A5 for antifungal activity against *Fusarium oxysporum* f. sp. *cubense*. Mar Drugs 19(9):516

Solter LF, Hajek AE, Lacey LA (2017) Exploration for entomopathogens. In: Lacey LA (ed) Microbial control of insect and mite pests: from theory to practice. Academic, pp 13–23

Son M, Yu J, Kim KH (2015) Five Questions about Mycoviruses. PLoS Pathog 11(11):e1005172

Song WY (2004) Identification and characterization of *Bdellovibrio bacteriovorus*, a predator of *Burkholderia glumae*. J Microbiol Biotechnol 14:48–55

Soo RM, Woodcroft BJ, Parks DH, Tyson GW, Hugenholtz P (2015) Back from the dead; the curious tale of the predatory cyanobacterium *Vampirovibrio chlorellavorus*. PeerJ 3:e968

Sparks TC, Nauen R (2015) IRAC: mode of action classification and insecticide resistance management. Pest Biochem Physiol 121:122–128

Sree KS, Padmaja V, Murthy YL (2008) Insecticidal activity of destruxin, a mycotoxin from *Metarhizium anisopliae* (hypocreales), against *Spodoptera litura* (lepidoptera: noctuidae) larval stages. Pest Manag Sci 64(2):119–125

Steller S, Vater J (2000) Purification of the fengycin synthetase multienzyme system from *Bacillus subtilis* b213. J Chromatogr B Biomed Sci Appl 737(1–2):267–275

Strobel G, Li JY, Sugawara F, Koshino H, Harper J, Hess WM (1999) Oocydin A, a chlorinated macrocyclic lactone with potent anti-oomycete activity from *Serratia marcescens*. Microbiology (Reading) 145(Pt 12):3557–3564

Su Y, Liu C, Fang H, Zhang D (2020) *Bacillus subtilis*: a universal cell factory for industry, agriculture, biomaterials and medicine. Microb Cell Factories 19:173

Suryawanshi RK, Patil CD, Koli SH, Hallsworth JE, Patil SV (2017) Antimicrobial activity of prodigiosin is attributable to plasmamembrane damage. Nat Prod Res 31:572–577

Szewczyk B, de Souza ML, de Castro MEB, Moscardi ML, Moscardi F (2011) In: Stoytcheva M (ed) Baculovirus biopesticides, pesticides-formulations, effects, fate. ISBN: 978-953-307-532-7. InTech. Available from: http://www.intechopen.com/articles/show/title/baculovirus-biopesticides

Tan S, Jiang Y, Song S, Huang J, Ling N, Xu Y, Shen Q (2013) Two *Bacillus amyloliquefaciens* strains isolated using the competitive tomato root enrichment method and their effects on suppressing *Ralstonia solanacearum* and promoting tomato plant growth. Crop Prot 43:134–140

Thakur N, Kaur S, Tomar P, Thakur S, Yadav AN (2020) Microbial biopesticides: current status and advancement for sustainable agriculture and environment. In: New and future developments in microbial biotechnology and bioengineering. Elsevier Inc., pp 243–282

Théatre A, Cano-Prieto C, Bartolini M, Laurin Y, Deleu M, Niehren J et al (2021) The surfactin-like lipopeptides from *Bacillus* spp.: natural biodiversity and synthetic biology for a broader application range. Front Bioeng Biotechnol 9:623701

Tilocca B, Balmas V, Hassan ZU, Jaoua S, Migheli Q (2019) A proteomic investigation of *Aspergillus carbonarius* exposed to yeast volatilome or to its major component 2-phenylethanol reveals major shifts in fungal metabolism. Int J Food Microbiol 306:108265

Tojo S, Kim JY, Tanaka Y, Inaoka T, Hiraga Y, Ochi K (2014) The mthA mutation conferring low-level resistance to streptomycin enhances antibiotic production in *Bacillus subtilis* by increasing the S-adenosylmethionine pool size. J Bacteriol 196(8):1514–1524

Trippe K, McPhail K, Armstrong D, Azevedo M, Banowetz G (2013) *Pseudomonas fluorescens* SBW25 produces furanomycin, a nonproteinogenic amino acid with selective antimicrobial properties. BMC Microbiol 13:1–10

Tyagi S, Kumar R, Kumar V, Won SY, Shukla P (2020) Engineering disease resistance plants through CRISPR-Cas9 technology. GM Crops Food 12(1):125–144

Utermark J, Karlovsky P (2007) Role of zearalenone lactonase in protection of *Gliocladium roseum* from fungitoxic effects of the mycotoxin zearalenone. Appl Environ Microbiol 73:637–642

Vachon V, Laprade R, Schwartz J-L (2012) Current models of the mode of action of *Bacillus thuringiensis* insecticidal crystal proteins: a critical review. J Invertebr Pathol 111(1):1–12

van der Meij A, Worsley SF, Hutchings MI, van Wezel GP (2017) Chemical ecology of antibiotic production by actinomycetes. FEMS Microbiol Rev 41:392–416

van Houte S, Ros VI, Mastenbroek TG, Vendrig NJ, Hoover K, Spitzen J, van Oers MM (2012) Protein tyrosine phosphatase-induced hyperactivity is a conserved strategy of a subset of baculoviruses to manipulate lepidopteran host behavior. PLoS One 7(10):e46933

Vero S, Garmendia G, Allori E, Sanz JM, Gonda M, Alconada T et al (2023) Microbial biopesticides: diversity, scope, and mechanisms involved in plant disease control. Diversity 15:457

Veselova SV, Burkhanova GF, Rumyantsev SD, Blagova DK, Maksimov IV (2019) Strains of *Bacillus* spp. regulate wheat resistance to greenbug aphid *Schizaphis graminum* Rond. Appl Biochem Microbiol 55(1):41–47

Viaene T, Langendries S, Beirinckx S, Maes M, Goormachtig S (2016) *Streptomyces* as a plant's best friend? FEMS Microbiol Ecol 92(8):fiw119

Vinale F, Sivasithamparam K, Ghisalberti EL, Marra R, Barbetti MJ, Li H, Woo SL, Lorito M (2008) A novel role for *Trichoderma* secondary metabolites in the interactions with plants. Mol Plant Pathol 72:80–86

Voisard C, Keel C, Haas D, Defago G (1989) Cyanide production by *Pseudomonas fluorescens* helps suppress black root rot of tobacco under gnotobiotic conditions. EMBO J 8:351–358

Wang J, Zhang H, Wang H, Zhao S, Zuo Y, Yang Y, Wu Y (2016) Functional validation of cadherin as a receptor of Bt toxin Cry1Ac in *Helicoverpa armigera* utilizing the CRISPR/Cas9 system. Insect Biochem Mol Biol 76:11–17

Wu L, Li X, Wu H, Gao X (2018) Research advances on bacilysin from Bacillus. J Nanjing Agric Univ 41:778-783

Wang F, Xiao J, Zhang Y, Li R, Liu L, Deng J (2021a) Biocontrol ability and action mechanism of *Bacillus halotolerans* against *Botrytis cinerea* causing grey mould in postharvest strawberry fruit. Postharvest Biol Technol 174:111456

Wang H, Peng H, Li W, Cheng P, Gong M (2021b) The Toxins of *Beauveria bassiana* and the strategies to improve their virulence to insects. Front Microbiol 12:705343

Wang J, Zheng X, Yuan J, Wang S, Xu B, Wang S, Zhang Y, Wu Q (2021c) Insecticide resistance monitoring of the diamondback moth (Lepidoptera: *Plutellidae*) populations in China. J Econ Entomol 114(3):1282–1290

Wang Q, Zou Q, Dai Z, Hong N, Wang G, Wang L (2022) Four novel mycoviruses from the hypovirulent *Botrytis cinerea* SZ-2-3y isolate from Paris polyphylla: molecular characterization and mitoviral sequence transboundary entry into plants. Viruses 14:151

Wang S, Fang X, Liang K, Li S, Han S, Zhu T (2023) Cloning, expression and antifungal effect of the recombinant chitinase from *Streptomyces sampsonii* KJ40. Ciência Rural 53(4):e20210663

Weller DM, Mavrodi DV, van Pelt JA, Pieterse CM, van Loon LC, Bakker PA (2012) Induced systemic resistance in *Arabidopsis thaliana* against *Pseudomonas syringae* pv. tomato by 2,4-diacetylphloroglucinol-producing *Pseudomonas fluorescens*. Phytopathology 102(4):403–412

Werner S, Polle A, Brinkmann N (2016) Belowground communication: impacts of volatile organic compounds (VOCs) from soil fungi on other soil-inhabiting organisms. Appl Microbiol Biotechnol 100(20):8651–8665

Williams T, Virto C, Murillo R, Caballero P (2017) Covert infection of insects by baculoviruses. Front Microbiol 8:1337

Wilson K, Grzywacz D, Curcic I, Scoates F, Harper K, Rice A, Paul N, Dillon A (2020) A novel formulation technology for baculoviruses protects biopesticide from degradation by ultraviolet radiation. Sci Rep 10(1):13301

Wu K, Su L, Fang Z, Yuan S, Wang L, Shen B, Shen Q (2017) Competitive use of root exudates by *Bacillus amyloliquefaciens* with *Ralstonia solanacearum* decreases the pathogenic population density and effectively controls tomato bacterial wilt. Sci Hortic 218:132–138

Xiao X, Cheng J, Tang J, Fu Y, Jiang D, Baker TS, Xie J (2014) A novel partitivirus that confers hypovirulence on plant pathogenic fungi. J Virol 88:10120–10133

Xu Y, Zhan J, Wijeratne EM, Burns AM, Gunatilaka AA, Molnár I (2007) Cytotoxic and Antihaptotactic beauvericin analogues from precursor-directed biosynthesis with the insect pathogen Beauveria bassiana ATCC 7159. J Nat Prod 70(9):1467–71

Xiao J, Guo X, Qiao X, Zhang X, Chen X, Zhang D (2021) Activity of fengycin and iturin A isolated from *Bacillus subtilis* Z-14 on *Gaeumannomyces graminis* var. *tritici* and soil microbial diversity. Front Microbiol 12:682437

Yadav A, Jaiswal N, Malviya S, Malhotra SK (2021) Entomopathogenic protozoa. In: Omkar (ed) Microbial approaches to insect pest management. Springer, Singapore, pp 337–384

Yang X, Hong C (2020) Biological control of Phytophthora blight by *Pseudomonas protegens* strain 14D5. Eur J Plant Pathol 156:591–601

Yip C-H, Mahalingam S, Wan K-L, Nathan S (2021) Prodigiosin inhibits bacterial growth and virulence factors as a potential physiological response to interspecies competition. PLoS One 16:e0253445

Youdkes D, Helman Y, Burdman S, Matan O, Jurkevitch E (2020) Potential control of potato soft rot disease by the obligate predators *Bdellovibrio* and like organisms. Appl Environ Microbiol 86(6):e02543–e02519

Zaczek-Moczydłowska MA, Young GK, Trudgett J, Plahe C, Fleming CC, Campbell K, Hanlon RO (2020) Phage cocktail containing Podoviridae and Myoviridae bacteriophages inhibits the growth of *Pectobacterium* spp. under in vitro and in vivo conditions. PLoS One 15:e0230842

Zeigler DR, Perkins JB (2008) The genus *Bacillus*. In: Goldman E, Green LH (eds) Practical handbook of microbiology, 2nd edn. CRC Press, Boca Raton, pp 309–339

Zeriouh H, de Vicente A, Pérez-García A, Romero D (2014) Surfactin triggers biofilm formation of *Bacillus subtilis* in melon phylloplane and contributes to the biocontrol activity. Environ Microbiol 16(7):2196–2211

Zhang J, Fu B, Lin Q, Riley IT, Ding S, Chen L, Cui J, Yang L, Li H (2020) Colonization of *Beauveria bassiana* 08F04 in root-zone soil and its biocontrol of cereal cyst nematode (*Heterodera filipjevi*). PLoS One 15(5):e0232770

Zhang D, Yu S, Zhao D, Zhang J, Pan Y, Yang Y et al (2021) Inhibitory effects of non-volatiles lipopeptides and volatiles ketones metabolites secreted by *Bacillus velezensis* C16 against *alternaria solani*. Biol Control 152:104421

Zhang R-S, Wang F-G, Qi Z-Q, Qiao J-Q, Du Y, Yu J-J et al (2022) Iturins produced by *Bacillus velezensis* Jt84 play a key role in the biocontrol of rice blast disease. Biol Control 174:105001

Zhang N, Wang Z, Shao J, Xu Z, Liu Y, Xun W et al (2023) Biocontrol mechanisms of Bacillus: improving the efficiency of green agriculture. Microb Biotechnol 16(12):2250–2263

Zhao P, Li P, Wu S, Zhou M, Zhi R, Gao H (2019) Volatile organic compounds (VOCs) from *Bacillus subtilis* CF-3 reduce anthracnose and elicit active defense responses in harvested litchi fruits. AMB Express 9(1):119

Chapter 10
Microbes as Natural Biocontrol Agents in Organic Farming

S. A. Ganiyu and M. Buba

Abstract Globally, pests and diseases often affect crop output and thereby impose serious challenges to sufficient food for the growing global populace. Therefore, because of constant crop yield loss through these pests, the necessity of protecting crops from pests and diseases, especially, in organic farming system cannot be overemphasized. This necessitates the search for a more practical and sustainable means of managing both insect pests and phytopathogens, aside from organic farming, that will dominate the most used and often abused synthetic pesticides. Using naturally occurring antagonistic microbes for biocontrol, and biological control agents (BCA), is a strategy that is safe for the environment to prevent plant diseases. Organic farming (OF), is a sustainable farming method that produces a steady supply of farm products while being mindful of the environment, economy, and society. The utilization of microorganisms as organic farming's natural biocontrol agents is highlighted in this chapter. Microbes are microscopic, unicellular organisms that are invisible to the naked eye but under a microscope. They are also known as microorganisms and care in numerous life forms, sizes, and features, which include bacteria, fungi, protists, viruses, and archaea. They could either be advantageous or harmful. Utilizing biological control agents relies on the "natural control" phenomenon to lower the quantity of deleterious plant diseases and pests. The purpose of introducing or maintaining living organisms as biocontrol agents is to damage organisms that cause diseases. More target specificity and ecological safety could be provided by microbes as a potential substitute for chemical insecticides.

Keywords Biocontrol · Diseases · Microbes · Natural enemies · Organic farming · Pathogens · Pests

---

S. A. Ganiyu (✉) · M. Buba
Department of Agronomy, Faculty of Agriculture, Federal University of Kashere, Kashere, Gombe State, Nigeria

O. O. Babalola et al. (eds.), *Ecofriendly Frontiers*,
https://doi.org/10.1007/978-3-031-98700-7_10

10.1 Introduction

Plant diseases are one the primary issues with crop production against achieving food security objectives for the world's expanding population. Evaluations showed that annual crop losses due to diseases are very huge compared to the growing population (Junaid and Gokce 2024). Several approaches, for instance, adoption of crop rotation and the use of resistant plant varieties, use of synthetic agrochemicals, among others, are being used to control infections in plants. Commonly, synthetic chemical insecticides are widely used in controlling plant pests and pathogens. However, when these chemicals are continually used, pests and diseases may evolve and become resistant, which has made their use flawed and necessitated the creation of sustainable and safe alternatives (Ganiyu et al. 2023). Due to their persistence and lack of biodegradability, these synthetic pesticides pose major threats to both human health and the agro-ecosystem, which can also cause resistance, toxicity to organisms that are not their intended targets, and biomagnification in the food chain (Buba and Salisu 2023). Chemical pesticide substitutes are prized for their effectiveness against a variety of crop diseases and pests, due to the fact that they are typically simple to apply, have a brief residual effect, and do not build up in the environment (Lale 2002), and attention on biocontrol has become strong owing to worries over the general practice of using chemical pesticides (Whipps 2001).

In addition to improving plant growth, biocontrol employing naturally occurring antagonistic bacteria, or biocontrol agents (BCAs), are not dangerous and ecologically sound technique that, in many situations, is the solitary method toward avoiding plant diseases and pests (Gamalero and Glick 2020). Moreover, BCAs have been reported to strengthen plants, enhance their capability to tolerate environmental pressure, aid in nutuirets uptake, cause disease resistance in plants, and act as biofertilizers. According to Gamalero and Glick (2020), biological control is the process by which diseases and dangerous pests are eliminated by natural enemies. Numerous microorganisms have been explored as possible means to manage diseases caused by bacteria, nematodes, viruses, fungi, and protozoa (Usta 2013).

Researchers have reported variable descriptions of the activities of biocontrol agents of pests, which are primarily regarded for their antagonistic qualities against phytopathogens on a broad spectrum of plants. *Erwinia carotovora*'s growth is inhibited by siderophores produced by antagonistic microorganisms, as stated by Antoun (2013). P and Q isolate of *Trichoderma virens* were identified by Oskiera et al. (2017). In contrast, strains from the Q group were able to both perform as agents of control of cotton seedling diseases and cause a progressive manifestation of phytoalexins. In contrast, strains from the P group were unable to achieve as control agents of cotton seedling diseases and could infect susceptible seeds (Junaid et al. 2013). Hence, the review delves into the use of microbes as natural biocontrol agents in organic farming for food security.

10.2 Concept of Organic Farming and Natural Enemies

10.2.1 Organic Farming

An agricultural practice with several facets, organic farming (OF) emphasizes the use of natural, biological methods as opposed to artificial chemicals. It's a system that combines biological, mechanical, and cultural activities to support biodiversity conservation, ecological balance, and resource cycling. It is not just the absence of synthetic materials but a knowledge and practice that appreciates the complications of ecosystems. According to Finckh and van Bruggen (2015) and van Bruggen and Finckh (2015), "an environmentally, economically, and socially responsible method of farming that produces a consistent supply of nutritious food and fibers while minimizing energy and nutrient losses and adverse environmental effects" is what is meant to be organic farming (OF). This cropping system is becoming more and more prevalent around the world, which strictly conforms with the idea and principle that neither synthesized pesticides nor chemical fertilizers should be used as the first option in crop production (Verma et al. 2022). The main characteristics of OF systems are (i) more plant variety, (ii) greater soil carbon-based content due to cover crops and rotations, and (iii) greater biomass, variety, and action of soil fauna and microbes. The others include better cation exchange capacity, increased internal cycling, and decreased nutrient loss; they minimize run-off and increase rooting depth, which improves water usage efficiency (Gomiero et al. 2011; Tuck et al. 2014).

Application of bioagents, which are economical and accessible to even small-scale farmers, is a feature of organic farming. These bioagents include biofertilizers, biopesticides, and biocontrol agents (BCAs). Organic farming relies on methods that integrate traditional agricultural techniques derived from naturally occurring biological processes with contemporary technology and ecological knowledge. The fundamental strategy used by farmers in organic farming (OF) to manage pests is to create an environment that supports beneficial creatures, use of biopesticides, which are pesticides made from natural sources like plants, animal tissue, or minerals, trap crops, or biocontrol agents (Wallace et al. 2021). The USDA National Organic Program states that organic production comprises an ecological production system that is adapted to site-specific conditions by integrating biological, mechanical, and cultural practices that support ecological balance, promote resource cycling, and maintain biodiversity (ECFR 2010; Greene et al. 2010). Accordingly, organic crop producers use methods that aim to maintain or improve the physical, chemical, or biological condition of the soil in order to reduce soil erosion and meet the natural dietary and behavioral needs of animals (Greene et al. 2010).

10.2.2 Natural Enemies

The term "natural enemy" in farming systems refers to naturally occurring creatures that lessen the impact of weeds, pests, or diseases in a particular ecosystem. These natural enemies are part of the same biological groups as pests and pathogens. But compared to pests, natural enemies typically occupy a higher trophic level (Neumeister 2010). The kinds of natural enemies might differ depending on the agroecosystem. Every common pest typically has a large number of natural adversaries.

Pathogens, parasites and parasitoids, and predators are the three functional classes into which natural enemies can be classified. Essentially, a fungus (a pathogen) that affects pupae or larvae might also be categorized as a parasite, but for the sake of simplification, microorganisms that impact and dwell in plant tissues are described as pathogens. This classification offers a more logical technique because it is impossible to draw a line between them. There's another crucial distinction between generalists and specialists. Neumeister (2010) states that a generalist has a wider range of interactions (a spider web does not select), but a specialist typically affects, parasitizes, or preys on one (or very few) unique species of animals. Once more, a generalist may focus on a certain creature, whereas a specialist may alter its specificity depending on the situation.

10.3 Microbes

10.3.1 Microbes and Their Compounds Produced for Plant Pathogens Suppression

Microbes are microscopic, unicellular organisms that are invisible to the naked eye. They are in numerous life forms, sizes, and features, and these include bacteria, fungi, protists, viruses, and archaea, which can either be advantageous or harmful. Some microbes cause severe infections on plants as well as causing food spoilage. Positively, microorganisms aid in the decomposition of chemicals in the environment, supporting biological processes and thus being crucial to the preservation of environmental equilibrium. A wealth of natural chemicals with the ability to treat plant diseases can be found in microorganisms that are beneficial. The natural substances produced by microbes that reduce plant pathogens include secondary metabolites (SMs) (Ayaz et al. 2021; Zubair et al. 2021).

10.3.2 Diversity of Opposing Microorganisms

Antagonistic microorganisms' effectiveness in plant disease management is a result of their biodiversity having been reported to be abundant (Negi et al. 2023) and in close association with both biotic and abiotic forms such as soil, plants, nematodes, and arthropods. Antagonistic microbes belong to different phyla and genera, which emphasizes their diversity. Different workers established an association between antagonistic microbes and other organisms. For instance, *Bacillus lentimorbus* which is known to associate with rhizospheric region of weeds was found to antagonize *Fusarium oxysporum* (Kang et al. 2005). Similarly, Ayyadurai et al. (2006) isolated *Pseudomonas aeruginosa*, an antagonist of *F. oxysporum*, from the rhizospheric zone of bananas. The growth of *Alternaria alternata* was inhibited by the soil bacterium *Pseudomonas putida* (Pandey et al. 2006). The growth of *Phytophthora citricola* was found to be inhibited by *Kitasatospora* from *Fagus sylvatica* (Haesler et al. 2008); the development of *Phytophthora* sp. was reported to be suppressed by *Streptosporangium* sp. from *Azadirachta indica* (Verma et al. 2009); *Verticillium dahlia* was found to be inhibited by endophytic bacteria (Aranda et al. 2011); and *Meloidogyne incognita* was found to be inhibited by *Fusarium oxysporum* from the rhizosphere of coffee (Freire et al. 2012). The biodiversity of some antagonistic microbes is shown in Table 10.1 below.

10.3.3 Microbes as Natural Biocontrol Agents

According to Heimpel and Mills (2017), the natural occurrence of bacteria or other living organisms lowering plant pathogen populations is known as the biological control of plant diseases. Carefully chosen isolates of helpful microorganisms that can regulate disease include those that are highly effective against pathogens and can be cultivated on artificial media. It is critical to understand that the virulence of the control agents employed determines the effectiveness of any biocontrol regime. Biocontrol agents are living things that humans introduce or maintain with the goal of harming organisms that cause diseases.

10.3.4 Limiting Factors for the Underutilization of Microbes as BCAs

These biological agents hold great potential for disease and pest control but remain largely untapped. There are two primary reasons for this: (1) the technical challenges associated with their use, stemming from absence of basic knowledge about the bioagents and their ecosystem; and (2) the expenses associated with production and the necessary permits for formulation, and application (Lahlali et al. 2022). Furthermore, shallow knowledge of these organisms. This is due to the fact that biocontrol is frequently more difficult to handle than physical and chemical approaches and always requires the understanding of interactions at the ecosystem. Compared to most physical or chemical controls, biological control is usually more durable and stronger but it is also probably less dramatic (Pandit et al. 2022). Another limiting factor is their need for food throughout their lifetime, which has posed a huge challenge to both farmers and crop protectionists. This implies that different food sources are required occasionally. Invertebrates that are not

Table 10.1 Diversity of antagonistic microorganisms from various sources (Negi et al. 2023)

Microbes	Source	Target pest/pathogen	References
Acinetobacter rhizosphaere	Wild olives	*Verticillium dahlia*	Aranda et al. (2011)
Advenella incenata	Tobbaco	*Phytophthora nicotianae*	Jin et al. (2011)
Alcaligenes faecalis	Soil	*Fusarium oxysporum*	Chowhan et al. (2023)
Aureobasidium pullulans	Soil	*Rhizoctonia solani*	Hilber-Bodmer et al. (2017)
Bacillus aerius	Soil	*Fusarium oxysporum*	Chowhan et al. (2023)
Bacillus amyloliquefaciens	Banana	*Fusarium oxysporum*	Wang et al. (2013)
Bacillus cereus	*Arabidopsis thaliana*	*Botrytis cinerea*	Hong et al. (2015)
Bacillus infantis	Soil	*Fusarium oxysporum*	Chowhan et al. (2023)
Bacillus lentimorbus	Weed	*Fusarium oxysporum*	Kang et al. (2005)
Bacillus licheniformis	Banana	*Fusarium oxysporum*	Yadav et al. (2021)
Bacillus methylotrophicus	Soil	*Sclerotinia minor*	Pane et al. (2012)
Bacillus paralicheniformis	Rice	*Xanthomonas oryzae*	Kannan et al. (2021)
Bacillus simplex	Soil	*Phakopsora pachyrhizi*	Twizeyimana et al. (2023)
Bacillus subtillis	Ginseng	*Sclerotinia sclerotiorum*	Lee et al. (2017)
Bacillus toyonensis	Tomato	*Alternaria alternate*	Pane and Zaccardelli (2015)
Bacillus velezensis	Rice	*Magnaporthe grisea*	Jing et al. (2020)
Bacillus weihenstephanensis	Soil	*Caenorhabditis elegans*	Rae et al. (2010)
Brevibacillus brevis	Tobacco	*Ralstonia solanacearum*	Liu et al. (2013)
Candida subhashii	Soil	*Rhizoctonia solani*	Hilber-Bodmer et al. (2017)
Cyberlindnera sargentensis	Soil	*Rhizoctonia solani*	Hilber-Bodmer et al. (2017)
Delftia tsuruhatensis	Tobacco	*Phytophthora nicotianae*	Jin et al. (2011)
Enterobacter cowanii	Tomato (surface)	*Botrytis cinerea*	Shi and Sun (2017)
Enterobacter hormaechei	Sweet pepper	*Ralstonia solanacearum*	Mamphogoro et al. (2021)
Fructobacillus fructosus	*Apis mellifera*	*Paenibacillus larvae*	Al-Ghamdi et al. (2020)
Fusarium oxysporum	Coffee	*Meloidogyne incognita*	Freire et al. (2012)
Jeotgalicoccus aerolatus	*Halodule uninervis*	*Pyricularia oryzae*	Bibi et al. (2018)

(continued)

Table 10.1 (continued)

Microbes	Source	Target pest/pathogen	References
Klebsiella oxytoca	Soil	*Fusarium oxysporum*	Chowhan et al. (2023)
Lysinabacillus fusiformis	Nasutitermes	*Phytophthora capsici*	Fitriana et al. (2022)
Moraxella osloensis	*Halodule uninervis*	*Phytophthora capsici*	Bibi et al. (2018)
Mortierella globalpina	*Caenorhabditis elegans*	*Meloidogyne chitwoodi*	DiLegge et al. (2019)
Myroides odoratimimus	Tobacco	*Phytophthora nicotianae*	Jin et al. (2011)
Ochrobactrum intermedium	Tomato	*Fusarium oxysporum*	Gowtham et al. (2016)
Paenibacillus alvei	Nasutitermes	*Phytophthora capsici*	Fitriana et al. (2022)
Paenibacillus jamilae	Cucumber	*Bipolaris sorokiniana*	Wang et al. (2019)
Paenibacillus polymyxa	*Morinda citrifolia*	*Aspergillus aculeatus*	Liu et al. (2018)
Penicillium oxalicum	Cereal	*Oncopeltus fasciatus*	Santamarina et al. (2002)
Proteus mirabilis	*Apis mellifera*	*Paenibacillus larvae*	Al-Ghamdi et al. (2020)
Pseudomonas aeruginosa	*Valeriana wallichii*	*Aspergillus flavus*	Chandra et al. (2020)
Pseudomonas brenneri	Potato	*Alternaria solani*	Caulier et al. (2018)
Pseudomonas fluorescens	Tomato	*Rhizoctonia solani*	Solanki et al. (2022)
Pseudomonas indica	Cabbage	*Pythium aphanidermatum*	Al-Daghari et al. (2023)
Pseudomonas protegens	Potato	*Alternaria solani*	Caulier et al. (2018)
Pseudomonas putida	Soil	*Alternaria alternate*	Pandey et al. (2006)
Pseudomonas reactans	Wild olives	*Verticillium dahlia*	Aranda et al. (2011)
Rahnella aquatilis	Wild olives	*Verticillium dahlia*	Aranda et al. (2011)
Serratia marcescens	Cabbage	*Pythium aphanidermatum*	Al-Daghari et al. (2023)
Staphylococcus epidermidis	*Halodule uninervis*	*Pyricularia oryzae*	Bibi et al. (2018)
Streptomyces rochei	Tobacco	*Ralstonia solanacearum*	Liu et al. (2013)
Sulftobacter dubius	*Halodule uninervis*	*Pythium ultimum*	Bibi et al. (2018)
Trichoderma asperellum	Rice	*Xanthomonas oryzae*	Kannan et al. (2021)
Trichoderma gamsii	Lentil	*Pythium afertile*	Rinu et al. (2014)
Trichoderma harzianum	Sugacane	*Colletotrichum falcatum*	Viswanathan et al. (2003)
Trichoderma reesei	Banana	*Fusarium oxysporum*	Damodaran et al. (2020)

considered pests, or alternative prey, might offer predators and parasitoids an additional food source. Additionally, biological control is becoming less dependable due to climate change. The naturally occurring harmonized timing of pest-predator sequences can remain upset by changes in weather patterns (Wallace et al. 2021).

10.3.5 Biological Control Agents: Current Situation and Future Prospects

According to Hajek et al. (2007), nematodes, viruses, and a variety of fungi are used to biologically manage arthropods. Thirty fungal species were used as biocontrol means against weeds by Schwartzländer et al. (2018), while more than two hundred microbial species/strains were used for augmentative biological control by van Lenteren et al. (2018). According to O'Brien (2017), about 38 species and/or strains have been used for the biological treatment of crop infections. van Lenteren et al. (2018) supplied a list of around 80 fungal strains, indicating 39 species, and 94 bacterial strains, representing 28 species. It was claimed that these strains were approved for use as products for plant protection. The record of pathogens listed in *Classical Biological Control of Insects and Mites: A Worldwide Compendium of diseases and nematode introductions* (Hajek et al. 2016) is one of the additional sources.

Unlike the chemical control approach, BCAs have evolved with their hosts, which ensures continued sensitivity of pests and pathogens to the antagonists (Roy et al. 2006; Wright and Bennet 2018). The effectiveness of BCA formulations has also been studied and reported by various researchers (Ambethgar 2009; Panebianco et al. 2015). Organic farmers maintain a balance between beneficial and detrimental elements in the environment. This method of managing pests needs a detailed knowledge of life cycles as well as eating patterns of many living creatures. The crucial mechanisms that bacteria, fungus, viruses, and archaea play in biological control were emphasized by Alamoudi (2024).

10.4 Soil-Dwelling Microbes

10.4.1 Soil-Dwelling Microbes as Antagonists to Pests and Pathogens

Soil microbial biomass abounds with both eukaryotic and prokaryotic organisms (Shannon et al. 2002). Literature abounds with the activity of soil-dwelling microbes, which exhibit antagonistic activity against plant pests and pathogens alike. As an illustration, more than 100 species of fungi are parasites, and many them feed on nematodes (Ahmad et al. 2021). Streptomycetes are well known for producing antibiotics and extracellular hydrolytic enzymes because of their capacity to colonize plants and minimize damage from a variety of pathogens. Accordingly, they could play a major role in an integrated disease control system in many crops (Samac et al. 2003).

Aside from soil-dwelling microbes, some pathogens live as endophytes (organisms that live in another plant) and serve as agents for biocontrol, and their main activities include the production of antibiotic agents (Lambert 2005), production of siderophores, exclusion of niches, promotion of systemic acquired host resistance, and nutritional competition (Antoun 2013). It is well documented that endophytes of bacteria contribute to pathogen control and complementary crop sequences that promote advantageous allelopathy.

10.5 Categories of Microbial Control Agents

Biological control can be classified into four main categories based on whether it is produced naturally (natural biological control), or through targeted human intervention (conservation biological control), or whether it is created temporarily (augmentative biological control) or permanently (classical biological control) through the introduction of BCAs into the target environment (Heimpel and Mills 2017; Stenberg et al. 2021).

10.5.1 Natural Biocontrol

According to Buitenhuis et al. (2023), natural biological control refers to the unassisted eradication of pests, pathogens or weeds using living organisms. It can only be imagined the magnitude of damage to agricultural produce by pests without the utilization of biocontrol agents to suppress them. It might be inferred that the lack of biocontrol agents would have facilitated the spread and establishment of pests and diseases in previously uninhabited areas. Natural biodiversity serves as the foundation for natural biological control, which emphasizes the vital significance of protecting it.

10.5.2 Conservation Biocontrol

A variety of techniques are used here, to maintain and increase the activity of natural enemies in order to increase current pest control levels and lessen the detrimental impacts of dangerous species (Eilenberg et al. 2001; Heimpel and Mills 2017). This strategy includes manipulating habitat and vegetation; using crop management techniques; directly providing resources.

Vegetation management practices often aim to preserve a diverse community of native natural enemies, especially plant pathogen antagonists or generalist predators, which can rapidly colonize the crop in the event of a pest outbreak or keep pest populations at a minimum (Costamagna et al. 2015; Jonsson et al. 2017; Zaviezo et al. 2021; Woltz and Landis 2013; Crowder and Jabbour 2014; Finke and Snyder 2008, 2010).

Crop management techniques make up the second group. Among these are soil management techniques, which are especially important for plant pathogen antagonists and soil entomopathogens. However, for these techniques to be consistently successful, detail knowledge of the biology of individual species and occasionally entire communities of microorganisms is required. It has been observed that increasing relative humidity by other crop-management approaches or by altering crop irrigation regimes or techniques increases the activities of fungi (Pell et al. 2010).

The direct supply of resources to predators is by providing protein supplies or pollen to arthropod predators, supplying sugar sprays to parasitoids, or adding organic amendments to the soil to function as an antagonist to microorganisms are just a few examples of the various ways this can be accomplished (Timper 2014).

10.5.3 Classical Biocontrol

Classical biological control entails the intentional introduction, release, and establishment of specialized natural enemies in previously uninhabited areas in order to suppress detrimental populations of invasive terrestrial arthropod pests, weeds, and, in rare cases, vertebrate pests (Sheppard et al. 2019). The first successful method for managing invasive pests was the use of natural enemies for arthropod pests, which can be divided into three main groups: diseases, parasitoids, and predators. Herbivorous insects, mites, and occasionally plant pathogens are the main tools used in biological management programs for weeds (Schultz et al. 2019). By searching the native range for natural enemies, this biocontrol method-sometimes known as "foreign exploration." A target pest's natural enemies, which were previously the focus of a traditional biocontrol program can be "borrowed" from collaborating scientists to be imported into a quarantine facility for safety testing before possibly being released and established in a new location. Entomopathogens are a group of pathogens that exist naturally as enemies of insects, and they include nematodes, bacteria, viruses, and fungi.

10.5.4 Augmentative Biocontrol

Natural enemies raised for commercial purposes can be released to supplement beneficial species found in natural communities when they are not large enough to control pests. To temporarily suppress arthropod pests and diseases, it involves raising in large quantities natural enemies of pests under carefully monitored circumstances, then releasing them (Heimpel and Mills 2017). Both inoculative and inundative releases are forms of augmentative release. Inoculative releases are usually made at the start of the cropping system's season to "inoculate" it, releasing only one or two natural enemies. Numerous natural enemies are repeatedly released during the growth season in a process known as "inundative release." Although the released natural enemies have the ability to multiply, they provide the majority of biocontrol in the short term.

10.6 Biological Control and Organic Farming

Although natural enemies of plant pathogens and pests are important, it is important to note that they typically must be applied in concert with other pertinent control measures, particularly appropriate crop control practices (Daniel et al. 2018). Altieri and Nicholls (2003) investigated controlling of insect infestations and soil fertility, among other works introducing biocontrol into organic farming, to balance health of plants and soil in agroecosystems. In practical terms, controlling pests, diseases, and weeds in organic cropping systems could be accomplished in three steps with different tools, according to Wyss (2011). Creating ideal growing conditions for plants is the first step toward increasing their resistance and resilience. In order to eliminate pests, diseases, or weeds with the least amount of environmental damage possible, direct control methods must be applied after the succeeding stage of fostering natural mechanisms of control through the development of natural antagonists. Put another way, cultural practices, habitat management, and direct control methods are some of the instruments used to manage diseases and pests (Vacante and Bonsignore 2018). Included are the use of mineral-based insecticides, pheromones, repellents, mass-trapping devices, physical barriers to keep pests out, and natural enemies' ability to control pest populations. Caldwell et al. (2013) offered a process that involves utilization of materials permitted in organic production as well as cultural control measures.

The structure and the size of the soil biomass are also impacted by soil management techniques that change the carbon concentration of the soil. Microorganisms have greater access to carbon when organic materials and cover crops are added to the soil (Schjonning et al. 2002). Crop wastes provide carbon that boosts the activity of soil microorganisms, and consequently, the possibility of competition in the soil. A pathogen's capacity to survive may be lowered if the residue is deposited in the soil and displaces the pathogen from its preferred niche (Bailey and Lazarovits 2003). Thus, land management that includes introducing organic materials results in changes to the biomass of soil microbes (Shannon et al. 2002). Applying organic matter and other treatments that increase the overall microbial activity in the soil can promote the general suppression of diseases by increasing competition for resources. Crop rotation may also benefit microbes in ways other than those frequently associated with disease host range and saprophytic survival. Rotation is the most effective strategy to lessen the effects of pathogens with a low potential for saprophytic survival or biotrophic infections, which require living host tissues (Bailey et al. 2001).

Biopesticides are significant enough to warrant increased production and marketing on a global scale, and it has been suggested that the rise in biological product registrations globally corresponds with the growing demand for more organic farming and the accomplishment of certain sustainable development goals (Alamoudi 2024). An assortment of microorganisms that have been successfully tested on a broad scale over many years has led to their mass production and patenting (see Table 10.2).

10.6.1 Soil-Borne Microorganisms to Manage Diseases of Organic Crop

With growing awareness of the harm that chemical pesticides cause to both individuals and the environment, there is an increasing interest in introducing biocontrol bacteria into the rhizosphere. According to van Bruggen et al. (2016) research, soils modified with organic matter had much higher concentrations of cellulolytic and hemicellulolytic actinomycetes than soils treated with chemical fertilizers. The advantage of using combinations of control microbes is that they can function broadly, which generally improves biological control's effectiveness and dependability and ensures that defensive enzymes are more strongly induced in comparison to single isolate (Latha et al. 2009). An efficient one ought to be cheap, retain good viability without the need for a specific storage system, and able to colonize and multiply on plant components in addition to competing and surviving in the environment (Harman 2010).

Trichoderma spp., for example *T. harzianum,* are considered to be good candidates for development as biocontrol agents since they can effectively suppress root-knot nematodes (Sharon et al. 2001). However, obstacles to their development have included variations in the effectiveness of isolates, their potential for biocontrol, and the reproducibility of results under various circumstances (Sharon et al. 2001). When combined with *Pseudomonas aeruginosa* or used alone on okra, endophytic *T. viride* was found to be efficacious in reducing the *F. solani, F. oxysporum*, and root-knot nematode (Afzal et al. 2013). According to Naraghi et al. (2010), *Talaromyces flavus*'s antagonistic actions were successful in biologically controlling tomato verticillium disease. *Chaetomium*-formulated bioproducts have demonstrated potential in field experiments as broad-spectrum myco-fungicides to manage a variety of infections (Soytong et al. 2001). Likewise, *Thielaviopsis* bud rot of *Hyophorbe lagenicaulis* is protected from and cured in Thailand by the application of bioproducts derived from *Chaetomium* (Soytong et al. 2001).

Paecilomyces lilacinus inoculation of the soil reduced nematode proliferation (Garcia et al. 2000). When *P. lilacinus* was combined with organic debris, its

Table 10.2 A few carefully chosen microbial management products and agents utilized to manage pathogens and pests that live in the soil (Stewart et al. 2010)

Pest control agent	Pest targets	Product name	Reference
Oryctes virus	*Oryctes rhinoceros*	–	Ramle et al. (2005)
Bacillus popilliae	*Popillia japonica*	Doom	Klein (1992)
Serratia entomophila	*Costelytra zealandica*	Invade/Bioshield	Jackson (2007)
Bacillus thuringiensis var. *Buibui*	*Popillia japonica*, *Anomala* spp.	–	Alm et al. (1997)
Beauveria bassiana	*Cornitermes cumulans*	Boverial	Alves et al. (1995)
Beauveria bassiana	*Heterotermes tenuis*	Termitrap	Almeida et al. (1997)
Beauveria bassiana	*Cosmopolites sordidus*	–	Badilla and Alves (1991)
Beauveria brongniartii	*Melolontha melolontha*	Engerlingspilz	Keller (1992)
Beauveria brongniartii	*Hoplochelus marginalis*	Betel	Vercambre et al. (1994)
Metarhizium anisopliae	*Adoryphorus couloni*	BiogreenChaferGuard®;	Rath (2002)
Metarhizium anisopliae	*Cornitermes cumulans*	Metarril	Alves et al. (1995)
Steinernema scapterisci	*Scapteriscus* spp.	–	Parkman and Smart (1996)
Heterorhabditis bacteriophora	*Phyllopertha horticola*	Nema-green	Sulistyanto and Ehlers (1996) www.e-nema.de
Heterorhabditis spp.	*Otiorhynchus* spp.	Dickmaulrüssler nematoden	Product literature, Andermatt Biocontrol AG
Heterorhabditis megidis	*Otiorhynchus* spp.	Larvanem	www.koppert.com
Thelohania solenopsae	*Solenopsis invicta*	–	Oi and Williams (2002)
Nematode control agent	**Nematode targets**	**Product name**	**Reference**
Paecilomyces lilicanus	*Meloidogyne*, *Globodera* spp	Bionemat	www.biotech-int.com
Bacillus firmus	*Meloidogyne* spp	Bionem WP	Blanchinsky et al. (2007)
Microbial control agent	**Microbial targets**	**Product name**	**Reference**

(continued)

Table 10.2 (continued)

Agrobacterium radiobacter	*Agrobacterium tumefaciens*	NoGall™	Ryder and Jones (1990), www.beckerunderwood.com
Bacillus subtilis	*Rhizoctonia*, *Fusarium* spp.	Kodiak®	Brannen and Kenney (1998)
Bacillus subtilis	*Fusarium*, *Verticillium*, *Pythium* spp	Biosubtilin	www.biotech-int.com
Pseudomonas fluorescens	*Sclerotinia*, *Rhizoctonia*, *Pythium* spp.	Biomonas	www.biotech-int.com
Streptomyces griseoviridis	*Fusarium*, *Phytophtora*, *Pythium*, *Rhizoctonia* spp.	Mycostop	www.verdera.fi
Coniothyrium minitans	*Sclerotinia* spp.	Contans®	www.prophyta.de
Pythium oligandrum	*Sclerotinia*, *Pythium*, *Alternaria* spp.	Polyversum	www.polyversum.cz
Trichoderma atroviride	*Sclerotium cepivorum*	Tenet®	Stewart and McLean (2007), www.agrimm.co.nz
T. harzianum T-22	*Pythium*, *Rhizoctonia*, *Fusarium* spp.	Rootshield®, PlantShield®	Harman and Bjorkman (1998), www.bioworksinc.com
T. virens	*Pythium*, *Rhizoctonia* spp.	SoilGard®	Lumsden and Knauss (2007)

capacity to control nematodes was enhanced. According to Prabhu et al. (2008), it is thought that fungal activity and persistence were enhanced by leftover organic matter, while nematicidal characteristics were released with the decay of living things. It is more effective to use a combination of *P. lilacinus* and *P. aeruginosa* to decrease the infection of root-infecting fungus and root-knot nematode on watermelon, chili, pumpkin, and guar (Ahmed and Monjil 2019). Apart from parasitizing the eggs of root-knot and cyst nematodes, *Paecilomyces lilacinus* also generated compounds that were nematicidal (Li and Zhang 2014).

Along with antagonistic fungi and bacteria, arbuscular mycorrhizal fungi (AMF) have been used to combat soil-borne diseases (Berta et al. 2005). Vesicular arbuscular endo-mycorrhizas are the most common form of mycorrhizal relationship, and they come from almost all cultivated plants, whether they are horticultural, fruit crops, or agricultural (Pfleger and Linderman 2000). This kind of symbiotic fungal infection may be one of the more beneficial biological methods of guaranteeing plant output because of its significance for plant nutrition and mineral uptake, as well as for plant health in general (Gupta 2003; Pfleger and Linderman 2000). By increasing nutrient intake, stress tolerance, and disease resistance, vesicular arbuscular mycorrhizae (VAM) promote crop development (Bouamri et al. 2006). According to researches, VAM lessens the intensity of plant-pathogenic fungus-induced diseases (Filion et al. 2003). This has been demonstrated to lessen damage from wilt of cotton caused by Verticillium, *Fusarium* crown and *Phytophthora* diseases on tomato (Arruda et al. 2021).

Plant Growth-Promoting Rhizobacteria (PGPR) are rhizospheric microorganisms that produce bioactive substances, promote plant growth, and protect plants from diseases. Plant growth-promoting rhizobacteria are bacteria that colonize roots and are known to help plants. They are said to enhance plant growth by either directly by generating growth regulators or by suppressing pathogens (Raaijmakers et al. 2002).

10.7 Potential of Entomopathogens in Organic Agriculture

Arthropods have a well-documented history of being frequently associated with microbes (van der Geest 2010; van der Geest et al. 2000; Bruin and van der Geest 2009). A well-known species in the *Bactericeae* family, *Bacillus thuringiensis* Berliner (as well as other species) is manufactured by industry today and sold to control a range of insect pests, comprising flies (Diptera), beetles (Coleoptera), as well as moths (Lepidoptera) (Gomis-Cebolla and Berry 2023). The subclassifications; Deuteromycotina, Mastigomycotina, and Ascomycotina contain the majority of pathogenic fungi (Bava et al. 2022). According to Chandler et al. (2000), at least 73 different types of mites have been found to be infected by roughly fifty-eight different fungal diseases, either in the laboratory or in the field. While certain species, like *Neozygites floridana* Weiser, Muma (Entomophthorales) and *H. thompsonii*, remain exclusive to mites and some species of fungal which can eradicate insects pests (Chandler et al. 2000).

10.8 Mechanisms Used by BCAs to Attack Phytopathogens

10.8.1 Colonization

In situations when there is no physical contact between the microorganisms, biological control agents can control plant pathogens indirectly. BCAs either directly or indirectly, control plant infections in most situations. One of the most important requirements for each BCA is the quick colonizatioon as well as occupying phyllosphere (indirect control). Preemptively occupying ecological niches (infection sites) is made possible by colonization, which also quickly consumes available substrates (plant exudates) to increase the difficulty of pathogen infections. In plant habitats like the phyllosphere, where nutrients are sparse, this capacity is especially crucial.

10.8.2 Antibiosis

The release of volatile organic compounds (VOCs) induces antibiosis, which is another method of plant defense. Di Francesco et al. (2015) discovered that *Aureobasidium pullulans* L1 and L8 produced 2-phenethyl alcohol, a volatile organic compound (VOC) that was very successful in suppressing post-harvest pathogens from germination.

10.8.3 Competition

BCAs' ability to inhibit crop diseases in the rooting zone is significantly influenced by competition for micronutrients. Iron is one of the rarest elements in the rhizosphere due to the ferric form's poor solubility. Thus, BCAs' ability to absorb iron effectively affects their ability to infiltrate the plant rhizosphere and repel plant pathogens from potential infection sites. Microorganisms chelate iron from their surroundings by generating and releasing secondary compounds called siderophores (Singh et al. 2020). In this respect, siderophores generated by bacteria that are helpful to plants exhibited a higher affinity for iron than siderophores originating from phytopathogenic fungus. In particular, it was discovered that *Pseudomonas* biocontrol bacteria were very successful in protecting plants by chelating iron in the root zone (Lemanceau et al. 2009).

10.8.4 Induced Systemic Resistance (ISR)

This enables plants to react quickly to pathogenic microbe and phytophagous insect attacks, which can also be triggered by some BCAs (Pieterse et al. 2014). It was found that *Pythium oligandrum* inhibited *B. cinerea* in grapevine and *Fusarium oxysporum* f. sp. *radicis-lycopersici* in tomato, suggesting that oomycete BCAs could also provide indirect plant protection. The ability of *P. oligandrum* to activate defense mechanisms in plants is associated with the production of oligandrin, a protein involved in the activation of ISR processes (Benhamou and Garand 2001; Kamle et al. 2020).

10.8.5 Parasitism and Production of Lytic Enzymes

According to Rey et al. (2008), *P. oligandrum* is a mycoparasite that can control plant diseases by interacting closely with the host pathogen and eliminating its living hyphae. Proteases, chitinases, glucanases, and cellulases are among the lytic enzymes that are released when the main components of fungal and oomycete cell walls are broken down, which is a characteristic of myco-parasitism in general. Similarly, by secreting a variety of chitinases and β-1,3-glucanases, as well as secondary metabolites that have toxic activity against pathogens, such as water-soluble substances like heptedelic acid, volatile antibiotics like 6-pentyl-a-pyrone, and toxic oligopeptides, *Trichoderma* species can parasitize and subsequently control a wide range of plant pathogenic fungi and oomycetes (Vinale et al. 2008). It is noteworthy that the pycnidia ascomycete *Ampelomyces quisqualis*, a mycoparasite that may multiply and infiltrate the mycelial structures (hyphae, conidiophores) of powdery mildew infections, does not produce antibiotics as part of its mechanism of action (Kiss et al. 2010).

10.8.6 Synthesis of Cell Wall Degrading Enzymes and Antibiotics

This is likewise the classic of BCAs from the *Lysobacter* species that prey on them (Puopolo et al. 2018). As a result, *Lysobacter capsici* AZ78 can use its twitching motility to colonize *Phytophthora infestans* mycelium and produce enzymes that break down cell walls to feed on the dead mycelium. Furthermore, *P. infestans*' apoptotic processes have been linked to *L. capsici* AZ78, most likely as a result of the latter's emission of a macrolactam that contains tetrameric acid (Lazazzara et al. 2017). *Lysobacter* BCAs can also suppress host defense reaction or discharge a particular group of antibiotic compounds into the surroundings to control plant pathogens without coming into close contact with them (Kilic-Ekici and Yuen 2003; Islam et al. 2012; Wan et al. 2008; Puopolo et al. 2014). One characteristic that has been thoroughly investigated in BCAs from *Pseudomonas* and *Bacillus* genera is the manufacture together with the release of antibiotics into the environment. Additionally, according to Thrane et al. (2000), pseudomonad BCAs may produce additional substances such as viscosinamide, a cyclic lipopeptide with antibacterial action, into the rhizosphere.

10.9 Disease and Pest Management in Organic Crop Production

10.9.1 Disease Management

High genetic diversity, closed systems for organic matter and nutrients, maintaining biological cycles on the farm, and preserving the environment as pure as possible are the core principles of organic crop production, according to van Bruggen and Finckh (2016). These theories suggest that systems of farming should be able to self-regulate through interrelated natural processes within agroecosystems, emphasizing problem avoidance rather than problem-solving with external inputs.

Smaller field plots, mixed crops, intricate rotation schemes, and tolerance for some weed levels in the crops are typical components of organic cropping systems that promote the development of numerous species of plants over time and place. Additionally, certain agronomic techniques, such as proper soil management, increase the number of soil organic matter, which subsequently turn increases microbial colonization and variety as well as plant canopies (Köhl 2022). Compared to conventional cropping systems, organic cropping systems pose unique challenges for disease control due to their inherent characteristics, which significantly impact the incidence and severity of plant diseases (van Bruggen and Finckh 2016). As such, bioprotectants play a different role in organic cropping systems than they do in conventional cropping systems when it comes to controlling plant diseases. Various strategies are utilized in organic crop production to prevent foliar infections. Examples include broad robustness against biotic stressors in cultivar selection, removing sources of inoculum to promote sanitation, and using mixed crops, intercropping, strip crops, and cover crops to boost habitat variety (Ditzler et al. 2021). The multicyclic foliar infections that cause issues for organic growers include those that van Bruggen and Finckh (2016) identified in their study.

10.9.2 Insect Management

In order to manage insect pests without using synthetic pesticides, augmentative release and the employment of biocontrol agents, like parasitoids, insect predators, and diseases, are very crucial. Should there be an unexpected surge in the number of insects, the previously indicated substances can be employed as therapeutic measures. In the cultivation of organic crops, Table 10.3 contains a list of some of the often utilized and possible biological control agents.

10.10 Merits and Weaknesses of Using Biocontrol Agents (BCAs)

10.10.1 Merits

Management of diseases and pests can be greatly enhanced by using biological control agents. Several studies conducted over multiple decades have confirmed the safety of BCAs and their isolates for the control of various organisms (Barratt and Mooed 2005; Tjamos et al. 2013; Hokkanen and Lynch 2003; Hokkanen et al. 2003;

Table 10.3 Possibile biocontrol agents in managing pests in organic farming (Kumaranag et al. 2013; Mohan et al. 2013)

Biocontrol agents	Pest	Crop
Bacteria *Bacillus thuringiensis*	Lepidopteran pests	Cotton, sunflower, and vegetables
Entomopathogenic fungi *Metarhizium anisopliae* *Beauveria bassiana* *Verticillium lecanii*	Coleopteran grubs Lepidopteran and coleopteran Aphids and whiteflies	Coconut, cotton, and greenhouse vegetables
Entomopathogenic nematodes *Heterorhabditis bacteriophora* *Steinernema carpocapsae*	Coleopteran and Lepidopterans	Sugarcane and plantation crops
Entomopathogenic Viruses Nuclear polyhedrosis virus (NPV), *Granulosis* virus	*Helicoverpa* and other Lepidopterans	Pulses and vegetables
Trichogramma brassiliensis—1.0 cc/ac. once in 10 days, (Egg parasitoid)	Lepidopteran, *Heliothis* sp.	Cotton, Tomato
Trichogramma chilonis—2 cc/ac once in 15 days	Borers	Sugarcane, paddy, pulses, Vegetables
Nuclear Polyhedrosis Virus (NPV) 100–200 LE/ac	*Spodoptera* sp. & *Heliothis* sp.	Vegetables
Chrysoperla sp. 5000–10000 eggs /ha, 3–4 times in 15 days, (Green lacewing)	Prudenia, Caterpillars, Whiteflies, thrips, aphids	Vegetables
Beauveria bassiana—1.0% It affects the young stage,	Helicoperva, Spodoptera, borers, hairy caterpillars, mites, scales, etc	Vegetables, cereals, fruits
Metarhizium anisopliae - 0.5–1.0% affects all stages	White grubs, Beetle grubs, caterpillars, Semiloopers, mealy bugs, BPH	Sugarcane, groundnut, rice, potato, cotton, cereals
Verticillium lecanii—0.5–1.0% Affects all stages	All sucking soft bodies insects	Sugarcane, groundnut, rice, potato, cotton, cereals
Phascilomycetes	Nematodes	All crops
Bacillus thuringiensis var *kustaki* 0.3–0.4%	Helicoperva, spodoptera, borers, hairy caterpillars, mites, scales, etc	Vegetables, cereals, fruits
NPV—*Nuclear Polyhedrosis* Virus of *Spodotera litura* 250–500 ml/ ha 2–3 times at 10-day intervals.	*Spodotera litura*	Cotton, groundnut, pulses, cabbage, chillies
NPV-*Nuclear Polyhedrosis* Virus of *Helicoverpa armigera* 250–500 ml/ha, 2–3 times at 10-day intervals.	*Helicoverpa armigera*	Cotton, groundnut, pulses, cabbage, chillies

Vurro 2007; Elnahal et al. 2022). The benefits and drawbacks of biocontrol agents were detailed by Sharma et al. (2017) and included the following:

- does not harm non-target organisms or the environment.
- no residual effect on the land, air, or water.
- being comparatively easier and cheaper.
- being able to mobilize plant nutrients for the benefit of plants.
- has great efficacy against a multitude of pathogens found in soil
- have capacity to eradicate targeted pathogens efficiently from the infection site.
- it can be combined with biofertilizers.
- they boost agricultural yields by promoting root and plant growth by fostering beneficial microorganisms in the rhizosphere, without posing any hazard to the plants. It facilitates the mobilization of plant nutrients and ensures that the plant has access to them.
- biocontrol agents both prevent resistance issues and help crop species develop systemic resistance.
- biocontrol maintains the ecosystem by itself, does not need complex management, and is self-regulating.
- because the effectiveness of biocontrol agents almost entirely depends on the surrounding environment, they operate more slowly and ineffectively than chemical pesticides.
- biocontrol agents are mostly utilized as a preventive rather than a therapeutic measure against particular diseases.
- agents for biocontrol have a limited shelf life and antagonism. *Trichoderma viride*, for instance, has a four-month shelf life, while *Pseudomonas fluorescens* has a three-month duration.
- it takes experienced individuals to proliferate and provide the biocontrol agents without contamination, and they are also necessary for frequent checking to preserve the amount of biocontrol agents that are effective in agricultural areas.
- biocontrol agents are currently scarce and limited in availability.

10.10.2 Weaknesses

The fact that microorganisms used as BCAs must be cultivated rather than synthesized and must continue to exist during the product's shelf life is one of their key disadvantages. In addition, microbial pesticides are likely to be vulnerable to adverse environmental conditions when released into the crop. The survival of microbial pathogens is a function of the availability of a minimum host population density and it should be noted, however, that individuals within a population might vary in their virulence towards the target pest. Another disadvantage is that they are likely to kill their hosts more slowly than when compared with synthetic chemicals, which are quick to act on pathogens. Furthermore, they require direct contact with the target species to be effective unlike some synthetic chemicals, which exhibit both contact and fumigant actions.

10.11 Conclusion

Microbial control agents include bacteria, fungi and viruses and their applications in organic farming can be in form of seed treatments, soil amendments and foliar sprays. Microbes are natural and sustainable way to control pests and diseases and are therefore environmentally friendly. As well, they can be specific to a particular pest or disease, hence, reducing harm to beneficial organisms. Moreover, their long-term efficacy cannot be overemphazised since they can persist in the environment. Also, microbes reduce the need for chemical pesticides, minimizing residues on crops. They enhance biodiversity by supporting beneficial organisms and maintaining ecosystem balance. However, developing cost-effective methods for mass-producing and formulating microbial biocontrol agents, establishing clear regulatory guidelines for the development and use of microbial agents as well as educating farmers, consumers, and policymakers about the benefits and potential of microbial agents would be of necessity.

References

Afzal S, Tariq S, Sultana V, Ara J, Ehteshamul-Haque S (2013) Managing the root diseases of Okra with endo-root plant growth promoting *Pseudomonas* and *Trichoderma viride* associated with healthy okra roots. Pak J Bot 45(4):1455–1460

Ahmad G, Khan A, Khan AA, Ali A, Mohhamad HI (2021) Biological control: a novel strategy for the control of the plant parasitic nematodes. Antonie Van Leeuwenhoek 114(7):885–912

Ahmed S, Monjil MS (2019) Effect of Paecilomyces lilacinus on tomato plants and the management of root knot nematodes: *Paecilomyces lilacinus* on tomato root knot disease. J Bangladesh Agric Univ 17(1):9–13

Alamoudi SA (2024) Using some microorganisms as biocontrol agents to manage phytopathogenic fungi: a comprehensive review. J Plant Pathol 106:3–21. https://doi.org/10.1007/s42161-023-01542-7

Al-Daghari DSS, Al-Sadi AM, Al-Mahmooli IH, Janke R, Velazhahan R (2023) Biological control efficacy of indigenous antagonistic bacteria isolated from the rhizosphere of cabbage grown in biofumigated soil against *Pythium aphanidermatum* damping-off of cucumber. Agriculture 13(3):626. https://doi.org/10.3390/agriculture13030626

Al-Ghamdi A, Al-Abbadi AA, Khan KA, Ghramh HA, Ahmed AM, Ansari MJ (2020) *In vitro* antagonistic potential of gut bacteria isolated from indigenous honey bee race of Saudi Arabia against *Paenibacillus larvae*. J Apic Res 59(5):825–833. https://doi.org/10.1080/00218839.2019.1706912

Alm SR, Villani MG, Yeh T, Shutter R (1997) *Bacillus thuringiensis* serovar. *japonensis* strain Buibui for control of Japanese and oriental beetle larvae (Coleoptera: Scarabaeidae). Appl Entomol Zool 32:477–484

Almeida JEM, Alves SB, Pereira RM (1997) Selection of *Beauveria* spp. isolates for control of the termite *Heterotermes tenuis* (Hagen, 1858). J Appl Entomol 121:539–543

Altieri MA, Nicholls CI (2003) Soil fertility management and insect pests: harmonizing soil and plant health in agroecosystems. Soil Tillage Res 72:203–211

Alves SB, Almeida JEM, Moino A Jr, Stimac JL, Pereira RM (1995) Use of *Metarhizium anisopliae* and *Beauveria bassiana* for control of *Cornitermes cumulans* (Kollar, 1832) in pastures. Ecossistema 20:50–57

Ambethgar V (2009) Potential of entomopathogenic fungi in insecticide resistance management (IRM): a review. J Biopest 2(2):177–193

Antoun H (2013) Plant-growth-promoting rhizobacteria. https://hdl.handle.net/20.500.11794/15642

Aranda S, Montes-Borrego M, Jimenez-Diaz RM, Landa BB (2011) Microbial communities associated with the root system of wild olives (*Olea europaea* L. subsp. *europaea* var. *sylvestris*) are good reservoirs of bacteria with antagonistic potential against *Verticillium dahliae*. Plant Soil 343(1):329–345. https://doi.org/10.1007/s11104-011-0721-2

Arruda B, George PB, Robin A, de LC Mescolotti D, Herrera WF, Jones DL, Andreote FD (2021) Manipulation of the soil microbiome regulates the colonization of plants by arbuscular mycorrhizal fungi. Mycorrhiza 31(5):545–558

Ayaz M, Ali Q, Farzand A, Khan AR, Ling H, Gao X (2021) Nematicidal volatiles from *Bacillus atrophaeus* GBSC56 promote growth and stimulate induced systemic resistance in tomato against *Meloidogyne incognita*. Int J Mol Sci 22:5049. https://doi.org/10.3390/ijms22095049

Ayyadurai N, Ravindra NP, Sreehari RM, Sunish KR, Samrat SK, Manohar M, Sakthivel N (2006) Isolation and characterization of a novel banana rhizosphere bacterium as fungal antagonist and microbial adjuvant in micropropagation of banana. J Appl Microbiol 100(5):926–937. https://doi.org/10.1111/j.1365-2672.2006.02863.x

Badilla F, Alves SB (1991) Control del picudo de la caña de azúcar *Sphenophorus levis* Vaurie (Col.: Curculionidae) con *Beauveria bassiana* y *Beauveria brongniartii* en condiciones de laboratorio y campo. 20–21:34–38

Bailey KL, Lazarovits G (2003) Suppressing soil-borne diseases with residue management and organic amendments. Soil Till Res 72:169–180

Bailey KL, Gossen BD, Lafond GP, Watson PR, Derksen DA (2001) Effect of tillage and crop rotation on root and foliar diseases of wheat and pea in Saskatchewan from 1991 to 1998: univariate and multivariate analyses. Can J Plant Sci 81(4):789–803

Barratt BIP, Mooed A (2005) Environmental safety of biological control: policy and practice in New Zealand. Biol Control 35(3):247–252

Bava R, Castagna F, Piras C, Musolino V, Lupia C, Palma E, Britti D, Musella V (2022) Entomopathogenic fungi for pests and predators control in beekeeping. Veterin Sci 9(2):95

Benhamou N, Garand C (2001) Cytological analysis of defense-related mechanisms induced in pea root tissues in response to colonization by nonpathogenic *Fusarium oxysporum* Fo47. Phytopathology 91(8):730–740

Berta G, Sampo S, Gamalero E, Massa N, Lemanceau P (2005) Suppression of *Rhizoctonia* root-rot of tomato by *Glomus mossae* BEG12 and *Pseudomonas fluorescens* A6RI is associated with their effect on the pathogen growth and on the root morphogenesis. Eur J Plant Pathol 111(3):270–288

Bibi F, Naseer MI, Hassan AM, Yasir M, Al-Ghamdi AAK, Azhar EI (2018) Diversity and antagonistic potential of bacteria isolated from marine grass *Halodule uninervis*. Biotech 8(1):48. https://doi.org/10.1007/s13205-017-1066-1

Blanchinsky D, Antonov J, Bercovitz A, Elad B, Feldman K, Husid A, Lazore M, Marcov N, Shamai I, Keren-Zur M (2007) Bionem WP: a unique tool for nematode control. Biological control of fungi and bacterial plant pathogens. IOBC/wprs Bull 30:23–26

Bouamri R, Dalpé Y, Serrhini MN, Bennani A (2006) Arbuscular mycorrhizal fungi species associated with rhizosphere of *Phoenix dactylifera* L. in Morocco. Afr J Biotechnol 5(6):510–516

Brannen PM, Kenney DS (1998) Kodiak Reg.-a successful biological-control product for Suppression of soil-borne plant pathogens of cotton. J Ind Microbiol Biotechnol 19:169–171

Bruin J, van der Geest L (2009) Diseases of mites and ticks. Reprinted from. Exp Appl Acarol 46(1–4), VI:350

Buba M, Salisu N (2023) Pesticidal potentials of ginger (*Zingiber officinale* Roscoe) powder on Red-rust Flour Beetle (*Tribolium castaneum* Herbst) in stored groundnut (*Arachis hypogaea* L.) in Kashere, Gombe State, Nigeria. Nigerian J Agric Food Environ 19(2&3):1–9

Buitenhuis R, Cock MJW, Colmenarez YC, De Clercq P, Edgington S, Gadaleta P, Gwynn R, Heimpel G, Hill M, Hinz HL, Hoddle MS, Jäkel T, Klapwijk JN, Leung K, Mc Kay F, Messelink GJ, Silvestri L, Smith D, Sosa A, Wäckers FL, Cabrera WG, Wyckhuys KAG, Zaviezo T (2023) Sustainable use and conservation of microbial and invertebrate biological control agents and microbial biostimulants. Background Study Paper No. 71. Commission on Genetic Resources for Food and Agriculture. FAO, Rome. https://doi.org/10.4060/cb3571en

Caldwell B, Sideman E, Seaman A, Shelton A, Smart C (2013) New York State Agricultural Experiment Station (NYSAES). Cornell University, Arnold Printing Corp, Ithaca, p 210

Caulier S, Gillis A, Colau G, Licciardi F, Liepin M, Desoignies N, Bragard C (2018) Versatile antagonistic activities of soil-borne *Bacillus* spp. and *Pseudomonas* spp. against *Phytophthora infestans* and other potato pathogens. Front Microbiol 9:143. https://doi.org/10.3389/fmicb.2018.00143

Chandler D, Davidson G, Pell JL, Ball BV, Shaw K, Sunderland KD (2000) Fungal biocontrol of Acari. Biocontrol Sci Tech 10:357–384

Chandra H, Kumari P, Bisht R, Prasad R, Yadav S (2020) Plant growth promoting *Pseudomonas aeruginosa* from *Valeriana wallichii* displays antagonistic potential against three phytopathogenic fungi. Mol Biol Rep 47(8):6015–6026. https://doi.org/10.1007/s11033-020-05676-0

Chowhan LB, Mir MI, Sabra MA, El-Habbab AA, Kumar BK (2023) Plant growth promoting and antagonistic traits of bacteria isolated from forest soil samples. Iran J Microbiol 15(2):278. https://doi.org/10.18502/ijm.v15i2.12480

Costamagna AC, Venables WN, Schellhorn NA (2015) Landscape-scale pest suppression is mediated by timing of predator arrival. Ecol Appl 25(4):1114–1130. https://doi.org/10.1890/14-1008.1

Crowder DW, Jabbour R (2014) Relationships between biodiversity and biological control in agroecosystems: current status and future challenges. Biol Control 75:8–17. https://doi.org/10.1016/j.biocontrol.2013.10.010

Damodaran T, Rajan S, Manoharan MRG, Yadav K, Kumar S, Ahmad I, Jha SK (2020) Biological management of banana *Fusarium* wilt caused by *Fusarium oxysporum* f.sp. *cubense* tropical race 4 using antagonistic fungal isolate CSR-T-3 (*Trichoderma reesei*). Front Microbiol 11:595845. https://doi.org/10.3389/fmicb.2020.595845

Daniel C, Barloggio G, Stoeckli S, Luka H, Niggli U (2018) Management of crops to prevent outbreaks. In: Vacante V, Kreiter S (eds) Handbook of Pest management in organic farming. CAB International, UK, pp 1–23

Di Francesco A, Ugolini L, Lazzeri L, Mari M (2015) Production of volatile organic compounds by *Aureobasidium pullulans* as a potential mechanism of action against postharvest fruit pathogens. Biol Control 81:8–14

DiLegge MJ, Manter DK, Vivanco JM (2019) A novel approach to determine generalist nematophagous microbes reveals *Mortierella globalpina* as a new biocontrol agent against *Meloidogyne* spp. nematodes. Sci Rep 9(1):7521. https://doi.org/10.1038/s41598-019-44010-y

Ditzler L, Van Apeldoorn DFV, Schulte RPO, Tittonell P, Rossing WAH (2021) Redefining the field to mobilize three-dimensional diversity and ecosystem services on the arable farm. Eur J Agron 122:126197

Eilenberg J, Hajek A, Lomer C (2001) Suggestions for unifying the terminology in biological control. BioControl 46:387–400

Electronic Code of Federal Regulations (ECFR) website (e-CFR) (2010) Title 7, Part 205, National Organic Program. Available at: http://ecfr.gpoaccess.gov/cgi/t/text/text-idx?c=ecfr&tpl=/ecfrbrowse/Title07/7cfr205_main_02.tpl. Accessed December 21, 2023

Elnahal ASM, El-Saadony MT, Saad AM, Desoky ESM, El-Tahan AM, Rady MM, AbuQamar SF, El-Tarabily KA (2022) The use of microbial inoculants for biological control, plant growth promotion, and sustainable agriculture: a review. Eur J Plant Pathol 162:759–792. https://doi.org/10.1007/s10658-021-02393-7

Filion M, St-Arnaud M, Jabaji-Hare SH (2003) Quantification of *Fusarium solani* f.sp. *phaseoli* in mycorrhizal bean plants and surrounding mycorrhizosphere soil using real-time polymerase chain reaction and direct isolations on selective media. Phytopathology 93(2):229–235

Finckh MR, van Bruggen AHC (2015) General introduction. In: Finckh MR, AHC VB, Tamm L (eds) Plant diseases and their management in organic agriculture. APS Press, St Paul, pp 3–11

Finke DL, Snyder WE (2008) Niche partitioning increases resource exploitation by diverse communities. Science 321(5895):1488–1490. https://doi.org/10.1126/science.1160854

Finke DL, Snyder WE (2010) Conserving the benefits of predator biodiversity. Biol Conserv 143(10):2260–2269. https://doi.org/10.1016/j.biocon.2010.03.022

Fitriana Y, Tampubolon DAT, Suharjo R, Lestari P, Swibawa IG (2022) *Lysinabacillus fusiformis* and *Paenibacillus alvei* obtained from the internal of nasutitermes termites revealed their ability as antagonist of plant pathogenic fungi. Plant Pathol J 38(5):449. https://doi.org/10.5423/PPJ.OA.03.2022.0031

Freire E, Campos V, Pinho R, Oliveira D, Faria M, Pohlit A, Noberto N, Rezende E, Pfenning L, Silva J (2012) Volatile substances produced by *Fusarium oxysporum* from coffee rhizosphere and other microbes affect *Meloidogyne incognita* and *Arthrobotrys conoides*. J Nematol 44(4):321

Gamalero E, Glick BR (2020) The use of plant growth-promoting bacteria to prevent nematode damage to plants. Biology 9(11):381

Ganiyu SA, Yassah RJ, Hamzat OTH, Popoola AR (2023) Assessing the efficacy of plant extract treatment for controlling tuber soft rot caused by *Pectobacterium carotovora* subsp. *carotovora* in sweet potato tubers. Indian Phytopathol 76:929–935. https://doi.org/10.1007/s42360-023-00657-2

Garcia I, Job D, Matringe M (2000) Inhibition of *p*-hydroxyphenylpyruvate dioxygenase by the diketonitrile of isoxaflutole:a case of half-site reactivity. Biochemistry 39(25):7501–7507

Gomiero T, Pimentel D, Paoletti MG (2011) Environmental impact of different agricultural management practices: conventional vs. organic agriculture. Crit Rev Plant Sci 30:95–124

Gomis-Cebolla J, Berry C (2023) *Bacillus thuringiensis* as a biofertilizer in crops and their implications in the control of phytopathogens and insect pests. Pest Manag Sci 79(9):2992–3001. https://doi.org/10.1002/ps.7560. Epub 2023 Jun 4. PMID: 37203311

Gowtham HG, Hariprasad P, Nayak SC, Niranjana SR (2016) Application of rhizobacteria antagonistic to *Fusarium oxysporum* f.sp. *lycopersici* for the management of *Fusarium wilt* in tomato. Rhizosphere 2:72–74. https://doi.org/10.1016/j.rhisph.2016.07.008

Greene C, Slattery E, McBride WD (2010) America's organic farmers face issues and opportunities. Amber Waves 8(2):34–39

Gupta SS (2003) Chemotactic response of plant-growth-promoting bacteria towards roots of vesicular-arbuscular mycorrhizal tomato plants. FEMS Microbiol Ecol 45(3):219–227

Haesler F, Hagn A, Frommberger M, Hertkorn N, Schmitt-Kopplin P, Munch JC, Schloter M (2008) *In vitro* antagonism of an actinobacterial *Kitasatospora* isolate against the plant pathogen *Phytophthora citricola* as elucidated with ultrahigh resolution mass spectrometry. J Microbiol Methods 75(2):188–195. https://doi.org/10.1016/j.mimet.2008.05.028

Hajek A, McManus ML, Delalibera I (2007) A review of introductions of pathogens and nematodes for classical biological control of insects and mites. Biol Control 41(1):1–13. https://doi.org/10.1016/j.biocontrol.2006.11.003

Hajek AE, Gardescu S, Delalibera I (2016) Classical biological control of insects and mites: a worldwide catalogue of pathogen and nematode introductions. FHTET-2016-06. The Forest Health Technology Enterprise Team, Morgantown

Harman GE, Bjorkman T (1998) Potential and existing uses of Trichoderma and Gliocladium for plant disease control and growth enhancement. In: Kubicek CP, Harman GE (eds). Trichoderma and Gliocladium. Taylor and Francis, London, 2:229–265

Harman GE (2010) *Trichoderma* for biocontrol of plant pathogens: from basic research to commercialized products. Cornell Community–Conference on Biological Control

Heimpel GE, Mills NJ (2017) Biological control: ecology and applications. Cambridge University Press, Cambridge, UK. https://doi.org/10.1017/9781139029117

Hilber-Bodmer M, Schmid M, Ahrens CH, Freimoser FM (2017) Competition assays and physiological experiments of soil and phyllosphere yeasts identify *Candida subhashii* as a novel antagonist of filamentous fungi. BMC Microbiol 17(1):4. https://doi.org/10.1186/s12866-016-0908-z

Hokkanen HM, Lynch JM (2003) Biological control: benefits and risks, vol 4. Cambridge University Press

Hokkanen HM, Zeng QQ, Menzler-Hokkanen INGEBORG (2003) Assessing the impacts of Metarhizium and Beauveria on bumblebees. In: Environmental impacts of microbial insecticides: need and methods for risk assessment. Springer Netherlands, Dordrecht, pp 63–71

Hong CE, Jo SH, Moon JY, Lee J-S, Kwon S-Y, Park JM (2015) Isolation of novel leaf-inhabiting endophytic bacteria in *Arabidopsis thaliana* and their antagonistic effects on phytopathogens. Plant Biotechnol Rep 9(6):451–458. https://doi.org/10.1007/s11816-015-0372-5

Islam MR, Jeong YT, Lee YS, Song CH (2012) Isolation and identification of antifungal compounds from *Bacillus subtilis* C9 inhibiting the growth of plant pathogenic fungi. Mycobiology 40(1):59–65

Jackson TA (2007) A novel bacterium for control of grass grub. In: Vincent C, Goettel MS, Lazarovits G (eds) Biological control: a global perspective. CABI Publishing, Wallingford, pp 160–168

Jin F, Ding Y, Ding W, Reddy MS, Fernando WGD, Du B (2011) Genetic diversity and phylogeny of antagonistic bacteria against *Phytophthora nicotianae* isolated from tobacco rhizosphere. Int J Mol Sci 12(5):3055–3071. https://doi.org/10.3390/ijms12053055

Jing R, Li N, Wang W, Liu Y (2020) An endophytic strain JK of genus *Bacillus* isolated from the seeds of super hybrid rice (*Oryza sativa* L. Shenliangyou5814) has antagonistic activity against rice blast pathogen. Microb Pathog 147:104422. https://doi.org/10.1016/j.micpath.2020.104422

Jonsson M, Kaartinen R, Straub CS (2017) Relationships between natural enemy diversity and biological control. Curr Opin Insect Sci 20:1–6. https://doi.org/10.10106/j.cois.2017.91.001

Junaid M, Gokce A (2024) Global agricultural losses and their causes. Bullet Biol Allied Sci Res 1:66. https://doi.org/10.54112/bbasr.v2024i1.66

Junaid JM, Dar NA, Bhat TA, Bhat AH, Bhat MA (2013) Commercial biocontrol agents and their mechanism of action in the management of plant pathogens. Int J Modern Plant Anim Sci 1(2):39–57

Kamle M, Borah R, Bora H, Jaiswal AK, Singh RK, Kumar P (2020) Systemic acquired resistance (SAR) and induced systemic resistance (ISR): role and mechanism of action against phytopathogens. In: Fungal biotechnology and bioengineering. Springer, Cham, pp 457–470

Kang S, Kim J, Joo G (2005) Isolation of antagonistic bacteria against *Fusarium oxysporum* and physicochemical properties of compost mixed with microbial formulation. Korean J Hortic Sci Technol 23(3):342–350

Kannan C, Mishra D, Rekha G, Maruthi P, Shaik H, Sundaram RM (2021) Diversity analysis of antagonistic microbes against bacterial leaf and fungal sheath blight diseases of rice. Egypt J Biol Pest Control 31(1):115. https://doi.org/10.1186/s41938-021-00462-x

Keller S (1992) The *Beauveria-Melolontha* project: experiences with regard to locust and Grasshopper control. In: Lomer CJ, Prior C (eds) Biological control of locusts and grasshoppers. CAB International, Wallingford, pp 279–286

Kilic-Ekici O, Yuen GY (2003) Induced resistance as a mechanism of biological control by *Lysobacter enzymogenes* strain C3. Phytopathology 93(9):1103–1110

Kiss L, Pintye A, Zséli G, Jankovics T, Szentiványi O, Hafez YM, Cook RT (2010) Microcyclic conidiogenesis in powdery mildews and its association with intracellular parasitism by Ampelomyces. Eur J Plant Pathol 126:445–451

Klein MG (1992) Use of *Bacillus popilliae* in scarab control. In: Jackson TA, Glare TR (eds) Use of pathogens in scarab pest management. Intercept, Andover, UK, pp 179–189

Köhl J (2022) The role of bioprotectants for disease control in integrated crop protection approaches. Burleigh Dodds Science Publishing Limited, pp 1–29. https://doi.org/10.19103/AS.2021.0093.09

Kumaranag KM, Kedar SC, Patil MD (2013) Insect pest management in organic farming systems. Popular Kheti 1(4):3

Lahlali R, Ezrari S, Radouane N, Kenfaoui J, Esmaeel Q, El Hamss H, Belabess Z, Barka EA (2022) Biological control of plant pathogens: a global perspective. Microorganisms 10(3):596

Lale NES (2002) Stored-product entomology and acarology in tropical Africa. Mole Publication, pp 15–203

Lambert PA (2005) Bacterial resistance to antibiotics: modified target sites. Adv Drug Deliv Rev 57(10):1471–1485

Latha P, Anand T, Ragupathi N, Prakasam V, Samiyappan R (2009) Antimicrobial activity of plant extracts and induction of systemic resistance in tomato plants by mixtures of PGPR strains and Zimmu leaf extract against *Alternaria solani*. Biol Control 50(2):85–93

Lazazzara V, Perazzolli M, Pertot I, Biasioli F, Puopolo G, Cappellin L (2017) Growth media affect the volatilome and antimicrobial activity against *Phytophthora infestans* in four Lysobacter type strains. Microbiol Res 201:52–62

Lee K, Jung J, Mo Y, Yoon Y, Kim S (2017) Isolation of antagonistic microbes for biological control of ginseng root rot. Planta Med Int Open 4(1):S1–S202. https://doi.org/10.1055/s-0037-1608502

Lemanceau P, Expert D, Gaymard F, Bakker PAHM, Briat JF (2009) Role of iron in plant–microbe interactions. Adv Bot Res 51:491–549

Li GH, Zhang KQ (2014) Nematode-toxic fungi and their nematicidal metabolites. In: Nematode-trapp fungi. Springer, Dordrecht, pp 313–375

Liu Y, Shi J, Feng Y, Yang X, Li X, Shen Q (2013) Tobacco bacterial wilt can be biologically controlled by the application of antagonistic strains in combination with organic fertilizer. Biol Fertil Soils 49(4):447–464. https://doi.org/10.1007/s00374-012-0740-z

Liu Y, Bai F, Li T, Yan H (2018) An endophytic strain of genus *Paenibacillus* isolated from the fruits of Noni (*Morinda citrifolia* L.) has antagonistic activity against a Noni's pathogenic strain of genus *Aspergillus*. Microb Pathog 125:158–163. https://doi.org/10.1016/j.micpath.2018.09.018

Lumsden RD, Knauss JF (2007) Commercial development of *Gliocladium virens* for damping-off disease. In: Vincent C, Goettel MS, Lazarovits G (eds) Biological control: a global perspective. CAB International, Kew, pp 203–210

Mamphogoro TP, Kamutando CN, Maboko MM, Aiyegoro OA, Babalola OO (2021) Epiphytic bacteria from sweet pepper antagonistic *in vitro* to *Ralstonia solanacearum* bd 261, a causative agent of bacterial wilt. Microorganisms 9(9):1947. https://doi.org/10.3390/microorganisms9091947

Mohan S, Devasenapathy P, Vennila C, Gill MS (2013) IPM Booklet for Pest and disease management: organic ecosystem. pp 21–22

Naraghi L, Heydari A, Rezaee S, Razavi M, Jahanifar H, Khaledi EM (2010) Biological control of tomato Verticillium wilt disease by *Talaromyces flavus*. J Plant Protect Res 50(3):360–365

Negi R, Sharma B, Kaur S, Kaur T, Khan SS, Kumar S, Ramniwas S, Rustagi S, Singh S, Rai AK, Kour D, Thakur N, Yadav AN (2023) Microbial antagonists: diversity, formulation and applications for management of pest–pathogens. Egypt J Biol Pest Control 33:105–126. https://doi.org/10.1186/s41938-023-00748-2

Neumeister L (2010) Climate change and crop protection-anything can happen. Pesticide Action Network Asia and the Pacific (PAN AP), Malasia, pp 1–42

O'Brien PA (2017) Biological control of plant diseases. Australas Plant Pathol 46:293–304. https://doi.org/10.1007/s13313-017-0481-4

Oi DH, Williams DF (2002) Impact of *Thelohania solenopsae* (Microsporidia: Thelohaniidae) on polygyne colonies of red imported fire ants (Hymenoptera: Formicidae). J Econ Entomol 95:558–562

Oskiera M, Szczech M, Stępowska A, Smolińska U, Bartoszewski G (2017) Monitoring of *Trichoderma* species in agricultural soil in response to application of biopreparations. Biol Control 113:65–72

Pandey A, Trivedi P, Kumar B, Palni LMS (2006) Characterization of a phosphate solubilizing and antagonistic strain of *Pseudomonas putida* (B0) isolated from a sub-alpine location in the Indian central Himalaya. Curr Microbiol 53(2):102–107. https://doi.org/10.1007/s00284-006-4590-5

Pandit MA, Kumar J, Gulati S, Bhandari N, Mehta P, Katyal R, Rawat CD, Mishra V, Kaur J (2022) Major biological control strategies for plant pathogens. Pathogens 11(2):273

Pane C, Zaccardelli M (2015) Evaluation of *Bacillus* strains isolated from solanaceous phylloplane for biocontrol of *Alternaria* early blight of tomato. Biol Control 84:11–18. https://doi.org/10.1016/j.biocontrol.2015.01.005

Pane C, Villecco D, Campanile F, Zaccardelli M (2012) Novel strains of *Bacillus*, isolated from compost and compost-amended soils, as biological control agents against soil-borne phytopathogenic fungi. Biocontrol Sci Tech 22(12):1373–1388. https://doi.org/10.1080/09583157.2012.729143

Panebianco S, Vitale A, Polizzi G, Scala F, Cirvilleri G (2015) Enhanced control of postharvest citrus fruit decay by means of the combined use of compatible biocontrol agents. Biol Control 84:19–27

Parkman JP, Smart GC (1996) Entomopathogenic nematodes, a case study: introduction of *Steinernema scapterisci* in Florida. Biocontrol Sci Tech 6:413–419

Pell JK, Hannam JJ, Steinkraus DC (2010) Conservation biological control using fungal entomopathogens. BioControl 55(1):187–198. https://doi.org/10.1007/s10526-009-9245-6

Pfleger FL, Linderman RG (2000) Mycorrhiza and plant growth. The American Phytopathological Society Press, St. Paul, p 360

Pieterse CM, Zamioudis C, Berendsen RL, Weller DM, Van Wees SC, Bakker PA (2014) Induced systemic resistance by beneficial microbes. Annu Rev Phytopathol 52:347–375

Prabhu S, Kumar S, Subramanian S, Kumar PS (2008) Mass production and commercial formulation of Paecilomyces lilacinus. Indian J Nematol 38(2):131–133

Puopolo G, Cimmino A, Palmieri MC, Giovannini O, Evidente A, Pertot I (2014) *Lysobacter capsici* AZ78 produces cyclo (l-Pro-l-Tyr), a 2, 5-diketopiperazine with toxic activity against sporangia of *Phytophthora infestans* and *Plasmopara viticola*. J Appl Microbiol 117(4):1168–1180

Puopolo G, Tomada S, Pertot I (2018) The impact of the omics era on the knowledge and use of Lysobacter species to control phytopathogenic micro-organisms. J Appl Microbiol 124(1):15–27

Raaijimakers JM, Vlami M, de Souza JT (2002) Antibiotic production by bacterial biocontrol agents. Antonie van Leeuwenkoek 81(1–4):537–547

Rae R, Iatsenko I, Witte H, Sommer RJ (2010) A subset of naturally isolated *Bacillus* strains show extreme virulence to the free-living nematodes *Caenorhabditis elegans* and *Pristionchus pacificus*. Environ Microbiol 12(11):3007–3021. https://doi.org/10.1111/j.1462-2920.2010.02278.x

Ramle M, Wahid MB, Norman K, Glare TR, Jackson TA (2005) The incidence and use of *Oryctes* virus for control of rhinoceros beetle in oil palm plantations in Malaysia. J Invertebr Pathol 89:85–90

Rath AC (2002) Ecology of entomopathogenic fungi in field soils. In: Proceedings of the meeting of the society for invertebrate pathology, Iguassu Falls, Brazil, pp 65–71

Rey P, Le Floch G, Benhamou N, Tirilly Y (2008) *Pythium oligandrum* biocontrol: its relationships with fungi and plants. In: Plant-microbe interactions, pp 43–57

Rinu K, Sati P, Pandey A (2014) *Trichoderma gamsii* (NFCCI 2177): a newly isolated endophytic, psychrotolerant, plant growth promoting, and antagonistic fungal strain. J Basic Microbiol 54(5):408–417. https://doi.org/10.1002/jobm.201200579

Roy HE, Steinkraus DC, Eilenberg J, Hajek AE, Pell JK (2006) Bizarre interactions and endgames: entomopathogenic fungi and their arthropod hosts. Annu Rev Entomol 51:331–357

Ryder MH, Jones DA (1990) Biological control of crown gall. In: Hornby D (ed) Biological control of soil-borne plant pathogens. CAB International, Wallingford, pp 45–63

Samac DA, Willert AM, McBride MJ, Kinkel LL (2003) Effects of antibiotic-producing Streptomyces on nodulation and leaf spot in alfalfa. Appl Soil Ecol 22:55–66

Santamarina MP, Rosello J, Llacer R, Sanchis V (2002) Antagonistic activity of *Penicillium oxalicum* Corrie and Thom, *Penicillium decumbens* Thom and *Trichoderma harzianum* Rifai isolates against fungi, bacteria and insects *in vitro*. Rev Iberoam Micol 19(2):99–103

Schjonning P, Elmholt S, Munkholm LJ, Debosz K (2002) Soil quality aspects of humid sandy loams as influenced by organic and conventional long-term management. Agric Ecosyst Environ 88:195–214

Schultz AN, Lucardi RD, Marisco TD (2019) Successful invasions and failed biocontrol: the role of antagonistic species interactions. Bioscience 69:711–724. https://doi.org/10.1093/biosci/biz075

Schwarzländer M, Hinz HL, Winston R, Day M (2018) Biological control of weeds: an analysis of introductions, rates of establishment and estimates of success, worldwide. BioControl 63:319–331. https://doi.org/10.1007/s10526-018-9890-8

Shannon D, Sen AM, Johnson DB (2002) A comparative study of the microbiology of soils managed under organic and conventional regimes. Soil Use Manag 18:274–283

Sharma M, Tarafdar M, Ghosh R, Gopalakrishanan S (2017) Biological control as a tool for eco-friendly management of plant pathogens. In: Adhya TK et al (eds) Advances in soil microbiology: recent trends and future prospects, microorganisms for sustainability 4. Springer Nature Singapore Pvt Ltd. https://doi.org/10.1007/978-981-10-7380-9_8

Sharon E, Bar-Eyal M, Chet I, Herrera-Estrella A, Kleifeld O, Spiegel Y (2001) Biological control of the root-knot nematode *Meloidogyne javanica* by *Trichoderma harzianum*. Phytopathology 91(7):687–693

Sheppard AW, Paynter Q, Mason P, Murphy S, Stoett P, Cowan P, Brodeur J, Warner K, Villegas C, Shaw R, Hinz H (2019) IUCN SSC invasive species specialist group. In: The application of biological control for the management of established invasive alien species causing environmental impacts. The secretariat of the convention on biological diversity technical series (No. 91). Secretariat of the Convention on Biological Diversity, Montreal

Shi J-F, Sun C-Q (2017) Isolation, identification, and biocontrol of antagonistic bacterium against *Botrytis cinerea* after tomato harvest. Braz J Microbiol 48:706–714. https://doi.org/10.1016/j.bjm.2017.03.002

Singh P, Khan A, Kumar R, Kumar R, Singh VK, Srivastava A (2020) Recent developments in siderotyping: procedure and application. World J Microbiol Biotechnol 36:1–15

Solanki MK, Solanki AC, Rai S, Srivastava S, Kashyap BK, Divvela PK, Kumar S, Yandigeri MS, Kashyap PL, Shrivastava AK, Ali B, Khan S, Jaremko M, Qureshi KA (2022) Functional interplay between antagonistic bacteria and *Rhizoctonia solani* in the tomato plant rhizosphere. Front Microbiol 13:990850. https://doi.org/10.3389/fmicb.2022.990850

Soytong K, Kanokmadhakul S, Kukongviriyapa V, Isobe M (2001) Application of *Chaetomium* species (Ketomium®) as a new broad-spectrum biological fungicide for plant disease control: a review article. Fungal Divers 7:1–15

Stenberg JA, Sundh I, Becher PG, Björkman C, Dubey M, Egan PA, Friberg H, Gil JF, Jensen DF, Jonsson M, Karlsson M (2021) When is it biological control? A framework of definitions, mechanisms, and classifications. J Pest Sci 94(3):665–676. https://doi.org/10.1007/s10340-021-01354-7

Stewart A, McLean KL (2007) Biological control of onion white rot. In: Mukerji KG, Chincholkar SB (eds) Biological control of plant disease. The Haworth Press, New York, pp 123–149

Stewart A, Brownbridge M, Hill R, Jackson T (2010) Utilizing soil microbes for biocontrol. In: Dixon R, Tilston E (eds) Soil microbiology and sustainable crop production. Springer, Berlin/Heidelberg, pp 315–371

Sulistyanto D, Ehlers RU (1996) Efficacy of the entomopathogenic nematodes *Heterorhabditis megidis* and *Heterorhabditis bacteriophora* for the control of grubs (*Phyllopertha horticola* and *Aphodius contaminatus*) in golf course turf. Biocontrol Sci Tech 6:247–250

Thrane C, Nielsen TH, Nielsen MN, Sørensen J, Olsson S (2000) Viscosinamide-producing Pseudomonas fluorescens DR54 exerts a biocontrol effect on Pythium ultimum in sugar beet rhizosphere. FEMS Microbiol Ecol 33(2):139–146

Timper P (2014) Conserving and enhancing biological control of nematodes. J Nematol 46(2):75–89. PMID: 24987159; PMCID: PMC4077175

Tjamos EC, Papavizas GC, Cook RJ (2013) Biological control of plant diseases: progress and challenges for the future, vol 230. Springer Science and Business Media

Tuck SL, Winqvist C, Mota F, Ahnström J, Turnbull LA, Bengtsson J (2014) Land-use intensity and the effects of organic farming on biodiversity:a hierarchical meta-analysis. J Appl Ecol 51:746–755

Twizeyimana M, Hammer PE, Gachango E, Craig K, Espejo B, Biggs MB, Kremer J, Ingham DJ (2023) Diverse environmental bacteria displaying activity against *Phakopsora pachyrhizi*, the cause of soybean rust. Front Plant Sci 14:1080116. https://doi.org/10.3389/fpls.2023.1080116

Usta C (2013) Microorganisms in biological Pest control-a review (bacterial toxin application and effect of environmental factors). In: Current progress in biological research. IntechOpen, pp 287–318. https://doi.org/10.5772/55786

Vacante V, Bonsignore CP (2018) Natural enemies and Pest control. In: Vacante V, Kreiter S (eds) Handbook of Pest management in organic farming. CAB International, UK

van Bruggen AHC, Finckh MR (2015) General principles of organic plant production. In: Finckh MR, van Bruggen AHC, Tamm L (eds) Plant diseases and their management in organic agriculture. APS Press, St Paul, pp 15–23

van Bruggen AHC, Finckh MR (2016) Plant diseases and management approaches in organic farming systems. Annu Rev Phytopathol 54:25–54

van Bruggen AH, Gamliel A, Finckh MR (2016) Plant disease management in organic farming systems. Pest Manag Sci 72(1):30–44

van der Geest LPS (2010) IPM potentials of microbial pathogens and diseases of mites. In: Ciancio A, Mukerji KG (eds) Integrated management of arthropod pests and insects borne diseases, vol 5. Springer Science+Business Media B.V, Dordrecht, pp 249–308

van der Geest LPS, Elliot SL, Breeuwer JAJ, Beerling EAM (2000) Diseases of mites. Exp Appl Acarol 24:497–560

van Lenteren JC, Bolckmans K, Köhl J, Ravensberg WJ, Urbaneja A (2018) Biological control using invertebrates and microorganisms:plenty of new opportunities. BioControl 63:39–59. https://doi.org/10.1007/s10526-017-9801-4

Vercambre B, Goebel O, Riba G, Marchal M, Neuvéglise C, Ferron P (1994) Success in microbial control of a soil pest, *Hoplochelus marginalis*, in Reunion Island; choice of a suitable fungus. In: Proceedings of the VIth international colloquium on invertebrate pathology and microbial control. SIP, Montpellier, France, pp 283–288

Verma VC, Gond SK, Kumar A, Mishra A, Kharwar RN, Gange AC (2009) Endophytic actinomycetes from *Azadirachta indica* A. Juss.: isolation, diversity, and anti-microbial activity. Microb Ecol 57(4):749–756. https://doi.org/10.1007/s00248-008-9450-3

Verma RK, Sankhla MS, Kesarwani S, Parihar K, Sonone SS, Jadhav EB, Bhardwaj L (2022) Role of organic crop production systems in biological diseases as control agents. Biointerf Res Appl Chem 12(3):3681–3695

Vinale F, Sivasithamparam K, Ghisalberti EL, Marra R, Woo SL, Lorito M (2008) Trichoderma–plant–pathogen interactions. Soil Biol Biochem 40(1):1–10

Viswanathan R, Sundar AR, Premkumari SM (2003) Mycolytic effect of extracellular enzymes of antagonistic microbes to *Colletotrichum falcatum*, red rot pathogen of sugarcane. World J Microbiol Biotechnol 19(9):953–959. https://doi.org/10.1023/B:WIBI.0000007329.68910.e5

Vurro M (2007) Benefits and risks of using fungal toxins in biological control. In: Novel biotechnologies for biocontrol agent enhancement and management. Springer, Netherlands, pp 53–74

Wallace J, Hammermeister A, Geldart E (2021) Biological pest control. Organic Agriculture Centre of Canada, Dalhousie University, Truro, N. S, pp 1–12

Wan M, Li G, Zhang J, Jiang D, Huang HC (2008) Effect of volatile substances of Streptomyces platensis F-1 on control of plant fungal diseases. Biol Control 46(3):552–559

Wang B, Yuan J, Zhang J, Shen Z, Zhang M, Li R, Ruan Y, Shen Q (2013) Effects of novel bioorganic fertilizer produced by *Bacillus amyloliquefaciens* W19 on antagonism of *Fusarium wilt* of banana. Biol Fertil Soils 49(4):435–446. https://doi.org/10.1007/s00374-012-0739-5

Wang X, Li Q, Sui J, Zhang J, Liu Z, Du J, Xu R, Zhou Y, Liu X (2019) Isolation and characterization of antagonistic bacteria *Paenibacillus jamilae* HS-26 and their effects on plant growth. Biomed Res Int. https://doi.org/10.1155/2019/3638926

Whipps J (2001) Microbial interactions and biocontrol in the rhizosphere. J Exp Bot 52:487–511

Woltz JM, Landis DA (2013) Coccinellid immigration to infested host patches influences suppression of *Aphis glycines* in soybean. Biol Control 64(3):330–337. https://doi.org/10.1016/j.biocontrol.2012.11.012

Wright MG, Bennett GM (2018) Evolution of biological control agents following introduction to new environments. BioControl 63:105–116

Wyss E (2011) African organic agriculture training manual Module 04 Pest, desease and weed management. FiBL, Research Institute of Organic Agriculture, Switzerland, p 33

Yadav K, Damodaran T, Dutt K, Singh A, Muthukumar M, Rajan S, Gopal R, Sharma PC (2021) Effective biocontrol of banana fusarium wilt tropical race 4 by a *Bacillus* rhizobacteria strain with antagonistic secondary metabolites. Rhizosphere 18:100341. https://doi.org/10.1016/j.rhisph.2021.100341

Zaviezo T, Grez A, Miall J, Mason P (2021) Conservation biological control. In: Mason PG (ed) Biological control: global impacts, challenges and future directions of pest management. CSIRO Publishing, Clayton South, pp 37–66. https://doi.org/10.1071/9781486309351

Zubair M, Farzand A, Mumtaz F, Khan AR, Sheikh TMM, Haider MS, Yu C, Wang Y, Ayaz M, Gu Q, Gao X (2021) Novel genetic dysregulations and oxidative damage in *Fusarium graminearum* induced by plant defense eliciting psychrophilic *Bacillus atrophaeus* TS1. Int J Mol Sci 22(22):12094. https://doi.org/10.3390/ijms222212094

Part IV
Microbial Innovations in Food Systems and Circular Economy

Chapter 11
Microbial Applications in Indigenous and Traditional Food Systems

Adewale Olusegun Obadina and Olusola Bandele Oyewole

Abstract Cultural heritage, biodiversity, and ecological sustainability form the foundations of indigenous and traditional food systems. These systems are intrinsically linked to local ecosystems and resources, incorporating microorganisms into the creation, storage, and enrichment of food products. Microorganisms play a vital role in food production, preservation, and improvement, through processes, such as fermentation, biopreservation, and bioprocessing. The utilization of these microorganisms in food systems contributes to improved food security, nutritional value, and health benefits for the native populations. This chapter delves into the significance of microorganisms in established food processing methods, such as fermentation, biopreservation, and bioprocessing. These techniques play a crucial role in enhancing food safety, increasing product longevity, and improving nutritional value. Probiotic microorganisms and bioactive substances in fermented foods promote digestive health, whereas biopreservation methods help reduce food spoilage and minimize waste. These practices are not only culturally significant but also have economic and environmental benefits, offering sustainable alternatives and helping to maintain biodiversity. Integrating traditional knowledge with modern science can foster innovation, thus making these practices relevant to contemporary global food challenges.

Keywords Fermented food · Probiotic microorganism · Food safety · Nutritional values · Health benefits

---

A. O. Obadina (✉) · O. B. Oyewole
Department of Food Science and Technology, Federal University of Agriculture, Abeokuta, Nigeria

O. O. Babalola et al. (eds.), *Ecofriendly Frontiers*,
https://doi.org/10.1007/978-3-031-98700-7_11

11.1 Introduction

Microbes, some of the earliest life forms, exist for approximately 3.8 billion years. Those associated with specific microorganisms form what is known as the microbiome, as defined by Martínez-Espinosa (2020) and Poli et al. (2017). These microbiomes are crucial to the physiological processes of the host and, support overall health. In food-producing organisms, they significantly affect both the yield and product quality. Recognized as one of the most widespread and enduring life forms on Earth, microbiomes are integral to various ecological systems (Timmis et al. 2017; Cavicchioli et al. 2019). Fermented foods are a major component of global consumption, with approximately one-third of all foods and beverages involved in fermentation processes. This includes over 5000 distinct products essential to global food systems (Vogel et al. 2011). Fermentation typically involves controlled microbial growth and enzymatic conversion of food components to enhance the nutritional, sensory, and functional attributes of the final products (Marco et al. 2017). A wide range of raw materials such as vegetables, cereals, legumes, bamboo shoots, milk, tubers, fish, and meat are used in fermentation. These processes rely on diverse microbes, including bacteria, yeasts, and molds, from various genera.

The role of microorganisms in food production was first recognized in 1837 with the discovery of the function of yeast in alcohol fermentation (Maicas 2020). Louis Pasteur later expanded this understanding in the 1860s by, attributing beer and vinegar production to microbial activity (Daly et al. 1996; Raman et al. 2014). The 1850s marked a turning point in food fermentation, with the Industrial Revolution prompting large-scale food production to meet the needs of the growing urban populations. Simultaneously, the emergence of microbiology as a scientific discipline has deepened the understanding of the biological mechanisms behind fermentation (Enthoven and Broeck 2021; Opal 2009). Through intentional microbial inoculation, raw materials are transformed into preserved foods with distinctive flavors including kimchi, sauerkraut, miso, tempeh, ogi, and kenkey (Mannaa et al. 2021; Capozzi et al. 2017; Siddiqui et al. 2023). Fermentation enhances food safety, shelf life, nutritional value, and sensory appeal, while also making certain inedible substances consumable and reducing energy demands during preparation (Dimidi et al. 2019; Mannaa et al. 2021). These techniques are central to many traditional and ceremonial dishes, particularly in Africa.

In developing regions, food systems predominantly depend on locally sourced resources, that are cultivated, processed, and consumed within nearby communities. These local foods are vital to both urban and rural diets Waltz 2010). A food system encompasses all processes and institutions involved in food production, processing, distribution, and consumption, and is responsible for managing the associated social, economic, and environmental impacts (Voidarou et al. 2020; Ruben et al. 2019). Efforts to address global hunger are increasingly adopting a Food Systems Approach (FSA), which emphasizes improving agricultural production, food processing, and nutritional outcomes (Berkum et al. 2018; Global Panel 2016; HLPE 2017; Keding et al. 2013). However, ensuring adequate nutrition and food safety remains a persistent global challenge (Waltz 2010). The widespread use of fermentation across cultures has fostered broad acceptance of microbes in food. The modern shift toward natural foods and bio-preservation over synthetic additives continues to boost consumer appreciation for fermented products (Pakseresht et al. 2017; Doyle et al. 2013).

11.2 Importance of Microbial Applications in Indigenous and Traditional Food Systems

11.2.1 Food Safety and Preservation

Indigenous cultures have long used microbial processes to preserve food without modern refrigeration or chemical additives. Fermentation, pickling, and curing utilize beneficial microorganisms to inhibit the growth of harmful bacteria and extend the shelf life of food (Behera et al. 2020). In particular, fermentation enhances acidity and/or produces alcohol, creating an environment that suppresses harmful microorganisms and increases the longevity of unprocessed materials (Siddiqui et al. 2023). This has led to safer products with extended storage times, as supported by various studies (Nout et al. 2005, 2007). Additionally, during fermentation, microorganisms can release compounds such as bacteriocins, which further inhibit harmful bacteria (Nout et al. 2005).

The fermentation of perishable raw materials, such as fresh milk, is especially advantageous (Schoustra et al. 2022). It also helps minimize food waste by preserving the products for longer periods. For example, Ogi, a Nigerian fermented maize product effectively ensures microbiological safety, with an unrefrigerated shelf life of approximately 7 days (Moonga et al. 2020). Other traditional West African fermented foods, such as Ogiri and Dawadawa, share similar microbial ecologies during processing, and are expected to offer comparable levels of safety and shelf life.

Many traditional fermented foods are rich in probiotic microorganisms, which offer health benefits such as improved gut health, enhanced digestion, and a strengthened immune system (Kechagia et al. 2013). These beneficial microbes are present in foods such as yogurt, kefir, kombucha, and fermented vegetables. Recent research has also highlighted those various African fermented foods contain probiotics and beneficial bacteria that promote overall well-being. However, the origin of the spontaneous microorganisms involved in fermentation remains unclear, and the probiotic characteristics of many locally produced foods are still being explored.

11.2.2 Nutritional Enhancement

Research has shown that fermented foods generally have superior nutritional content compared with their unfermented counterparts (Hasan et al. 2014). Microbial activity during fermentation enhances nutritional value. Sharma et al. (2020) identified three specific microbial fermentation methods for improving nutrition. Microorganisms use catabolic and anabolic processes to decompose complex molecules and generate essential vitamins and growth factors (Kennedy 2016). The extraction of nutrients from indigestible plant components is crucial, especially for certain grains and seeds. Milling breaks down hemicellulosic and cellulosic structures surrounding the endosperm, which is rich in proteins and digestible carbohydrates, to enable nutrient extraction. In less industrialized regions, basic milling techniques are used to obtain nutrients from plants (Lovegrove et al. 2015). However, this method often fails to extract all nutritional benefits, as some nutrients remain bound and inaccessible even after cooking (Nkhata et al. 2018). Certain fungi, bacteria, and yeasts can overcome this by breaking down the indigestible protective layers and cellular boundaries (Hasan et al. 2014).

Individuals with limited financial resources often rely on grain-based foods, which can lack adequate nutrition. Fermentation with yeast and Lactic Acid Bacteria (LAB) can enhance the nutritional quality and digestibility of foods. Mokoena (2005) found that fermentation benefits from increased microbial enzyme activity in an acidic environment at temperatures between 22 and 25 °C. During this process, enzymatic hydrolysis reduces anti-nutrients such as phytic acid and tannins through phytase activity, producing nutrients such as simple polysaccharides, proteins, iron, and fatty acids more accessible (Sharma et al. 2020). Microorganisms secrete hydrolytic enzymes that break down complex molecules into absorbable forms, which are then metabolized into secondary metabolites and structural components that are essential for microbial growth (Song et al. 2022).

Lactic acid fermentation is an age-old process that involves various bacterial strains that convert carbohydrates into lactic acid. It plays a crucial role in the manufacture of diverse food products such as dairy, meat, sauerkraut, bread, and silage (Zapaśnik et al. 2022), and is widely used in the global food industry. Traditional fermentation practices across regions have utilized lactic acid bacteria for the solid-state fermentation of legumes, cereals, and starchy substances (Fukushima 1985). As fermentation progresses, lactic acid builds up, reducing dry matter yield and increasing acidity. Research suggests that fermented legumes tend to have higher pH values than other substrates, likely because of their elevated protein content (Zamora and Fields 1979; Paredes-López et al. 1991).

Certain nutritious foods may also contain anti-nutritional factors such as phytic acid, trypsin inhibitors, flatus factors, and lectins, which interfere with digestion and nutrient absorption (Singh et al. 2023; Salim et al. 2023). For instance, phytic acid binds to minerals and enzymes, thereby limiting their bioavailability. Lectins attach similarly to the intestinal walls and, disrupt nutrient uptake. However, fermentation helps degrade these compounds via microbial hydrolysis, enhancing nutrient assimilation.

The use of appropriate starter cultures for fermentation improves both nutritional content and flavor. Flavor development results from biochemical reactions initiated by microbial enzymes and metabolites (Sharma et al. 2020). These compounds enhance taste and sensory quality, often boosting commercial appeal (Thierry et al. 2015). Erkus et al. (2013), Thierry et al. (2015), and Phiri et al. (2019) showed that diverse microbial communities produce a broader range of aromas. Traditional fermentation methods often yield richer and more nuanced flavors than industrial processes. Materia et al. (2021) found that consumers prefer traditionally fermented foods. These methods are rooted in indigenous food systems, preserve cultural heritage, and foster a sense of communal identity (Swiderska et al. 2022).

11.2.3 The Ethnobiological and Cultural Importance of Fermented Foods

Fermented foods have been essential to human diets for millennia, offering more than nutrition by holding significant ethnobiological and cultural roles (Leeuwendaal et al. 2022). Microbial metabolic processes have produced items that influenced the traditions, healthcare practices, and cultural identities of indigenous communities globally. Fermented foods can be examined from various angles to comprehend their significance, including their roles in food preservation, biodiversity conservation, health enhancement, and cultural tradition maintenance (Cuamatzin-García et al. 2022).

These studies illustrate the intricate relationship between native cultures and their environments, as fermentation methods are tied to the distinct microbial landscapes of each region. The mastery of traditional fermentation requires an understanding of microbes thriving in specific climates. This expertise enables communities to preserve food and enhance its nutritional value (Leeuwendaal et al. 2022). Many societies view fermentation not just as a preservation method but also as a means to improve health. Microbial fermentation produces biologically active substances such as B vitamins, antimicrobial peptides, and antioxidants, which are vital for health, especially in resource-limited settings (Dimidi et al. 2019).

In addition to their biological benefits, fermented foods are embedded in cultural practices and social structures. They hold symbolic importance in ceremonies, festivities, and expressions of cultural identity, with specific fermentation methods closely tied to local customs and historical memories (Cuamatzin-García et al. 2022). In Africa, Asia, and Latin America, fermented foods are integral to social gatherings, spiritual events, community-building activities, and blending of production and consumption into these occasions (LeBlanc et al. 2024).

Fermented foods also aid in the conservation of biodiversity, particularly in area where traditional culinary practices are vital. The microbial communities involved are often unique to local environments, producing distinctive products that cannot be replicated by industrial methods cannot replicate (Valentino et al. 2024). These microorganisms make traditional fermentation crucial for preserving the cultural heritage and microbial ecosystems (Lugo-Morin 2020). As global food systems move toward uniformity and synthetic methods, protecting microbial communities is critical for maintaining cultural uniqueness and ecological diversity.

11.2.4 Economic and Social Aspects of Microbially Fermented Indigenous Foods

Fermented indigenous foods are vital to the socioeconomic fabric of many societies, especially those where traditional food preparation methods have been passed down through generations (Cuamatzin-García et al. 2022). Products such as yogurt, kimchi, tempeh, sourdough bread, and fermented fish result from microbial activity and often utilize local resources and labor, thereby supporting regional agriculture and artisanal industries. Traditional beverages such as chicha in South America and tella in Ethiopia benefit small-scale farmers (Tafere 2015). Manual and scalable fermentation methods create job opportunities, especially for women and marginalized groups, and promote sustainable livelihoods (Adesulu-Dahunsi et al. 2022). For instance, tempeh production in Indonesia provides substantial employment, enhancing the economic well-being of both the rural and urban populations.

The global demand for traditional fermented foods is rising owing to interest in their unique flavors, health benefits, and probiotic properties. For example, Korean kimchi and Japanese miso, have achieved international acclaim, positively impacting their respective economies (Wicaksono et al. 2024). Fermented foods also address nutritional deficiencies among low-income populations by enhancing access to proteins and nutrients, such as vitamin B12, and combating malnutrition (Melini et al. 2019). They are integral to cultural and social identities, reflecting local traditions and strengthening community bonds, as seen in the preparation of ogi, a West African fermented cereal pudding (Cuamatzin-García et al. 2022).

The health benefits of fermented foods include improved digestive health and immune function. Studies indicate that fermented dairy consumption in African and Asian communities reduces lactose intolerance and enhances nutrition (Valentino et al. 2024). Traditional fermentation practices also preserve cultural identity and culinary knowledge through skill transfer (D'Andrea and D'Ulizia 2023). Furthermore, fermentation extends the shelf life of perishable foods, boosting food security in regions that lack refrigeration (Terefe 2016).

Global interconnectedness has significantly affected the production, use, and cultural relevance of indigenous fermented foods. These foods, deeply rooted in local cultural practices and ecosystems, now face pressure from international trade, marketing, and global taste convergence (Mendez and Popkin 2004). While globalization increases awareness and consumption of fermented foods such as kimchi, miso, and kefir (Siddiqui et al. 2023; Muhialdin et al. 2021), it also poses challenges. Large-scale commercialization may lead to homogenization and loss of traditional microbial strains and preparation methods essential for authentic flavor and nutrition (Voidarou et al. 2020; Akinola et al. 2020). As a result, the cultural significance of these foods may be diluted, reducing them to commercial products stripped of their original meanings (Cuamatzin-García et al. 2022).

Indigenous fermented foods contain diverse microbial communities with valuable applications in food production, pharmaceuticals, agriculture, and environmental management (Valentino et al. 2024). Distinctive strains such as *Lactobacillus*, *Saccharomyces*, and *Aspergillus* exhibit resilience and the ability to produce beneficial compounds such as enzymes, vitamins, and antimicrobial peptides (Sharifi-Rad et al. 2020; Dimidi et al. 2019). Lactic acid bacteria from these foods are used to develop probiotics, natural preservatives, and flavor enhancers to improve food quality and shelf life (Zapaśnik et al. 2022). Microbial metabolites such as exopolysaccharides and bioactive peptides offer antioxidant, anti-inflammatory, and anticarcinogenic properties (Rani et al. 2021). Additionally, microbial cultures from fermented foods are being explored for waste processing and biofertilizer production to support sustainable agriculture through organic matter decomposition and plant growth promotion (Sharma and Tong 2025).

11.2.5 Biodiversity Preservation

Beverages and fermented foods have been staples of the human diet for a considerable period, serving vital nutritional, cultural, and economic roles across human societies. Although the functional and microbial varieties of most fermented foods are still largely unknown, they have ancient roots, are widely present, and offer notable health benefits. Recently, there has been growing interest in the emergence of tools to study microbial diversity in these foods.

Although lifestyle changes have contributed to the decline of some fermented foods, possibly impacting gastrointestinal health, they continue to be significant in many cultures. Indigenous communities often preserve traditional food practices involving locally sourced ingredients and wild microbial strains, helping to conserve biodiversity and maintain an ecological balance. Fermented foods are also crucial for preserving various microorganisms within the diet and are produced through controlled microbial growth and food component transformation, as described by. Archaeological evidence has confirmed the global prevalence of fermentation.

Indigenous food production methods are generally more environmentally friendly than industrialized food systems. Currently, the global food system faces two major challenges: operating within ecological limits to ensure sustainability and mitigate climate change while remaining economically efficient (Springmann et al. 2018), and providing health-promoting diets, particularly in regions resistant to genetically modified foods due to cultural or health concerns.

A viable solution is the expansion of the food fermentation technology. Fermentation is a biological process in which microorganisms, primarily bacteria or fungi, derive energy from carbohydrate-rich substances, supporting growth and metabolite production that is beneficial to humans (Paramithiotis et al. 2022). During fermentation, microbial populations such as temperature and pH are monitored to maintain optimal conditions. In spontaneous fermentations involving cereals, pulses, medicinal plants, and dairy products, lactic acid bacteria (LAB)—such as Streptococcus, Leuconostoc, Lactococcus, and Lactobacillus—typically dominate. This group has been widely studied by researchers including, with lactic acid production as a key trait.

In alcoholic fermentation, yeasts like *Saccharomyces cerevisiae* are the primary agents, whereas acetic acid fermentation involves fungi from the genera *Gluconobacter* and *Acetobacter*. Cheese fermentation, is characterized by the presence of *Penicillium*, *Propionibacterium* spp., and LAB. Fermentation and other microbial processes are environmentally advantageous because of their low energy demands and minimal waste generation.

Indigenous food systems often exhibit greater resilience to climate change because of their diversity and adaptability. Microbial applications contribute to this resilience by offering preservation and storage methods that are suited to adverse conditions. They are vital not only for food preservation and flavor but also for cultural heritage, biodiversity, and sustainable production.

Tailoring microbial processes to local ingredients allows communities to continue producing nutritious food despite the shifting environmental conditions. Fermented foods are particularly beneficial as they do not require refrigeration, which is a critical advantage in areas with unreliable energy access owing to climate impacts (Steinkraus 1997).

Ultimately, fermented foods strengthen food systems by extending shelf life, enhancing nutrition, promoting local production, and reducing refrigeration requirement. Localized microbial applications reduce dependency on global food chains, fostering food sovereignty and resilience (FAO 2011).

11.3 Future Directions in Microbial Research in Indigenous Foods

Future research should aim to deepen our understanding of microbes, use them sustainably, and tackle ethical bioprospecting, scalability, and technological challenges. The full extent of microbial diversity in traditional fermented foods remains largely unexplored, primarily because of the limitations of conventional culture-based methods. However, the advent of sophisticated omics technologies has revolutionized our ability to analyze microbial communities and their functional capabilities. These state-of-the-art techniques have led to the discovery of previously unknown microbial strains and have revealed their genetic potential for synthesizing bioactive compounds (Langari and Mantzouridou 2024), paving the way for innovative probiotics, enzymes, and metabolites.

CRISPR-Cas systems and RNA sequencing techniques have enabled researchers to explore microbial gene expression and regulation, unveiling the pathways responsible for probiotic effectiveness, flavor generation, and antimicrobial substance synthesis (Xu and Li 2020). Single-cell sequencing and enrichment culture techniques help isolate rare microbial strains with unique attributes that are valuable for specific applications (Satam et al. 2023).

Synthetic biology supports the development of specialized microbial consortia that emulate or enhance traditional fermentation, leading to improved efficiency, modified flavor profiles, or the production of targeted functional compounds (Lv et al. 2021). Research on the relationship between fermented food microbes and the human gut microbiome is also gaining momentum, examining their effects on gut health, metabolic functions, and the potential for personalized nutrition (Leeuwendaal et al. 2022).

Future endeavors should focus on extracting, analyzing, and integrating unique probiotic strains from traditional fermented foods into functional food and beverage products. Improved encapsulation techniques are expected to enhance the resilience and efficacy of probiotics (Valentino et al. 2024). By leveraging advanced technologies, promoting sustainability, and adhering to ethical standards, researchers can unlock the potential of microbial ecosystems while respecting their cultural and ecological significance.

11.4 Conclusion

Microbial applications play a vital role in preserving and enhancing the nutritional value of indigenous and traditional food systems as well as in preserving cultural heritage. These practices have demonstrated remarkable resourcefulness and inventiveness for utilizing natural processes for sustainable food production. Maintaining biodiversity, cultural heritage, and food security in a rapidly changing world requires understanding and preservation of traditional methods. Researchers can tap into the potential of microbial ecosystems while preserving their cultural and ecological significance by employing cutting-edge technologies, addressing sustainability issues, and maintaining ethical standards.

References

Adesulu-Dahunsi AT, Dahunsi SO, Ajayeoba TA (2022) Co-occurrence of lactobacillus species during fermentation of African indigenous foods: impact on food safety and shelf-life extension. Front Microbiol 13. https://doi.org/10.3389/fmicb.2022.684730

Akinola R, Pereira LM, Mabhaudhi T, de Bruin F-M, Rusch L (2020) A review of indigenous food crops in Africa and the implications for more sustainable and healthy food systems. Sustainability 12(8):3493. https://doi.org/10.3390/su12083493

Behera SS, El Sheikha AF, Hammami R, Kumar A (2020) Traditionally fermented pickles: how the microbial diversity associated with their nutritional and health benefits? J Funct Foods 70(1):103971. https://doi.org/10.1016/j.jff.2020.103971

Berkum SV, Dengerink J, Ruben R (2018) The food systems approach: sustainable solutions for a sufficient supply of healthy food. Wageningen Economic Research, The Hague

Capozzi V, Fragasso M, Romaniello R, Berbegal C, Russo P, Spano G (2017) Spontaneous food fermentations and potential risks for human health. Fermentation 2017(3):49

Cavicchioli R, Ripple WJ, Timmis KN, Azam F, Bakken LR, Baylis M, Chrenfeld MJ, Boetius A, Boyd PW, Classen AT (2019) Scientists warning to humanity: microorganisms and climate change. Nat Rev Microbiol 17:569–586

Cuamatzin-García L, Rodríguez-Rugarcía P, El-Kassis EG, Galicia G, de Meza-Jiménez ML, del Baños-Lara MR, Zaragoza-Maldonado DS, Pérez-Armendáriz B (2022) Traditional fermented foods and beverages from around the world and their health benefits. Microorganisms 10(6):1151. https://doi.org/10.3390/microorganisms10061151

D'Andrea A, D'Ulizia A (2023) Preserving local food traditions: a hybrid participatory approach for stimulating transgenerational dialogue. Societies 13(4):95. https://doi.org/10.3390/soc13040095

Daly C, Fitzgerald GF, Davis R (1996) Biotechnology of lactic acid bacteria with special reference to bacteriophage resistance. Antonie Van Leeuwenhoek 1996(70):99–110

Dimidi E, Cox SR, Rossi M, Whelan K (2019) Fermented foods: definitions and characteristics, impact on the gut microbiota and effects on gastrointestinal health and disease. Nutrients 11:1806

Doyle MP, Steenson LR, Meng J (2013) Bacteria in food and beverage production. In: Rosenberg E, DeLong EF, Lory S, Stackebrandt E, Thompson F (eds) The prokaryotes: applied bacteriology and biotechnology. Springer, Berlin/Heidelberg, pp 241–256, ISBN 978-3-642-31331-8

Enthoven L, Van den Broeck G (2021) Local food systems: reviewing two decades of research. Agric Syst 193:103226. https://doi.org/10.1016/j.agsy.2021.103226

Erkus O, de Jager VC, Spus M, van Alen-Boerrigter IJ, van Rijswijck IM, Hazelwood L, Janssen PW, van Hijum SA, Kleerebezem M, EJ S (2013) Multifactorial diversity sustains microbial community stability. ISME J 7:2126–2136

FAO (2011) Fermented fruits and vegetables: a global perspective. Food and Agriculture Organization of the United Nations

Fukushima D (1985) Fermented vegetable protein and related foods of Japan and China. Food Rev Intl 1:149–209

Global Panel (2016) Food systems and diets: facing the challenges of the 21st century. In: G.P.o.A.a.F.S.f (ed) Nutrition. http://glopan.org/sites/default/files/ForesightReport.pdf, London, UK

Hasan MN, Sultan MZ, Mar-E-Um M (2014) Significance of fermented food in nutrition and food science. J Sci Res 2014(6):373–386

HLPE (2017) Nutrition and food systems, a report by the high-level panel of experts on food security and nutrition of the Committee on World Food Security, Rome

Jay JM (1996) Modern food microbiology, 5th edn. Chapman and Hall, New York

Kechagia M, Basoulis D, Konstantopoulou S, Dimitriadi D, Gyftopoulou K, Skarmoutsou N, Fakiri EM (2013) Health benefits of probiotics: a review. ISRN Nutri 2013(1):1–7. https://doi.org/10.5402/2013/481651

Keding GB, Schneider K, Jordan I (2013) Production and processing of foods as core aspects of nutrition-sensitive agriculture and sustainable diets. Food Secur 5:825–846

Kennedy DO (2016) B vitamins and the brain: mechanisms, dose, and efficacy—a review. Nutrients 2016(8):68

Langari D, Mantzouridou FT (2024) Analytical aspects of metagenomic and volatilomic approaches that advance table olive integrity research. Explor Foods Foodomics 2(6):570–598. https://doi.org/10.37349/eff.2024.00052

LeBlanc KE, Baer-Sinnott S, Lancaster KJ, Campos H, Hei K, Tucker KL, Kushi LH, Willett WC (2024) Perspective: beyond the Mediterranean Diet -- exploring Latin American, Asian, and African heritage diets as cultural models of healthy eating. Adv Nutr 15(5):100221–100221. https://doi.org/10.1016/j.advnut.2024.100221

Leeuwendaal NK, Stanton C, O'Toole PW, Beresford TP (2022) Fermented foods, health and the gut microbiome. Nutrients 14(7):1527. https://pubmed.ncbi.nlm.nih.gov/35406140/

Lovegrove A, Edwards CH, De Noni I, Patel H, El SN, Grassby T, Zielke C, Ulmius M, Nilsson L, Butterworth PJ, Ellis PR, Shewry PR (2015) Role of polysaccharides in food, digestion, and health. Crit Rev Food Sci Nutr 57(2):237–253. https://doi.org/10.1080/10408398.2014.939263

Lugo-Morin DR (2020) Indigenous communities and their food systems: a contribution to the current debate. J Ethnic Foods 7(1). https://doi.org/10.1186/s42779-019-0043-1

Lv X, Wu Y, Gong M, Deng J, Gu Y, Liu Y, Li J, Du G, Ledesma-Amaro R, Liu L, Chen J (2021) Synthetic biology for future food: research progress and future directions. Future Foods 3:100025. https://doi.org/10.1016/j.fufo.2021.100025

Maicas S (2020) The role of yeasts in fermentation processes. Microorganisms 8(8):1142. https://doi.org/10.3390/microorganisms8081142

Mannaa M, Han G, Seo Y-S, Park I (2021) Evolution of food fermentation processes and the use of multi-omics in deciphering the roles of the microbiota. Foods 10(11):2861. https://doi.org/10.3390/foods10112861

Marco ML, Heeney D, Binda S, Cifelli CJ, Cotter PD, Foligné B, Gänzle M, Kort R, Pasin G, Pihlanto A (2017) Health benefits of fermented foods: microbiota and beyond. Curr Opin Biotechnol 2017(44):94–102

Martínez-Espinosa RM (2020) Microorganisms and their metabolic capabilities in the context of the biogeochemical nitrogen cycle at extreme environments. Int J Mol Sci 21:4228

Materia VC, Linnemann AR, Smid EJ, Schoustra SE (2021) Contribution of traditional fermented foods to food systems transformation: value addition and inclusive entrepreneurship. Food Secur 13(5):1163–1177. https://doi.org/10.1007/s12571-021-01185-5

Melini F, Melini V, Luziatelli F, Ficca AG, Ruzzi M (2019) Health-promoting components in fermented foods: an up-to-date systematic review. Nutrients 11(5):1189. https://doi.org/10.3390/nu11051189

Mendez MA, Popkin BM (2004) Globalization, urbanization and nutritional change in the developing world. Electron J Agric Dev Econ 1(2):220–241

Mokoena MP, Chelule PK, Gqaleni N (2005) Reduction of fumonisin B1 and zearalenone by lactic acid bacteria in fermented maize meal. J Food Prot 68:2095–2099

Moonga HB, Phiri S, Schoustra SE, Chileshe J, Chirwa-Moonga T, Shindano J (2020) The Munkoyo root: traditional uses, biochemistry, fermentation, and potential cultivation. In: African natural plant products, volume III: discoveries and innovations in chemistry, bioactivity, and applications. American Chemical Society, pp 81–99

Muhialdin BJ, Zawawi N, Abdull Razis AF, Bakar J, Zarei M (2021) Antiviral activity of fermented foods and their probiotics bacteria towards respiratory and alimentary tracts viruses. Food Control 127:108140. https://doi.org/10.1016/j.foodcont.2021.108140

Nkhata SG, Ayua E, Kamau EH, Shingiro J-B (2018) Fermentation and germination improve nutritional value of cereals and legumes through activation of endogenous enzymes. Food Sci Nutr 6(8):2446–2458. https://doi.org/10.1002/fsn3.846

Nout MJR, De Vos WM, Zwietering MH (2005) Food fermentation. Wageningen Academic Publishers

Nout MJR, Darkar PK, Beuchat LR (2007) Indigenous fermented foods. In: Doyle MP, Beuchat LR (eds) Food microbiology: fundamentals and frontiers, 3rd edn. ASM Press, Washington DC, pp 817–835

Opal SM (2009) A brief history of microbiology and immunology. In: Vaccines: a biography, pp 31–56. https://doi.org/10.1007/978-1-4419-1108-7_3
Pakseresht A, McFadden BR, Lagerkvist CJ (2017) Consumer acceptance of food biotechnology based on policy context and upstream acceptance: evidence from an Artefactual field experiment. Eur Rev Agric Econ 44:757–780
Paramithiotis S, Das G, Shin HS, Patra JK (2022) Fate of bioactive compounds during lactic acid fermentation of fruits and vegetables. Foods 11(5):733. https://doi.org/10.3390/foods11050733. PMID: 35267366; PMCID: PMC8909232
Paredes-López O, Gonzalez-Castaneda J, Carabez-Trejo A (1991) Influence of solid substrate fermentation on the chemical composition of chickpeas. J Ferment Bioeng 71:58–62
Phiri S, Schoustra SE, van den Heuvel J, Smid EJ, Shindano J, Linnemann A (2019) Fermented cereal-based Munkoyo beverage: processing practices, microbial diversity and aroma compounds. PLoS One 14:e0223501
Poli A, Finore I, Romano I, Gioiello A, Lama L, Nicolaus B (2017) Microbial diversity in extreme marine habitats and their biomolecules. Microorganisms 5:25
Raman S, Rogers JK, Taylor ND, Church GM (2014) Evolution-guided optimization of biosynthetic pathways. Proc Natl Acad Sci USA 111:17803–17808
Rani A, Saini K, Bast F, Mehariya S, Bhatia S, Lavecchia R, Zuorro A (2021) Microorganisms: a potential source of bioactive molecules for antioxidant applications. Molecules 26(4):1142. https://doi.org/10.3390/molecules26041142
Ruben R, Verhagen J, Plaisier C (2019) The challenge of food systems research: what difference does it make? Sustainability 11:171
Salim R, Nehvi IB, Mir RA, Tyagi A, Ali S, Bhat OM (2023) A review on anti-nutritional factors: unraveling the natural gateways to human health. Front Nutr 10. https://doi.org/10.3389/fnut.2023.1215873
Satam H, Joshi K, Mangrolia U, Waghoo S, Zaidi G, Rawool S, Thakare RP, Banday S, Mishra AK, Das G, Malonia SK (2023) Next-generation sequencing technology: current trends and advancements. Biology 12(7):997. https://doi.org/10.3390/biology12070997
Schoustra S, van der Zon C, Groenenboom A, Moonga HB, Shindano J, Smid EJ, Hazeleger W (2022) Microbiological safety of traditionally processed fermented foods based on raw milk, the case of Mabisi from Zambia. LWT 113997:113997. https://doi.org/10.1016/j.lwt.2022.113997
Sharifi-Rad J, Rodrigues CF, Stojanović-Radić Z, Dimitrijević M, Aleksić A, Neffe-Skocińska K, Zielińska D, Kołożyn-Krajewska D, Salehi B, Milton Prabu S, Schutz F, Docea AO, Martins N, Calina D (2020) Probiotics: versatile bioactive components in promoting human health. Medicina 56(9). https://doi.org/10.3390/medicina56090433
Sharma P, Tong YW (2025) Food waste-derived biofertilizers for agriculture sustainability. In: Waste-To-Energy. Elsevier, pp 195–215. https://doi.org/10.1016/b978-0-443-22356-3.00009-9
Sharma R, Garg P, Kumar P, Bhatia SK, Kulshrestha S (2020) Microbial fermentation and its role in quality improvement of fermented foods. Fermentation 6(4):106. https://doi.org/10.3390/fermentation6040106
Siddiqui SA, Erol Z, Rugji J, Taşçı F, Kahraman HA, Toppi V, Musa L, Di Giacinto G, Bahmid NA, Mehdizadeh M, Castro-Muñoz R (2023) An overview of fermentation in the food industry - looking back from a new perspective. Bioresour Bioprocess 10(1):85. https://doi.org/10.1186/s40643-023-00702-y
Singh P, Pandey VK, Sultan Z, Singh KN, Dar AH (2023) Classification, benefits, and applications of various anti-nutritional factors present in edible crops. J Agric Food Res 14:100902–100902. https://doi.org/10.1016/j.jafr.2023.100902
Song Y, Sun L, Zhang S, Fan K, Wang H, Shi Y, Shen Y, Wang W, Zhang J, Han X, Mao Y, Wang Y, Ding Z (2022) Enzymes and microorganisms jointly promote the fermentation of rapeseed cake. Front Nutr 9. https://doi.org/10.3389/fnut.2022.989410
Springmann M, Clark M, Mason-D'Croz D, Wiebe K, Bodirsky BL, Lassaletta L, Willett W (2018) Options for keeping the food system within environmental limits. Nature 562(7728):519–525

Steinkraus KH (1997) Classification of fermented foods: worldwide review of household fermentation techniques. Food Control 8(5–6):311–317

Swiderska K, Argumedo A, Wekesa C, Ndalilo L, Song Y, Rastogi A, Ryan P (2022) Indigenous peoples' food systems and biocultural heritage: addressing indigenous priorities using Decolonial and interdisciplinary research approaches. Sustainability 14(18):11311. https://doi.org/10.3390/su141811311

Tafere G (2015) A review on traditional fermented beverages of Ethiopian. J Nat Sci Res 5:94–102

Terefe NS (2016) Emerging trends and opportunities in food fermentation. In: Reference module in food science. Elsevier. https://doi.org/10.1016/b978-0-08-100596-5.21087-1

Thierry A, Pogačić T, Weber M, Lortal S (2015) Production of flavor compounds by lactic acid bacteria in fermented foods. In: Biotechnology of lactic acid bacteria: novel applications. Wiley, pp 314–340

Timmis K, de Vos WM, Ramos JL, Vlaeminck SE, Prieto A, Danchin A, Verstraete W, de Lorenzo V, Lee SY, Brüssow H (2017) The contribution of microbial biotechnology to sustainable development goals. Microb Biotechnol 10:984–987

Valentino V, Magliulo R, Farsi D, Cotter PD, O'Sullivan O, Ercolini D, De Filippis F (2024) Fermented foods, their microbiome and its potential in boosting human health. Microb Biotechnol 17(2):e14428. https://doi.org/10.1111/1751-7915.14428

Vogel RF, Hammes WP, Habermeyer M, Engel K-H, Knorr D, Eisenbrand G (2011) Microbial food cultures—opinion of the Senate Commission on Food Safety (SKLM) of the German Research Foundation (DFG). Mol Nutr Food Res 2011(55):654–662

Voidarou C, Antoniadou M, Rozos G, Tzora A, Skoufos I, Varzakas T, Lagiou A, Bezirtzoglou E (2020) Fermentative foods: microbiology, biochemistry, potential human health benefits and public health issues. Foods 10(1):69. https://doi.org/10.3390/foods10010069

Waltz CL (2010) Local food systems: background and issues. Incorporated: Nova Science Publishers

Wicaksono WA, Akinyemi OE, Wassermann B, Bickel S, Suwanto A, Berg G (2024) The terroir of tempeh: strong region-specific signatures in the bacterial community structures across Indonesia. Curr Res Microb Sci 7:100287–100287. https://doi.org/10.1016/j.crmicr.2024.100287

Xu Y, Li Z (2020) CRISPR-Cas systems: overview, innovations and applications in human disease research and gene therapy. Comput Struct Biotechnol J 18:2401–2415. https://doi.org/10.1016/j.csbj.2020.08.031

Zamora A, Fields ML (1979) Nutritive quality of fermented cowpeas and chickpeas. J Food Sci 44:234–237

Zapaśnik A, Sokołowska B, Bryła M (2022) Role of lactic acid bacteria in food preservation and safety. Foods 11(9):1283. https://doi.org/10.3390/foods11091283

Chapter 12
Microbial Biotechnology for Food Waste Reduction and Upcycling

Elijah Ige Ohimain

Abstract Food waste generation is increasing globally due to differing causes, either because of excessive consumption in developed countries or lack of postharvest and processing facilities in developing countries. Every year, about 6.3 billion ton agricultural residues and food wastes are generated globally. Food waste is a major concern for the environment especially because of greenhouse gas emissions, and they provide habitat for breeding of pathogens and their vectors. Food wastes also result in the loss of natural resources and farm input including land, water, energy, fertilizer and efforts. Recycling and upcycling of food wastes conserve resources, mitigate environmental impacts and global hunger and help to meet other sustainable development goals (SDGs). The microbial upcycling of food waste into value-added products including food and beverages, single-cell proteins, bioplastic, biochemicals, enzymes, compost and biofertilizers, mushroom, probiotics and prebiotics, and various biofuel carriers, can lead to a multiproduct value chain spurring biomanufacturing economy. Food upcycling provides a credible route for harnessing microbial potentials for enhancing food security. It also contributes to circular economy, potentially resulting in net-zero wastes. The suite of available technology for the upcycling of food wastes can be grouped into biochemical and thermochemical conversions and their hybrid. The future of wastes-to-products is quite promising, but there are logistical, infrastructural, technological, and policy hurdles that need to be addressed for it to come to reality.

Keywords Biochemical conversion · Ecofriendly approach · GHG · SGDs · Thermochemical conversion · Upscaling · Valorisation · Waste management · Waste-to-products

E. I. Ohimain (✉)
Department of Microbiology, Niger Delta University, Wilberforce Island, Bayelsa State, Nigeria
e-mail: elijahohimain@ndu.edu.ng

O. O. Babalola et al. (eds.), *Ecofriendly Frontiers*,
https://doi.org/10.1007/978-3-031-98700-7_12

12.1 Introduction

As the world's population exceeded 8 billion people in November 2022, wastes generation and management had become a challenge for humanity. Excessive production and consumption in developed countries and lack of postharvest and processing facilities in developing countries, have led to the generation of excessive wastes, which had become a challenge for environmental sustainability. Available development indices from the World Bank show that the world's Gross domestic product (GDP), which grows at 3.1% annually hit $101.3 trillion, hence a GDP per capita of $12.74 in 2022 with per capita CO_2 emissions of 4.29 metric tonnes (World Bank 2022). Globally, about 2.1 billion tonnes of municipal solid wastes (MSW) are generated annually, which was predicted to reach 3.8 billion tonnes by 2050 (UNEP 2024). The cost of waste management is also staggering. Year 2020 estimates from the United Nations Environmental Programme (UNEP) show that the direct cost of MSW management is $252 billion, which rose to $361 billion when the hidden costs of poor waste management are considered such as environmental pollution, climate change and human health impacts (UNEP 2024). They also estimated that under the current scenario i.e., without additional action, the annual waste management cost could reach $640.3 billion by 2050, but by taking proper action, could reduce the annual cost to $270.2 billion and even predicted a net gain of $108.5 billion per year in a circular economy model when the wastes are managed from end to end.

Food wastes, which lead to loss of valuable natural resources, also cause environmental pollution, present opportunity for bio-manufacturing due to their proximate composition especially because of the presence of proteins, carbohydrates and lipids (Sufficiency et al. 2022). Sustainable waste management approach could involve waste avoidance, minimization, valorisation and upcycling to value-added products. Waste valorisation and upcycling of bioproducts is a recent idea (Sufficiency et al. 2022), which presents enormous opportunities for bio-based manufacturing sector. Valorisation of wastes can become an opportunity for employment and wealth creation. Valorisation will strengthen food supply chains and reduce the cost and environmental impacts of food wastes (Esparza et al. 2020). Food, agricultural and forestry wastes are major components of MSW, hence, this chapter is focused on microbial biotechnology and other emerging approaches for their reduction, recycling and upcycling.

12.2 Current State of Food Wastes

The United Nations Environmental Programme defined wastes as unintended by-product of production and consumption, which include food wastes and their packaging materials especially plastics (UNEP 2024). Piercy et al. (2024) defined food wastes as the organic fraction of MSW, which comprise of organic wastes from various sources including domestic, food industry/manufacturing and agricultural residues. Food wastes accounted for about 60% of MSW, but the proportion varies among countries. For example, USA discard 60 million tons of food wastes annually, which accounts for about 22% of MSW (Mahabeer and Jin 2024). Food wastes are commonly generated along the global supply value chain including crop production, post-harvest and storage, food processing, distribution and retail, consumption and disposal (Lee et al. 2024).

The world food production stands at 9.5 billion tonnes in 2021 (FAO 2022), of which about 1.3 billion tonnes are wasted annually (Piercy et al. 2024), which comprise of grains, tubers, pulses, fruits, vegetables, meat, eggs, milk and seafood (Pleissner and Lin 2013). In addition, agricultural residues were estimated at 5.0 billion tonnes per year (Cherubin et al. 2018), which put the total biomass lost at 6.3 billion tonnes per year excluding food packaging wastes (Table 12.1). Majority of the food wastes generated annually comes from the Asian continent, particularly the industrialised Asian countries (China, Japan and South Korea), South/ Southeast Asia and Western/ Central Asia with North Africa generating 357, 275 and 97 million tons respectively. Food wastes generation from other regions including Sub-Saharan Africa 127 million tons, Europe 205 million tons, Latin America 126 million tons and North America /Oceania 108 million tons (Pleissner and Lin 2013). In the European Union, food industry processing by-products, which comprise of forestry, agriculture and fishing wastes account for about 1.4% of total waste production (Sabater et al. 2020). Esparza et al. (2020) reported that 30–50 percent of global food produced is either lost or wasted, but fruits and vegetables account for about 38% of food wastes (Lee et al. 2024). About a third of global food crop production is lost along the supply chains (Pleissner and Lin 2013; FAO 2019; UNEP 2024), which also results in the loss of farm input including human efforts, energy, water and fertilizer, while polluting the environment. This volume of food losses can feed a population of 1.26 billion people annually (UNEP 2024). Hence, hunger can be reduced by minimizing food wastage, redistribution of surplus food and valorisation of food waste (UNEP 2024).

Lee et al. (2024) presented the scale of losses in fruit and vegetable value chain, which are 10–20% during farming, 4–10% postharvest, 20–25% processing, 8–17 distribution, 15–35% retail and 5–12% consumption for developing countries. The values are the same for developed countries except for processing and consumption

Table 12.1 Global food wastes generation

Types of wastes	Description	Volume, billion ton/yr	References
Food wastes	Postharvest losses, food leftover, food industry processing wastes	1.3	UNEP (2024), FAO (2019)
Agricultural residues	Leaves, stem, straw, stover, stalks, husks, pods	5.0	Cherubin et al. (2018)
Food packaging wastes	Plastics, paper, carton		
Total		6.3	

which were estimated as 2% and 15–28% respectively. Food wastes are generally high in resource poor countries due to poor storage and handling infrastructure (Said et al. 2023). Fruits and vegetable losses during crop production and postharvest are mostly due to pest and diseases, unfavourable weather, poor handling and inadequate infrastructure for cold storage and processing especially in resource poor communities under tropical climates (Lee et al. 2024); but excessive consumption causes food wastes in developed countries (Esparza et al. 2020). In terms of sector, majority of food wastes are generated from households accounting for about 61%, while food services and retail accounted for about 26% and 13% respectively (Said et al. 2023). Therefore, there exists a considerable potential for the upscaling or upcycling of unavoidable food wastes and agricultural residues to reduce hunger and support environmental sustainability.

The food industry is largely associated with the generation of copious amount of wastes. The on-farm, domestic and industrial processing of certain crops especially sugarcane, maize, oil palm, cocoa, rice, cassava, coffee, wheat etc. generate diverse waste streams, which disposal present considerable challenge to environmental sustainability. For instance, about 70–80% of cocoa fruit in dry weight, which comprise of cocoa bean shell and pulp, cocoa pod husks, and cocoa sweatings are considered as farm wastes (Vásquez et al. 2019; Anoraga et al. 2024). Food industry processing wastes including lipid-rich and carbohydrate-rich effluents that are typically discharged into the environment, has been reviewed as substrates for microbial valorisation (Lad et al. 2022). Slaughterhouses wastes are rich in proteins, fruits and food processing wastes are rich in carbohydrates, vegetable oil processing wastes are rich in lipids, while agricultural residues contain high levels of lignocellulose.

Food crops come with natural covering or packaging, which are variously referred to as epicarp, peelings, cob, husk, shell, coir etc. Most of these natural coverings are typically not consumed, but form a major component of food processing wastes both domestically and industrially. Take for example, the processing of raw cassava tubers to garri by smallholders in Nigeria, only 34% is typically converted to the product that is consumed (Fig. 12.1), while the rest end up as various waste streams including 19.8% air emissions, 16.2% effluents and 30% solid wastes (Ohimain et al. 2013a). The proportion of solid wastes increases to nearly 50%, when mechanical peelers are used for cassava processing. Since Nigeria is the largest producer of cassava tubers in the world, and produces about 60 million tons of cassava per annum (Ikuemonisan et al. 2020), the waste generation is therefore huge. Waste generation from raw food during processing of many food substances results in the generation of 50–85% of wastes, which can be polluting to the environment if not well managed.

The proximate composition of food wastes including food processing wastes and agricultural residue is widely varied depending on the type of biomass, pretreatment and processing methods. They can contain variable amounts of carbohydrates, protein, lipids, fibres, ash and moisture. Pleissner and Lin (2013) reported that food waste is composed (w/w) of 30–60% carbohydrates, 10–40% lipids and 5–10% proteins. Agricultural residue consists of 29.4% cellulose, 20.1% hemicelluloses and 14.6% lignin (Piercy et al. 2024). Bontaş et al. (2023) reported a range of

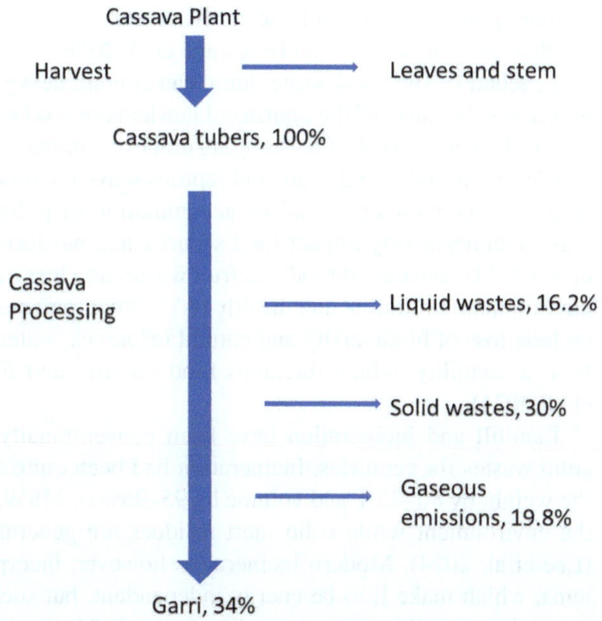

Fig. 12.1 Waste generation from processing cassava tubers to garri. (Source: modified from Ohimain et al. 2013a)

30–35% cellulose, 15–35% hemicellulose and 20–35% of lignin in lignocellulosic biomass. The proximate composition of food wastes suggests that they can be utilized for the production of various value-added products, which is the main thrust of this chapter.

12.3 Environmental Impacts and Bioremediation of Food Wastes

Wastes are challenges for environmental sustainability. Food wastes which result in loss of resources, pose serious environmental challenges. Food wastes especially when disposed carelessly, can be nauseating, unsightly and emit foul odours, which affect aesthetics. Food wastes abandoned on roadsides, have caused vehicles to skid leading to road accidents, block drainage canals causing flooding, and often attract and become breeding grounds for disease vectors particularly rodents and insects especially flies, roaches and mosquitoes. Microbes including pathogens are common in food waste dumps (Sufficiency et al. 2022). Emissions from decaying food waste can contain carbon dioxide, methane, SOx and NOx, which can contribute to climate change. Fruit and vegetable wastes were estimated to cause a per capita emission of about 1.36 kg CO_2 equivalent (Lee et al. 2024). In South Africa for example, crop production resulted in the emission of a total of 5.2 million tonnes of CO_2 equivalent in 2012 with the highest emissions of 57% followed by 30% caused by the application of synthetic fertiliser and lime respectively, while agricultural residue accounted for 13% (Tongwane et al. 2016).

Leachates from food waste dump can contain heavy metals that can pollute soil and rivers. Because of the unutilized nutrients in food wastes, it can cause eutrophication (Lee et al. 2024). Free-ranging domestic animals and wildlife including avian species, mammals, land snail, and reptiles scavenge food waste dumps, which often results in their infection and bioaccumulation of pollutants in food chains. Food waste can negatively impact food security and nutrition (Esparza et al. 2020). Lee et al. (2024) considered food and fruit wastes as a loss of valuable resources causing environment pollution and health risk. Other potential impacts of food wastes include loss of biodiversity and natural resources, water pollution, and reduction in food availability, which threatens food security and food supply chains (Mahish et al. 2024).

Landfill and incineration have been conventionally used for the treatment of solid wastes for centuries. Incineration had been quite effective in the reduction of the weight by 80–85% and volume by 95–96% of MSW, generating air emissions to the environment while solid inert residues are generally disposed-off in landfills (Lee et al. 2024). Modern incinerators however, incorporate waste-to-energy systems, which make it to be energy independent, but such systems are complex and expensive to build and manage (Lee et al. 2024). Incinerators are not common in resource-limited countries, where MSW is typically disposed off by open dumping and burning, which is quite polluting to the environment. But the drawback of these conventional methods is that they often lead to the loss of valuable biomass and also cause air, water and soil pollution, hence not an ecofriendly approach. Besides, landfill and incineration treatment of wastes, generate products that are either harmful to the environment or of less economic value, whereas upcycling, which involves the production of higher value substances from food wastes are now trending.

The remediation of food waste follows the six-tier inverted pyramidal hierarchical waste management priorities, which involves waste prevention, reduction, reuse, recycling, recovery and dumping (Axil 2024), applicable to food wastes (Fig. 12.2). This is a top-down approach where highest priority is given to waste treatment methods that have relative least environmental impacts. This approach results in

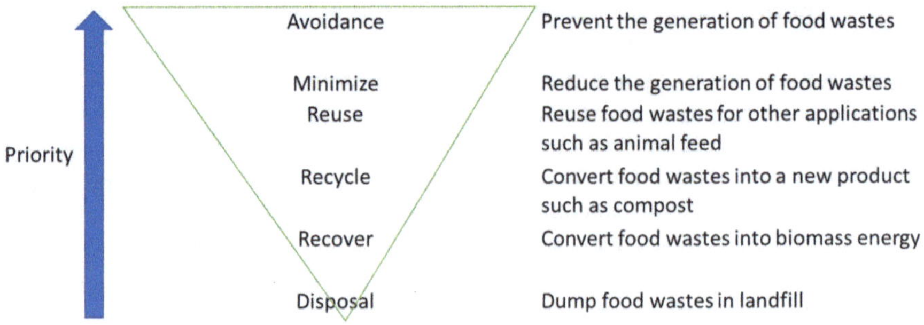

Fig. 12.2 Food waste management hierarchy and priorities

maximum recovery of biomass, with only a small fraction or no waste at all reaching the landfill. These processes support circular economy and zero discharge. For example, food wastes can be avoided or prevented by not using the raw material at all. This approach can be impractical, because humans need to eat food for survival. But a fruit such as carrot can be eaten completely without leftovers, which avoids waste generation. Though, carrot used in the preparation of fruit salad and juice are often deskinned or expressed, which generates fruit wastes. Reduction strategy involves minimizing the use of the food item or maximizing the raw material when used, which will lead to the generation of the least amount of wastes. For instance, raw apples can be consumed almost 100%, but some people optionally deskin it before consumption, which creates some wastes. Reuse options involve repurposing of the wastes to be reused either by the waste generator in other process or other users or turned into a new product, which is referred to as recycling. For instance, food processing wastes like egg shell can be repurposed or reused as bio-calcium supplements or compost. Another option is recovery, where material or energy is recovered from the food waste. For example, various energy careers including electricity, bioethanol and biogas can be derived from sugarcane bagasse and molasses. In addition, molasses can be used as medium for the production of single cell proteins, enzymes and microalgae, which can be used for the production of biodiesel, bioethanol and other biochemicals. When the waste management hierarchy is followed, only minimal amount of food waste can end in the landfill. There can even be zero food waste if the final residue returns to the farm as compost or manure. The role of microbes in waste degradation, mitigating climate change, nutrient cycling, environmental remediation and enhancing food security is presented in a recent review (Malkawi and Kapiel 2024).

Waste reuse, recycling and recovery minimize the use of virgin material for production and therefore maximizes farm input such as energy, water, fertilizer and human effects, while minimizing emissions. Waste reduction approaches could involve crop breeding, food processing options, waste repurposing and recycling, energy generation from food wastes etc. These approaches can create value for wastes and opportunities for business and employment. Food waste generation can be reduced through breeding and raw material selection. Take for example, the conversion of full fresh bunch (FFB) into palm oil. The material balance for the conversion of FFB show that waste generated is depended on the breed of the FFB (Fig. 12.3). Only 9.4–12.8% palm oil and 5.7–7.2% kernel was produced but accompanied by various waste streams; 0.8–2.4% chaff, 23.7–32.4% empty fruit bunch (EFB), 10.0–18.8% palm kernel shell (PKS), 23.2–28.1% palm press fibre (PPF), and 11.4–15.4% water, with Dura variety of oil palm. But when the improved Tenera variety was used as raw material palm oil yield double to 26.0–28.2% oil while kernel reduced slightly to 5.5–5.6% kernel, but accompanied by less wastes; 0.9–1.4% chaff, 25.7–28.2% EFB, 6.8–7.5% kernel shell, 19.1–20.3% palm press fibre, and 12.0–12.5% water (Ohimain et al. 2013b). Hence, oil palm processing generates copious amounts of diverse waste streams including palm oil mill effluents (POME), solid wastes (chaff, EFB, PPF, PKS), and gaseous emissions (Figure 12.3), which pollutes the environment if not valorised. Since, Nigeria

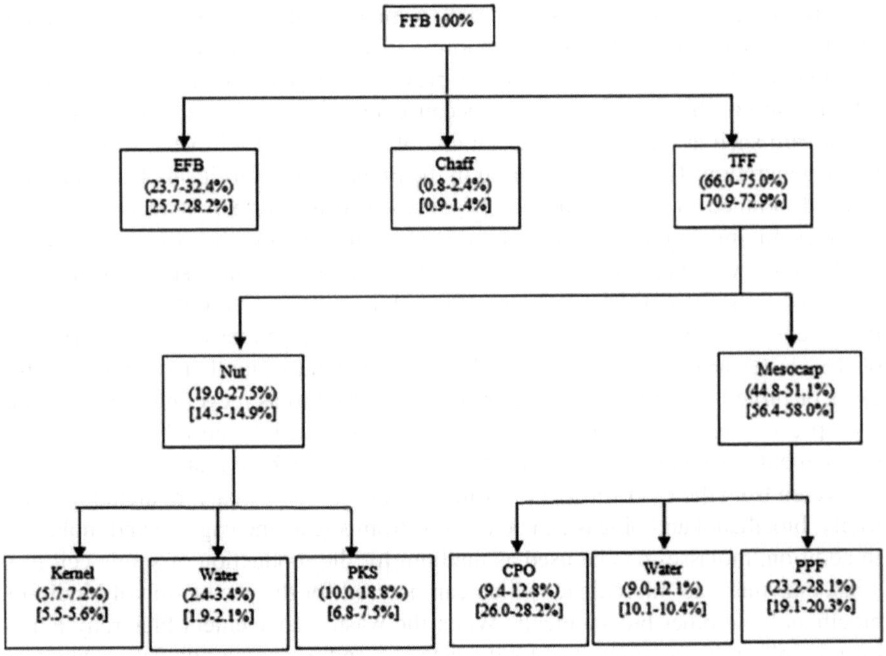

Fig. 12.3 Waste generation by smallholder oil palm processing in Nigeria. (Dura variety), [*Tenera* variety]. (Source: Adapted from Ohimain et al. 2013b)

produces about 1.4 million tons of palm oil annually, the generation of waste biomass is huge (Ohimain and Izah 2014a). However, it has been demonstrated that the diverse waste stream from oil palm processing can be recovered and used for energy and other applications. The energy self-sufficiency of small-scale and medium-scale oil palm processing in Nigeria have been demonstrated through the conversion of the wastes to energy (Ohimain and Izah 2014b, 2015a, 2015b). It should be noted that energy recovery is the last stage of the waste management hierarchy before disposal.

12.4 Microbial Enzymes in Food Waste Degradation and Utilization

Food wastes typically contain mixtures of different types of food substances containing carbohydrates, proteins, fibres, lipids and ash, hence a cocktail of microbial enzymes such as cellulases, amylases, glucoamylases, proteases, lipases, and phosphatases is involved in their degradation and subsequent mineralization (Pleissner and Lin 2013; Lee et al. 2024). Several hydrolytic enzymes including amylases, deaminases proteases, lipases, and disaccharidases such as β-galactosidase can play major roles in the degradation of various components or functional substances in food wastes. These enzymes are mostly produced by bacteria and fungi. Microbes and their enzymes have been used for the recovery of nutrients from food wastes. For instance, fungi such as *Aspergillus oryzae* and *Aspergillus awamori* are known for their secretion of hydrolytic enzymes including protease, glucoamylase, and phosphatase, which are used for the degradation of food wastes leading to the recovery of nitrogen, phosphate and glucose (Pleissner and Lin 2013).

Enzymes are organic catalysts that are frequently used in industrial processes. Enzymes have diverse applications in different sectors especially food and beverages, laundry, bioenergy, pharmaceutical and biomedical purposes. On the other hand, because of their high nutrients, food wastes can be used as excellent media for the cultivation of diverse enzyme-producing microbes leading to the expression of various enzymes that have industrial applications. For example, amylases are produced by *Rhizopus*, *Aspergillus* and *Bacillus*, which have applications in bakery, textiles, beverages and pharmaceutical industries. Proteases used in the meat industry is produced by *Streptomyces*, *Bacillus* and *Aspergillus*. Lipases produced by *Rhizopus* and catalases produced by *Aspergillus* are important in the biofuel and food industry. For instance, carbohydrate-rich food waste can be subjected to enzymatic valorisation processes involving the use of hydrolase and isomerase which can be used for the production of biochemicals such as prebiotics, biofuel and sweeteners (Sufficiency et al. 2022). In addition, diverse enzymes will be needed in the valorisation of food wastes and agricultural residues for the production of biofuels and other biochemicals. The global market for industrial enzyme has more than doubled from \$2.8 billion in 2008 to \$6 billion in 2016 (Sufficiency et al. 2022). But production of industrial enzymes can be challenged by the availability of good feedstock. Hence, food wastes and agricultural residues present excellent substrates for enzyme production.

12.5 Microbial Upcycling of Agricultural Residues and Food Wastes

Piercy et al. (2024) considered food wastes as sources of carbon that need to be harnessed. They considered microbial upcycling of food wastes as opportunity to support circular economy and recover valuable resources. Upcycling prevent food from becoming a waste, preserving them in diverse forms suitable for human consumption and other uses. Upcycling can conserve farm inputs such as efforts, time, fertilizer, water, pesticides etc. and prevent air pollution impacts of food wastes. Said et al. (2023) considered bioprocessing as the most cost-effective and sustainable management of food wastes because it often does not lead to the generation of toxic substances. Rakesh and Mahendran (2023) considered upcycling of food waste, which is in line with circular economy, as a sustainable approach for minimizing waste generation and greenhouse emissions.

Recently, attention has focused on the microbial valorisation of agro-processing wastes for the production of high-value-adding substances with potential applications in diverse industries such as food and feed, medical and pharmaceutical and cosmetic (Vásquez et al. 2019). Lenkiewicz (2023) opined that circular economy can mitigate climate change, reduce energy shortage, protect public health, and strengthen food supply chains. The challenge posed by food wastes can become an opportunity for business especially for the production of diverse products such as compost and manure, flours, fermented foods and beverages.

The production of compost and manure from farm residues and food waste is fairly conventional, as such will be briefly discussed here. Lad et al. (2022) opined that traditional approaches of food waste management including use in animal feeds, incineration, composting and fertilizer, extract little value from the wastes. Whereas, microbial upcycling into diverse products presents avenue for the valorisation of wastes.

The process of valorisation of food waste can involve first, the biochemical hydrolysis leading to the production of basic substances including sugars, fatty acids, amino acids, nitrogen and phosphate, which can be recovered or utilised by diverse microbes for the production of diverse products (Pleissner and Lin 2013).

12.5.1 Composting and Vermicomposting

Synthetic fertilizers used in modern agriculture negatively affect soil microbiome causing near irreversible damage to the soil ecosystem (Lenkiewicz 2023). Hence, compost and manure appear to be credible alternative to synthetic fertilizers. Composing is an aerobic microbial system for the degradation of organic matter. Traditionally, microbes use various enzymes to breakdown and mineralise organic materials such as food wastes and convert them to compost or manure, which are used as biofertilizers for soil enhancement. Recently, vermicomposting, which involves the use of earthworms have been incorporated into conventional composting systems. Earthworms enhance microbial degradation process by physically breaking through the organic wastes, aerating them and digesting them using various enzymes, which often lead to the production of higher quality compost, with less pathogens and heavy metals (Lee et al. 2024).

12.5.2 Fermented Food and Beverages

Many food substances especially fruits contain fermentable sugars, while tubers such as yam, cassava and most grains including rice, maize, wheat, rye etc contain starch, which can be broken down to sugars upon hydrolysis. Therefore, wastes from these products can be converted by yeasts and some bacteria to alcoholic beverages such as beer, stout, wine, rum, etc. Fruit wastes like citrus (Figure 12.5), pineapples, apples, bananas can be converted to juices. Upcycled foods, which prevent wastage, can enhance food security, and return food stuff to the value chain for human consumption (Lee et al. 2024).

12.5.3 Flours

Edible flours can be produced from tubers such as cassava, yam, cocoyam and used for baking purposes or eaten as meals. These food substances are typically fermented, dried and milled into flours. Wastes from food processing such as peelings, sievings, etc are also dried, milled and used as composite flour. Flours made from food processing wastes are rich in fibres and are used as fibre supplements. This practice is common in tropical countries.

12.5.4 Animals Feeds

Various agricultural residues, food and fruit processing wastes and food wastes have been traditionally processed, stabilized and used as animal feeds or as feed substituents. Food wastes present a cost-effective and alternative feed for animals (Sufficiency et al. 2022). Literature are many on the use of different food wastes and agricultural residues for the feeding of pets such as dogs and cats, and farm animals including poultry, pigs, lamb, rabbits, snails, cattle, and fishes.

12.6 Innovative Microbial Technologies

There are several innovative approaches for the microbial utilization of food wastes for the production of various substances including industrial enzymes, single cell proteins (SCP), yeast and mushrooms, algae, probiotics and prebiotics, renewable energy including biogas, bioethanol and biodiesel. Esparza et al. (2020) discussed emerging approaches for the valorization of food wastes, which involved the production of value-added products such as bioplastics, exopolysaccharides and other biopolymers, enzymes, and biofuels.

12.6.1 Biofertilizers

Beside the traditional production of compost from food waste, the production of biofertilizers have gained considerable interest in recent years. Mahish et al. (2024) reviewed literature on the conversion of food waste derived from various sources such as fruits, sugar crops, starch crops, vegetable oils and proteins; dairy, poultry, fish to biofertilizers by diverse species of fungi, bacteria and actinomyces. Food wastes are used as substrates for the cultivation of microbial guilds that are used for the production of biofertilizer, which are not only rich in organic matter like traditional compost, but also rich in phosphorus, nitrate and potassium. Some of these microbes play diverse roles as nitrogen fixers, nitrifiers, ammonifiers, phosphate solubilizers, while some produce plant growth hormone like indole acetic acid (IAA). Ghoreishi et al. (2023) demonstrated the production of IAA from pruning and grass clipping wastes via solid-state fermentation by *Trichoderma harzianum*.

Mahish et al. (2024) compiled from primary research works, a list of food wastes valorised by specific microbes that were used as fertilizers to boost plant yields. For instance, they reported food leftovers, fruit peels, rice bran, vegetable wastes, rapeseed mill cake, fish wastes and used vegetable oil were utilized for the production of biofertilizers by diverse bacteria and fungi species including *Pseudomonas aeruginosa*, *Bacillus thermoamylovorans*, *Bacillus stearothermophilus*, *Streptomyces thermonitrificans*, *Brevibacillus borstelensis*, *Lactobacillus* sp, and *Aspergillus niger* for the production of biofertilizer that enhanced the yield of tomato, cucumber, lettuce, cabbage, *Brassica rapa*, maize, *Ipomoea aquatica*, alfalfa.

12.6.2 Single Cell Protein

Utilization of food wastes for the production of single cell protein (SCP), used as food and feed is another way of upcycling food waste. SCP can be produced through the fermentation activities of diverse microbes including yeast, fungi, bacteria and microalgae. Food wastes present a credible alternative media for the production of SCP as opposed to the traditional expensive media. Although, both solid-state and submerged fermentation are used for the production of SCP, semi-solid fermentation is emerging especially for the cultivation of some microalgae heterotrophically. Piercy et al. (2024) reviewed and presented 82 genera of microbes that can use food wastes substrate for the production of SCP, though with emphasis on mixed cultures over pure cultures. Sekoai et al. (2024) listed some SCP that have been commercialized including Marmite produced by Unilever using brewers' yeast *Saccharomyces cerevisiae*, Quorn™ produced by Marlow Foods using the fungus *Fusarium venenatum* and *Spirulina*, a microalga produced by Cyanotech Corporation. Sekoai et al. (2024) presented some advantages of SCP for the production of protein-rich food including minimal requirements for natural resources particularly land and water with relatively low carbon footprint and short production cycle. Moreover, the high protein composition (30–70%) of SCP is comparable with that of conventionally reared animals such as cattle and poultry. They reviewed food waste as a sustainable substrate for the production of SCP.

12.6.3 Mushroom Production

Food wastes, including agricultural residues and forestry residues have been commonly used as substrate for the production of culinary and medicinal mushrooms. This is one of the well-established methods of waste biomass valorisation. Most of the commercialized mushrooms have been produced by solid state fermentation of sawdust and dead wood such as shitake, oyster mushroom, and *Ganoderma* species. Mushroom are healthy human food that contain diverse bioactive substances that have exhibited several activities such as antioxidant, anti-inflammatory, antimicrobial, anticancer, antidiabetic, antihypertensive and protective of key organs such as heart, liver and kidneys (Ohimain 2020; Yu et al. 2023; Poyeri and Ohimain 2024).

Mushroom production can contribute beyond the direct management of waste food biomass, into other types of solid wastes generated from food packaging. Take for instance, small-scale mushroom production makes use of plastic and glass bottles or containers and metallic drums that were recovered from waste dumps and repurposed for mushroom cultivation (Figure 12.6). For instance, plastics and mayonnaise/jam bottles are cleaned and sterilized and used for fruiting. Mayonnaise/jam bottles are also used for mushroom liquid culture and solid spawn production. Recovered iron and steel including drums are used for the local fabrication of boilers used for substrate sterilization. Scrap metals are also used for the fabrication of milling equipment and fruiting shelves. Reuse of metal wastes is driving the valorisation of scrap metals in Nigeria (Ohimain and Jenakumo 2013). Waste biomass are used as fuel for the boiler and substrate for the production of fruiting bodies. Small-scale mushroom production provide means for the management not only of food wastes but also food packaging wastes as well.

12.6.4 Probiotics and Prebiotics

Food wastes, because of their nutrient content, are used as media for the cultivation of yeasts and lactic acid bacteria (LAB). Cultures of yeasts particularly *Saccharomyces cerevisiae* and various species of LAB have been consumed directly as probiotics, used in food manufacturing processes including the fermentation of biomass leading to the production of various biochemicals, which have been used as prebiotics among other uses. Sabater et al. (2020) discussed the LAB species frequently used in the dairy manufacturing processes including *Lactobacillus* spp., *Lactococcus lactis*, *Streptococcus thermophilus*, *Leuconostoc* spp. while members of the genera *Oenococcus*, *Pediococcus*, and *Weissella* have been used in the fermentation of plant-based biomass. Upcycled products that can be produced from LAB fermentation of waste food biomass include yoghurt, cheese, lactic acid, sauerkraut, and pickles.

12.6.5 Biomass Energy

Biomass is among the most important renewable energy and material sources, which can become more relevant especially when food wastes are used as feedstock sustainably, which support circular economy and resource conservation (Huang et al. 2017). Utilizing food wastes for biofuel production addresses some concerns in biofuel economy particularly food versus fuel conflicts and cheaper feedstock, which can conserve resources and make the overall biofuel production to become economically competitive. Diverse energy carriers can be potentially produced from food wastes including ethanol, methanol, hydrogen, biodiesel, bio-oil, biochar, briquettes, bioelectricity, methane, and biogas (Mahish et al. 2024). Miran et al. (2016) generated bioelectricity from the conversion of orange peeling by *Pseudomonas* sp in a microbial fuel cell.

Biodiesels are fatty acid ethyl esters (FAEEs) or fatty acid methyl esters (FAMEs) produced by esterification or transesterification of vegetable oils in the presence of a catalyst (ethanol/ methanol and sodium hydroxide or sulphuric acid). Used vegetable oils, oily wastewater from vegetable oil production and refining are commonly used for the production of biodiesel. Microalgae grown autotrophically or heterotrophically on food wastes have emerged as credible source of oils for the production of biodiesel.

Microbial hydrolysis and fermentation of the carbohydrate fraction of food wastes is a common pathway for the production of bioethanol that is used not only as biofuel, but as chemical feedstock. Different types of food wastes containing carbohydrates have been demonstrated for ethanol production including sugarcane molasses, corn cob and Stover, cassava peelings and sieving, and fruit wastes such as papaya, mango and citrus.

12.6.6 Microalgal Production

Algal processes have the added advantage of trapping and using carbon dioxide, which present a positive impact on climate and they can be cultured on nutrient poor soils or even bioreactors. Open ponds or closed bioreactors are commonly used for microalgal cultivation. Pleissner and Lin (2013) discussed the fungal enzymatic hydrolysis of food wastes, which produced the nutrient required for the cultivation of microalgae. Lee et al. (2024) described the potential use of fruit wastes as media for the production of microalgae, which can be useful industrially for the production of commercially important substances such as lipids especially unsaturated fatty acids, proteins, carbohydrates, vitamins, coenzymes, and pigments, with potential industrial applications in food, feed and beverages, textiles, biofuels and other bioenergy applications, bioplastics and synthetic fibres, pharmaceutics and nutraceuticals, biochemicals and cosmetics. Pleissner and Lin (2013) listed products from microalgae produced from food wastes that can be used as feedstock for the production of value-added substances such as lipids, carbohydrates and proteins used for the production of plasticiser, biodiesel, polyurethane, organic acids, biosurfactants, polyols and polyhydroxyalkanoates. Food wastes are increasingly used to grow microalgae heterotrophically for the production of diverse bio-pigments including phycobiliproteins and xanthophylls such as zeaxanthin, lutein, neoxanthin, and violaxanthin, which are used in the food industry.

12.6.7 Plastics and Bioplastics

Plastics especially polyethyleneterephthalate (PET) are among the commonest materials used for the packaging of food. Hence, plastic wastes comprise a major portion of MSW disposed in landfills. While the organic fractions of food wastes are easily biodegradable, plastic wastes are not and become persistent even in landfills, which are mostly treated via open incineration, causing air pollution and has become a challenge for environmental sustainability. Data from a recent editorial showed that since commercial production of plastics started in the 1950s, its production, which is mostly from petroleum sources has steadily grown annually reaching 400 million tonnes in 2022, which had reached 10 billion tonnes cumulatively. Of these, 22% littered the environment particularly soil and waterbodies, 18% are incinerated, only a small fraction of about 10% is recycled, while the bulk 50% is disposed in landfills. It is estimated that about 10 million tons of plastic waste reaches the ocean annually (Pereyra-Camacho and Pardo 2024).

Microbial biotechnology presents a twin approach to solving the global food and plastic wastes challenge. Food wastes and plastics can be used as substrates for the production of biodegradable plastics (bioplastics). Pereyra-Camacho and Pardo (2024) reviewed the recycling of fossil plastics and upcycling of plastic wastes to valuable products by microbes. They reviewed microbial enzymes such as PET hydrolases, urethanases and nylonases that have been recently demonstrated to breakdown plastic polymers especially PET, polyurethanes and polyamides respectively, with potential of recovery and reuse of constituent monomers. However, due to limitations in bioconversion of plastics, integration with thermochemical methods have been advocated for the upcycling of plastic wastes (Tiso et al. 2022; Pereyra-Camacho and Pardo 2024).

Tiso et al. (2021) demonstrated the conversion of fossil PET waste into two bioplastics, polyhydroxyalkanoate (PHA) and poly(amide urethane). The process involved, first the breakdown of PET by hydrolase into pure terephthalate and ethylene glycol, which are used as substrate by *Pseudomonas umsongensis* GO16, for the production of PHA and hydroxyalkanoyloxy-alkanoates (HAAs), which are in turn used for synthesis of poly(amide urethane. Other studies have demonstrated the microbial depolymerization and upcycling of PET by *P. umsongensis* GO16 (Narancic et al. 2021) and engineered *P. putida* (Brandenberg et al. 2022). The use of food wastes as substrate for the production of biodegradable plastic especially PHA had been demonstrated. Mahabeer and Jin (2024) compiled a list of bacteria species that have been demonstrated to produce and accumulate PHA from the degradation of various types of food wastes; *Curpriavidus necator*, *Burkholderia cepacian*, *Burkholderia thailandensis*, *Bacillus firmus*, *Bacillus cepacia*, *Bacillus megaterium*, *Ralstonia eutropha*, *Halogeometricum borinquense*, *H. mediterranei*, *Paracoccus* sp and *Alcaligenes latus*. Recombinant *E. coli* have been used for upcycling of various food wastes such as corn, soya, coffee, malt, cheese whey etc. for bioplastic production. Mahish et al. (2024) discussed the conversion of sugarcane molasses to polyhydroxybutyrate (a bioplastic) by *Bacillus cereus*.

12.6.8 Biochemicals

Microbial fermentation processes, in addition to the production of ethanol have been utilized for the production of other biochemicals such as acetic acid, vinegar, citric acid. Sufficiency et al. (2022) listed several organic acids that can be produced from food wastes including succinic, malic, fumaric, glutamic, itaconic acids, and 3 -hydroxy propionic acid. Many other substances with specialized applications have also been produced from food wastes. For instance, *Clostridium* species can ferment food waste to produce biohydrogen, butanol, isopropanol, butyric acid and acetone. *Enterobacter* can produce ethanol, formic acid, butanediol, and biohydrogen from waste biomass.

Studies and reviews have discussed the potentials of converting food wastes into other useful materials. Mahabeer and Jin (2024) reviewed recent literature pertaining to the upcycling of food waste into value-added biomaterials with potential applications in the fabrication of medical products such as biodegradable polymers, cellulose, pectin, collagen, lignin, chitosan, and hydroxyapatite. Through systematic review, Zamil et al. (2022) identified some bioactive substances and by-products from rice processing wastes including squalene, betaine, tricin, phenolic compounds. These substances exhibited various properties such as anti-inflammatory, antiaging, whitening, moisturizing photoprotective, and melanin synthesis, which have potential applications in the manufacturing of cosmetics and wound healing products. Similarly, Chen et al. (2024) produced a fermented substance via *Aspergillus oryzae* Fermentation of red rice that exhibited bioactive properties including anti-ageing and antioxidative properties, regenerative, moisturising and whitening effects, with potential applications in skincare/cosmetics. Mahabeer and Jin (2024) reviewed the methods and collated basic research including the roles of microbes, the state of technology for the extraction and upcycling of various food wastes into diverse biomaterials and their medical applications. Mahish et al. (2024) also reviewed literatures on important substances such as proteins and amino acids, lipids and fatty acids, polysaccharides, disaccharides and monosaccharides, cellulose, hemicellulose, lignin, pectin, and fibre that could be recovered from different types of food wastes. Anoraga et al. (2024) reported some value-added substances derivable from cocoa pod husk including dietary fibre, pectin, xylitol (an alternative sweetener), fragrance, encapsulation material, and contain bioactive substances that has exhibited antibacterial and antioxidant effects with potential applications in the production of black soap and cosmetics.

12.6.9 Biopesticides

Pesticides are chemicals designed to kill nuisance organisms such as insects including mosquitoes, roaches and ticks, weeds, rodents and pathogens. They are important in disease prevention, agricultural production and protection of biomaterials especially furniture and food. However, most of the pesticides in use are made from fossil sources, which consumed much energy and release tonnes of GHGs. Besides, pesticides can pollute the environment, bioaccumulate in the food chain, become toxic to non-target organisms and potentially affecting humans (Šunjka and Mechora 2022). Many pesticides, especially chlorinated pesticides, tend to be persistent in the environment due to their resistance to microbial degradation. Hence, attention is now focused on the production of alternative pesticides from biological sources, which are less toxic and amenable to biodegradation. Biopesticides can be produced from fungi, bacteria and viruses. For instance, *Bacillus thuringiensis* has been used as insecticide since 1920. The biopesticide market has grown exponentially in recent years reaching $2.5 billion (Balasubramanian and Tyagi 2017). One of the major constraints for the production of microbial pesticides is the cost of the substrate. Hence, food wastes present a cheap alternative for the production of biopesticides, which can reduce the cost of production by 35–59% (Balasubramanian and Tyagi 2017). Ghoreishi et al. (2023) demonstrated the production of biopesticides from grass pruning and clipping wastes through *Trichoderma harzianum* via solid-state fermentation.

12.7 Microbial Biotechnology and Thermochemical Processes for Food Waste Valorization

Existing and emerging technologies for the upcycling of food wastes can be broadly be divided into biochemical and thermochemical. Biochemical processes, which are mostly facilitated by microbes and/or their enzymes, include composting, fermentation, anaerobic digestion, esterification and transesterification, while thermochemical processes include combustion, briquetting, pyrolysis, liquefaction, gasification and Fischer-Tropsch synthesis (Fig. 12.7). Besides, energy recovery from waste heat arising from both biochemical and thermochemical processes are being harnessed and used beneficially either as process heat or for the generation of green electricity via steam turbines. A biorefinery approach driven by microbial processes in combination with various processes such as biochemical, thermochemical, and physical methods for the conversion of biomass especially food wastes into diverse

Fresh oil palm fruits | Sliced empty fruit bunch | Oil palm fruit threshing | Palm fruits with empty fruit bunch in the background | Heap of palm kernel | Semi-mechanized palm oil extraction

Fig. 12.4 Oil palm processing generates diverse waste stream

products (Fig. 12.4), which promote circular economy leading to net zero- wastes is being advocated (Sekoai et al. 2024; Lee et al. 2024). However, Esparza et al. (2020) suggested the integration of conventional with emerging methods for a more holistic approach.

12.7.1 Fermentation

Fermentation has been an old technology for the cultivation of microbes especially bacteria, fungi and algae and/or the production of biochemicals. Fermentation processes reroute materials especially carbon into high-value biomolecules (Lad et al. 2022). Two types of fermentation methods are commonly used, solid state (SSF) and submerged fermentation (SF), but semi-solid fermentation is also becoming popular. For instance, both SSF and SF have been used for the production of mushroom fruiting bodies using agro-wastes and food wastes as substrates. Hydrolysis and fermentation have been used for the production of alcoholic beverages for millennia. For instance, practically any fruit, food crop or their wastes containing glucose can be fermented to produce ethanol by the yeast *Saccharomyces cerevisiae* and the bacteria *Zymomonas mobilis* whereas other microbes can produce other biochemicals such as lactic acid by *Lactobacillus* sp, citric acid by *Aspergillus niger* and phenol by *Rhizopus oligosporus* (Mahish et al. 2024).

Peelings account for a major proportion of food and fruit wastes. For example, peelings account for 30–40% of bananas, 40–50% of tangerine, 50% of pomegranates (Mahish et al. 2024). Peels are generally not edible and are sometimes used as animal feed. Recently, peels have been upcycled to other useful substances via microbial fermentation processes. Studies have listed diverse peelings that have been valorised by various microbes to useful products. For instance, biosorbents have been successfully produced from the peelings of pomegranate, tangerines, and orange (Mahish et al. 2024).

12.7.2 Anaerobic Digestion

Methane and biogas are renewable energy sources that has been produced from microbial conversion of practically any biomass including sewage, fruit and food wastes including production effluents via anaerobic digestion, which results in the production of nutrient-rich solid residue or digestate that is used as biofertilizer. Anaerobic digestion is a complex process involving several microbial processes including biodegradation, hydrolysis, fermentation, acidogenesis, acetogenesis, methanogenesis, decarboxylation, volatilization etc leading to the production of methane-rich biogas and biohydrogen by consortia of microbes under anaerobic conditions. Biogas is a mixture of methane and carbon dioxide and can also contain some impurities such as water vapour, carbon monoxide and hydrogen sulphide, hence will require treatment prior to use. Biogas and biomethane are commonly used as fuel for cooking, heating and power generation and can be further purified and compressed and used as automotive fuel. Piercy et al. (2024) discussed the use of anaerobic digestion methods for the production of biogas from food wastes. Benefits of biogas production from food wastes can include reduction in air emissions, especially of greenhouse gasses particularly carbon dioxide, methane, and nitrous oxide (Sufficiency et al. 2022).

Ohimain and Izah (2017) described several configurations of bioreactors for the production of biogas from palm oil mill effluents (POME). Ohimain and Izah (2015a, b) projected that about 75 billion Litres of biohydrogen can be produced from POME by mixed culture of anaerobes in Nigeria by 2029. Piercy et al. (2024) estimated that biogas generated from UK food wastes can power 26,500 households with green electricity. Energy recovery from food wastes supports sustainable development goals especially the global transition to renewable energy.

12.7.3 Thermochemical Conversion

Thermochemical methods especially pyrolysis, gasification, liquefaction, briquetting and pelleting can convert biomass including food wastes to diverse biofuel carriers (Fig.12.5). Briquetting is a thermochemical process for the conversion of loose biomass material to high density/ energy solid biofuel. It involved the application of compressive pressures for the densification and compaction of biomass. Both mechanical and hydraulic press are used, with or without binders. Many types of wastes can be converted to briquettes such as MSW, agricultural residues and

Fig. 12.5 Spoilt orange dumped along a Nigerian highway

food processing wastes such as corn wastes, straw, oil palm wastes, sugarcane bagasse, husks of various crops and timber residues particularly sawdust. Biomass pelleting is another method of compressing biomass to form pellets, which are similarly used for cooking and heating. Pellets are made from similar biomass like briquettes.

Pyrolysis involves the heating of biomass above 500 °C in the absence of air or oxygen, which result in breakdown of food wastes leading to the production of gases (synthetic gas i.e., syngas (H_2, CO), CO_2, light hydrocarbons), biochar, and bio-oil, which can be refined to biofuel. Under optimized conditions i.e. at high heating rate of 1000 °C/s (fast pyrolysis), yields (w/w) of 60–70% bio-oil, 10–15% syngas and 15–25% bio-char can be obtained from food wastes, which can be self-sustaining because a fraction of these energy carriers can drive the process (USDA 2024).

Liquefaction is a thermochemical process carried out in water or other solvents at moderate temperature of 250–400 °C and pressures of 5–25 MPa resulting in the breakdown of biomass into bio-oil, biochar, and gases (Huang et al. 2017). Hence, biomass liquefaction results in the production of similar products as fast pyrolysis. Bontaş et al. (2023) described the depolymerisation of biomass such as food wastes via liquefaction processes for the production of polyols, which are used for the synthesis of polyurethane, which is a bioplastic.

Gasification is a thermochemical process that converts biomass such as food wastes at a temperature of about 900 °C in a gasifier in the presence of limited amount of oxygen leading to the production of syngas and other gases (CO_2, CH_4, light hydrocarbons). Syngas has flexibility, because it can be utilized directly or used for the synthesis of diverse fuels in the presence of appropriate catalysts especially when integrated with Fischer-Tropsch Synthesis (FTS). Hydrogen, heat, electricity and many carbon-containing chemicals can be produced from biomass via gasification integrated with FTS including ethanol and methanol, methane, ethers, diesel, gasoline, kerosene. Both solid and liquid wastes can be valorised via gasification. Gasification is a thermochemical breakdown or partial oxidation or incomplete combustion of biomass. More information on gasification can be gotten from Mishra and Upadhyay (2021).

Biomass combustion is another thermochemical method of generating heat and power from biomass in the presence of air (biomass burning) at temperature of around 500 °C. It is fairly conventional because it has been in use for millennia and might possibly be the oldest form of energy conversion. It involved the complete oxidation of biomass. Fuel wood, charcoal and agricultural residues has been traditionally used for cooking in many rural communities. In larger scale, biomass can be used as boiler fuel, which can drive steam turbines for electricity. There are now combined heat and power systems. Dua (2024) reported the conversion of banana peelings to bioenergy, biofertilizer, and textile fibres in Pakistan. Annually, about 80 million tonnes of banana wastes are generated in the country, which was estimated to yield nearly 58 million cubic meters of syngas that can supply electricity to 50% of rural Pakistan and generate biofertilizers for soil replenishment.

The efficiency and diversity of value-added substances that can be produced from food wastes can be increased when thermochemical methods are integrated with biochemical methods. For instance, syngas produced via pyrolysis, liquefaction, and gasification can be used as feedstock for microbial fermentation leading to the production of many value-added biochemicals (Fig. 12.6). Syngas fermentation utilizes obligately anaerobic or acetogenic autotrophic bacteria such as *Clostridium* species including *Cl. ljungdahlii*, *Cl. Carboxidivorans, Cl. ragsdalei* and *Cl. autoethanogenum* to produce alcohols and other biochemicals from syngas (Henstra et al. 2007; Phillips et al. 2017). Rückel et al. (2021) demonstrated the production of ethanol, 1-hexanol and 1-butanol by *Clostridium carboxidivorans* through syngas and carbon dioxide fermentation in a stirred-tank reactor even in the presence of impurities. The technology for the production of ethanol via syngas fermentation route is currently available commercially (Henstra et al. 2007) (Figs. 12.7, 12.8, and 12.9).

Fig. 12.6 Culinary and medicinal mushrooms produced from food processing wastes, agricultural and forestry residues using recycled plastics, glass bottles and fabricated metal shelves

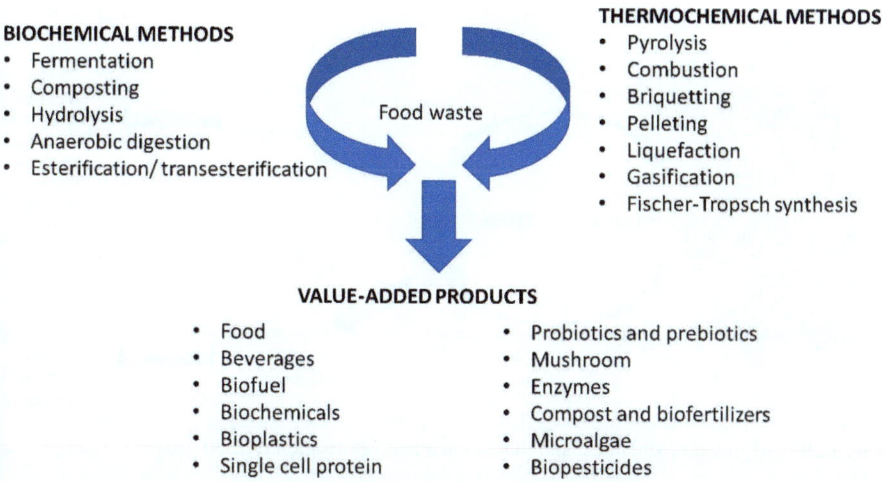

Fig. 12.7 Biochemical and thermochemical routes for food waste valorisation

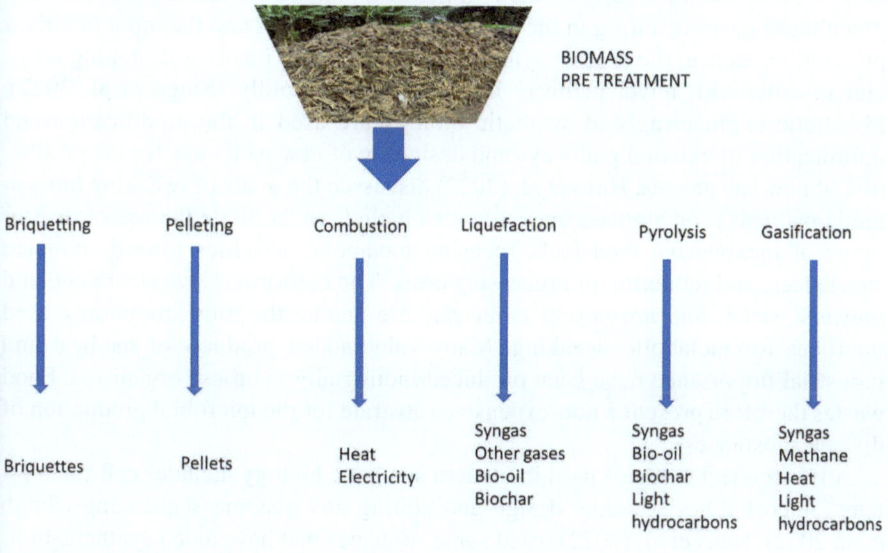

Fig. 12.8 Thermochemical conversion of food waste to bioenergy

12.8 Advancements in Synthetic Biology for Food Waste Utilization

Microbes have diverse metabolic activities, which makes them flexible for use in the valorisation and upcycling of waste biomass and other complex wastes. But the development and production of synthetic molecules, has created a challenge

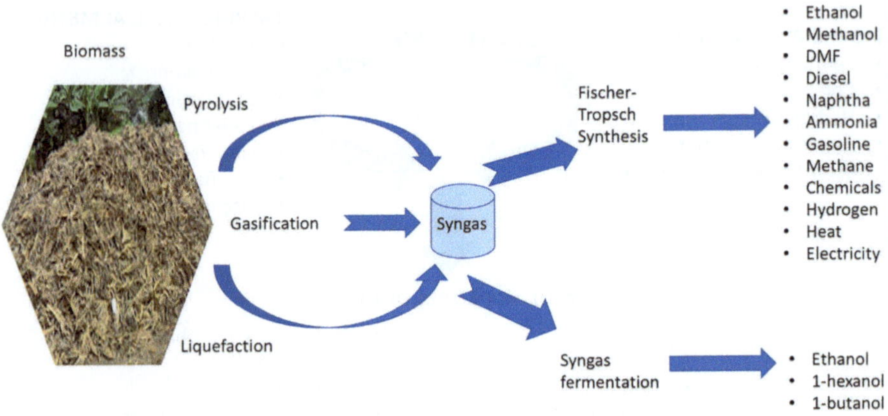

Fig. 12.9 Integrated biochemical and thermochemical valorisation of food waste

for the management of their wastes. Besides, there is an increasing demand of value-added products with novel functionalities. The complete microbial degradation of food wastes results in their mineralization, which emits pollutants and greenhouse gases resulting in the loss of natural resources and the input into their production. Hence, the field of synthetic biology, which involves designing artificial systems with novel pathway is now growing rapidly (Singh et al. 2022). Metabolic engineering and synthetic biology are used in the modification and optimization of existing pathways and designing of new pathways for the production of new substances. Hans et al. (2022) discussed the goals of tweaking biosynthetic pathways for biomass upcycling can include to facilitate the use of diverse range of inexpensive feedstock, focusing metabolic activities towards targeted substances, and reduction of processing costs. The coliform, *Escherichia coli* and brewers' yeast, *Saccharomyces cerevisiae* are among the most commonly used microbes for metabolic tweaking. Many value-added products of medical and industrial importance have been produced industrially via these organisms. Food wastes therefore present a non-expensive substrate for the microbial production of diverse substances.

Advanced technologies used in modern synthetic biology includes cell-free systems, microfluidics, genome design and editing and genome sequencing (Singh et al. 2022). Hans et al. (2022) listed some strategies that have aided synthetic biology for biomass upcycling including in-silico pathway prediction, de-novo pathway design, and metabolic engineering. Cell-free systems for example have found wide applications in the production of biofuel and other biochemicals. Maurya et al. (2023) discussed the production of biofuels including ethanol, hydrogen, butanol, bisabolene, isoprene, limonene, farnesene and pinene using cell-free systems.

Liu et al. (2022) discussed advances in protein engineering and synthetic biology with potential applications in food science and biofuel production. They observed that recent technological advances in recombinant DNA, CRISPR-Cas and "omics," when combined with high-throughput screening infrastructure can lead to major improvement in the tweaking of proteins particularly enzymes for improved performance and novel applications.

12.9 Policy and Economic Aspects of Microbial Food Waste Management

Food waste policy can interact with other sectors especially agriculture, nutrition and health, food processing and manufacturing, product consumer and domestic sector, waste management and environmental protection, and energy sector especially renewables. About 98% of global hunger population dwell in developing countries of which around 15% suffer from various nutritional challenges such as nutritional deficiency and malnutrition (Mahish et al. 2024). The microbial valorisation of food wastes supports circular economy, preservation of natural resources, waste-to-wealth including renewable energy and contribute to meeting most of the United Nations Sustainable Development Goals (SGDs) (Table 12.2). For instance, SDG 2 advocated for an end to hunger by 2030, which calls for increased food production, whereas SDG 12 calls for the halving of food wastes before 2030. It has been predicted that agricultural production needs to increase by about 70% to be able to feed the teaming population of people by 2050 (Šunjka and Mechora 2022). Hence, the microbial valorisation of food wastes could drastically minimize or potentially eliminate food wastes, which can reduce world hunger considerably. Lenkiewicz (2023) opined that the management of food wastes following waste hierarchy and circular economy ideologies can contribute to achieving SGDs.

The effective management of food wastes, will require a shift in human approach and lifestyles especially consumption pattern. Choices that are sustainable need to be made in key sectors of human endeavour. The manner in which wastes are perceived need to change. Sustainable waste management could involve actors at different levels such as individual, families, community, state or region, national and international. It could involve governments, community-based organizations and non-governmental organizations including international organizations and development banks. Said et al. (2023) advocated for government support in valorisation of food wastes especially in the formulation of policies backed with sanctions and incentives. Additional policies are needed on food, renewable energy, green and circular economy, environmental and resource conservation, and waste management and valorisation. Ecological economics can play a role in the valorisation of food wastes. For instance, if the cost of avoided environmental pollution of food wastes are considered along with the value-added products, can make food waste upcycling to be more economically competitive. Discussions on the full extent of applicable policies is beyond the scope of this chapter.

Table 12.2 Possible contribution of food valorisation to sustainable development goals

SGDs	Goals	Interface with food waste
Goal 1	No poverty	Reduction of food wastes especially among rural farmers in resource limited countries could contribute to poverty reduction
Goal 2	Zero hunger	Valorization of food wastes and sustainable agricultural practices could contribute to ending hunger, enhance nutrition and food security
Goal 3	Good health and well-being	Valorization of food waste, which could lead to reduction in hunger and enhancement of nutrition can contribute to good health and reduce morbidity and mortality
Goal 4	Quality Education	Waste management education at all levels can led to behavioural change towards sustainability
Goal 5	Gender equality	Food wastes generated at home are mostly managed by women and their daughters. Education of women on food waste avoidance, reduction and management can empower them
Goal 6	Clean water and sanitation	Food wastes valorization will prevent their dumping into waterbodies and prevent water pollution.
Goal 7	Affordable and clean energy	Food and other agricultural wastes can be utilized for the generation of renewable energy including biogas, bioethanol, biodiesels and briquettes to enhance energy access and security
Goal 8	Decent work and economic growth	Waste recycling especially in resource poor countries are done by the informal sector and need to be integrated into the formal sector
Goal 9	Industry, innovation and infrastructure	Innovations, conversion technology and infrastructure are key in the valorization of food wastes at different scales.
Goal 10	Reduced inequalities	Upcycling of food wastes could result in waste-to-wealth, which could reduce inequality within and between countries
Goal 11	Sustainable cities and economies	Upcycling of food wastes could reduce emissions and diseases, which could enhance economic activities and contribute to sustainable cities
Goal 12	Responsible consumption and production	Valorization of food wastes could contribute to sustainable production and responsible consumption. Reduction of food wastes at the supply chain including postharvest losses can lead to more efficient use of natural resources and sustainable production
Goal 13	Climate action	Upcycling of food wastes could contribute to reduction of greenhouse gas from landfills. Educate people on climate actions including adaptation and mitigation
Goal 14	Life below water	Food wastes valorization will prevent their dumping into waterbodies and prevent water pollution.
Goal 15	Life on land	Food wastes valorization will prevent their dumping indiscriminately on land leading to air emissions. Compost from recycling of food wastes can be used to increase the fertility of degraded lands
Goal 16	Peace, justice and strong institutions	Food wastes upscaling, which can provide wealth and reduce inequality can contribute to peace
Goal 17	Partnership for the goals	The successful valorization of food wastes will require partnerships at all levels; individuals, families, government and industry

12.10 Future Perspectives and Conclusion

The goal of circular economy is to recycle and upcycle wastes to prevent their disposal in dumpsites and landfills. Upcycling food wastes not only create value-added products, but mitigate against air emissions and has several other environmental, economic and social benefits. The future of wastes-to-wealth is quite promising, but there is policy, technological, logistical and infrastructural challenges that need to be addressed before it can be actualized. Lee et al. (2024) noticed that the upcycling and commercialization of food wastes has been slow due to technical, operational, economic, and regulatory challenges.

Logistical and infrastructural including energy challenges are major barriers for the upcycling of food wastes. This challenge is especially prevalent in resource-limited countries where there is lack of adequate infrastructure and supply chains for the collection, storage and processing of food wastes in a timely manner. Esparza et al. (2020) recognized the various challenges in the valorization of food wastes such as logistics, bulkiness, high moisture content, perishability and heterogeneity. The food waste situation is most problematic in tropical countries with poor economies, due to the high ambient temperatures that accelerates food spoilage because of higher microbial activities and the presence of tropical flies that lay eggs, which can hatch into maggots within 24 hours. Besides, sanitation is poor, access road for the evacuation of farm products and food wastes is inadequate, electricity and cold chain facilities are limited in poor countries. Hence, it is being advocated in tropical countries that food wastes should be collected frequently i.e. within 24 hours in order to prevent fly infestation (UNEP 2024). This can increase the cost of food waste handling compared to other types of wastes. Besides, food wastes are biochemically unstable and therefore require considerable energy for their storage, pretreatment and processing. Most of the food upcycling methods require energy, which may be lacking, unstable or insufficient in resource poor countries. Said et al. (2023) suggested the use of emerging information technologies such as internet of things, artificial intelligence and machine learning especially by enhancing the supply chain and predicting both the quality and quantity of food wastes, which could enable real-time decision-making and optimization of the valorisation process.

Technology limitation is another challenge for food waste valorisation especially among poor countries. Food waste can emerge as mixed waste stream, which can present considerable technical challenges in their handling. For instance, when food wastes are mixed in MSW, they could become contaminated with other substances containing heavy metals, which potentially reduces their value for valorisation (UNEP 2024) and increase the cost of their handling. Food wastes can also emerge in different states such as solid, semi-solid and liquids or even their mixtures containing diverse substances like oil, wastewater, packaging materials, household items. Besides, food wastes have varied physicochemical properties (Sufficiency et al. 2022), which present considerable technological challenges in their handling and valorisation. However, Lad et al. (2022) opined that due to mixture of substances and their complexity, food processing wastes and other industrial effluents including sewage are better handled by microbial valorisation processes than physicochemical processes. Technological advances in food waste valorisation including biochemical, thermochemical and synthetic biology is not widespread and the developing countries are lagging behind.

Public acceptability of upcycled food poses a challenge to their advancement. Waste recycling in developing countries are mostly operated by the informal sector, which are hardly regulated. UNEP (2024) emphasized the need to remove barriers militating against food waste valorisation by providing the necessary infrastructure, establish or strengthen food waste collection value chain and incentivise wastes. Hence, policies pertaining to food, wastes, and environmental protection, similarly need to be upgraded to make the sector economically sustainable. The informal waste recyclers should be recognized, formalized and rewarded for their ecological role in pre-empting environmental pollution arising from waste food disposal. Such policy shift could encourage behavioural changes that could encourage domestic gathering, sorting, packaging and preservation that could strengthen food wastes supply chain. Public awareness is key especially on consumption choices, lifestyle pattern, domestic waste handling, and acceptability of upcycled products. These could increase market value for food wastes and spur private sector investment in the subsector. Regulatory supervision of food wastes value chain can boost consumer confidence in the adoption of upcycled products.

References

Anoraga SB, Shamsudin R, Hamzah MH, Sharif S, Saputro AD (2024) Cocoa by-products: a comprehensive review on potential uses, waste management, and emerging green technologies for cocoa pod husk utilization. Heliyon 10(16):e35537

Axil (2024) What is a waste management hierarchy? https://axil-is.com/blogs-articles/waste-management-hierarchy/. Accessed 9 Aug 2024

Balasubramanian S, Tyagi RD (2017) Biopesticide production from solid wastes. In: Wong JWC, Tyagi RD, Pandey A (eds) Current developments in biotechnology and bioengineering. Elsevier, pp 43–58

Bontaş MG, Diacon A, Călinescu I, Rusen E (2023) Lignocellulose biomass liquefaction: process and applications development as polyurethane foams. Polymers 15(3):563

Brandenberg OF, Schubert OT, Kruglyak L (2022) Towards synthetic PETtrophy: engineering Pseudomonas putida for concurrent polyethylene terephthalate (PET) monomer metabolism and PET hydrolase expression. Microb Cell Factories 21(1):119. https://doi.org/10.1186/s12934-022-01849-7

Chen M, Sun Y, Zhu L, Li L, Zhao Y (2024) Study on the skincare effects of red Rice fermented by *Aspergillus oryzae* in vitro. Molecules (Basel, Switzerland) 29(9):2066. https://doi.org/10.3390/molecules29092066

Cherubin MR, Oliveira DMDS, Feigl BJ, Pimentel LG, Lisboa IP, Gmach MR et al (2018) Crop residue harvest for bioenergy production and its implications on soil functioning and plant growth: a review. Sci Agric 75:255–272

Dua S (2024) New method turns banana waste into green textiles, energy. https://interestingengineering.com/energy/banana-waste-green-textiles-energy. Assessed 27 Aug 2024

Esparza I, Jiménez-Moreno N, Bimbela F, Ancín-Azpilicueta C, Gandía LM (2020) Fruit and vegetable waste management: conventional and emerging approaches. J Environ Manag 265:110510
FAO (2019) The state of food and agriculture: moving forward on food loss and food waste reduction. Food and Agriculture Organization of the United Nations, Rome. https://www.fao.org/3/ca6030en/ca6030en.pdf
FAO (2022) Agricultural production statistics 2000–2021 FAOSTAT analytical brief 60. Food and Agricultural Organization of the United Nations, Rome
Ghoreishi G, Barrena R, Font X (2023) Using green waste as substrate to produce biostimulant and biopesticide products through solid-state fermentation. Waste Manag 159:84–92
Hans S, Kumar N, Gohil N et al (2022) Rebooting life: engineering non-natural nucleic acids, proteins and metabolites in microorganisms. Microb Cell Factories 21:100. https://doi.org/10.1186/s12934-022-01828-y
Henstra AM, Sipma J, Rinzema A, Stams AJ (2007) Microbiology of synthesis gas fermentation for biofuel production. Curr Opin Biotechnol 18(3):200–206
Huang H, Yuan X, Wu G (2017) Liquefaction of biomass for bio-oil products. In: Singh L, Kalia V (eds) Waste biomass management – a holistic approach. Springer, Cham. https://doi.org/10.1007/978-3-319-49595-8_11
Ikuemonisan ES, Mafimisebi TE, Ajibefun I, Adenegan K (2020) Cassava production in Nigeria: trends, instability and decomposition analysis (1970-2018). Heliyon 6(10):e05089. https://doi.org/10.1016/j.heliyon.2020.e05089
Lad BC, Coleman SM, Alper HS (2022) Microbial valorization of underutilized and nonconventional waste streams. J Ind Microbiol Biotechnol 49(2):kuab056
Lee A, Lan JCW, Jambrak AR, Chang JS, Lim JW, Khoo KS (2024) Upcycling fruit waste into microalgae biotechnology: perspective views and way forward. Food Chem Mol Sci 8:100203
Lenkiewicz Z (2023) The circular economy. In: Ghosh SK, Eduljee G (eds) Meeting sustainable development goals, vol 51, chapter 10. Royal Society of Chemistry, pp 224–251
Liu R, Liang L, Lacerda MP, Freed EF, Eckert CA (2022) Advances in protein engineering and its application in synthetic biology. In: New Frontiers and applications of synthetic biology. Academic, pp 147–158
Mahabeer G, Jin S (2024) Upcycling food waste into biomaterials applicable to medical products. Sustainability 16(11):4473
Mahish PK, Verma DK, Ghritlahare A, Arora C, Otero P (2024) Microbial bioconversion of food waste to bio-fertilizers. Sustain Food Technol 2(3):689–708
Malkawi HI, Kapiel TYS (2024) Microbial biotechnology: a key tool for addressing climate change and food insecurity. Eur J Biol Biotechnol 5(2):1–15
Maurya R, Chaudhari H, Mansuri A, Gohil N, Bhattacharjee G, Khambhati K, Singh V (2023) Cell-free systems for sustainable production of biofuels. In: Singh V, Show PL (eds) Biomanufacturing for sustainable production of biomolecules. Springer Nature Singapore, Singapore, pp 331–348. https://doi.org/10.1007/978-981-19-7911-8_17
Miran W, Nawaz M, Jang J, Lee DS (2016) Conversion of orange peel waste biomass to bioelectricity using a mediator-less microbial fuel cell. Sci Total Environ 547:197–205. https://doi.org/10.1016/j.scitotenv.2016.01.004
Mishra S, Upadhyay RK (2021) Review on biomass gasification: gasifiers, gasifying mediums, and operational parameters. Mater Sci Energy Technol 4:329–340
Narancic T, Salvador M, Hughes GM, Beagan N, Abdulmutalib U, Kenny ST, Wu H, Saccomanno M, Um J, O'Connor KE, Jiménez JI (2021) Genome analysis of the metabolically versatile Pseudomonas umsongensis GO16: the genetic basis for PET monomer upcycling into polyhydroxyalkanoates. Microb Biotechnol 14(6):2463–2480. https://doi.org/10.1111/1751-7915.13712
Ohimain EI (2020) Possible use of Ganoderma mushroom extracts for the treatment of some chronic risk factors associated with severe Covid-19. Virol Immunol J 4(3):000254

Ohimain EI, Izah SC (2014a) Possible contributions of palm oil mill effluents to greenhouse gas emissions in Nigeria. Br J Appl Sci Technol 4(33):4705–4720

Ohimain EI, Izah SC (2014b) Energy self-sufficiency of smallholder oil palm processing in Nigeria. Renew Energy 63:426–431

Ohimain EI, Izah SC (2015a) Estimation of potential biohydrogen from palm oil mills' effluent in Nigeria using different microorganisms under light independent fermentation. J Environ Treat Techniq 3:97–104

Ohimain EI, Izah SC (2015b) Energy self-sufficiency of semi-mechanized oil palm processing: a case study of Bayelsa Palm Mill, Elebele, Nigeria. Energy Econom Lett 2(3):35–45

Ohimain EI, Izah SC (2017) A review of biogas production from palm oil mill effluents using different configurations of bioreactors. Renew Sust Energ Rev 70:242–253

Ohimain EI, Jenakumo CB (2013) Scrap metal recycling and valorization in Bayelsa State, Nigeria. J Mater Sci 119:137–147

Ohimain EI, Silas-Olu DI, Zipamoh JT (2013a) Biowastes generation by small scale cassava processing centres in Wilberforce Island, Bayelsa State, Nigeria. Greener J Environ Manag Public Safe 2(1):51–59

Ohimain EI, Izah SC, Obieze FA (2013b) Material-mass balance of smallholder oil palm processing in The Niger Delta, Nigeria. Adv J Food Sci Technol 5(3):289–294

Pereyra-Camacho MA, Pardo I (2024) Plastics and the sustainable development goals: from waste to wealth with microbial recycling and upcycling. Microb Biotechnol 17(4):e14459

Phillips JR, Huhnke RL, Atiyeh HK (2017) Syngas fermentation: a microbial conversion process of gaseous substrates to various products. Fermentation 3(2):28

Piercy E, Ellis PR, Guo M (2024) Microbial upcycling of food waste using anaerobic digestion for energy and single-cell protein productio. Curr Opin Food Sci 57:101173

Pleissner D, Lin CSK (2013) Valorisation of food waste in biotechnological processes. Sustain Chem Proc 1:1–6

Poyeri WR, Ohimain EI (2024) Cultivation, molecular identification and productivity assessment of a commercial strain and two wild strains of *Ganoderma* mushroom harvested from tree stumps in a rainforest at Ugbowo, Benin City, Nigeria. J Appl Sci Environ Manag 28(7):2207–2215

Rakesh B, Mahendran R (2023) Upcycling of food waste and food loss–a sustainable approach in the food sector. Trends Food Sci Technol 143:104274

Rückel A, Hannemann J, Maierhofer C, Fuchs A, Weuster-Botz D (2021) Studies on syngas fermentation with Clostridium carboxidivorans in stirred-tank reactors with defined gas impurities. Front Microbiol 12:655390

Sabater C, Ruiz L, Delgado S, Ruas-Madiedo P, Margolles A (2020) Valorization of vegetable food waste and by-products through fermentation processes. Front Microbiol 11:581997. https://doi.org/10.3389/fmicb.2020.581997

Said Z, Sharma P, Nhuong QTB, Bora BJ, Lichtfouse E, Khalid HM et al (2023) Intelligent approaches for sustainable management and valorisation of food waste. Bioresour Technol 377:128952

Sekoai PT, Roets-Dlamini Y, O'Brien F, Ramchuran S, Chunilall V (2024) Valorization of food waste into single-cell protein: an innovative technological strategy for sustainable protein production. Microorganisms 12(1):166

Singh V, Bhattacharjee G, Gohil N, Maurya R, Lam NL, Alzahrani KJ (2022) An introduction to advanced technologies in synthetic biology. In: New Frontiers and applications of synthetic biology. Academic, pp 1–9

Sufficiency E, Qamar SA, Ferreira LFR, Franco M, Iqbal HM, Bilal M (2022) Emerging biotechnological strategies for food waste management: a green leap towards achieving high-value products and environmental abatement. Energy Nexus 6:100077

Šunjka D, Mechora Š (2022) An alternative source of biopesticides and improvement in their formulation-recent advances. Plants (Basel, Switz) 11(22):3172. https://doi.org/10.3390/plants11223172

Tiso T, Narancic T, Wei R, Pollet E, Beagan N, Schröder K, Honak A, Jiang M, Kenny ST, Wierckx N, Perrin R, Avérous L, Zimmermann W, O'Connor K, Blank LM (2021) Towards bio-upcycling of polyethylene terephthalate. Metab Eng 66:167–178. https://doi.org/10.1016/j.ymben.2021.03.011

Tiso T, Winter B, Wei R, Hee J, de Witt J, Wierckx N, Quicker P, Bornscheuer UT, Bardow A, Nogales J, Blank LM (2022) The metabolic potential of plastics as biotechnological carbon sources - review and targets for the future. Metab Eng 71:77–98. https://doi.org/10.1016/j.ymben.2021.12.006

Tongwane M, Mdlambuzi T, Moeletsi M, Tsubo M, Mliswa V, Grootboom L (2016) Greenhouse gas emissions from different crop production and management practices in South Africa. Environ Dev 19:23–35

UNEP (2024) Global Waste Management Outlook 2024: beyond an age of waste – turning rubbish into a resource. United Nations Environment Programme, Nairobi. https://wedocs.unep.org/20.500.11822/44939

USDA (2024) What is Pyrolysis? https://www.ars.usda.gov/northeast-area/wyndmoor-pa/eastern-regional-research-center/docs/biomass-pyrolysis-research-1/what-is-pyrolysis/

Vásquez ZS, de Carvalho Neto DP, Pereira GVM, Vandenberghe LPS, de Oliveira PZ, Tiburcio PB, Rogez HLG, Góes Neto A, Soccol CR (2019) Biotechnological approaches for cocoa waste management: a review. Waste Manag (New York, NY) 90:72–83. https://doi.org/10.1016/j.wasman.2019.04.030

World Bank (2022) World Development Indicators. https://datatopics.worldbank.org/world-development-indicators/. Accessed 26 Aug 2024

Yu F, Teng Y, Li J, Yang S, Zhang Z, He Y, Zhou P (2023) Effects of a Ganoderma lucidum Proteoglycan on type 2 diabetic rats and the recovery of rat pancreatic islets. ACS Omega 8(19):17304–17316

Zamil DH, Khan RM, Braun TL, Nawas ZY (2022) Dermatological uses of rice products: trend or true? J Cosmet Dermatol 21(11):6056–6060. https://doi.org/10.1111/jocd.15099

Chapter 13
Microbial Bioconversion of Agro-industrial Wastes into Value-Added Products

Chioma Christiana Obi and Olukayode Oladipo Amund

Abstract Agro-industrial wastes and by-products are organic materials that come from a variety of processes in the agricultural and agro-processing industries. Globally, agro-industrial wastes remain under-exploited and poorly managed; many current approaches in managing agro-industrial wastes especially in third world countries culminate in environmental pollution, emission of greenhouse gases and a public health risk. Therefore, the use of microorganisms to convert these wastes into value added products (VAPs) such as biopolymers, biofuels, and enzymes, microbial protein, organic acids, antibiotics is a sustainable renewable approach to handle the above-mentioned environmental concerns. Agro-industrial wastes include field crop wastes, crop leftovers, harvest wastes, animal dung or manure, wastes from slaughterhouses and meat processing, and wastes from food preparation and consumption. Because these wastes are rich in nutrients that support microbial growth, they can be employed as substrates in different bioprocesses such as fermentation (solid state and submerged) for the production of value added products that have numerous applications in chemical, cosmetics, food, textile and pharmaceutical industries. In this chapter, we considered agro-industrial wastes and their sources, different strategies for microbial bioconversion of these wastes into VAPs, valorization pathways for diverse VAPs and examples of value-added products with their possible industrial applications.

Keywords Bioconversion · Agro-industrial wastes · Valorization · Fermentation · VAPs

C. C. Obi (✉)
Department of Biological Sciences, Bells University of Technology, Ota, Nigeria
e-mail: ccobi@bellsuniversity.edu.ng

O. O. Amund
Department of Microbiology, Faculty of Science, University of Lagos, Akoka, Lagos State, Nigeria

O. O. Babalola et al. (eds.), *Ecofriendly Frontiers*,
https://doi.org/10.1007/978-3-031-98700-7_13

13.1 Introduction

The need for food and agro-industrial products due to the growing worldwide population has forced agriculture to evolve into a sustained mode of production, generating massive amounts of food wastes, agro-industrial wastes, and agricultural residues every year (Yafetto et al. 2023). Researchers from all around the world have recently focused on the financial losses, environmental contamination, and health risks brought on the ecosystem by improper handling of excess agro-industrial wastes (Sadh et al. 2018). It is estimated that about 700 million tonnes of agro-industrial wastes are generated annually worldwide with Nigeria accounting for about 183 ± 8.9 metric tonnes or 26% (Oluwasola et al. 2021; Fritsch et al. 2017). Globally, agro-industrial wastes remain under-exploited and poorly-managed. Many current approaches in managing agro-industrial wastes especially in third world countries constitute environmental pollution and a public health risk (Oluwasola et al. 2021), making the development of more eco-friendly means of managing this menace imperative. One of such ways is the use of microorganisms (bioconversion) through several bioprocesses to produce biological products such as bioethanol (Singh et al. 2022; Huzir et al. 2018), biodiesel (Devi et al. 2022; Martínez et al. 2015), single cell protein (Mondal et al. 2012), antibiotics (El-Housseiny et al. 2021; Shata 2018), organic acids (Torrado et al. 2011; Sukan et al. 2014), biopolymers (Yadav et al. 2020), enzymes (Tuysuz et al. 2020; Paul et al. 2020) that will not only reduce agro-industrial wastes but increase economic growth and human well-being.

Wastes from slaughterhouses and meat processing, animal manure or dung, field crop wastes, crop leftovers, harvest wastes, and wastes from food preparation and consumption are examples of agro-industrial wastes (Rasool et al. 2023; Yafetto et al. 2023; Diaz-Vazquez et al. 2021). These wastes possess significant nutritional value due to their elevated levels of proteins, minerals and sugars, thereby promoting microbial growth. Therefore, they can serve as substrates for the production of value added products through different fermentation strategies (Ravindran et al. 2018). A vast population of naturally occurring microbes including bacteria (*Bacillus subtilis, Lactobacillus, Candida tropicalis*), fungi (*Penicillium* sp., *Aspergillus* sp., *Rhizopus oryzae*) and algae (*Chlorella* sp., *botryococcus braunii, Crypthecodinium cohnii, Spirulina* sp. etc.) are capable of transforming agro-industrial wastes into plant nutrients which help in reducing the carbon-nitrogen ratio in the ecosystem thereby drastically minimizing ecological imbalance and enhance soil fertility and productivity (Aslam et al. 2023). This could be achieved through several strategies such as composting; an aerobic solid phase breakdown of organic materials generally lignocellulosic contents (lignin, cellulose and hemicellulose) of agricultural and industrial wastes. Furthermore, microorganisms possess the capability to produce value added products such as biodiesel, biogas and bioethanol via anaerobic digestion of agro-industrial wastes under controlled conditions. Bioconversion of agro-industrial wastes by pure cultures or consortia using solid state fermentation (SSF) or submerged fermentation processes has also produced value-added products like organic acids, antibiotics, amino acids, enzymes, biopolymers, and bioactive compounds (phenolics). This chapter is therefore aimed at providing an overview of agro-industrial wastes, the different strategies for microbial bioconversion of these wastes into value-added products, microbial bioconversion pathways and inhibitors, and the possible value-added products that could be generated from them.

13.2 Agro-industrial Wastes: An Overview

Agro-industrial wastes refer to biodegradable wastes generated from agricultural and industrial practices, including crop residues (leaves, stalks, seeds, stems and pods), food wastes, and by-products (molasses, bagasse, husks, pulps etc.) from processing industries. Agro-industrial waste is produced during the industrial processing of agricultural or livestock products (Naik et al. 2023). Agro-wastes make up around 30% of the agricultural products produced worldwide. These wastes can be grouped into solid, liquid, or slurry forms (El-Ramady et al. 2022). However, International Energy Agency (IEA) (2017) grouped agro-industrial waste into two: namely primary and secondary agro-residues. Wastes originating directly from crop and animal production activities, such as stover, stalk, bagasse, cob, pruning, straw, leaves, and animal dung or manure, are classified as primary residues. On the other hand, wastes originating from agro-allied industrial processing, such as pit, shell, peels, husk, cake, slurry, and slaughterhouse wastes, are classified as secondary wastes (IEA 2017). Several studies have used digital agricultural technologies and sustainable management to quantify the production of agricultural wastes and their environmental effects, including global nutritional and environmental losses, with a focus on bioconversion of agricultural wastes as a more cost-effective and environmentally friendly alternative to treatment (Ogunmoroti et al. 2022).

In this chapter, we categorised agro-industrial wastes based on the origin of the wastes into two groups namely; crop residues and livestock based agro-industrial wastes (Fig. 13.1). Plant based (crop residues) agro-waste is further classified into agricultural field generated wastes (leaves, stems, pods, seeds, stalks, straws etc.) and post-harvest wastes arising from processing industries which include fruit residues, pulp, husks, molasses, bagasse. Similarly, agrowastes from livestock is classified into wastes generated from animal husbandry which include cow dung, poultry litters, diary shed effluent, shells, spilled feed, residual milk, bedding materials, renderings etc. and wastes from meat processing industries such as liquid cheese whey, oil/fat residues, packing materials, bones, offal, waste water (effluents containing blood). The fruit, juice, chip, meat, and confectionery industries, as well as other food processing sectors, generate enormous amounts of organic wastes and associated effluents annually. According to Sadh et al. (2018), these organic wastes can be used for a variety of energy sources (biofuels). Manikandan et al. (2022) reported a specific waste index for animals to be 0.56, 0.87, 0.2 and 0.1 for cow, calf, pig and sheep respectively. The calculation of these indexes involves dividing

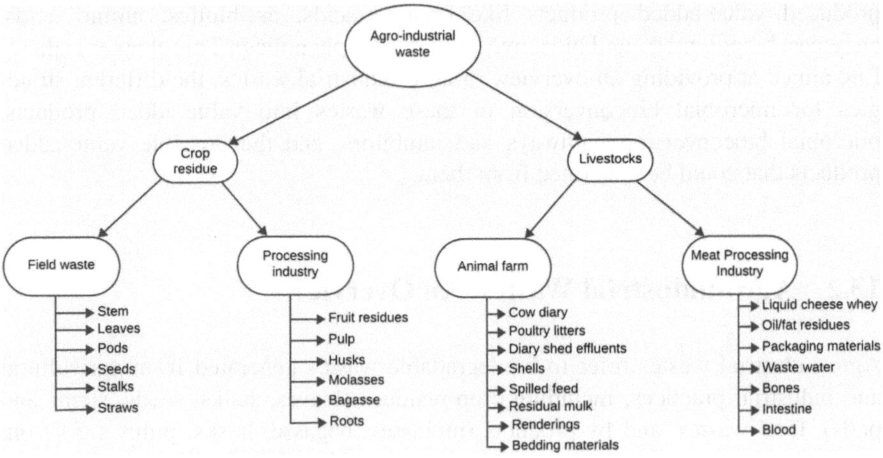

Fig. 13.1 Types of Agro-industrial wastes

the total mass of wastes by the mass of product that can be sold. However, other agricultural waste, such as rice and wheat straw waste were estimated to have been 709 and 673 metric tons in the previous century, respectively. In contrast, there are currently 147 metric tonnes of fiber sources worldwide (Manikandan et al. 2022).

13.3 Strategies for Microbial Bioconversion of Agro-industrial Wastes

The process of turning agro-industrial wastes into value-added products by microbial bioconversion has garnered a lot of attention recently, especially in developing countries (Manikandan et al. 2022). Depending on the waste type and the microorganism producing it, these processes can be either solid state fermentation (SSF) which involves growing microorganisms on moist solid substrate without the presence of free-flowing water or submerged fermentation (SmF), which involves growing microorganisms in liquid nutrient media. Both processes take place under controlled environmental conditions. Complexes of proteins, carbohydrates, and polyphenolic compounds are among the many components found in agro-industrial wastes that support microbial growth. However, microbes are prone to microbial inhibitions that could arise from pretreatment steps as well as during fermentation process, which can have negative effects on the rate of bioconversion due to their inhibitory effect on the microbial active enzymes. These wastes can therefore be used as inexpensive, sustainable sources of nutrients for fermentation processes that produce value-added products, generating mass and energy balances from cultivation, extraction, processing, and disposal (Panesar et al. 2015). Since the nutritional components of a culture medium make up between 38% and 72% of the total production cost, the availability and affordability of agro-industrial wastes are essential for the commercially viable production of microbial products (Buraimoh et al. 2021; Astudillo et al. 2023). The most commonly reported microbial products that can be produced via fermentation processes whether SSF or SmF include biogas, biofuel, enzymes, pigments, bioethanol, biopolymers, organic acids, single cell protein, biosurfactants, antibiotics, and phenolic compounds used for different industrial applications.

13.3.1 Substrate Pretreatment Methods

Agro-industrial wastes are known to be resistant to microbial breakdown as a result of their complex physical and chemical structure. As a result of this, few procedures are applied to the biomass in order to overcome the above mentioned limitation. Pretreatment procedures are actions used on biomass to get over the earliest obstacles to its simple conversion. These processes increase the conversion efficiency by changing the biomass's components. The complex molecules in the biomass are broken down into hemicellulose, cellulose and lignin during the pretreatment. The goal of cellulose breakdown is to decrease the crystallinity of the cellulosic moiety while increasing its porosity and surface area which enhances microbial/substrate interaction. Reduced energy consumption during conversion, cost savings, and the direct generation of sugars from the biomass are the primary goals of pretreatment of agricultural wastes and other lignocellulosic biomass (Awogbemi and Von Kallon 2022). Pretreatment techniques that are commonly employed include chemical, biological, physical, physicochemical, and green solvent-based procedures (Areepak et al. 2022). To reduce undesirable inhibitors, several physical, chemical, and biological steps must be taken in an integrated manner rather than using a single pretreatment process. One of the most promising and successful methods for pretreatment of biomass on an industrial scale is chemical pretreatment. Alkaline pretreatment has been widely employed as a successful delignification technique. In this processs, NaOH is typically used to remove lignin by dissolving the ester bonds that bind the lignin and hemicellulose biomass (Awogbemi and Von Kallon 2022). This causes the polysaccharide-lignin connections to dissociate, which delignifies the biomass at the temperatures between 50–180 °C. The majority of traditional alkaline pretreatments are carried out at a low solid:liquid ratio (1:5–1:20) with excessive soaking that produces wastewater; as a result, a solid-liquid separation step is necessary following the pretreatment which in turn increases the bioprocess time and cost (Kim et al. 2020). The alkaline pretreatment may also cause microstructural swelling, which makes the biomass more vulnerable to enzymatic degradation. Some physical characteristics of the lignocellulosic biomass, such as the degree of polymerization and the crystallinity of cellulose, nevertheless, may continue to impede the biological conversion phase. To get over these restrictions, cellulose's crystalline area and depolymerization have been reduced by mechanical size reduction, which speeds up the enzymatic hydrolysis process that releases fermentable sugars. Physical pretreatment methods are employed to increase the agricultural wastes surface area and reduce their particle size. More accessibility to enzymes and microbes during processing, a decrease in cellulose crystallinity, improved conversion efficiency, and ultimately a higher yield of product are all guaranteed by the size reduction during microbial bioprocessing (Awogbemi and Von Kallon 2022). Biological pretreatment involves the use of aerobic bacteria, enzymes, or fungi to break down complex carbohydrates such as lignin and facilitate the conversion of lignocellulose material into useful products. This method of pretreatment is environmentally benign, generates no harmful substances into the environment, uses little energy, and produces little to no inhibitors that could hamper successful bioconversions (Farkas et al. 2019). As a result, it takes a long time and is slow, which results in a large residence time. Additionally, there is a chance that microbial instability will cause unpredictable reaction outcomes.

13.3.2 Solid-State Fermentation (SSF)

Solid state fermentation is a process that involves the growth of microbes on moist solid material particles in beds where the spaces between the particles are filled with a continuous gas phase (Mitchell et al. 2011; Belorkar and Jogaiah 2022).). This fermentation occurs in the lack or near absence of free water. Vital enzymes (lipase, protease, cellulose, xylanase), secondary metabolites, organic acids, and their possible uses as biocontrol agents, bioremediation agents, and biofuels have all been produced using SSF as a biotechnological process (Das and Kumar 2018; Al Mousa et al. 2022; Basto et al. 2022). Solid-state fermentation offers an attractive perspective and is strongly tied to the concept of turning waste into value-added goods since it increases biomass value and reduces the need for downstream liquid volume treatment (Chilakamarry et al. 2022).

Recent years have seen the publication of several theme reviews that focus on the available SSF methods for the reprocessing and reuse of various agricultural and food industry wastes, especially lignocellulosic and vegetable wastes (Mitchell et al. 2011). This is mostly because of its benefits for producing high-value products, managing solid waste, and conserving biomass energy. Solid materials can be subjected to SSF as long as their surface is moist and reasonably solid. Opportunities offered by SSF include increased fermentation efficiency, enhanced product stability, and increased concentrations of final products, cultivation of water-insoluble substrates, decreased catabolic suppression, and combination agriculture of different fungal species (Yafetto 2022). Lower water activity-adapted microorganisms such as filamentous fungi and some yeast species flourish in specific SSF processes and hence minimize the number of upstream bioprocessing operations. SSF supports a more environmentally-friendly manufacturing process due to a very high reduction of wastewater levels (Chowdhary et al. 2021). The majority of microorganisms utilized in solid-state fermentation are filamentous fungi belonging to the *Penicillium, Rhizopus, Aspergillus, Fusarium*, and *Trichoderma* genera. Moreover,

Actinobacteria species such as *Streptomyces thermonitrificans, Streptomyces chattanoogensis* and some yeasts e.g. *Saccharomyces cerevisiae, Saccharomyces boulardii, Candida* sp. are also used in solid-state fermentation (Yafetto 2022). SSF has the advantage that majority of microorganisms employed in the process, especially filamentous fungi and yeasts, are typically thought to be safe mostly since the products are usually free of toxins, which makes them suitable for consumption by both humans and animals (Yafetto et al. 2019). SSF is basically a multistep process involving the following steps.

1. **Pre-treatment step**: In order to improve the availability of the bound nutrients and to decrease the size of the components, substrates for SSF will go through pre-treatment through mechanical, chemical, or biological processing. For instance, crushing straw and shredding vegetable components might enhance the physical features of the procedure. This stage which is also known as "upstream bioprocess," entails preparing growth media and substrates as well as isolating the microorganisms needed for fermentation (Yafetto 2022).
2. **Hydrolysis of primarily polymeric substrates, e.g., polysaccharides and proteins and their utilization (fermentation).** This step is also known as the midstream bioprocess i. e. the heart of any bioprocess where the bioconversion of substrates that leads to the production of VAPs takes place
3. **Separation and purification of end products:** This step involves the downstream processes where the biological products are recovered and purified to meet required quality and standard. It is worthy of note that the sequential steps involved in SSF are also generally used in many industries though with little modifications in methods involved in obtaining the final expected product (Nambi et al. 2017).

Solid-state fermentation has striking advantages as a biotechnological process for microbial conversion which include the following:

1. The process generates very little waste or liquid effluent, making it sustainable and friendly to the environment.
2. SSF uses natural solids as the medium because of its affordability, ease of use, and minimal energy and capital requirements.
3. Application of SSF for enzyme production provides higher productivity, high enzyme titre and stability
4. The bioreactor design, aeration process, and effluent treatment are relatively quite simple.
5. Wastes from homes, industries and farms can all be productively utilized in SSF

13.3.3 Submerged Fermentation (SmF)

The most widely used technique for producing enzymes in large scale is submerged fermentation, which entails inoculating microbial cultures into a liquid medium to produce desired products. Large-scale use of the submerged fermentation method necessitates more advanced infrastructure and exact control. Conversely, it offers a number of benefits like increased yields and productivity, efficient control of process parameters, lower fermentation period, reduced labour expenses, and a decreased danger of contamination (Luft et al. 2020). Batch, fed-batch, and continuous systems are the most commonly ways to perform submerged fermentation; the batch method is most frequently used to synthesize various compounds, including citric acid. Under either anaerobic or aerobic conditions, submerged fermentation can occur. For instance, butanol is created by anaerobic fermentation, whereas antibiotics and enzymes are produced by aerobic fermentation, which incorporates oxygen into the fermentation broth through stirring aeration. However, facultative anaerobic organisms like *Saccharomyces cerevisiae,* which can grow in the presence of oxygen and increase cell biomass, are used in other fermentation processes, like the production of ethanol. During the process of producing ethanol, such organisms may switch to anaerobic SmF process has been used in the production of many biological products from agro-industrial wastes such as enzymes/peptides (e.g. α-amylase, lipase, protease, xylanase, cellulose, laccase), industrial pigments (e.g. lycopene, carotenoids, astaxanthin), biosurfactants (rhamnolipids, lipopeptides), phenolic compounds (ellagic acid), antibiotics and some organic acids (Astudillo et al. 2023). Table 13.1 shows the list of VAPs that had been reported to be produced through submerged fermentation process using agro-industrial wastes by both bacteria and fungi.

Submerged fermentation process as an effective method for microbial conversion of agro-industrial wastes into VAPs has been reported extensively over the last decades (Das and Kumar 2018; Basto et al. 2022; Astudillo et al. 2023). Nonetheless, this biotechnological process has several drawbacks, such as high energy consumption, high wastewater generation, the necessity to regulate foam creation, and high capital investment required for stirred bioreactors (scale-up) (Asghari et al. 2020; Astudillo et al. 2023). Nevertheless, adequate evaluation and technological advancement of this process could be essential to achieve production at an industrial level.

13.4 Value-Added Bioproducts from Agro-industrial Wastes

Compounds of significant relevance for the pharmaceutical, food and beverage, textile and energy production industries had been obtained from agro-industrial residues through biotechnological processes that involve microbial transformation of these wastes (Dos Santos et al. 2024). In this section, some examples of VAPs, the

Table 13.1 VAPs from agro-industrial wastes through submerged fermentation (SmF)

Microorganism	Substrate	VAP	References
Bacteria			
Peudomonas azotoformans	Bagasse Potato peels	Biosurfactants	Das and Kumar (2018)
Bacillus subtilis M-9	Corn stover/banana stalk	Enzyme (protease)	Javed et al. (2013)
Bacillus subtilis ICA 56	Sugarcane molasses and corn steep solids (CSS)	Acetolactate decarboxylase (ALDC)	Oliveira et al. (2022)
Lactobacillus rhamnosus ATCC 10863	Oil palm trunk sap	Lactic acid	Saelee (2022)
Pseudomonas aeruginosa LBI2A1	Guava seed oil and peel	Rhamnolipids (biosurfactant)	Yoshimura et al. (2022)
Bacillus haynesii	Orange peel	Lipopeptide	Rastogi et al. (2021)
Rhodopseudomonas faecalis	Soybean meal	Lycopene	Patthawaro et al. 2020
Bacillus stearothermophilus	Corn steep liquor-soya bean meal	α-amylase	Basto et al. (2022)
Fungi			
Mucor circinelloides and *M. hiemalis*	Tangerine peel pomegranate peel and wheat straw	Enzyme (Xylanase)	Al Mousa et al. (2022)
Penicillium multicolour, P. canescens, P. herquie, Talaromyces verruculosus and Fusarium solani	Mixture of green waste and whey	Pigments	Molelekoa et al. (2021)
Trichoderma harzianum	Sugarcane bagasse, castor oil cake	Lipase	Coradi et al. (2013)
Pleurotus citrinopileatus	Olive-oil mill wastewater	Antityrosinase Compounds	Zerva et al. (2021)
Penicillium brevicompactum	Cheese-whey (CW) and cornsteep liquor (CSL)	Pigments (yellow, orange, and red)	Basto et al. (2022)
Candida tropicalis	Olive mill waste	d-limonene methyl butanoate	Guneser et al. (2017)
Aspergillus fumigatus	Orange peel waste	Ellagic acid	Sepúlveda et al. (2020)
Microalgae			
Phormidium autumnale	Slaughterhouse Wastewater	Carotenoids	Rodrigues et al. (2014)

substrate for their production, the bioprocess involved, the microorganisms and their industrial applications has been described.

13.4.1 Organic Acids

The term “organic acids” refers to polyfunctional acids, which include the core metabolic pathway, that have one or more carboxyl (-COOH), hydroxyl (-OH), thiol (-SH), etc. groups. Organic acids are among the most promising future products of industrial biotechnology, despite the fact that these carboxylates had less of an obvious effect on humans in the past. Recently, there is a great increase in the global demand for organic acids because of their several applications in industries such as pharmaceutical, cosmetics, cleaning, food and chemical feedstocks (Sharma et al. 2021). Notably, one approach that has been seen as promising for obtaining precursor chemicals from renewable feedstocks, such lignocellulosic agro-industrial wastes, is the microbial synthesis of organic acids through biotechnological processes. However, the implementation of this approach is still greatly hampered by the need to identify an appropriate microbial source and ideal growing conditions for maximum organic acid production (Dezam et al. 2017; Kaur et al. 2022). Examples of organic acids that could be produced by microorganisms from agro-industrial wastes are include lactic acid (El-Sheshtawy et al. 2022), citric acid (Egbe et al. 2022), gluconic acid (Singh and Singh 2006), succinic acid (Nagime et al. 2023; Qin et al. 2024), acetic acid (Roda et al. 2017) etc. (Table 13.2). Fungal species are the mostly reported group of organisms known for their ability to produced organic acids such as gluconic acid, succinic acid, citric acid, kojic acid e.t.c. Furthermore, microbial production of organic acids can be carried out by bacteria (*Bacillus licheniformis, Corynebacterium spp. Arthrobacter paraffineus*) and yeasts (*Candida tropicalis, C. citroformans, C. oleophila*) but the fungal groups are well known for their production of high amounts of various organic acids (Dezam et al. 2017; Sharma et al. 2021).

Submerged fermentation process has been widely reported as the best microbial cultivation method for the large scale production of organic acids (Liu et al. 2019). Nevertheless, the cultivation of filamentous fungi under solid-state fermentation (SSF) has received growing attention due to the intrinsic advantages of this cultivation method (Niego et al. 2023; Liu et al. 2019). These organic acids (acetic acid, citric acid, oxalic acids, lactic acid etc. have a wide range of applications in industries. For example, formic or acetic acids are used for oil and gas well stimulation treatment while others such as citric acid and oxalic acid are applied for rust removal. Similarly, organic acids are used in food preservation as a result of their effects on bacteria. They also form basic component of animal feeds.

Table 13.2 Organic acids from microbial conversion of agro-industrial wastes and industrial applications

Organic acid	Substrates	Bioprocess involved	Microorganism	Industrial application	References
Lactic acid	Cotton textile and coffee wastes	Submerged fermentation (SmF)	*Kosakonia cowanii*	Control of pathogens, shelf life increase, and food preservation, flavouring factor	El-Sheshtawy et al. (2022)
	Molasses cassava starch hydrolyzate	SmF, SSF	*Lactobacillus delbrueckii* NCIMB 8130		John et al. (2007)
Citric acid	Wheat straw, beet molasses, potato peel powder, corn steep liquor, rice straw	SmF	*Aspergillus niger*	Flavour enhancement, emulsification, preservation, anti-oxidation, food, chemical, beverage, cosmetic and pharmaceutical for stabilization, acidulation, as synergistic agent and plasticizer	Egbe et al. (2022)
Gluconic acid	Grape must Bagasse	SmF Solid state fermentation (SSF)	*Aspergillus niger* ORS-4.410	Beverages, pharmaceutical, food, cement, metal chelating, textile agent, leather, and dairy sectors	Singh and Singh (2006)
Hyaluronic acid (HA)	Corn steep liquor (CSL) and sugarcane molasses	SmF	*Streptococcus zooepidemicus*	Cosmetics industry	Amado et al. (2016)
Succinic acid	Cane molasses Palm oil mill effluent Sugar cane bagasse	SmF, batch fermentation	*Actinobacillus succinogenes* CGMCC1593 *Enterococcus gallinarum* (MW931746) *Actinobacillus succinogenes* GXAS137	Food and feed additives, pharmaceuticals, perfumes production of plasticizers, solvents, polymers, inks, clothing fibers, paints	Nagime et al. (2023), Qin et al. (2024)

(continued)

Table 13.2 (continued)

Organic acid	Substrates	Bioprocess involved	Microorganism	Industrial application	References
Acetic acid	Cereals and their brans Pineapple (*Ananas comosus*) peels Apple pomace Corn cob	Solid state fermentation Batch fermentation	*Aspergillus*, *Rhizopus* *Monascus* and *Saccharomyces cerevisiae* *Acetobacter aceti*	Preparation of metal acetates, printing processes, plastics photographic films and textiles, cleaning agent, flavouring and preservatives	Roda et al. (2017), Parmar and Rupasinghe (2013)

13.4.2 Antibiotics

Microbial production of antibiotics from agro-wastes is of enormous interest globally mainly from medical and economical point of view. It was recently reported that the annual consumption of antibiotics worldwide is estimated to have reached over 200,000 tons (Kashif and Shahid 2022). Therefore, an alternative and cost-effective source for the production of antibiotics has become imperative. Worthy of note is that one of the most cost effective ways of meeting this demand for novel and effective antibiotics is through the use of agro-industrial wastes as cheap substrates for microbial production of these compounds. Antibiotics are made via microbial secondary metabolic pathways, which use common metabolites in less accurate and sometimes more complicated ways than primary metabolism. The abundance of bioactive chemicals in agricultural wastes, such as apple pomace, rice bran,, green grain husk, banana peel, wheat bran, pomegranate peel, black grain husk sawdust, rice hulls, pineapple peel, orange peel, peanut shells, and maize cobs, makes them suitable substrates for the production of antibiotics through various bioprocesses (Al-Farraj et al. 2020; Kalaiyarasi et al. 2020). The yield and quality of antibiotics produced from agro-industrial wastes are determined by the processing procedures as well as the raw material or substrate used.

Traditionally, liquid culture (SmF) has been used to manufacture antibiotics, but in recent years, SSF has also been used (Kalaiyarasi et al. 2020). Using agricultural waste as the substrate, solid state fermentation (SSF) has been widely employed to manufacture a number of antibiotic classes, including neomycin, oxytetracycline, and rifamycin, according to recent publications (Al-Farraj et al. 2020). Because submerged fermentation is easier to automate, scale up, and monitor, it is widely preferred in industrial operations. Contrary to solid-state fermentation, large-scale fermentation avoids the dynamic gradients in temperature, pH, and O_2 concentration that are frequently required for industrial viability in large bioreactors (Correia et al. 2023). Additionally, the by-products become easier to purify, although this process produces a lot of wastewater. Since liquid fermentation typically occurs in 1–2 weeks, it is notable for its rapidity and necessitates a variety of apparatus for real-time management (Correia et al. 2023).

The primary classes of antibiotics from microbial origin are the cephalosporins, penicillins, and tetracyclines. Different species of microorganisms are known to have the capability to produce antibiotics using agro-industrial wastes as substrates especially bacteria of the class actinomycetes (*Streptomyces* sp., Actinoplanes, Micromonospora, Amycolatopsis, Saccharopolyspora etc.) and many filamentous fungal species including the *Penicillium* spp., *Fusarium* spp. (Hutchings et al. 2019). Vastrad and Neelagund (2011) reported the production of Neomycin, an important aminoglycoside via SSF using apple pomace and cotton seed meal.

13.4.3 Bioethanol

Bioethanol over the years has attracted great interest as an important alternative fuel for the replacement of gasoline. Bioethanol is by far the most popular biofuel utilized for transportation globally due to its positive impact on the environment. This is because it is an oxygenated biofuel that is renewable, nontoxic and biodegradable and these characteristics help to minimize particle emissions from compression-ignition engines. These days, the majority of alcohol is produced using raw materials that contain sugar and starch. However, a range of easily obtainable lignocellulosic biomass sources, such as agricultural and forestry wastes and herbaceous energy crops could be utilized as feedstocks to create bioethanol and other valuable chemicals. Despite financial and technological obstacles, renewable lignocellulosic raw materials are cheap feedstocks that support sustainability by not competing with the food and feed chain (Bušić et al. 2018).

An estimated 442 billion litres of bioethanol are produced annually from lignocellulosic biomass, according to a previous study by Sarkar et al. (2012). Lignocellulosic biomass is a cheap, renewable, and widely accessible agro-industrial waste. Cellulose (50%) lignin (25%), hemicellulose (22%), and other constituents like protein, ash, organic acids, and other non-structural elements are the main constituents of these complex carbohydrate polymers (Carrillo-Nieves et al. 2019). Linear and crystalline is cellulose. It is a homopolymer composed of repeated glucose sugar units connected by β-1,4 glycosidic bonds. Hemicellulose is a short polymer with a lot of branches. Hemicellulose is composed of D-xylose, D-arabinose, D-glucose, D-galactose, and D-mannose. These two carbohydrate polymers are closely linked to lignin, which is hydrophobic in nature. Consequently, it shields these polymers from microbial attacks/breakdown. Prior to bioethanol production, Lignocellulosics are processed through three major operations which include:

1. Pre-treatment for delignification: This initial step is necessary in order to liberate cellulose and hemicellulose. The main processing step in the production of bioethanol is the process that involves the solubilization and separation of one or more of these biomass components. There are four types of pretreatments: physicochemical (like liquid hot water, hydrothermal, steam explosion, CO_2 explosion, ammonia recycling percolation, and ammonia fiber explosion); biological

(like brown rot, white rot, and soft rot fungi); chemical (like acids and bases combined with oxidizing agents, organic solvents, ozonolysis, and ionic liquids); and physical (like grinding, chipping, or milling) (Zabed et al. 2017; Carrillo-Nieves et al. 2019). It is noteworthy that most often than not, pre-treatment of lignocellulosic wastes involves the combination of above-mentioned techniques (Talebnia et al. 2010; Sarkar et al. 2012). However, each of this pre-treatment method has its advantages and disadvantages, hence necessitating the adoption of combination technique. For example, acid pre-treatment gives a better yield but results in the production of various inhibitors like acetic acid, furfural and 5 hydroxymethylfurfural which are known growth inhibitors of microorganisms

2. Hydrolysis: In order to create fermentable sugars such as glucose, galactose, mannose, xylose, arabinose, and fermentation of reducing sugars, cellulose and hemicellulose must be hydrolyzed. This procedure, which is also referred to as saccharification, can be carried out by enzymatic hydrolysis or acid hydrolysis using biological enzymes such xylanase, hemicellulases, cellulase (endo and exoglucanase and β-glucosidases).
3. Fermentation: This technique uses a variety of microorganisms to ferment the saccharified biomass, but its commercial application is limited by the high costs of these fermentation technologies and the variability of the substrate composition. Specifically, the absence of capable bacteria that can effectively ferment both hexose and pentose sugars is a barrier to the industrial use of lignocelluloses for the production of bioethanol (Carrillo-Nieves et al. 2019). The optimum option for producing ethanol on a big industrial scale would therefore be a microbe with a wide utilization range, a high ethanol yield, and productivity. Such microbial cells ought to be resistant to high ethanol concentrations, high temperatures, and the inhibitors found in hydrolysate. A number of biotechnological bioprocessing methods have been previously described for the production of bioethanol from agroindustrial wastes. These methods include pre-saccharification followed by simultaneous saccharification and fermentation (PSSF), consolidated bioprocessing (CBP), simultaneous saccharification and fermentation (SSF), simultaneous hydrolysis and co-fermentation (SHCF), and simultaneous saccharification and fermentation (SSF) (Carrillo-Nieves et al. 2017, 2019). Because of its inherent benefits—such as a higher ethanol yield, a faster conversion time, lower costs, and the ability to use the same reactor for all steps—SSF, SSCF, and PSSF are the most often used methods for producing bioethanol. Nevertheless, a pre-treatment phase is necessary (Carrillo-Nieves et al. 2019). Contrariwise, CBF uses a single microbe that can produce all of the enzymes needed to produce sugars and ethanol. This approach reduces the overall process cost, develops all steps in a single reactor, and eliminates the need for a pre-treatment step (Wilkinson et al. 2016). However, the conversion time is typically rather high, and the productivity and ethanol yield are low, making this method unsuitable for commercial production.

Bacteria, yeasts, and fungus are among the microorganisms that have been widely employed recently to produce bioethanol. *Saccharomyces cerevisiae* is the most frequently reported bacterium producing bioethanol from agro-industrial wastes, particularly sugar-containing feedstocks, because of its capacity to break down sucrose into hexoses (fructose and glucose). It has been reported that other organisms such as *Streptomyces, Zymomonas mobilis, Thermoanaerobacter ethanolicus, Pichia stipitis, Klebsiella oxytoca, Escherichia coli, Aspergillus niger, Candida shehatae, Mucor indicus*, and others are also used (Wilkinson et al. 2016; Bušić et al. 2018; Buraimoh et al. 2021).

13.4.4 Biogas

Biogas is a renewable energy source produced by the anaerobic digestion of organic materials (biomass) by bacteria consortia (Gupta et al. 2022). It mainly consists of a mixture of methane (60%), carbon dioxide (35–40%) and small amount of hydrogen, ammonia, hydrogen sulphide etc. It is worthy to note that the component mixture of biogas after production depends on the source and type of the organic waste being used (Khalil et al. 2019). Researchers over the years have reported that biogas can be produced from agricultural wastes (crop residues), industrial wastes, food wastes, animal dung, manure and sewage sludge (Scano et al. 2021). For better biogas production, a formulation comprising of mixtures of agro-industrial wastes gives better yield than when used individually. This has the advantages such as (1) provision of a better nutrient balance, and (2) dilution of possible toxic compounds, such as ammonia in manure resulting in increased biogas production (Barros et al. 2023). In the last decade, it was reported that biogas production under anaerobic condition is the most adopted method used due to its inherent advantages such as low cost and environmental friendliness, pathogens' reduction, and the production of organic fertilizer as by-product (Abdeshahian et al. 2016).

Biogas is produced by five main processes: pre-treatment, hydrolysis, methanogenesis, acetogenesis, and acidogenesis (Devi et al. 2022). It is produced through a purification and upgrading process, a variety of metabolic reactions involving different microorganisms, and the removal of CO_2 from the gas mixture to improve the energy content. A range of bioreactors, such as continuous stirred tank reactors, batch bioreactors (one stage), continuously fed systems (two stages), multi-stage continuously fed systems, and up-flow anaerobic sludge blankets, can be used for production when utilizing agro-industrial waste as the organic substrate. The most basic type of bioreactor is the batch type, which is filled with feedstock and then discharged after a length of time known as the hydraulic retention time. Anaerobic batch reactors can carry out quick digestion using low-cost, basic equipment and make it simple to gauge the rate of digestion (CompAretti et al. 2011). Depending on the procedure and the kind of bioreactor, the operating temperature for anaerobic digestion can be kept in either the mesophilic or thermophilic range. Nevertheless, the results showed that whereas the mesophilic digestion produces up to 20% more methane than the thermophilic digestion, the thermophilic digestion produces up to 47% more biogas (Barros et al. 2023). Furthermore, the anaerobic digestion of organic wastes into biogas is known to involve a number of microorganisms. These include, among others, Desulfovibrio, Desulfomicrobium, Archaeon, Methanoculleus, Methanosarcina, and Thermogymnomonas. The growth and activity of anaerobic microbes, and thus the effectiveness and capacity of anaerobic digestion for the production of biogas and digestate yields, are influenced by a number of significant parameters: raw material concentration, type, chemical composition of substrate, and pH, oxygen concentration, microbial composition, temperature, presence of inhibitors and amounts (like ammonia), toxic compounds, intermediate product concentration (like volatile fatty acids), bioreactor design, and stirring speed (CompAretti et al. 2011; Devi et al. 2022).

In conclusion, though producing biogas from animal waste has a lot of potential as a sustainable energy resource substitute and as a way to manage animal waste; there is still a lack of use of this technology in many parts of the world especially in third world countries. The execution of biogas operations, whether at small or industrial scale is often hampered by a number of obstacles including technical and financial difficulties which directly affects operating costs.

13.4.5 Biodiesel

Biodiesel has gained much attention as one of the most promising biofuels for global fuels market as a result of gradual depletion of fossil based petroleum products and rapid increase in costs as well as environmental impact (pollution, climate change and release of greenhouse gases) (Torres-Rodrìguez et al. 2016). Thus, production of biodiesel from oleaginous microorganisms grown on agro-industrial substrates which is economically viable and eco-friendly technology for renewable energy is highly sustainable. Montero-Rodríguez et al. (2016) described biodiesel as a blend of fatty acid methyl esters (FAMEs) that are produced by transesterification of triglycerides (TAG) primarily from vegetable and animal fats with alcohols, primarily methanol or ethanol. Biodiesel has physical and chemical characteristics that are comparable to those of diesel made from non-renewable petroleum and is extremely biodegradable and non-toxic. Numerous microorganisms, such as bacteria, fungi, yeasts, and microalgae, can collect oils under particular growth circumstances, which make them suitable as feedstock for the synthesis of biodiesel. The use of Algae biomass as a feedstock for the production of many value-added products has attracted much attention over the years. This could be attributed to the fact Algal extracts are excellent sources of biotechnologically viable chemicals with a variety of medicinal and restorative qualities, including antimicrobial, antioxidative, anti-inflammatory, and antibacterial characteristics (Uma et al. 2023). Research on the bioenergy potential of algal biomass is now a global topic because algal fuels are effective, develop quickly, and provide high-quality fuel.

Utilizing microbial oil as feedstock for biodiesel production has several benefits, including a lower labour demand for oil harvesting, shorter culture growth period, reduced reliance on climatic conditions (weather) and season, and ease of scaling up. In addition, low cost agro-industrial wastes (corn stover) are readily available. As a result, microbial-derived oil is garnering more interest as a vital component in the production of biodiesel. When a chemical catalyst (acids or bases) is present e.g. sulphuric acid, short-chain alcohols like methanol are frequently utilized in the transesterification reaction to produce biodiesel. Alkali catalysts are mostly preferred because of their higher reactivity and milder process conditions such as low temperature requirement (Brahma et al. 2022). An ideal biodiesel according to US Department of Energy should consist of higher percentage of monounsaturated fatty acids compared to less and polyunsaturated fatty acids. This is because high levels of polyunsaturated fatty acids will have negative effects on the oxidative stability of the biodiesel thereby leading to increase of nitrogen oxide exhaust emissions not suitable for biodiesel engines.

13.4.6 Biopolymers

Plastics are durable, affordable, and versatile materials that have raised living standards and economic activity all over the world. It is essential to many different operational sectors, including building materials, textiles, electronics, automotive, fiber and industrial and food packaging because of their characteristic stability and durability properties (Gutierrez-Macias et al. 2017). Nevertheless, due to growing concern about the environmental damage and health risks associated with petroleum-based plastics, which are resistant to microbial degradation, the production of bioplastics utilizing biopolymers produced by microbes is gaining attention over the years. In addition to being complex molecules made by bacteria, biopolymers can also be discovered in or derived from a variety of wastes, including agro-industrial wastes. They are separated into several kinds according to their chemical makeup, such as polypeptides, polyesters, polynucleotides, and polysaccharides (Horue et al. 2021). Alginate, gellan, kefiran, bacterial cellulose, levan, polyhydroxyalkanoates, ß-glucans, pullulan, and xanthan gums are classic examples of biopolymers.

Wastes from agricultural, domestic and industrial origin serve as substrates for fermentation by microbial cultures (bacteria and fungi) for biopolymer production. Paddy, wheat straw, beet molasses and cane, corn cob, wheat bran, bagasse, whey, chicken feathers, glycerol and waste frying oil are examples of agro-industrial wastes that could be used as substrates for cultivation of microorganisms capable of biopolymer accumulation (Naitam et al. 2022). However, the primary disadvantages of using renewable polymers are the high expense of waste-derived polymer purification procedures and the need for scaling up when producing biopolymers. Nonetheless, the commercial application of biopolymers in the creation of high-value products, such as coating materials, medical or pharmaceuticals and packaging materials, might offset the initial expenses of producing biopolymers while simultaneously enhancing local economies and environmental sustainability.

Notably, wide ranges of microorganisms (bacteria and fungi) are known for their inherent capability to accumulate biopolymers under high carbon concentration when grown under some nutritional stress such as low nitrogen, phosphorus, and other elements etc. Members of the genus *Bacillus* (*Bacillus megaterium, B. cereus, B. sphaericus, B. firmuss*), *Lactobacillus* (*L. confuses*), *Pseudomonas* (*Pseudomonas oleovorans, Pseudomonas putida*), *Burkholderia* (*B. cepacia, B. sacchari*) *Cupriavidus necator* etc. were reported to possess the capacity to produce biopolymers (Gutierrez-Macias et al. 2017). Poly-hydroxyalkanoates (PHA), which are highly biodegradable and produce CO_2, water, and humic material as byproducts of microbial degradation, are the most researched microbial biopolymers with plastic characteristics (Naitam et al. 2022). The literature claims that Poly-3-Hydroxybutyrate (PHB) is the best-characterized biopolymer among PHAs because of its thermochemical characteristics, which include optical activity, partial crystallinity, high melting temperature, and low permeability for water and atmospheric gases. These properties have indicated that PHB has the potential to be produced on a large scale in industry (Gomaa 2014; Horue et al. 2021).Although biopolymers (PHBs) can be produced from readily available inexpensive substrates (lignocellulosics), the cost of production is still not sustainable as a result of several pretreatment (saccharification) steps involved in making the substrates available in a form that will support microbial growth for maximum biopolymer accumulation is quite challenging.

13.4.7 Amino Acids

Globally, the requirement for protein sources is steadily rising as a result of population expansion, changing dietary choices, and the growing trend toward a high-protein diet. The recommended dietary allowance (RDA) for protein is 0.8 g/kg of body weight, according to the Food and Nutrition Board. The Food and Agriculture Organization of the United Nations (FAO) has predicted that by 2050, there would be 9.7 billion people on the planet (Bajić et al. 2023). Therefore, to be able to sustain a continuous supply of food rich in protein for general human well-being and as supplement for animal feeds, there is need for a sustainable source of this requirement and microbial protein is an alternative. Microbial proteins, often referred to as single-cell proteins (SCP) or microbial biomass, are rich in protein and a good source of lipids, vitamins, minerals, carbs, and amino acids. They can be cultivated for both human consumption and animal feed.

The building blocks of proteins are amino acids, and it is through the synthesis of new structural and functional proteins—like hormones and enzymes—that living things are able to construct and regenerate their cells. Interestingly, the biological quality of proteins varies depending on the ratios and concentrations of the amino acids that comprise them, i.e., if a protein contains a higher proportion of necessary acids. Conventionally, humans and animals derive protein, which is a macronutrient from two major sources namely; plant based protein (beans, grains, seeds, nuts etc.) and animal protein (muscles and tissues). Though protein sources are readily available, but with the constantly growing world population, the demand for protein-rich foods will rise exponentially leading to food crisis. SCPs are dead, dry microbial cells or total proteins obtained from pure microbial cell cultures. Microorganisms including bacteria (*Cellulomas*, *Methylococcus capsulatus*, *Alcaligenes* etc.), yeasts (*Candida*, *Hansenula*, *Pichia*, *Torulopsis*, *Saccharomyces* etc.), algae (*Spirullina*, *Chlorella*, *Scenedesmus* etc.) and filamentous fungi (*Fusarium*, *Rhizopus*) are usually used in the large scale production of microbial proteins (Bogale 2020). Nonetheless, there are numerous benefits to using yeasts as microbial protein over other species, including their greater size, which facilitates harvesting, high lysine content, lower nucleic acid concentration, and capacity to thrive at acidic pH. Additionally, due to its long history of usage in conventional fermentations, its most significant advantage is familiarity and acceptability (Bajić et al. 2023).

A few intrinsic characteristics are considered before choosing a microbe to produce SCP. According to Raziq et al. (2020), these include the growth morphology, requirement for heat and oxygen production during fermentation, growth rate foam properties, productivity, and/or yield of specific low-cost substrates, genetic stability, tolerance to pH and temperature, end product composition and structure, and the simplicity of protein recovery and purification. When opposed to conventional animal farming and crop production, the production of microbial proteinaceous biomass offers a number of advantages. Compared to farm animals or plants, which have doubling times of about 1–2 years and a few months, respectively, microbes make protein far more efficiently—algae and molds take 2.6 h, while bacteria and yeasts takes approximately 0.33–2 h). Additionally, depending on the microbe utilized, the dry mass basis protein content of microorganisms ranges from 30% to 80% w/w, and the protein has a excellent dietary value. When producing SCP, a variety of raw materials can be used as substrates, including low-value agro-industrial wastes and by-products. These substrates include renewable plant resources (cellulose, starch, and sugar), high-energy resources (ethanol, methanol, n-alkanes, gas oil, natural gas, and acetic acid), various agro processing wastes (sugar cane bagasse, molasses, milk whey, fruit waste, waste liquor,), and carbon dioxide etc. (Raziq et al. 2020; Saleh et al. 2020). With relatively miniature continuous fermentation processes (solid state and submerged) and a relatively limited land area, microbial proteins can be generated in large quantities. Furthermore, climate, as well as seasonal and climatic fluctuations, has little effect on the production of SCP. Also, compared to plants and animals, microorganisms are easier to manipulate genetically for improved characteristics. In addition, SCP satisfies the FAO/WHO's recommended essential amino acid requirements for human nutrition and animals (Bajić et al. 2023). Animals and humans mostly use single-cell proteins as dietary supplements. In addition, there are numerous other uses for single cell proteins. It is consumed as a health meal to reduce obesity and give athletes rapid energy particularly males. It has been reported to lower blood sugar in diabetic patients, improves skin and eye health, increases lactation, and reduces cholesterol and stress. It is also utilized in therapeutic and natural remedies. It is also utilized as a foam stabilizer and in the processing of paper and leather in industrial applications.

13.4.8 Enzymes

Enzymes are biological catalysts that are used in a variety of industries, including paper, pulp, detergents, baking, and brewing. In many cases, they are favoured over chemical catalysis because of their strong operating parameters and high degree of substrate specificity. With the development and widespread use of recombinant DNA technology, scientists are now able to adopt biotechnological approaches for large synthesis of enzymes from any source in microorganisms to meet the needs of a wide range of industries. Furthermore, another technique that is frequently used in industry to raise the yield of enzyme synthesis is strain improvement through induced mutation (Ravindran et al. 2018). Presently, current global production of industrial enzymes can be estimated to represent a value between 4 and 5 billion USD.

Enzymes are comparatively expensive reagents, which raises the operational costs of processes that use them despite their obvious need in industrial processes. A thorough examination of the process plant economics for the manufacture of enzymes shows that capital expenditures account for about half of production costs, with raw material costs making up the other third. Because of this, producing enzymes from inexpensive raw materials like agro-industrial wastes will inevitably lower operating costs for the companies that use enzymes in their production, which will significantly lower the cost of the finished product (Mojumdar and Deka 2019). Nevertheless, scientists started investigating practical ways to use agricultural wastes and lignocellulosic residues as a source of raw materials for the synthesis of different enzymes. Using agro-industrial wastes such corn cob, rice bran, sugarcane bagasse, potatoes and banana peels etc. as substrates, a wide variety of microorganisms, such as bacteria (*Bacillus amyloliquefaciens, Bacillus subtilis, Streptomyces)*, yeasts (*Geotrichum candidum*, *Saccharomyces cerevisiae*) and fungus (*Aspergillus spp,* F*usarium spp and Pleurotus ostreatus*, create various types and classes of enzymes (Parmjit et al. 2016; Panesar et al. 2016; Putri et al. 2020; Legodi et al. 2023) through biotechnological process that involves several unit operations. The majority of industrial enzymes that are now in use are the hydrolases, which include lipases and proteases and are widely employed in the chemical, detergent, and dairy sectors while the second largest group of enzymes is made up of various carbohydrases, most notably cellulases and amylases (Putri et al. 2020). The main area of enzyme application is the food and feed sector e.g. *a*-amylase, xylanse, protease which consumes over one-third of enzyme's production (Javed et al. 2013; Mojumdar and Deka 2019; Al Mousa et al. 2022), closely followed by detergents and cleaners.

13.5 Challenges, Future Perspectives and Conclusion

Microbial bioconversion of agro-industrial wastes undoubtedly is a promising way to handle the problems associated with indiscriminate disposal of agrowastes that leads to environmental degradation. Nevertheless, because agro-industrial waste is complex in nature which makes them resistance to microbial attack, there is need for pretreatment process such as enzymatic saccharification and biomass pretreatment. These two important bioconversion processes influence the techno-economic viability of producing VAPs on a wide scale from agro-industrial wastes. The cost-effectiveness and quality of fractionated products must be maximized as the environmentally benign and sustainable principles of bioconversion are implemented. Even though some pretreatment techniques have improved and now give high sugar yields with fewer drawbacks, there is still more space for improved lignocellulosic biomass pretreatment and developments based on detoxification. Therefore, it is necessary to design suitable and economical pretreatment techniques that may maximize biomass degradation while using less energy and chemicals and enhancing the enzymatic hydrolysis that follows. Another major challenge in pretreatment step is how to tackle side reactions that produce by-products from lignocellulose that block subsequent biochemical processes, though its goals are to separate the major components of the feedstock and reveal the cellulose for enzymatic saccharification. Thus, with respect to certain well-known inhibitors, the selection and engineering of biocatalysts with enhanced resistance to these inhibitors can provide solutions, though; further work is required to cover all classes of inhibitory chemicals that affect biotechnological processes.

In conclusion, this chapter highlights the potential of various agro-industrial residues as substrates for microbial production of different value-added products of known industrial applications. Notably, using agro-industrial wastes for microbial fermentation might lessen the environmental problems associated with their disposal while also improving process economics. However, there is still need for innovative research and effective technology that must be used to address the conversion process's technological constraints in order to effectively develop and optimize VAPs production from agricultural wastes in the near future. Also, pretreatment techniques that release inhibitors that impact bioconversion still require significant development to achieve optimal utilization of agro-industrial wastes for higher yields of VAPs.

References

Abdeshahian P, Lim JS, Ho WS, Hashim H, Lee CT (2016) Potential of biogas production from farm animal waste in Malaysia. Renew Sustain Energy Rev 60:714–723

Al-Farraj DA, Varghese R, Vágvölgyi C, Elshikh MS, Alokda AM, Mahmoud AH (2020) Antibiotics production in optimized culture condition using low cost substrates from Streptomyces sp. AS4 isolated from mangrove soil sediment. J King Saud Univ Sci 32(2):1528–1535

Al Mousa AA, Abo-Dahab NF, Hassane AMA, Gomaa AEF, Aljuriss JA, Dahmash ND (2022) Harnessing Mucor spp. for Xylanase production: statistical optimization in submerged fermentation using agro-industrial wastes. BioMedical Res Int 3816010:1–17

Amado IR, Vázquez JA, Pastrana L, Teixeira JA (2016) Microbial production of hyaluronic acid from agro-industrial by-products: molasses and corn steep liquor. Biochem Eng J 117A:181–187

Areepak C, Jiradechakorn T, Chuetor S, Phalakornkule C, Sriariyanun S, Raita M, Champreda V, Laosiripojana N (2022) Improvement of lignocellulosic pretreatment efficiency by combined chemo – mechanical pretreatment for energy consumption reduction and biofuel production. Renew Energy 182:1094–1102

Asghari M, Jahadi M, Hesam F, Ghasemi-Sepro N (2020) Optimization of Monascus pigment production on date waste substrates using solid state fermentation. Appl Food Biotechnol 8:247–254

Aslam AA, Shamim M, Nazir MS, Kokab O, Miran M, Akhtar MN, Ali Z, Tahir Z, Abdullah MA (2023) Microbial degradation of agricultural and food wastes into value-added products. Sustain Mater 2:157–198

Astudillo Á, Rubilar O, Briceño G, Diez MC, Schalchli H (2023) Advances in agroindustrial waste as a substrate for obtaining eco-friendly microbial products. Sustainability 15:3467

Awogbemi O, Von Kallon DV (2022) Pretreatment techniques for agricultural waste. Case Stud Chem Environ Eng 6:100229

Bajić B, Vučurović D, Vasić Đ, Jevtić-Mučibabić R, Dodić S (2023) Biotechnological production of sustainable microbial proteins from agro-industrial residues and by-products. Foods 12:107

Barros RS, Contreras MD, Morris FR, Chamorro MV, Arrieta AA (2023) Evaluation of the methanogenic potential of anaerobic digestion of agro-industrial wastes. Heliyon 9:14317

Basto B, da Silva NR, Teixeira JA, Silvério SC (2022) Production of natural pigments by *Penicillium brevicompactum* using agro-industrial by-products. Fermentation 8:536

Belorkar SA, Jogaiah S (2022) Solid-state fermentation and submerged fermentation for enzyme production. In: Protocols and applications in enzymology. Elsevier, pp 71–90

Bogale TT (2020) Microbial protein production from agro-industrial wastes as food and feed. Am J Life Sci 8(5):121–126

Brahma S, Nath B, Basumatary B, Das B, Saikia P, Patir K, Basumatary S (2022) Biodiesel production from mixed oils: a sustainable approach towards industrial biofuel production. Chem Eng J Adv 10:00284

Buraimoh M, Ogunyemi AK, Isanbor C, Aina OS, Amund OO, Ilori MO, Familoni OB (2021) Sustainable generation of bioethanol from sugarcane wastes by *Streptomyces coelicolor* strain COB KF977550 isolated from a tropical estuary. Sci Afr 11:1–7

Bušić A, Marđetko N, Kundas S, Morzak G, Belskaya H, Ivančić ŠM, Komes D, Novak S, Šantek B (2018) Bioethanol production from renewable raw materials and its separation and purification: a review. Food Technol Biotechnol 56(3):289–311

Carrillo-Nieves D, Ruiz HA, Aguilar CN, Ilyina A, Parra-Saldivar R, Torres JA et al (2017) Process alternatives for bioethanol production from mango stem bark residues. Bioresour Technol 239:430–436

Carrillo-Nieves D, Alanísa MJR, Quiroza R, Ruizb HA, Iqbala HMN, Parra-Saldívar R (2019) Current status and future trends of bioethanol production from agroindustrial wastes in Mexico. Renew Sust Energ Rev 102:63–74

Chilakamarry CR, Sakinah AMM, Zularisam AW, Sirohi R, Khilji IA, Ahmad N, Pandey A (2022) Advances in solid-state fermentation for bioconversion of agricultural wastes to value-added products: opportunities and challenges. Bioresour Technol 343:126065

Chowdhary P, Gupta A, Gnansounou E, Pandey A, Chaturvedi P (2021) Current trends and possibilities for exploitation of Grape pomace as a potential source for value addition. Environ Pollut 278:116796

CompAretti A, Febo P, Greco C, Orlando S (2011) Current state and future of biogas and digestate production. Bulgarian J Agr Sci 19(1):1–14

Coradi GV, da Visitação VL, de Lima EA, Saito LYT, Palmieri DA, Takita MA, Neto P, de Lima VMG (2013) Comparing submerged and solid-state fermentation of agro-industrial residues for the production and characterization of lipase by *Trichoderma harzianum*. Ann Microbiol 63(2):533–540

Correia J, Borges A, Simões M, Simões LC (2023) Beyond penicillin: the potential of filamentous fungi for drug discovery in the age of antibiotic resistance. Antibiotics (Basel) 12(8):1250

Das AJ, Kumar R (2018) Utilization of agro-industrial waste for biosurfactant production under submerged fermentation and its application in oil recovery from sand matrix. Bioresour Technol 260:233–240

Devi MK, Manikandan S, Oviyapriya M, Selvaraj M, Assiri MA, Vickram S, Awasthi MK (2022) Recent advances in biogas production using agro-industrial waste: a comprehensive review outlook of techno-economic analysis. Bioresour Technol 363:127871

Dezam APG, Vasconcellos VM, Lacava PT, Farinas CS (2017) Microbial production of organic acids by endophytic fungi. Biocatal Agric Biotechnol 11:282–287

Diaz-Vazquez D, Carrillo-Nieves D, Orozco-Nunnelly DA, Senes-Guerrero C, Gradilla-Hernandez MS (2021) An integrated approach for the assessment of environmental sustainability in agro-industrial waste management practices: the case of the Tequila industry. Front Environ Sci 9:682093

Dos Santos FKF, Barcellos-Silva IGC, Leite-Barbosa O, Ribeiro R, Cunha-Silva Y, Veiga-Junior VF (2024) High added-value by-products from biomass: A case study unveiling opportunities for strengthening the agroindustry value chain. Biomass 4:217–242

Egbe NE, Ihediwa L, Abdulsalami MS, Adebayo A, Onuh K (2022) Citric acid production from agricultural wastes using *Aspergillus niger* isolated from some locations within Kaduna metropolis, Nigeria. J Appl Sci Environ Manage 26(9):1607–1614

El-Housseiny G, Ibrahim A, Yassien M, Aboshanab K (2021) Production and statistical optimization of paromomycin by Streptomyces rimosus NRRL 2455 in solid state fermentation. BMC Microbiol 21:34

El-Ramady H, Brevik EC, Bayoumi Y, Shalaby TA, El-Mahrouk ME, Taha N et al (2022) An overview of agro-waste management in light of the water-energy-waste Nexus. Sustainability 14:15717

El-Sheshtawy HS, Fahim I, Hosny M, El-Badry MA (2022) Optimization of lactic acid production from agro-industrial wastes produced by Kosakonia cowanii. Curr Res Green Sustain Chem 5:100228

Farkas C, Rezessy-Szabó JM, Gupta VJ, Duy HT, Friedrich L, Felföldi J, Nguyen QD (2019) Microbial saccharification of wheat bran for bioethanol fermentation. J Cleaner Prod 240:118269

Fritsch C, Staebler A, Happel A, Cubero M, Miguel A, Aguiló-Aguayo I, Abadias M et al (2017) Sustainability in food supply chain and food industry. Sustainability 9(8):1492

Gomaa EZ (2014) Production of polyhydroxyalkanoates (PHAs) by Bacillus subtilis and Escherichia coli grown on cane molasses fortified with ethanol. Arch Biol Technol 57:145–154

Guneser O, Demirkol A, Yuceer YK, Togay SO, Hosoglu MI, Elibol M (2017) Production of flavour compounds from olive mill waste by *Rhizopus oryzae* and *Candida tropicalis*. Braz J Microbiol 48:275–285

Gupta J, Kumari M, Mishra A, Swati A, M., & Thakur, I.S. (2022) Agro-forestry waste management – a review. Chemosphere 287:132321

Gutierrez-Macias P, Hernandez De Jesus M, Barragan-Huerta BE (2017) The production of biomaterials from agro-industrial waste. Fresenius Environ Bull 26(6):4128–4152

Horue M, Berti IR, Cacicedo ML, Castro GR (2021) Microbial production and recovery of hybrid biopolymers from wastes for industrial applications – a review. Bioresour Technol 340:125671

Hutchings MI, Truman AW, Wilkinson B (2019) Antibiotics: past, present and future. Curr Opin Microbiol 51:72–80

Huzir NM, Aziz MMA, Ismail SB, Abdullah B, Mahmood NAN, Umor NA, Muhammad SAFS (2018) Agro-industrial waste to biobutanol production: eco-friendly biofuels for next generation. Renew Sustain Energy Rev 94:476–485

IEA (International Energy Agency) (2017) Mobilization of agricultural residues for bioenergy and higher value bio-products: resources, barriers and sustainability. IEA Bioenergy Task 43:1

Javed S, Meraj M, Bukhari SA, Irfan R, Mahmood S (2013) Hyper-production of alkaline protease by mutagenic treatment of *Bacillus subtilis* M-9 using agro-industrial wastes in submerged fermentation. Microb Biochem Technol 5:074–080

John RP, Nampoothiri KM, Pandey A (2007) Production of L (+) lactic acid from cassava starch hydrolysate by immobilized Lactobacillus delbrueckii. J Basic Microbiol 47:25–30

Kalaiyarasi M, Ahmad P, Vijayaraghavan P (2020) Enhanced production antibiotics using green gram husk medium by *Streptomyces* sp. SD1 using response surface methodology. J King Saud Univ Sci 32:2134–2141

Kashif A, Shahid MK (2022) Agricultural wastes as an alternative source for the production of antibiotics: recent developments and future perspectives. In: Nayak SK, Baliyarsingh B, Mannazzu I, Singh A, Mishra BB (eds) Advances in agricultural and industrial microbiology. Springer, Singapore

Kaur N, Panesar PS, Ahluwalia S (2022) Production of organic acids from agro-industrial waste and their industrial utilization. In: Valorization of agro-industrial byproducts. CRC Press, pp 227–264

Khalil M, Berawi MA, Heryanto R, Rizalie A (2019) Waste to energy technology: the potential of sustainable biogas production from animal waste in Indonesia. Renew Sust Energ Rev 105:323–331

Kim S, Kim S, Sohn SY (2020) Evaluation of the wastewater generated during alkaline pretreatment of biomass for feasibility of recycling and reusing. Renew Energy 155:1156–1164

Legodi LM, La Grange DC, van Rensburg ELJ (2023) Production of the cellulase enzyme system by locally isolated *Trichoderma* and *Aspergillus* species cultivated on banana pseudostem during solid-state fermentation. Fermentation 9:412

Liu J, Lv Y, Qian X, Jiang Y, Wu M, Zhang W, Zhou J, Dong W, Xin F, Jiang M (2019) Current advances in organic acid production from organic wastes by using microbial co-cultivation systems. Biofuels Bioprod Biorefin 14:1648–1660

Luft L, Confortin T, Todero I, Zabot G, Mazutti M (2020) An overview of fungal biopolymers: bioemulsifiers and biosurfactants compounds production. Crit Rev Biotechnol 40:1059–1080

Manikandan A, Muthukumaran P, Poorni S, Priya M, Rajeswari R, Kamaraj M, Aravind J (2022) Microbial approaches for bioconversion of agro-industrial wastes: a review. In: Aravind J et al (eds) Strategies and tools for pollutant mitigation, pp 151–182

Martínez EJ, Raghavan V, González-Andrés F, Gómez X (2015) New biofuel alternatives: integrating waste management and single cell oil production. Int J Mol Sci 16(5):9385–9405

Mitchell D, de Lima DA, Luz LF, Krieger N (2011) Bioreactors for solid-state fermentation. In: Moo-Yong M (ed) Comprehensive biotechnology. Elsevier, pp 347–360

Mojumdar A, Deka J (2019) Recycling agro-industrial waste to produce amylase and characterizing amylase–gold nanoparticle composite. Int J Recycl Org Waste Agric 8:263–269

Molelekoa TBJ, Regnier T, da Silva LS, Augustyn W (2021) Production of pigments by filamentous fungi cultured on agro-industrial by-products using submerged and solid-state fermentation methods. Fermentation 7(4):295

Mondal AK, Sengupta S, Bhowal J, Bhattacharya DK (2012) Utilization of fruit wastes in producing single cell protein. Int J Sci Environ Technol 1(5):430–438

Montero-Rodríguez D, Andrade R, Lima R, Silva G, Rubio-Ribeaux D, Silva T, Takaki GC (2016) Conversion of agro-industrial wastes by *Serratia marcescens* UCP/WFCC 1549 into lipids suitable for biodiesel production. Chem Eng Trans 49:307–312

Nagime PV, Upaichit A, Cheirsilp B, Boonsawang P (2023) Bio-succinic acid production from palm oil mill effluent using *Enterococcus gallinarum* with sequential purification of biogas. Fermentation 9:369

Naik B, Kumar V, Rizwanuddin S, Chauhan M, Gupta AK, Rustagi S, Kumar V, Gupta S (2023) Agro-industrial waste: a cost-effective and eco-friendly substrate to produce amylase. Food Prod Process Nutr 5:30

Naitam MG, Singh TG, Kaushik R, Singh S, Nain L (2022) Agro-industrial waste as potential renewable feedstock for biopolymer poly-hydroxyalkanoates (PHA) production. Enzyme Eng 11:190

Nambi VE, Chandrasekar V, Karthikeyan S (2017) Value addition of grains using solid state fermentation. Int J Nutr Food Sci 3(4):555619

Niego GAT, Lambert C, Mortimer P, Thongklang N, Rapior S, Grosse M, Schrey H, Charria-Girón E, Hyde KD, Stadler M (2023) The contribution of fungi to the global economy. Fungal Divers 121:95–137

Ogunmoroti A, Liu M, Li M, Liu W (2022) Unraveling the environmental impact of current and future food waste and its management in Chinese provinces. Resour Environ Sustain 9:100064

Oluwasola OD, Afolabi SA, Leonard EN, Osei KBB (2021) Country-level assessment of agri-food waste and enabling environment for sustainable utilisation for bioenergy in Nigeria. J Environ Manage 294:112929

Panesar PS, Kaur R, Singla G, Sangwan RS (2016) Bio-processing of agro-industrial wastes for production of food-grade enzymes: progress and prospects. Appl Food Biotechnol 3(4):208–227

Panesar R, Kaur S, Panesar P (2015) Production of microbial pigments utilizing agro-industrial waste: a review. Curr Opin Food Sci 1:70–76

Parmar I, Rupasinghe HV (2013) Bio-conversion of apple pomace into ethanol and acetic acid: enzymatic hydrolysis and fermentation. Bioresour Technol 130:613–620

Parmjit SP, Kaur R, Singla G, Sangwan RS (2016) Bio-processing of agro-industrial wastes for production of food-grade enzymes: progress and prospects. Appl Food Biotechnol 3(4):208–227

Patthawaro S, Lomthaisong K, Saejung C (2020) Bioconversion of agro-industrial waste to value-added product lycopene by photosynthetic bacterium *Rhodopseudomonas faecalis* and its carotenoid composition. Waste Biomass Valorization 11:2375–2386

Paul J, Beliya E, Tiwari S, Patel K, Gupta N, Jadhav S (2020) Production of biocatalyst-amylase from agro-waste 'rice bran' by using Bacillus tequilensis TB5 and standardizing its production process. Biocatal Agric Biotechnol 26:101648

Putri DN, Khootama A, Perdani MS, Utami TS, Hermansyah H (2020) Optimization of *Aspergillus niger* lipase production by solid state fermentation of agro-industrial waste. Energy Rep 6(1):331–335

Oliveira RC, Maciel VMM, Hissa DC, França ÍW, Gonçalves LRB (2022) Production of the food enzyme acetolactate decarboxylase (ALDC) from *Bacillus subtilis* ICA 56 using agro-industrial residues as feedstock. Fermentation 8:675

Qin Y, Li Y, Liang G, Shen N, Xian L, Wang Q (2024) Efficient production of succinic acid from sugarcane bagasse hydrolysate by *Actinobacillus succinogenes* GXAS137. Fermentation 10:22

Rasool K, Hussain S, Shahzad A, Miran W, Mahmoud KA, Ali N (2023) Comprehensive insights into sustainable conversion of agricultural and food waste into microbial protein for animal feed production. Rev Environ Sci Biotechnol 22:527–562

Rastogi S, Tiwari S, Ratna S, Kumar R (2021) Utilization of agro-industrial waste for biosurfactant production under submerged fermentation and its synergistic application in biosorption of Pb^{2+}. Bioresour Technol Rep 15:100706

Ravindran R, Hassan SS, Williams GA, Jaiswal AK (2018) A review on bioconversion of agro-industrial wastes to industrially important enzymes. Bioengineering (Basel) 5(4):93

Raziq A, Lateef M, Ullah A, Ullah H, Khan MW (2020) Single cell protein (SCP) production and potential substrates: a comprehensive review. Pure Appl Biol 9:1743–1754

Roda A, Lucini L, Torchio F, Dordoni R, De Faveri DM, Lambri M (2017) Metabolite profiling and volatiles of pineapple wine and vinegar obtained from pineapple waste. Food Chem 229:734–742

Rodrigues DB, Flores ÉMM, Barin JS, Mercadante AZ, Jacob- Lopes E, Zepka LQ (2014) Production of carotenoids from microalgae cultivated using agro-industrial wastes. Food Rev Int 65:144–148

Sadh PK, Duhan S, Duhan JS (2018) Agro-industrial wastes and their utilization using solid state fermentation: a review. Bioresour Bioprocess 5:1

Saelee N (2022) Lactic acid production from old oil palm trunk sap in the open batch, open repeated batch, fed-batch, and repeated fed-batch fermentation by *Lactobacillus rhamnosus* ATCC10863. Fermentation 8:430

Saleh F, Hussain A, Younis T, Ali S, Rashid M, Ali A, Mustafa G, Jabeen F, Al-Surhanee AA, Alnoman MM et al (2020) Comparative growth potential of thermophilic amylolytic *Bacillus* sp. on unconventional media food wastes and its industrial application. Saudi J Biol Sci 27:3499–3504

Sarkar N, Ghosh SK, Bannerjee S, Aikat K (2012) Bioethanol production from agricultural wastes: an overview. Renew Energy *37*:19–27

Sepúlveda L, Laredo-Alcalá E, Buenrostro-Figueroa J, Ascacio-Valdés J, Genisheva Z, Aguilar C, Teixeira J (2020) Ellagic acid production using polyphenols from orange peel waste by submerged fermentation. Electron J Biotechnol 43:1–7

Scano EA, Grosso M, Pistis A, Carboni G, Cocco D (2021) An in-depth analysis of biogas production from locally agro-industrial by-products and residues. An Italian case. Renew Energy 179:308–318

Sharma N, Sahota P, Singh M (2021) Organic acid production from agricultural waste. In: Kashyap BK, Solanki MK, Kamboj DV, Pandey AK (eds) Waste to energy: prospects and applications. Springer, Singapore, pp 415–438

Shata H (2018) Statistical optimization of erythromycin production by *Saccharopolyspora erythraea* under solid state fermentation of agro-industrial materials using response surface methodology. J Microbiol Biotechnol Food Sci 8:692–697

Singh OV, Singh RP (2006) Bioconversion of grape must into modulated gluconic acid production by *Aspergillus niger* ORS4410. J Appl Microbiol 100(5):1114–1122

Singh S, Kumar A, Sivakumar N, Verma JP (2022) Deconstruction of lignocellulosic biomass for bioethanol production: recent advances and future prospects. Fuel 327:125109

Sukan A, Roy I, Keshavarz T (2014) Agro-industrial waste materials as substrates for the production of poly (3-hydroxybutyric acid). J Biomater Nanobiotechnol 5:229–240

Talebnia F, Karakashev D, Angelidaki I (2010) Production of bioethanol from wheat straw: an overview on pretreatment, hydrolysis and fermentation. Bioresour Technol 101(13):4744–4753

Torrado AM, Cortés S, Salgado JM, Max B, Rodríguez N, Bibbins BP, Converti A, Domínguez JM (2011) Citric acid production from orange peel wastes by solid-state fermentation. Braz J Microbiol 42(1):394–409

Torres-Rodrìguez DA, Romero-Ibarra IC, Ibarra IA, Pfeiffer H (2016) Biodiesel production from soybean and Jatropha oils using cesium impregnated sodium zirconate as a heterogeneous base catalyst. Renew Energy 93:323–331

Tuysuz E, Gonul-Baltaci N, Omeroglu M, Adiguzel A, Taskin M, Ozkan H (2020) Co-production of amylase and protease by locally isolated thermophilic bacterium *Anoxybacillus rupiensis* T2 in sterile and non-sterile media using waste potato peels as substrate. Waste Biomass Valorization 11:6793–6802

Uma VS, Usmani Z, Sharma M, Diwan D, Sharma M, Guo M, Tuohy MG, Makatsoris C, Zhao X, Thakur VK, Gupta VK (2023) Valorisation of algal biomass to value-added metabolites:emerging trends and opportunities. Phytochem Rev 22:1015–1040

Vastrad BM, Neelagund SE (2011) Optimization and production of neomycin from different agro industrial wastes in solid state fermentation. Int J Pharma Sci Drug Res *3*(2):104–111

Wilkinson S, Smart KA, James S, Cook DJ (2016) Bioethanol production from brewers spent grains using a fungal consolidated bioprocessing (CBP) approach. Bioenergy Res 2016:1–12

Yadav B, Chavan S, Atmakuri A, Tyagi RD, Drogui P (2020) A review on recovery of proteins from industrial wastewaters with special emphasis on PHA production process: sustainable circular bioeconomy process development. Bioresour Technol 317:124006

Yafetto L, Odamtten GT, Adu SK, Ekloh E (2019) The biotechnological potential of using mono- and co-cultures of *Aspergillus niger* van Tieghem and *Trichoderma viride* Pers ex Fr. to enhance the protein content of cassava (*Manihot esculenta* Crantz) peels by solid substrate fermentation. Studies Fungi 4(1):111–122

Yafetto L (2022) Application of solid-state fermentation by microbial biotechnology for bio-processing of agro-industrial wastes from 1970 to 2020: a review and bibliometric analysis. Heliyon 8:09173

Yafetto L, Odamtten GT, Wiafe-Kwagyan M (2023) Valorization of agro-industrial wastes into animal feed through microbial fermentation: a review of the global and Ghanaian case. Heliyon 9:14814

Yoshimura I, Salazar- Bryam AM, Faria AU, Leite LP, Lovaglio RB, Contiero J (2022) Guava seed oil: potential waste for the rhamnolipids production. Fermentation 8:379

Zabed H, Sahu JN, Suely A, Boyce AN, Faruq G (2017) Bioethanol production from renewable sources: current perspectives and technological progress. Renew Sustain Energy Rev 71:475–501

Zerva A, Tsafantakis N, Topakas E (2021) Evaluation of Basidiomycetes wild strains grown in agro-industrial residues for their anti-tyrosinase and antioxidant potential and for the production of biocatalysts. Fermentation 7:19

Chapter 14
Genome Editing as Genetic Engineering Tool: Potential Applications in Production of Microbial Foods for Food Security

Oluwafemi Sunday Obayori and Oluwaseun Oyetunji Olasunkanmi

Abstract From the ancient times, microorganisms have been at the centre of production of safe, nutritious, and delicious foods with enhanced shelf life. Improvement of food quality via fermentation by microorganisms is as old as civilization. Industrialization brought forth the era of use of microorganisms for production of proteins, supplements, feeds or foods in the form of single-cell proteins (SCPs). Thus, selected bacteria, fungi and algae have been used as feeds for animals and part of human food as supplements since the 20th century. However, limited numbers of these microbes met the conditions for such applications. The advent of gene manipulation brought the prospect of engineering of SCP microbes for improved qualities related to organoleptic properties, safety and nutritive values and possibilities of using microbes as food for man on a large scale. Classical tools of genetic engineering such as electrophoresis, gene cloning, mutagenesis, Southern blotting, sequencing, and later polymerase chain reaction (PCR), were central to this revolution in bioscience, enabling untold breakthroughs in medicine, agriculture, food industry and other human endeavours. This is happening at the same time as global population explosion, climate change induced scarcity of arable land and food security concerns continue to drive the quest for new more sustainable food sources. The last two decades have seen an explosion in high throughput technologies, enabling unprecedented feats in the biosciences. Gene Editing stands out as a technology which when combined with the classical genetic engineering techniques provides handy but superb toolkit sets for precise, accurate and massive fine-tuning and repurposing of microbes for desired ends. This chapter proposes to navigate the landscapes of potential applications of this revolutionising technology for discovery and repurposing of existing SCPs and other microbes as microbial foods.

Keywords Food security · Cloning · Genetic engineering · Microbial food · Single cell protein · CRISPR-Cas

O. S. Obayori (✉) · O. O. Olasunkanmi
Department of Microbiology, Faculty of Science, Lagos State University, Ojo, Lagos, Nigeria

O. O. Babalola et al. (eds.), *Ecofriendly Frontiers*,
https://doi.org/10.1007/978-3-031-98700-7_14

14.1 Introduction

Since the advent of food industries and food microbiology, shelf-life, taste, aroma, quantity, biological and physico-chemical properties of foods produced have been greatly improved upon due to the applications of various microorganisms in the production processes as well as elimination of some microorganisms.

Several groups of microorganisms from yeasts, bacteria and fungi are useful to humans and have made major contributions in the food industry. From time immemorial, scientists all over the world have used various microorganisms to achieve different feats in the food industry, which include fermentation and preservation, and have also developed and engineered some microorganisms into foods such as microbial protein. Yeast is the most widely used microbe in both food- and feed-processing industry, thereafter, bacteria (Tamang et al. 2020). Microorganisms can be genetically manipulated to make food products with the essential properties such as high nutritional contents on a commercial scale, with safety as a key factor as well.

Microorganisms have been involved in the preparations of some products such as bread, cheese, beverages and yoghurt even when the understanding of what the microbes do in the production processes was vague. The first instance of fermented milk was probably manufactured accidentally by nomads, as it became acidic and clotted under the influence of certain microbes. By chance, the bacteria used were harmless, acidophilic and non-toxic (Johnson and Steele 2014).

Other more recent advances in the application of microorganisms in the food industry include: using fermentation to produce microbial biomass as a nutritional food source is starting to be adopted in both animal feed and human foods (Shurson 2018; Jones et al. 2020; Banks et al. 2022). Microalgae such as *Spirulina* and *Chlorella*, which are single-cell proteins, highly rich in proteins, vitamins and phytonutrients, are consumed as food and health supplements (Janssen et al. 2022). Microbes are now potential animal meat alternatives. Single-cell protein obtained from filamentous fungi, resembles meat product from chicken nuggets to beef mince (Whittaker et al. 2020). Microorganisms such as bacteria and yeast have been used to create dairy products (Lv et al. 2021). Human milk oligosaccharides (HMOs) have been produced in both *Saccharomyces cerevisiae* and *Bacillus subtilis* (Lu et al. 2021) and human milk fats in the oleaginous yeast *Yarrowia lipolytica* (Bhutada et al. 2022). Microbes have also been engineered to produce sweeteners such as stevia-derived molecules, xylitol or erythritol (Moon et al. 2010; Mohamad et al. 2015; Xu et al. 2022). More exotic, hoppy flavours have been engineered into yeast to make tastier beer (Denby et al. 2018).

In this chapter, the historical background to the application of microorganisms in food production will be briefly elucidated and due consideration given to single-cell-proteins (SCP). The role of climate change and population explosion in driving the quest for novel food sources will be highlighted. The emerging concept of Microbial Food will be considered as distinct from mere application of microorganisms in food production. Furthermore, we will explore the various technologies used in genetic engineering from the early days in the 1960s to the present, emphasising the latest high throughput dimensions. The landscapes of potential applications of the latest revolutionising technologies for discovery and repurposing of existing SCPs and other microbes as microbial foods will the explored and future directions highlighted. Thus, rather than detailing applications, the possibilities open to applications based on description of the available technologies are brought to the fore.

14.2 Climate Change, Food Security and the Search for Novel Foods

Historically, there has always been a correlation between human population growth and the ability to gather or produce food in functional quantities and qualities. Not only the evolution of man as a social and rational being but his gross anatomy, physiology and psychosocial existence bear astounding correlation with his level of mastery of access to nutriments. From mere population of about 50 million in the pre-Christian era, today the world population is over 8.1 billion (Worldometer 2024). Much of this growth has taken place in the twenty-first century as a result of mastery over the elements, technological advancement and applications in agriculture and medicine. Albeit, food security remains a daunting challenge in the twenty-first century with a significant number of the world population considered food insecure by the Food and Agricultural Organisation (FAO). This has become exercabated as a result of global emergencies such as the COVID-19 pandemic (FAO et al. 2021). According to an FAO report, in 2022 29.6% of the world population is either moderately or severely food insecure (FAO et al. 2023).

Food security is a broad concept with several component parts encompassing not only physical availability but also access, utilization and consistent or stable supply. Food security is defined as "a situation that exists when all people, at all times, have physical, social and economic access to sufficient, safe and nutritious food that meets their dietary needs and food preferences for an active and healthy life" (FAO et al. 2018.

Food insecurity has become exacerbated by a number of factors, chief among which is competition for agricultural products and farmland by emerging industries that rely on biobased products such as biofuels, increasingly tenacious and unrelenting climate change induced loss of arable land to desertification and flooding, receding capacity in the face of economic crisis, especially in the Global South but also in the rurality of the Global North (Raj et al. 2022). Heat stress, for instance, affects not only ecological stability but has overall deleterious effect on productivity and wellbeing (Kuhla et al. 2021).

Climate change drives change in food security as there could be reduced agricultural production, altered suitability of land for crop production, changes in crop production and change in sustainability of rain-fed agriculture as a result of disruption in the traditionally predictable precipitation cycle in agricultural systems (Ayo et al. 2014; Kalele et al. 2021; World Bank 2022). Flooding of farmlands and grazing land or pasture could reduce productivity and also cause mortality of food animals. Climate affects food production systems, food quality and safety, lowers nutrient content of crops and significantly alters the disease and health environment (Ritala et al. 2017). It also causes disruption of supply chain leading to food price volatility and increases number of people at risk of malnutrition. Vulnerable population on the margins of the social sphere are often more at-risk. Lower income, lower physical access to nutrition food and discrimination are but some of the factors that exacerbate the effect for people on the lower rung of the social ladder and or margins of society. (O'Neill et al. 2022).

Livestock emission is estimated at 8.1 gigaton of CO_2—eq per annum, which represents 14.5% of total anthropogenic emission (Cheng et al. 2022), even though recent studies have shown that on a global scale it might be difficult to correctly correlate results from emission with climate change (Scoones 2023).

Microbial foods can be defined as the biomass of cultured microorganisms or the culture broth which can be used as ingredients for cooking and consumed in the forms of typical, contemporary foods cooked with diverse recipes. They have been described as whole-cell-foods, distinguishing them from single cell proteins obtained from algae such as *Spirulina* or other algae and bacteria which are used as supplements in the form of tablets or ingredients (Choi et al. 2022). Microbial foods are likely to gain popularity and acceptance in future for a number of reasons. One is the growing consciousness of animal rights and drive towards elimination of unethical methods of raising livestock and preparing meat. There are also increasingly prohibitive costs of meat production. There has also been concern about climate change as the livestock farm remains a major source of methane, a greenhouse gas (Cheng et al. 2022; Choi et al. 2022). Global warming will in the nearest future make conventional plant or crop production difficult as plants wither as result of high temperature. A recent report noted that the creation of microbial biomass with wastewater resulted in reduction of climate change impact by 96%, reduced freshwater usage by 85% and improved land use footprint by 99% (Durkin et al. 2022).

14.3 Single Cell Proteins and Microbial Foods

14.3.1 Single Cell Proteins

The first food and feed sources that could be regarded as microbial foods of sort were the single-cell proteins (SCP). Single cell proteins (SCPs) are proteins produced from microorganisms and used as supplements or additives for human food and animal feeds. The attraction to single cell proteins stems from the fact that apart from being rich sources of protein they also contain other essential nutrients such as vitamins, amino acids nucleic acids, carbohydrates, bioactive substances such as quinones and are low in fat. They have been in production for more than 100 years, through the years of economic depression and two world wars which, needless to say, play pivotal roles in focusing attention on them (Hamdan and Senez 1992; Ritala et al. 2017). But it was not until the late 1960s that the concept single-cell-proteins was adopted to describe this source of nutrient.

Single-Cell-Proteins can be defined as protein produced by growing single-celled organisms. The organisms used include species used includes of bacteria, moulds, yeasts and algae. The literature is replete with species that have been used for this purpose and more are being added. (Zepka et al. 2008; Gao et al. 2012; Liu et al. 2014; Patelski et al. 2015; Yunus et al. 2015; Putra et al. 2020; Zha et al. 2021; Khan et al. 2022; Wada et al. 2022; Jain et al. 2023; Sekoai et al. 2024). They include bacteria—*Bacillus cereus, Bacillus subtilis, Escherichia coli, Rhodobacter capsulatus, Streptomyces tuirus*; Fungi—*Penicillium citrinum, Cladosporium cladosporioides, Trichoderma viride, Geotrichum candidum. Candida guilliermondii, Candida tropicalis, Candida utilis* and *Saccharomyces cerevisiae,* Algae—*Aphanothece microscopica* Nägeli, *Chlorella pyrenoidosa, Dunaliella*, sp. and *Spirulina* spp.. The fungi, however, constitute the most widely used group, with *Saccharomyces cerevisiae* as most common species, followed by *Candida* spp. (Ribeiro et al. 2023).

Generally, organisms to be considered for SCP production must possess the following characteristics:

- Low nucleic acid content
- Rapid growth
- Ease of recovery
- Not pathogenic
- Digestible

Additionally, they must have high protein yield and the growth conditions must not be in the extremes (Salazar-López et al. 2022; Thiviya et al. 2022).

Substrates used in production of SCPs include carbon dioxide, hydrogen, ethanol, methanol, n-alkanes, starch, sugar and cellulose, waste stream carbon and nitrogen sources such as sulfite waste liquor, molasses, whey, milk and fruit waste (Gao et al. 2012; Wada et al.2022; Bratosin et al. 2021; Bajic et al. 2022; Thiviya et al. 2022; Jain et al. 2023). Thus, apart from their use as sources of human and animal supplements, they provide a sustainable way to mop up biomass and reduce carbon footprint. They have a number of advantages over other sources of protein, namely:

- Microorganisms grow at a fast rate compared to protein-rich food crops
- They produce proteins of higher quality and quantity
- A wide range of inexpensive raw materials are available as substrates
- The production process is easy and simple
- The microorganisms are more amenable to genetic manipulation
- They can utilize a wide range of substrates.
- The production of process is SCP is eco-friendly, cost-effective, and energy efficient and generally sustainable

Mainly used as food additives, starter cultures, sources of surfactants in cosmetics, antifoaming agents in paper and leather industry and as animal feeds supplements, SCPs have never really gained ascendancy to be considered as whole food until recently. Partly responsible for this are a number of factors:

- The capital cost of production due to sophisticated machinery requirement
- High level of nucleic acid in biomass that may lead to gastrointestinal problems or lead to high level of uric acid in consumer
- It does not contain all the essential amino acids; hence it is mainly used as a supplement to other protein sources
- It may trigger an allergic reaction in some consumers.
- Production of SCP requires strict control over various environmental factors and there could be challenges with optimization
- There is public concern over food produced with microorganism as many people see microbes as more of disease agents

Single-cell protein process can be summarized in the following steps: Strain selection, fermentation, harvesting, treatment after harvesting, and single-cell protein food processing. These are steps that include biotechnologically intensive procedures and design of appropriate bioreactors.

14.3.2 Microbial Foods

Single cell proteins having evolved and become established through the years is readily taken for granted whenever microbial food is mentioned. However, strictly speaking the microbial foods should be restricted to use of microbes as whole food. Jahn et al. (2022) identified three main categories of microbial foods:

(i) Fermented foods—traditionally produced through the controlled transformation by microorganisms.
(ii) Microbial biomass is produced from a defined sugar source and possibly from different waste streams or atmospheric gases, followed by the harvesting of microbial biomass and its processing into palatable and safe foods.
(iii) Cell factories—which are microorganisms that are selected and (genetically) optimized to produce certain compounds.

Notwithstanding, microbial biomass produced for human consumption and the cell factories appear to fit more into the emerging conception of microbial foods. Thus, lab-grown meats from microorganism, other edible biomasses as well as sauces, additives and condiments made from whole-cells of microorganisms can be broadly accepted and categorised as microbial foods.

Microbial foods of the future are most likely to come in form of lyophilized inoculum to be cultured and consumed at home or cultured and sold as ready-to-eat microbial food by outlets and companies (Choi et al. 2022). Such foods would be consumed whole as part of food ingredient. It is projected that food microbes would become a major part of the food supplies of spacecraft during inter-planetary voyage as the systems impose limits on amount of food cargo that could be carried and engineered food microbe, that would utilize organic and inorganic wastes, including carbon dioxide to produce biomass, using sunlight energy (Choi et al. 2022; Llorente et al. 2022).

Domestication, adaptive laboratory evolution and screening of natural variations as well as random mutagenesis and careful selection have been at the centre of efforts to enhance desired traits in organisms (Jahn et al. 2022). In the last 50 years, the most remarkable outcomes have largely been the result of genetic engineering. However, as far as the use of SCP for human foods is concerned—microbial foods—the use of microbes is still skewed towards production of supplements and condiments. Advancement towards increased use of microbes for food can leverage on this.

14.4 Genetic Engineering

Genetic engineering is the alteration of genetic make-up of an organism through deletion, insertion or silencing of genes. It is defined by the National Human Genome Institute (web) as a process that uses laboratory-based technologies to alter the DNA makeup of an organism. Also known as gene manipulation or recombinant DNA technology, it allows genes from any source, irrespective of its origin, to be introduced into a recipient for the purpose of expression of information encoded in the transferred gene in the new entity. In the classical sense, genetic engineering was almost entirely dependent on cloning and other technologies associated with the development of the field of genetics from the early 1950, through early twenty-first century (Brown 2010). Today, even newer technologies have been thrown up, with gene editing standing tall among these and poised to remain in the limelight for some time, thanks to the immense possibilities it offers.

Classical genetic engineering involves the identification of gene of interest, cutting and joining of DNA fragments, transfer of DNA from source to recipient and identification of clones with correct or expected inserts. All of these objectives determined the tools employed in recombinant DNA technology. It is therefore important to do a succinct overview of the molecular tools, and technologies employed in the process without delving into detailed protocols.

The DNA lends itself to manipulation and engineering largely on account of its composition and structure, which essentially consists of four-letter words (monomers adenine (A), cytosine (C) guanine (g) and thymine (T) (Fig. 14.1). Being made up of two complementary strands that run in opposite direction each running 5′–3′, with a sticky overhang which can be denatured, cut and rejoined under appropriate condition and the presence of the right enzyme is the underlining molecular principle that makes manipulation of genome of organisms possible—it is at the basis of all brilliant and ingenious revolutions in recombinant DNA technology and gene manipulation.

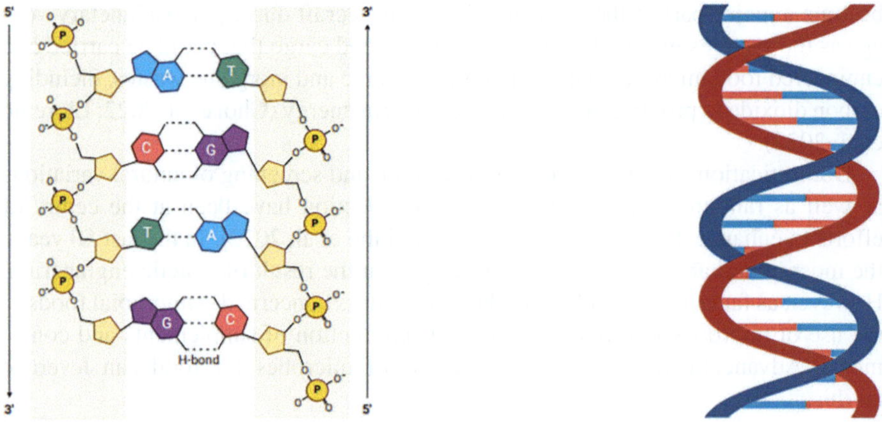

Fig. 14.1 Structure of the DNA is double stranded antiparallel helix made up of four types of nucleotides. This is the basis of all molecular genetic interventions and processes. (Figure was prepared using Biorender objects (www.biorender.com))

Genetically engineered bacterium (GEB)is a bacterium with the ability to efficiently express heterologous proteins or molecular compounds for a specific purpose after genetic engineering (Liu et al. 2022a, b). Bacteria that have gained reckoning as very useful tool of GEB include: *E. coli, Lactobacillus, Salmonella* among others (He et al. 2017; Kawaguchi et al. 2018; Liu et al. 2018; Christofi et al. 2019; Alfano et al. 2020; Oh et al. 2020; Terol et al.2021).

14.4.1 Gene Cloning

The first step in classical cloning is usually the identification and isolation of the DNA of interest or source DNA. This is followed by incorporation of the source DNA into a vector (that is a plasmid, virus or a construct made up of plasmid and viral components). The source DNA bearing vector, now known as the recombinant DNA, is incorporated into a recipient (host) cell that can replicate the vector.

Subsequently, cells that are transferred are selected and further screened for detection of the recombinant DNA. The cultures of cells containing the cloned DNA can now be grown. These steps are illustrated in Fig. 14.2. To achieve this, certain biological tools are required, named enzymes vectors and competent cells. The key enzymes and their roles are outlined below:

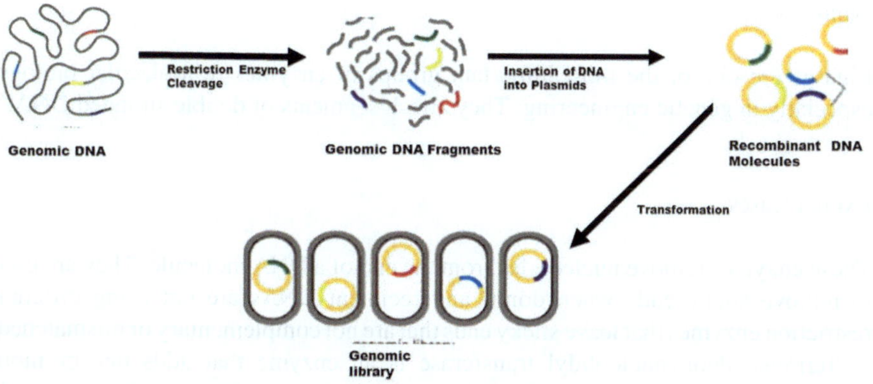

Fig. 14.2 Steps in gene cloning

14.4.1.1 Some of the Key Enzymes Used in Cloning

Some of the key enzymes used in cloning include: Restriction endonucleases, Terminal deoxynucleotidyl transferase, Exonucleases and Ligases. They are used for variety of purposes including cutting trimming and modification and joining of the source DNA and the vector.

Endonucleases

These enzymes are those that catalyze internal breaks in single or double strands. The S1 Endonuclease obtained from *Aspergillus oryzae* is an example of one that cuts single strands. DNAase 1 cuts both single and double strands. However, the most important and most commonly used endonucleases are the restriction endonucleases (RE), which cut double strands at specific location, after identifying specific sites, called restriction sites. The enzymes play important role in removing DNA of invading phages from the bacterial genome. Type I RE cut while the DNA randomly at locations outside the recognition site, Type II cut at the recognition site, usually at a palindrome sequence, while Type III cut the DNA at a specific location away from the recognition site. The palindromes are inverted repeats of 4–8 pairs of bases. The cleavage by Type II often results in sticky OH′ overhang or at times blunt end depending on the enzyme. Type II are the most commonly used RE and include EcoRI from *E. coli* BamHI from *Bacillus amyloliqouefaciens*, Hind III from *Haemophilus influenzae* d among others. HpaI and HaeIII and example of R.E. that cuts DNA leaving blunt ends.

Ligases

Ligases are one of the most important groups of enzymes in molecular biology, especially in genetic engineering. They attach segments of double stranded DNA.

Exonucleases

These enzymes remove nucleotides from the end of a DNA molecule. They are used to remove sticky ends when donor and recipient DNAs are cut using different restriction enzymes that leave sticky ends that are not complementary or mismatched.

Terminal deoxynucleotidyl transferase is an enzyme that adds one or more deoxynucleotide to the 3′ end of a DNA molecule. It is used for homopolymer tailing of blunt ended fragment by adding poly A (polyadenosine or poly T (polythymidine).

Alkaline phosphatase is used to remove phosphate group from the 5′ end of a DNA. It is usually obtained from *E. coli*, calf intestine tissue or arctic shrimp. Another modifying enzyme used is genetic engineering is polynucleotide kinase, it is used for adding phosphate group 5′ to end of a polynucleotide.

Polymerases

Polymerases are some of the most basic enzymes of genetic engineering—catalyzing the synthesis of a polynucleotide strand using a complementary template. DNA polymerase synthesises DNA molecule using a single-stranded DNA template and requires a primer and OH' and (the so-called sticky end). They are easily obtained from *E. coli*.

Klenow is an *E. coli* DNA polymerase I which has been modified by removing its exonuclease activity (5′-3′). It is used in filling in 3′ ends created by restriction enzyme digestion, randomly labelling of primer, DNA sequencing and second strand DNA synthesis in cDNA cloning. Other important DNA polymerases include T4 DNA polymerase, T7 DNA polymerase which is obtained from the thermophilic bacterium *Thermus aquaticus* and owes its allure to its heat stability. It is used in PCR.

Reverse transcriptases are polymerases that synthesize DNA using ssRNA as template. They are commonly obtained from Avian myoblastosis virus or monkey murine leukaemia virus. RNA polymerases synthesise RNA from DNA template. They are obtained from phages such as SP6, T3, and T7.

14.4.1.2 Vectors

The vector is required to transfer the DNA of interest into the host/recipient and protect it from degradation. It is a molecular vehicle used to transfer DNA of interest into a host cell. The vectors commonly used are outlined in Table 14.1. The

Table 14.1 Some vectors traditionally used in cloning

Type	Description
Plasmids	They are the most commonly used but are limited in the amount of DNA insert they can carry
Phages	They are viral particles or transducing phages. They have the advantage of high efficiency and ability to packages large inserts (~24 kb)
Cosmids	Vectors made from a plasmid and the cohesive end sites (cos) of phage and can package larger inserts (~50 kb) than phages and easier to handle
Bacteria Artificial Chromosomes (BACs)	They are large DNA constructs used for cloning in bacteria cells. They package large inserts (˃100 kb) and more stable than plasmids but may be more difficult to handle
Yeast Artificial Chromosomes (YACs)	YACs are similar to BACs but are specifically designed for transfer of foreign DNA into yeast cells

preferences depend on a number of factors including the efficiency, the purpose, the size of DNA molecule being transferred amongst others. However, a good vector must fulfill a number of conditions and properties including:

- It should be small and manageable.
- The ability to replicate in the recipient cell
- Ability to carry the DNA of interest with it into the host
- Presence of genes that enable selection of cells that successfully picked up the vector, that is, the transformant.
- Amenability to mechanism that allows for quick and reliable identification of clones with correct source DNA insert.

The plasmid pBR322 (Fig. 14.3a) is an example of commonly used cloning vector. It is large and uses resistance genes for both selection and screening Another classic example is the pUC19 (Fig. 14.3b) which is smaller, has a higher copy number, is more flexible and uses the antibiotic resistance gene for ampicillin (*AmpR*) and the *lacZ* gene for selection and screening respectively. A plasmid vector that is used for cloning and expression of a foreign DNA in a host of recipient is called an expression vector. Such plasmids must be constructed to enable the transcription of the cloned gene and expression of the functional product of interest. Thus, they usually have as part of their construct a strong promoter, a termination codon, a translation initiation sequence and a transcription termination sequence.

14.4.1.3 Transfer of Recombinant DNA to a Host Cell

Once the donor and recipient DNAs have been successfully ligated to form a recombinant, the next step is to transfer the recombinant into the host cell of interest. In bacteria this is usually done by transformation, a process by which competent cells uptake the DNA across the cell envelope. There are naturally competent strains of

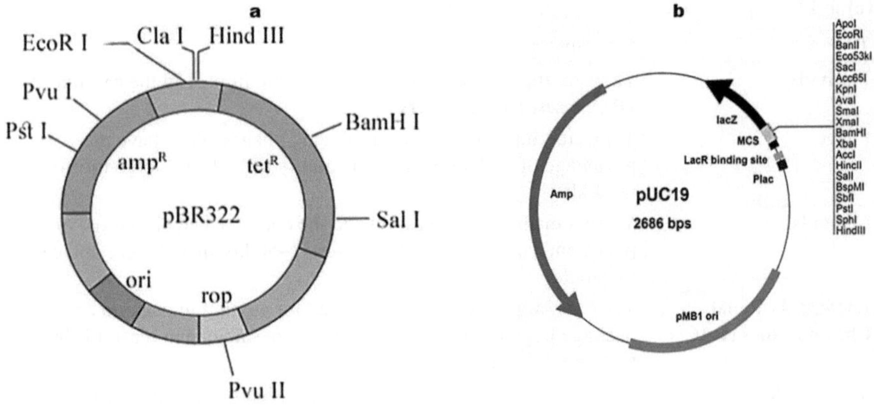

Fig. 14.3 Structure of pBR322 plasmid (**a**) and the puC19 plasmid (**b**). Ori - origin of replication, rop - repressor of primer

species such as *Streptococcus*, *Bacillus subtilis* and so forth, but the rate of natural transformation often does not meet the desired end and in many cases the host cell of interest may not be competent. Thus, in most laboratory experiments and studies, commercially available competent cells of strains of *E. coli* are used. This procedure uses a combination of cooling to 0 °C and treatment with calcium chloride or rubidium chloride and subsequent heating to 42 °C and cooling again (Glazer and Nokaido 2007; Glick et al. 2010).

Transformed cells are selected using at least one resistance marker on the plasmid vector. The transformants are thereafter screened for the presence of correct insert using a second marker as in the case of pBR322 or *lacZ* gene as in the case of pUC19.

When handling eucaryotic cells transfer of foreign DNA, especially when working with mammalian cells or plant protoplasts, methods other than transformation or use of phages are used and referred to as Transfection. These include electroporation which relies on the use of short pulses of high voltage to create temporary pores in the cell membrane, allowing DNA to move downstream the electrical potential across the cell envelope (Potter 2003; Sherba et al. 2020). This process requires pretreatment of the cells and stringent ionic conditions, temperature, cell/DNA ratio and is carried out in special equipment—the electroporator. DNA or gene guns are also used to achieve effective transfection. In this procedure, DNA coated gold or tungsten nanoparticles are fired at the cells as projectiles. Other procedures use conjugation. For example, the tumorigenic bacterium *Agrobacterium tumefaciens* is used in transfection of plants, while various viruses are used as vectors in both bacteria and animal cells to effect injection. The large sizes of mammalian and plant cells also allows micro injection procedure in which the recombinant DNA is directly injected into the cell. Some of the approaches to handling transfer of large recombinant DNA to host cell have been well elucidated by Bai et al. (2023).

14.4.2 Some Classical Genetic Engineering Tools

From the early days in the 1970s through the 1990s genetic engineering benefited from quite an array of methods and technologies which even as some of them are surpassed or improved upon still remain important reference points and milestones in our understanding of biotechnological applications. Worthy of note among these are Gel electrophoresis, Southern blotting, Sangers sequencing and Polymerase chain reaction (PCR).

14.4.2.1 Agarose Gel Electrophoresis

Agarose gel electrophoresis provides a simple method to separate and size DNA fragments, relying on the force of electric field. Gel electrophoresis has a long history predating the introduction of cloning, and agarose gel electrophoresis had become common in nucleic acid works towards the late 1960s. In what has come to be established as perhaps the most enduring workhorse of molecular biology, a simple equipment consisting of gel tank, electrodes and agarose slab immersed in a buffered solution (tris acetate or tris borate EDTA—ethylenediaminetriacetate) is layered with the DNA sample and run under sets of conditions including temperature, gel concentration, buffer ionic condition, specific voltage and duration. The agarose acts as a sieve that impedes the movement of DNA as it is driven by the electric field from the anode to the cathode. Because the amount of charge per molecule is proportional to its size the rate of movement is inversely proportional to the logarithm of the DNA fragment, fragments move at different rates. Visualisation is achieved by staining commonly by staining with ethidium bromide (Lee et al. 2012). Separated DNA can then be extracted for sequencing, amplification or used for hybridization procedures on the gel. Since the 1980s there have been other electrophoretic procedures following similar principles but with modifications to address particularly issues related to efficiency in the light of varying DNA sizes, and using another substratum such as polyacrylamide. The denaturing gradient electrophoresis (DGGE) is used to separate DNA fragments of the same sizes but different nucleotide sequences (Fischer and Lerman 1983; Muyzer et al. 1993) while the pulsed field gel electrophoresis (PFGE) is used to separate large DNA fragment (ranging from 30 to 2000 kb) difficult to separate by conventional agarose gel electrophoresis (Schwartz and Cantor 1984; Sharma-Kuinkel et al. 2016).

14.4.2.2 Southern Blotting

Southern blotting came into existence at a time when cloning was beginning to gain currency in molecular biology and hybridization of separated DNAs for identification purpose was rather challenging as a result of the fragility of the agarose gel, necessitating a procedure to transfer the separated DNA to a firm substratum that

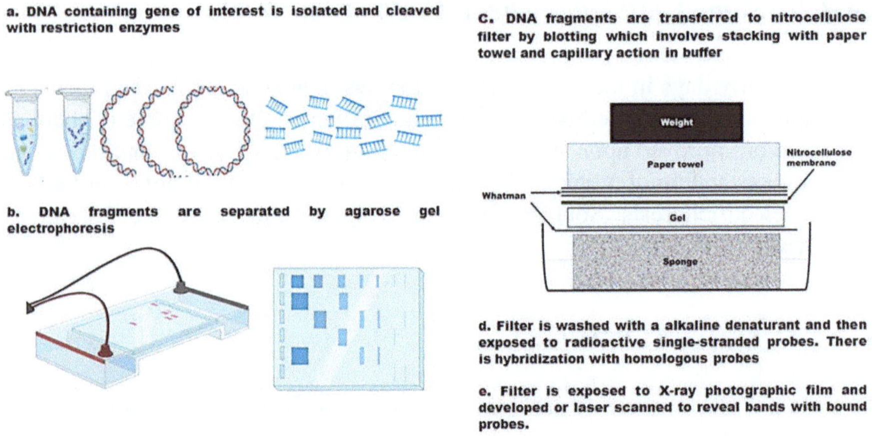

Fig. 14.4 Southern blotting procedure. (Figure was prepared partly from Biorender objects (www.Biorender.com))

could withstand the rigour of treatment steps. The development of the procedure by Edwin Mellor Southern (Southern 1975) infused a new life to the work of molecular biologists. In cloning, it is useful not only in the handling of DNA fragments but also in confirming clones that carry the inserts in the plasmid vector. Key steps in this procedure includes isolation of DNA of interest, cleavage restriction enzyme fragmentation of the DNA, followed by electrophoresis and transfer to a nitrocellulose filter by capillary action. The DNA is then denatured and washed with radiolabelled probe. Thereafter it is read in a reader to detect found bands. The protocol is outlined in Fig. 14.4.

14.4.2.3 Sangers Sequencing

Sequencing, which has been defined as the reading and recording of the nucleotide sequence of a DNA or RNA, came in the nick of time to address an important aspect of genetic engineering: the need to know the precise composition and ordering of nucleotides of a nucleic acid of interest. Developed about the same time as the Maxam-Gilbert chemical synthesis, the Sangers chain termination method soon gained ascendancy for a number of reasons. It uses a single-stranded DNA template and a mixture of the four deoxyribonucleotides and their respective dideoxyribonucleotides in ratios 4:1 to initiate synthesis catalysed by DNA polymerase. The dideoxyribonucleotides lack the OH’ overhang, hence when incorporated during synthesis the process is automatically terminated. Since the dideoxyribonucleotides are labelled, such insertion is detectable. With proportional chances of incorporation of normal and dideoxyribonucleotides, different chain lengths are produced such that when the synthesis products are electrophoresed and aligned based on the terminating dideoxyribonucleotide the sequence of nucleotide in the DNA can be determined.

14.4.2.4 Polymerase Chain Reaction

Polymerase chain reaction (PCR), developed by Kary Mullis in 1985, is a technique used to amplify designated segment of DNA. It was a fortuitous but rather simple technique which has come to stay as a corner stone of molecular genetics and the most valuable tool in the biosciences (Zhu et al. 2020). For many molecular biology procedures, and cloning especially, obtaining the appropriate concentration of the gene of interest could be challenging, often requiring laborious and expensive cloning procedure. PCR enables amplification of DNA via a three-step procedure carried out in a thermo cycler (Fig. 14.5). As Kary Mullis himself simply put it, the whole idea was to pick a piece of DNA and make as much of it as possible (Mullis 1990) The thermocycler allows repeated cycles of temperature changes over specified time duration resulting in geometrical increase in amount of DNA fragments. Typically, the steps include Denaturation 94 °C, Annealing at 50–60 °C and Elongation at 72 °C. The thermostable enzyme Taq polymerase from *Thermus aquaticus* or others such as Pfu polymerase from *Pyrococcus furiosus* is used. Following amplification, the PCR product is run in agarose gel electrophoresis and preparatory to visualization and other procedures. The annealing step requires two primers which are oligonucleotides complementary to the DNA strands.

Although there are variations and different types for different purposes, including quantitation, identification, diagnosis and fingerprinting, the standard PCR has retained its iconoclastic position as a clean, seamless procedure with varied application (Lorenz 2012; Garibyan and Avashia 2013).

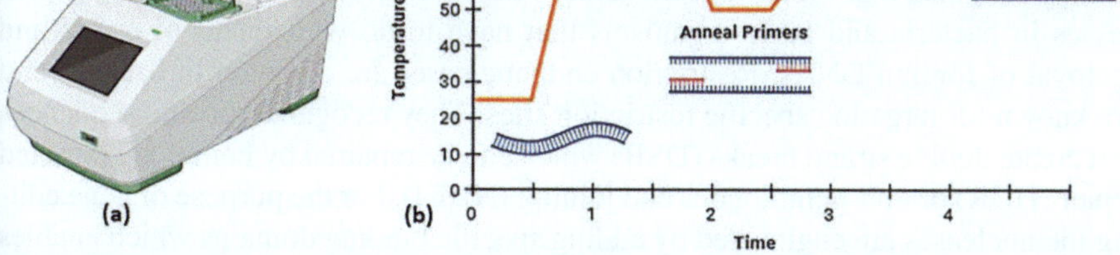

Fig. 14.5 PCR (**a**) A Thermocycler (Biorad T100) (**b**) Steps in PCR: Denaturation 94 °C, Annealing at 50–60 °C and Elongation at 72 °C. (Figure was prepared using Biorender objects (www.biorender.com))

14.5 New Era of Throughput Technologies in Gene Manipulation

The first ten years of the twenty-first century have brought on board developments in molecular biology, bioinformatics and nanotechnology that put genetic engineering unimaginably at the frontlines of addressing myriads of life's challenges. The classical tools of fledgling genetic engineering of the late twentieth century that awed the scientific world, though still very relevant, have become dwarfed by the high-throughput, ever-shrinking tools of modern biotechnology. Next generation sequencing, gene editing and synthetic biology are but some of the most awesome of these technologies. Even as more technologies come to fore from the frontiers of scientific innovation, it appears that in time to come gene editing poises to drive developments in genetic engineering.

14.5.1 Gene Editing

Classical genetic engineering modifies organisms by identifying genes of interest and inserting the genes into the genome of the target organism. Basically, it alters the genome by cutting it and adding a foreign entity. This comes with a number of implications, including the insertion of exogenous DNA that might not be desired or might have deleterious effect on the modified organism or consumers. For example, in genetically modified plants there may be the need to use certain agrochemicals in conjunction with herbicide tolerant plants or there could be carryover of antibiotic resistance genes from the production line (Zaidi et al. 2020). Genome editing makes it possible to alter the genome without introducing foreign genes (He and Zhao 2019; Zhang et al. 2023).

Genome editing is simply the re-writing of the genetic code of an organism to remove or introduce desired information using engineered nucleases. Nucleases are natural enzymes that break nucleic acids. They are involved in a number of processes in bacteria and other organisms that have to do with repair of errors and removal of foreign DNAs. Restriction endonucleases are common in bacteria and are known for targeting specific restriction sites. They recognize specific sequences and create double strand breaks (DSB) which can be repaired by homology directed repair (HDR) or non-homologous end joining (NHEJ). For the purpose of gene editing the nucleases are engineered by adding specific binding domains which enables targeted cutting on sites in the DNA longer than the sort sequence that make up the restriction sites of organism's genome.

There are basically four types of such molecular scissors, namely Meganucleases, Zinc-finger nucleases (ZFNs), TALENs (Transcription Activator-Like Effector Nucleases) and CRISPR-Cas (Gaj et al. 2016). A diagrammatic representation of the components of the various systems is shown in Fig. 14.6. They all have in common their ability for targeted cutting and having the binding domain and cleavage

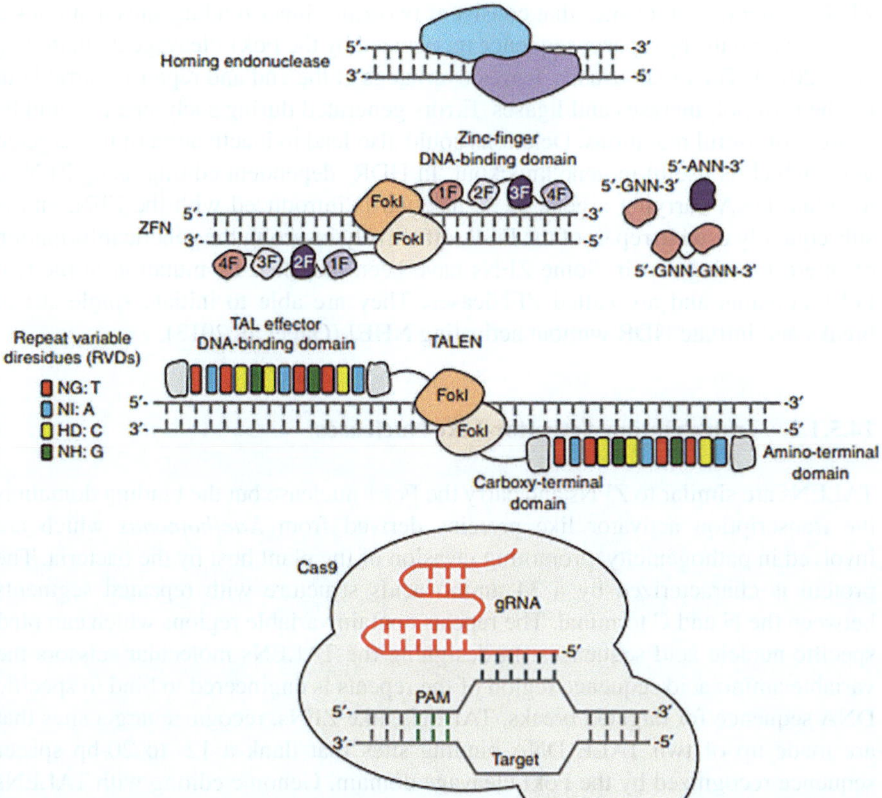

Fig. 14.6 Nucleases used in gene editing

domain but differ on a number of scores including design, efficiency, cost and easeof application among others.

The Homing endonucleases are large, constituting a major hinderance to their use; Zinc Fingers and TALENs both use the Fokl nuclease and are highly specific because of dimeric binding but differ in binding domains or effectors, while CRISPR Cas system only bind a strand of the DNA fragment (Gaj et al. 2016).

14.5.1.1 Zinc Finger Nucleases

ZFNs have zinc finger repeats as their binding domain. It is a protein motif made up of Histidine and Cysteine residues held together by zinc between two B-sheets and an a-helix. The cleavage domain is Fokl, a restriction enzyme from the bacterium *Flavobacterium okeanokoites*. Fusion of the zinc finger with the nuclease produces efficient molecular scissors able to bind long sequences and make targeted break.

ZFNs recognize target sites that consist of two zinc-finger binding sites that flank a 5- to 7-base pair (bp) spacer sequence recognized by the Fok1 cleavage domain (Gaj et al. 2016). The break usually leaves overhangs at the end and repair is carried out by the host polymerases and ligases. Errors generated during such repairs could be sources of useful mutations. Deletions could also lead to inactivation of the targeted gene, which is useful in gene knockout. In HDR- dependent editing using ZFNs a template DNA carrying a gene to be inserted is introduced with the ZFN, and is subsequently used in repair of the DSB, effectively inserting the genetic information of interest during repair. Some ZFNs have been modified by mutation in the two Fok1 domains and are called ZFNicases. They are able to initiate single strand breaks and initiate HDR without activating NHEJ (Gaj et al. 2013).

14.5.1.2 Transcription Activator-Like Nucleases

TALENs are similar to ZFNs and carry the Fok1 nuclease but the binding domain is the transcription activator like proteins derived from *Xanthomonas* which are involved in pathogenicity, promoting invasion of the plant host by the bacteria. The protein is characterized by a 34 amino acids structure with repeated segments between the N and C terminal. The repeats contain variable regions which can bind specific nucleic acid sequences. In designing the TALENs molecular scissors the variable amino acid sequence region of the repeats is engineered to bind to specific DNA sequence for targeted breaks. TALENs, like ZFNs, recognize target sites that are made up of two TALE DNA-binding sites that flank a 12- to 20-bp spacer sequence recognized by the FokI cleavage domain. Genome editing with TALENs and ZFNs use diametric molecules. This makes for low rates of off-target breaks.

TALENs is more specific than ZFNs as each TALE is specific for particular DNA triplet and it produces less off-target breaks, which may be deleterious, but it may be difficult to deliver to the host cell because of its large size (Gaj et al. 2013; Pattanayak et al. 2014). ZFNs on the other hand is difficult to produce because of the difficulty in constructing the zinc finger array. Today, both TALENs and ZFNs have been overtaken by the CRISPR-Cas system as the gene editing tool of interest perhaps because of its malleability, ease of design and implementation and greater editing efficiency outcome and possibility of programming for multiplexing (Stella and Montoya 2016; Zhang et al. 2023). The CRISPR-Cas9 system, unlike the ZFN and TALENs, is also more tolerant of positional multiple consecutive mismatching (Saifaldeen et al. 2020).

14.5.1.3 CRISPR-Cas Systems

The Clustered Regularly Interspaced Palindromic Repeat (CRISPR) system of the bacteria is a defence system of bacteria and archaea against bacteriophages, discovered back in the 1980s when its role was still unknown (Horvath and Barrangou 2010; Al-Attar et al. 2011). It was brought back to reckoning in 2007 when it was found and revealed to be linked to resistance to phage infection in *Streptococcus thermophilus.* (Barrangou et al. 2007). It was observed that in the genome of the organism there were regions of inverted repeated sequences separated by sequences that are complementary to sequences in viral genome. This region together with a leader sequence constitute the CRISPR array. The CRISPR-associated genes which encode specific proteins of what would later be known as CRISPR-associated proteins (CAS) form the next region after the leader.

The operational mechanism of the CRISPR-Cas formation is kickstarted when a specific sequence of an invading foreign DNA is integrated into the host at the CRISPR locus in the repeat closest to the leader (Xiao et al. 2017). Transcripts derived from the spacer in CRISPR locus via a pre-CRISPR RNA (pre-crRNA) are processed by specific Cas proteins to form CRISPR-RNAs (crRNA). The crRNAs are assembled with Cas proteins to form an effector search complex which begins to scavenge the cytoplasm for complementary DNA (Jiang and Doudna 2015). There is a third component which is the tracrRNA (trans-activating CRISPR). Once the components are assembled the scavenging for foreign DNA begins. They do this by base pairing with the crRNA guide sequence, followed by Cas protein mediated cleavage. The crRNA contains 20 nucleotides similar to the DNA sequence. This seed sequence and a conserved protospacer adjacent motif (PAM) upstream of the crRNA binding site are required for target recognition by Cas9 protein (Gaj et al. 2016). The Cas proteins have nuclease or helicase activities depending on their specific roles in the CRISPR-Cas system. This mechanism is illustrated in Fig. 14.7 (Jiang and Doudna 2015).

The CRISPR Cas9 is engineered to form a single unit by combining the crRNA and the traRNA to give the single-guide RNA (sg-RNA) which combines with the Cas9 protein (Liu et al. 2022b). The cleavage efficiency of CRISPR-Cas9 is affected by sg-RNA properties, G+C content of PAM proximal and distal regions among others (Jung et al. 2024). Further modifications to the CRISPR-Cas 9 have sought to reduce the high frequency of off-target cleavage. One of such is the CRISPR-Fok1dead Cas 9 (fdCas9). It is constructed by linking a Fok1 endonuclease catalytic domain to an inactive Cas9 protein. It uses two sgRNA to bind to the two strands of DNA in a protospacer adjacent motif (PAM)-out orientation. They have enhanced specificity and can easily form complexes but are more difficult to deliver due to their sizes (Saifaldeen et al. 2020).

There are six types of CRISPR-Cas system, different in the mechanism of processing of the pre-crRNA, the type of nucleic acid targeted—RNA or DNA, and efficiency of cleavage. The best studied and most commonly used CRISPR-Cas system is the CRISPR-Cas 9, but other CRISPR-Cas systems have been developed with prospect of higher efficiency and adaptability. Of particular interest are Cas-12 and Cas-13.

A typical CRISPR locus consists of an AT-rich leader sequence upstream of the CRISPR array (gray box), a series of repetitive DNA sequences called repeats (brown diamond) and non-repetitive spacers (colored boxes), as well as a set of CRISPR-associated (cas) genes (shown as arrows). In the acquisition phage, new spacer (highlighted in green) derived from the viral DNA or other invasive genetic

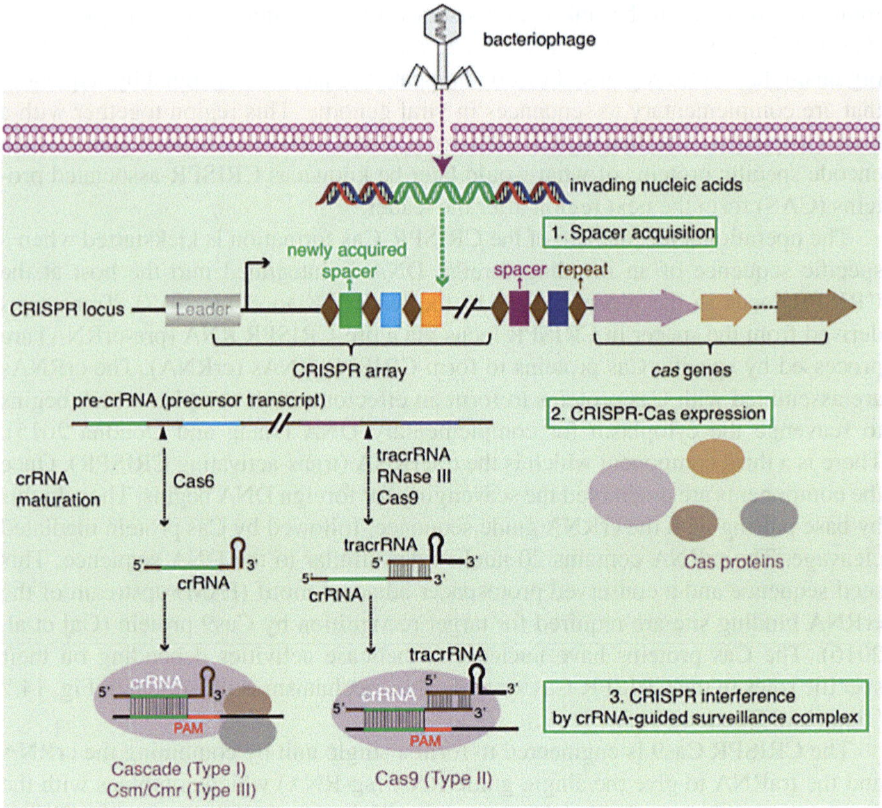

Fig. 14.7 Overview of CRISPR-Cas adaptive immune systems

elements is excised and inserted into the CRISPR array. During the expression stage, *cas* genes encoding a diverse set of Cas proteins are transcribed and translated, and the CRISPR array is transcribed into a long precursor CRISPR RNA (pre-crRNA). In Type I and III systems, precrRNA is further cleaved into small mature crRNAs by Cas6 endoribonucleases, whereas in Type II systems this processing involves a small trans-activating crRNA (tracrRNA) to form Fig. 14.7 (Jiang and Doudna 2015).

CRISPR-Cas tendency for off-target DSBs is a major drawback. There are several methods for screening and validation include but not limited to: Cas-OFFinder (Bae et al. 2014), CIRCLE-seq (Tsai et al. 2017) which uses small double stranded oligonucleotides to tag DSBs, amplify then followed by NGS analysis; BLISS (Yan et al. 2017) in which DSBs are biochemically labelled and the downstream sequences amplified by PCR, then analysed by NGS; SITE-seq (Cameron et al. 2017) and DIG-seq (Kim and Kim 2018).

When combined with other tools CRISPR-Cas system becomes a very powerful tool. For instance, next generation sequencing (NGS) is useful in screening mutants and for analyzing mutants and engineered products for exactitude of modification. This makes for easy screening and selection and obviate laborious process since large numbers of products are massively sequenced in parallel. It is useful in assessing the accuracy and safety of CRISPR-editing as it is able to detect on-target and off-target modifications, quantify editing efficiency and has proven to be more accurate for validation of CRISPR-Cas edited products than several other methods (Frock et al. 2015; Tsai et al. 2015; Sentmanat et al. 2018; Connelly and Pruett-Miller 2019).

14.5.1.4 Fanzor

There are various types CRISPR-Cas systems including CRISPR-associated transposons (CAST), a Tn7 associated CRISPR-Cas system (Peters et al. 2017). But there are other systems that have emerged outside the classical CRISPR-Cas, namely the obligate mobile genetic element guided activity (OMEGA) RNAs such as the TnpB found in prokaryotes and the Fanzor found in eukaryotes (Altae-Tran et al. 2021; Karvelis et al. 2021; Saito et al. 2023; Song and Bae 2024). Fanzor is the eukaryotic equivalent of the CRISPR-Cas system. It is a component of the transposable element in the eucaryotic cell and it is a TnpB′ homolog (Bao and Jurka 2013). The TnPB and Fanzor effectors contain a CRISPR-Cas 12 effector-like domain (RuvC) as their RNA- guided endonucleases (Badon et al. 2024; Wiegand et al. 2024). The Fanzor system is found in a wide range of eukaryote including algae, protozoan, fungi, arthropods and plants (Bao and Jurka 2013; Schargel et al. 2025). Also, it showcases diverse adjacent motif preferences, extended ωRNA structure and RuvC domain rearrangements (Jiang et al. 2025).

One important advantage of Fanzor over CRISPR-Cas system is its small size, which supports easy delivery and diverse applications (Awan et al. 2023) and the fact that it is less prone to off-target cleavage as it has high specificity for the transposon adjacent motif (TAM) where it binds Schargel et al. 2025).

14.5.2 Applications of Genome Editing

In its relatively short span of existence gene editing, especially with the CRISPR-Cas toolkit, has come to reckoning as revolutionizing tool for genome engineering in various field of human endeavours (Wright et al. 2016; Pickar-Oliver and Gersbach 2019; Boti et al. 2023) with immense resources and toolkits at hand (Jinek et al. 2012; Adli 2018; Hanna and Doench 2020). It is presently at the forefronts of genomic surveillance and precision medicine. For example, CRISPRCas9 has been projected for direct targeting of tumor genes in cancer cells to eliminate genetic changes that lead to tumor growth and cancer and in nervous diseases (Jiang et al. 2024).

Technologies for crop improvement include genotyping, marker-assisted selection and high throughput phenotypic among others (Hickey et al. 2019). But today genome editing is at the forefront of plant breeding, with focus on disease resistance, climate change and stress resilience (Krishnappa et al. 2021; Gao 2021; Zha et al. 2021; Mohan et al. 2022; Tian et al. 2024). Disease resistant plants can be engineered using CRISPR-Cas to generate insertions or deletions in coding sequences, leading to gene disruption or even in the promoter region by deletion. Multiplex sgRNA could also be used to achieve gene deletion or homology directed repair (Zaidi et al. 2020). A recent review detailed successes recorded in Africa concerning applications of CRISPR-Cas system to engineer crops, especially cereals resilient to climates change, other abiotic stressors and biotic stress (Ndudzo et al. 2024). Other applications include engineering of plant to resist insect pest via gene knockout to remove susceptibility targets in the plant, colour change and volatile substance gene to discourage disrupt attraction of insect to the plant, and gene knockout to increase susceptibility of insect larvae to biopesticide like the *cry* protein of *B. thuringiensis* (Xu et al. 2019; Tyagi et al. 2020).

14.5.3 Applications of CRISPR-Cas in Genetic Engineering for Microbial Foods

The use of microorganisms or microbial biomass as whole meal food has not enjoyed patronage and acceptance on a large scale because of potential side effects or sensory factors and public perception. To be able to use them as substitute to farm-grown foods like plants and animals, the negating properties have to be addressed. Many of the of the drawbacks of SCPs and the limitations that discourage their use as human foods are ones that could be addressed by biotechnological interventions. Areas of intervention include: redesign of proteins for quality and nutrition profile, programing of protein for overexpression, engineering for production of nutritional and functional compounds, improvement in organoleptic properties and texture. Another area of focus could also be to engineer microbes to use a broad range of substrates for growth. Moreover, genetic engineering tools specifically developed for microbial engineering such as promoters, transcription factors ribosomal binding sites and synthetic biosensors are transforming this prospect to reality as an unheard-of rate.

The CRISPR/Cas system could be a powerful tool for mining for organisms with potential for safe application as microbial food. Recently Liu et al. (2024) developed a method to randomly activate endogenous genes and mine for key targets associated with desired phenotypes using *Yarrowia lipolytica*.

The industrial strain of the mould *Aspergillus oryzae* has been extensively modified using TALENs and CRISPR Cas9 systems, achieving gene knockout and myriad other modifications (Maruyama 2021). *Aspergillus oryzae* has a long history of use in the Japanese food industry and a subject of several biotechnological applications and modification particularly because it is considered safe and has ability for heterologous protein production. Its large genome is also an added advantage. The CRISPR Cas system allows for the modification of *A. oryzae* without need to construct complex host/vector system. It allows for introduction of mutation to several genes at the same time.

Engineering SCPs for consumer acceptability could benefit immensely from gene knockout, knockin or knockdown as the case may be. Previously, flavour had been imparted on plant-based meat by engineering haemoglobin protein into the plant (Fraser et al. 2018). Indeed, the production of microbial food imitation of meat has a long history dating as far back as the late 1980s with the production of *Fusarium venenatum*-based 'meat' (Sadler 1988; Wackett 2020), the age of high-throughput biotech offers cheaper and more precise engineering options. Another approach might involve the use of gene knockout to inactivate the genes encoding the pathways responsible for the off-flavour impacting metabolites in fungi or bacteria intended for use as microbial food (Jahn et al. 2022).

CRISPR interference (CRISPRi) uses a catalytically inactive Cas9 for gene editing. It is useful in genome-scale functional screening. It is a simple and highly efficient method, which uses the class 2 type II-A dCas9. It has found application in all groups of organisms with potential application in food industry, including yeasts, moulds, Actinobacteria, Gram-positive and Gram-negative bacteria and algae (Sun et al. 2023). This can be very useful in both selection and modification of microbial food candidates.

Engineering the microbial food microorganism for flavour imparting metabolism could help reduce cost by obviating the need for additives and artificial flavouring. CRISPR-Cas9 technology has been used to construct *Saccharomyces cerevisiae* to produce linalool and geraniol for flavour improvement in beer (Denby et al. 2018). This helped to obviate the use of hops thus cutting cost. Strain construction was done with Cas9 plasmid ADE2Λ plasmid. ADE2 strain was co-transformed with Cas9/sgRNA plasmid and repair template leading to double break in the ADB23' sequence. The repair template contained ADE2 gene, templated DSB repair resulted in a white colony phenotype. Upon screening of colonies, stable integrators could be differentiated from heterologous integrators which had variegated colonies. This also highlights the continuous relevance of classic tools of genetic engineering (cloning, transformation etc) in combination with novel tools such as CRIPR-Cas9.

Meng et al. (2023) used CRISPR-Cas technology in the development of a *Pichia pastoris* with ability to use methanol on an industrial scale. The improvement achieved include efficient utilization of methanol, tolerance to higher temperature and higher protein yield. The engineering of the organism was done using CRISPR-Cas mediated genome editing derived-lactic acid by integrating the D-lactate dehydrogenase gene (d-LDH) into the non-transcribed spacers of ribosomal DNA (rDNA) locus followed by transformational amplification.

More and more novel tools and methods are being churned out and continue to enrich the available toolkits of CRISPR-Cas gene editing, with increasing efficiency and accuracy of target hits. A system for single-nucleotide modification of the chromosome and plasmids in gram-negative bacteria was developed by Kozaeva et al.

(2024). Named pAblo.pCasso, the system is able to carry out seamless cytidine and adenine base-editing with remarkable precision. As new tools and protocols emerge, it is simply a matter of time for the biotechnological barriers to effectively harnessing potential resources for microbial foods to be surmounted, barring ethical considerations and public acceptability.

While research on applications of Fanzor system has focused mainly on genome editing of human and animals specially for cancer treatment and stem cell engineering (Saito et al. 2023) its unique potentials for use in large genomes could be relevant to engineering of yeasts and moulds for microbial food. For instance, it could be programmed for increased protein expression in fungi and microalgae.

14.5.4 Potential Benefits of Engineered Microbial Foods for Food Security

One of the major reasons it has been difficult to use microorganisms as food on a large scale is finding organisms with full complement of characteristic features that enhance safety, flavour and aroma, and support high productivity. Combining next generation sequencing with gene editing is going to pave way for new sources and greater possibilities to engineer desired product.

The fact that microbial food production has requirement for smaller land area than conventional farming, makes use of less water resources and is able to generate biomass from industrial waste streams and biogas makes it more attractive than animal husbandry which today is a major driver of greenhouse gas production and global warming. It makes for conservation of water resources, sustainable production processes and waste management.

Increased use of genetically modified microorganisms as food will pave way for food production to meet the growing demand resulting from global population explosion while circumventing the ethical concerns associated with genetic modification of conventional food crops and food animals. These concerns include replacement of natural or existing breed by genetically modified varieties and ultimate loss of biodiversity. At the same time, it will provide a wider range of options of proteins and nutriment for consumers to choose from, with reduced pressure on the natural ecosystem.

Equally, those whose concerns and opposition to livestock farming and raising of food animals is centred around cruelty against animals and animal rights are more likely to embrace engineered microbial foods as replacement for conventional meats or animal-based GMO meat. It would be more acceptable for addressing looming insecurity as it would not involve any form of slaughtering or cruelty against animal.

14.6 Conclusion

The use of microorganisms as food has been practiced for decades but has not assumed centre stage in solving the challenges of food security for a plethora of reasons. One of these being the lack of popular acceptability and inability of scientists to effectively modify such potential microbes to remove identified negative properties and introduce new qualities. However, unfolding realities of increasing world population, climate change impact on agricultural productivity as well as concerns with modified plant and food animals have led to renewed interest in microbial foods. This challenge equally coincides with the emergence in the last two decades of new tools of genetic engineering such as gene editing and synthetic biology. Combining these with classical tools of genetic engineering like cloning, electrophoresis, southern blotting, amplicon sequencing, PCR to mention but a few, would go a long way in enabling discovery of microbes with potential for use as microbial foods and engineering them for yield, quality and acceptability.

References

Adli M (2018) The CRISPR tool kit for genome editing and beyond. Nat Commun 9:1911. https://doi.org/10.1038/s41467-018-04252-2

Al-Attar S, Westra ER, van der Oost J, Brouns SJ (2011) Clustered regularly interspaced short palindromic repeats (CRISPRs): the hallmark of an ingenious antiviral defense mechanism in prokaryotes. Biol Chem 392:277–289. https://doi.org/10.1515/BC.2011.042

Alfano A, Perillo F, Fusco A, Savio V, Corsaro MM, Donnarumma G et al (2020) *Lactobacillus brevis* CD2: fermentation strategies and extracellular metabolites characterization. Probiotics Antimicrob Proteins 12:1542–1554. https://doi.org/10.1007/s12602-020-09651-w

Altae-Tran H, Kannan S, Demircioglu FE, Oshiro R, Nety SP, McKay LJ, Dlakić M, Inskeep WP, Makarova KS, Macrae RK et al (2021) The widespread IS200/IS605 transposon family encodes diverse programmable RNA-guided endonucleases. Science 374:57–65

Awan MJA, Awan MRA, Amin I, Mansoor S (2023) Fanzor: a compact programmable RNA-guided endonuclease from eukaryotes. Trends Biotechnol 11:1332–1334. https://doi.org/10.1016/j.tibtech.2023.08.003

Ayo JA, Omosebi MO, Suleiman A (2014) Effect of climate change on food security in Nigeria. J Environ Sci Comput Sci Eng Technol 3:1768–1778

Bae S, Park J, Kim J-S (2014) Cas-OFFinder: a fast and versatile algorithm that searches for potential off-target sites of Cas9 RNA-guided endonucleases. Bioinformatics 30:1473–1475

Bajic B, Vucurovic D, Vasic D, Jevic-Mucibabic R, Dodic S (2022) Biotechnological production of sustainable microbial proteins from agro-industrial residues and by-products. Foos 12(1):107

Bai S, Luo H, Tong H, Wu Y (2023) Application and technical challenges in design, cloning, and transfer of large DNA. Bioengineering (Basel) 10:1425. https://doi.org/10.3390/bioengineering10121425

Badon IW,Oh Y, Kim HJ, Lee SH (2024) Recent application of CRISPR-Cas 12 and OMEGA system for genome editing. Mol Ther 32(1):32–43

Banks M et al (2022) Industrial production of microbial protein products. Curr Opin Biotechnol 75:102707

Bao W, Jurka J (2013) Homologues of bacterial TnpB_IS605 are widespread in diverse eukaryotic transposable elements. Mob DNA 4:12. https://doi.org/10.1186/1759-8753-4-12

Barrangou R et al (2007) CRISPR provides acquired resistance against viruses in prokaryotes. Science 315:1709–1712

Bratosin BC, Darjan S, Vodnar (2021) Single cell protein: a potential substitute in the human and animal nutrition. Sustainability 13(16): 9284

Bhutada G et al (2022) Production of human milk fat substitute by engineered strains of *Yarrowia lipolytica*. Metab Eng Commun 14:e00192

Boti MA, Athanasopoulou K, Adamopoulos PG, Sideris DC, Scorilas A (2023) Recent advances in genome-engineering strategies. Genes 14:129. https://doi.org/10.3390/genes14010129

Brown TA (2010) Gene cloning and DNA analysis: an introduction, 6th edn. Wiley, Winchester, p 319

Cameron P et al (2017) Mapping the genomic landscape of CRISPR-Cas9 cleavage. Nat Methods 14:600–606. https://doi.org/10.1038/nmeth.4284. Erratum in: Nat Methods (2023) 20:2068. https://doi.org/10.1038/s41592-023-02114-4

Cheng M, McCarl B, Fei C (2022) Climate change and livestock production: a literature review. Atmos 13:140. https://doi.org/10.3390/atmos13010140

Choi KR, Yu HE, Lee SY (2022) Microbial food: microorganisms repurposed for our food. Microb Biotechnol 15(1):18–25. https://doi.org/10.1111/1751-7915.13911

Christofi T, Panayidou S, Dieronitou I, Michael C, Apidianakis Y (2019) Metabolic output defines Escherichia coli as a health-promoting microbe against intestinal *Pseudomonas aeruginosa*. Sci Rep 9:14463. https://doi.org/10.1038/s41598-019-51058-3

Connelly JP, Pruett-Miller SM (2019) CRIS.py: a versatile and high-throughput analysis program for CRISPR-based genome editing. Sci Rep 9:4194

Denby CM, Li RA, Vu VT et al (2018) Industrial brewing yeast engineered for the production of primary flavor determinants in hopped beer. Nat Commun 9:965. https://doi.org/10.1038/s41467-018-03293-x

Durkin A, Finnigan T, Johnson R, Kazer J, Yu J, Stuckey D, Guo M (2022) Can closed-loop microbial protein provide sustainable protein security against the hunger pandemic? Curr Res Biotechnol 4:365–376

FAO, IFAD, UNICEF, WFP, WHO (2021) The state of food security and nutrition in the world, 2021. Transforming food systems for food security, improved nutrition and affordable healthy diets for all. FAO, Rome. https://doi.org/10.4060/cb4474en

FAO, IFAD, UNICEF, WFP, WHO (2023) In brief to the state of food security and nutrition in the world 2023. Urbanization, agrifood systems transformation and healthy diets across the rural–urban continuum. FAO, Rome. https://doi.org/10.4060/cc6550e

FAO et al. (2018) The state of food security and nutrition in the world: building climate resilience for food security and nutrition. Food and Agriculture Organization of the United Nations. FAO, Rome

Fischer SG, Lerman LS (1983) DNA fragments differing by single base-pair substitutions are separated in denaturing gradient gels: correspondence with melting theory. Proc Natl Acad Sci USA 80:1579–1583. https://doi.org/10.1073/pnas.80.6.1579

Fraser RZ, Shitut M, Agrawal P, Mendes O, Klapholz S (2018) Safety evaluation of soy leghemoglobin protein preparation derived from *Pichia pastoris*, intended for use as a flavor catalyst in plant-based meat. Int J Toxicol 37:241–262. https://doi.org/10.1177/1091581818766318

Frock RL, Hu J, Meyers RM, Ho YJ, Kii E, Alt FW (2015) Genome-wide detection of DNA double-stranded breaks induced by engineered nucleases. Nat Biotechnol 33:179–186. https://doi.org/10.1038/nbt.3101

Gaj T, Mercer AC, Sirk SJ, Smith HL, Barbas CF III (2013) A comprehensive approach to zinc-finger recombinase customization enables genomic targeting in human cells. Nucleic Acids Res 41:3937–3946

Gaj T, Sirk SJ, Shui SL, Liu J (2016) Genome-editing technologies: principles and applications. Cold Spring Harb Perspect Biol 8(12):a023754. https://doi.org/10.1101/cshperspect.a023754

Gao C (2021) Genome engineering for crop improvement and future agriculture. Cell 184:1621–1635. https://doi.org/10.1016/j.cell.2021.01.005

Gao Y, Li D, Liu Y (2012) Production of single cell protein from soy molasses using *Candida tropicalis*. Ann Microbiol 62:1165–1172

Garibyan L, Avashia N (2013) Polymerase chain reaction. J Invest Dermatol 133:1–4. https://doi.org/10.1038/jid.2013.1

Glazer AN, Nokaido H (2007) Microbial biotechnology: fundamentals of applied microbiology, 2nd edn. Cambridge University Press, Cambridge

Glick BR, Pasternak JJ, Patten CL (2010) Molecular biotechnology: principles and applications of recombinant DNA, 4th edn. ASM Press, Washington, DC

Hamdan I, Senez J (1992) Protein (SCP) production in the twenty-first century. In: DaSilva EJ, Ratledge C (eds) Biotechnology: economic and social aspects: issues for developing countries. Cambridge University Press, pp 142–164. https://www.cambridge.org/core/books/biotechnology-economic-and-social-aspects/1EE6FBE4E3C1028E2CF7D0C47640D7BD

Hanna RE, Doench JG (2020) Design and analysis of CRISPR-Cas experiments. Nat Biotechnol 38:813–823

He Y, Zhao Y (2019) Technological breakthroughs in generating transgene-free and genetically stable CRISPR-edited plants. aBIOTECH 1:88–96

He L, Yang H, Liu F, Chen Y, Tang S, Ji W et al (2017) Escherichia coli Nissle 1917 engineered to express Tum-5 can restrain murine melanoma growth. Oncotarget 8:85772–85782. https://doi.org/10.18632/oncotarget.20486

Hickey LT, Hafeez AN, Robinson H, Jackson SA, Leal-Bertioli SCM, Tester M, Gao C, Godwin ID, Hayes BJ, Wulff BBH (2019) Breeding crops to feed 10 billion. Nat Biotechnol 37:744–754. https://doi.org/10.1038/s41587-019-0152-9

Horvath P, Barrangou R (2010) CRISPR/Cas, the immune system of bacteria and archaea. Science 327:167–170

Jahn LJ, Rektal VM, Sommer MOA (2022) Microbial foods for improving human and planetary health. Cell 186:469–478. https://doi.org/10.1016/j.cell.2022.12.002

Janssen M, Wijffels RH, Barbosa MJ (2022) Microalgae based production of single-cell protein. Curr Opin Biotechnol 75:102705

Jain S, Hefferman J, Joshi J, Watt T, Marcellin E, Greening C (2023) Microbial conversion of waste gases into single-cell protein. Microbiol Aust 44(1):27–30

Jiang F, Doudna JA (2015) The structural biology of CRISPR-Cas systems. Curr Opin Struct Biol 30:100–111. https://doi.org/10.1016/j.sbi.2015.02.002

Jiang HB, Tang MY, Xu ZD, Wang YN, Li MP, Zheng SY, Zhu JH, Lin ZL, Zhang M (2024) CRISPR/Cas9 system and its applications in nervous system diseases. Genes Dis 11:675–686. https://doi.org/10.1016/j.gendis.2023.03.017

Jiang K, Gootenberg JS, Abudayyeh OO (2025) Fanzors, a family of eukaryotic RNA-guided DNA endonucleases. FEBS Lett 599:1089–1093. https://doi.org/10.1002/1873-3468.70038. Epub ahead of print

Jinek M, Chylinski K, Fonfara I, Hauer M, Doudna JA, Charpentier E (2012) A programmable dual-RNA-guided DNA endonuclease in adaptive bacterial immunity. Science 337:816–821. https://doi.org/10.1126/science.1225829

Johnson ME, Steele JL (2014) Fermented dairy products. ASM Press eBooks, pp 823--839. https://doi.org/10.1128/9781555818463.ch32

Jones SW et al (2020) Recent advances in single cell protein use as a feed ingredient in aquaculture. Curr Opin Biotechnol 61:189–197

Jung WJ, Park SJ, Cha S, Kim K (2024) Factors affecting the cleavage efficiency of the CRISPR-Cas9 system. Anim Cells Syst 28:75–83. https://doi.org/10.1080/19768354.2024.2322054

Kalele DN, Ogara WO, Oludhe C, Onono JO (2021) Climate change impacts and relevance of smallholder farmers' response in arid and semi-arid lands in Kenya. Sci African 12:e00814. https://doi.org/10.1016/j.sciaf.2021.e00814

Karvelis T, Druteika G, Bigelyte G, Budre K, Zedaveinyte R, Silanskas A, Kazlauskas D, Venclovas C, Siksnys V (2021) Transposon-associated TnpB is a programmable RNA-guided DNA endonuclease. Nature 599:692–696

Kawaguchi K, Miyake K, Zhao M et al (2018) Tumor targeting salmonella typhimurium A1-R in combination with gemcitabine (GEM) regresses partially GEM-resistant pancreatic cancer patient-derived orthotopic xenograft (PDOX) nude mouse models. Cell Cycle 17:2019–2026. https://doi.org/10.1080/15384101.2018.1480223

Khan MKI, Asif M, Razzaq ZU et al (2022) Sustainable food industrial waste management through single cell protein production and characterization of protein enriched bread. Food Biosci 46:101406

Kim D, Kim JS (2018) DIG-seq: a genome-wide CRISPR off-target profiling method using chromatin DNA. Genome Res 28:1894–1900. https://doi.org/10.1101/gr.236620.118

Kozaeva E, Nielsen ZS, Nieto-Domínguez M, Nikel PI (2024) The pAblo·pCasso self-curing vector toolset for unconstrained cytidine and adenine base-editing in gram-negative bacteria. Nucleic Acids Res 52:e19. https://doi.org/10.1093/nar/gkad1236

Krishnappa G, Savadi S, Tyagi BS et al (2021) Integrated genomic selection for rapid improvement of crops. Genomics 113:1070–1086. https://doi.org/10.1016/j.ygeno.2021.02.007

Kuhla K, Willner SN, Otto C, Wenz L, Levermann A (2021) Future heat stress to reduce people's purchasing power. PLoS One 16:e0251210. https://doi.org/10.1371/journal.pone.0251210

Liu B, Li Y, Song J, Zhang L, Dong J, Yang Q (2014) Production of single-cell protein with two-step fermentation for treatment of potato starch processing waste. Cellulose 21:3637–3645

Lee PY, costumbrado J, Hsu CY, Kim YH (2012) Agarose gel electrophoresis for the separation of DNA fragments. J Vis Exp (62):3923

Liu M, Li S, Zhang Q et al (2018) Oral engineered Bifidobacterium longum expressing rhMnSOD to suppress experimental colitis. Int Immunopharmacol 57:25–32. https://doi.org/10.1016/j.intimp.2018.02.004

Liu Y, Feng J, Pan H, Zhang X, Zhang Y (2022a) Genetically engineered bacterium: principles, practices, and prospects. Front Microbiol 13:997587. https://doi.org/10.3389/fmicb.2022.997587

Liu Y, Pinto F, Wan X et al (2022b) Reprogrammed tracrRNAs enable repurposing of RNAs as crRNAs and sequence-specific RNA biosensors. Nat Commun 13:1937. https://doi.org/10.1038/s41467-022-29604-x

Liu X, Deng J, Zhang J, Cui Z, Qi Q, Hou J (2024) Genome-scale transcriptional activation by non-homologous end joining-mediated integration in *Yarrowia lipolytica*. Biotechnol Biofuels Bioprod 17:24. https://doi.org/10.1186/s13068-024-02472-x

Llorente B, Williams TC, Goold HD et al (2022) Harnessing bioengineered microbes as a versatile platform for space nutrition. Nat Commun 13:6177. https://doi.org/10.1038/s41467-022-33974-7

Lorenz TC (2012) Polymerase chain reaction: basic protocol plus troubleshooting and optimization strategies. J Vis Exp 63:e3998. https://doi.org/10.3791/3998

Lu M, Mosleh I, Abbaspourrad A (2021) Engineered microbial routes for human milk oligosaccharides synthesis. ACS Synth Biol 10:923–938

Lv X et al (2021) Synthetic biology for future food: research progress and future directions. Future Foods 3:100025

Maruyama JI (2021) Genome editing technology and its application potentials in the industrial filamentous fungus *Aspergillus oryzae*. J Fungi (Basel) 7:638. https://doi.org/10.3390/jof7080638

Meng J, Liu S, Gao L, Hong K, Liu S, Wu X (2023) Economical production of Pichia pastoris single cell protein from methanol at industrial pilot scale. Microb Cell Factories 22:198. https://doi.org/10.1186/s12934-023-02198-9

Mohamad NL, Mustapa KSM, Mokhtar MN (2015) Xylitol biological production: a review of recent studies. Food Rev Int 31:74–89

Mohan C, Satish L, Muthubharathi BC et al (2022) CRISPR-Cas technology: a genome-editing powerhouse for molecular plant breeding. In: Arora S, Kumar A, Ogita S, Yau YY (eds) Biotechnological innovations for environmental bioremediation. Springer, Singapore. https://doi.org/10.1007/978-981-16-9001-3_32

Moon HJ et al (2010) Biotechnological production of erythritol and its applications. Appl Microbiol Biotechnol 86:1017–1025

Mullis KB (1990) The unusual origin of the polymerase chain reaction. Sci Am 262:56–61

Muyzer G, de Waal EC, Uitterlinden AG (1993) Profiling of complex microbial populations by denaturing gradient gel electrophoresis analysis of polymerase chain reaction-amplified genes coding for 16S rRNA. Appl Environ Microbiol 59:695–700. https://doi.org/10.1128/aem.59.3.695-700.1993

Ndudzo A, Sibanda A, Moyo S, Bobo ED (2024) CRISPR-Cas9 genome editing in crop breeding for climate change resilience: implications for smallholder farmers in Africa. J Agric Food Res 16:101132. https://doi.org/10.1016/j.jafr.2024.101132

O'Neill B, van Aalst M, Zaiton Ibrahim Z et al (2022) Key risks across sectors and regions. In: Pörtner HO, Roberts DC, Tignor M et al (eds) Climate change 2022: impacts, adaptation, and vulnerability. Contribution of Working Group II to the sixth assessment report of the Intergovernmental Panel On Climate Change. Cambridge University Press, Cambridge/New York, pp x-y. xxx

Oh JH, Schueler KL, Stapleton DS et al (2020) Secretion of recombinant interleukin-22 by engineered *Lactobacillus reuteri* reduces fatty liver disease in a mouse model of diet-induced obesity. mSphere 5:e00183–e00120. https://doi.org/10.1128/msphere.00183-20

Patelski P, Berlowska J, Dziugan P et al (2015) Utilisation of sugar beet bagasse for the biosynthesis of yeast SCP. J Food Eng 167:32–37

Pattanayak V, Guilinger JP, Liu DR (2014) Determining the specificities of TALENs, Cas9, and other genome-editing enzymes. Methods Enzymol 546:47–78. https://doi.org/10.1016/B978-0-12-801185-0.00003-9

Peters JE, Makarova KS, Shmakov S, Koonin EV (2017) Recruitment of CRISPR-Cas systems by Tn7-like transposons. Proc Natl Acad Sci USA 114:E7358–E7366

Pickar-Oliver A, Gersbach CA (2019) The next generation of CRISPR–Cas technologies and applications. Nat Rev Mol Cell Biol 20:490–507

Potter H (2003) Transfection by electroporation. Curr Protoc Mol Biol Chapter 9:Unit 9.3. https://doi.org/10.1002/0471142727.mb0903s62

Putra MD, Abasaeed AE, Al-Zahrani SM (2020) Prospective production of fructose and single cell protein from date palm waste. Electron J Biotechnol 48:46–52

Raj S, Roodbar S, Brinkley C, Wolfe DW (2022) Food security and climate change: differences in impacts and adaptation strategies for rural communities in the Global South and North. Front Sustain Food Syst 5:691191. https://doi.org/10.3389/fsufs.2021.691191

Ritala A, Häkkinen ST, Toivari M, Wiebe MG (2017) Single cell protein-state-of-the-art, industrial landscape and patents 2001–2016. Front Microbiol 8:2009

Ribeiro GO, Rodrigues LAP, Santos TBS, Alves JPS, Oliveira RS, Nery TBR, Barbosa JDV, Soares MBP (2023) Innovations and developments in single cell protein: Bibliometric review and patents analysis 13:1093464

Sadler M (1988) QUORN. Nutr Food Sci 88:9–11. https://doi.org/10.1108/eb059180

Saifaldeen M, Al-Ansari DE, Ramotar D, Aouida M (2020) CRISPR FokI dead cas9 system: principles and applications in genome engineering. Cells 9:2518. https://doi.org/10.3390/cells9112518

Sun L, Zheng P, Sun J, Wendisch VF, Wang Y (2023) Genome-scale CRISPRi screening: a powerful tool in engineering microbiology. Eng Microbiol 3(3):100089

Saito M, Xu P, Faure G, Maguire S, Kannan S, Altae-Tran H, Vo S, Desimone A, Macrae RK, Zhang F (2023) Fanzor is a eukaryotic programmable RNA-guided endonuclease. Nature 620:660–668

Salazar-López NJ, Barco-Mendoza GA, Zuñiga-Martínez BS et al (2022) Single-cell protein production as a strategy to reincorporate food waste and agro by-products back into the processing chain. Bioengineering 9:623. https://doi.org/10.3390/bioengineering9110623

Schargel RD, Qayyum MZ, Tanwar AS, Kalathur RC, Kellogg EH (2025) Structure of Fanzor2 reveals insights into the evolution of the TnpB superfamily. Nat Struct Mol Biol 32(2):243–246. https://doi.org/10.1038/s41594-024-01394-4

Schwartz DC, Cantor CR (1984) Separation of yeast chromosome-sized DNAs by pulsed field gradient gel electrophoresis. Cell 37:67–75. https://doi.org/10.1016/0092-8674(84)90301-5
Scoones I (2023) Livestock, methane, and climate change: the politics of global assessments. Wiley Interdiscip Rev Clim Chang 14:e790. https://doi.org/10.1002/wcc.790
Selvakumar SC, Preethi KA, Ross K, Tusubira D, Khan MWA, Mani P, Rao TN, Sekar D (2022) CRISPR/Cas9 and next generation sequencing in personalized treatment of cancer. Mol Cancer. 21(1):83
Sekoai PT, Roets-Dlamini Y, O'Brien F, Ramchuran S, Chunilall V (2024) Valorization of food waste into single-cell protein: an innovative technological strategy for sustainable protein production. Microorganisms 12:166. https://doi.org/10.3390/microorganisms12010166
Southern EM (1975) Detection of specific sequences among DNA fragments separated by gel electrophoresis. J Mol Biol 98(3):503–517
Sentmanat MF, Peters ST, Florian CP, Connelly JP, Pruett-Miller SM (2018) A survey of validation strategies for CRISPR-Cas9 editing. Sci Rep 8:888. https://doi.org/10.1038/s41598-018-19441-8
Sharma-Kuinkel BK, Rude TH, Fowler VG Jr (2016) Pulse field gel electrophoresis. Methods Mol Biol 1373:117–130. https://doi.org/10.1007/7651_2014_191
Sherba JJ, Hogquist S, Lin H, Shan JW, Shreiber DI, Zahn JD (2020) The effects of electroporation buffer composition on cell viability and electro-transfection efficiency. Sci Rep 10(1):3053. https://doi.org/10.1038/s41598-020-59790-x
Shurson GC (2018) Yeast and yeast derivatives in feed additives and ingredients: sources, characteristics, animal responses, and quantification methods. Anim Feed Sci Technol 235:60–76
Song B, Bae S (2024) Genome editing using CRISPR, CAST, and Fanzor systems. Mol Cells 47(7):100086. https://doi.org/10.1016/j.mocell.2024.100086
Stella S, Montoya G (2016) The genome editing revolution: a CRISPR-Cas TALE off-target story. BioEssays 38:S4–S13
Tamang JP et al (2020) Fermented foods in a global age: east meets west. Compr Rev Food Sci Food Saf 19:184–217. https://doi.org/10.1111/1541-4337.12520
Thiviya P, Gamage A, Kapilan R, Merah O, Madhujith T (2022) Single cell protein production using different fruit waste: a review. Separations 9:178. https://doi.org/10.3390/separations9070178
Tian J, Tang Z, Niu R et al (2024) Engineering disease-resistant plants with alternative translation efficiency by switching uORF types through CRISPR. Sci China Life Sci 67(8):1715–1726. https://doi.org/10.1007/s11427-024-2588-9
Tsai SQ, Zheng Z, Nguyen NT, Liebers M, Topkar VV, Thapar V, Wyvekens N, Khayter C, Iafrate AJ, Le LP, Aryee MJ, Joung JK (2015) GUIDE-seq enables genome-wide profiling of off-target cleavage by CRISPR-Cas nucleases. Nat Biotechnol 33:187–197. https://doi.org/10.1038/nbt.3117
Terol GL, Gallego-Jara J, Martinez RAS, Vivancos AM, Diaz MC, de Diego Puente (2021) Impact of the expression system on recombinant protein production in Escherichia coli BL21. Front Microbiol 12.001
Tsai SQ, Nguyen NT, Malagon-Lopez J et al (2017) CIRCLE-seq: a highly sensitive in vitro screen for genome-wide CRISPR-Cas9 nuclease off-targets. Nat Methods 14:607–614. https://doi.org/10.1038/nmeth.4278
Tyagi S, Kesiraju K, Saakre M et al (2020) Genome editing for resistance to insect pests: an emerging tool for crop improvement. ACS Omega 5:20674–20683
Wackett LP (2020) Microbial meat substitutes: an annotated selection of World Wide Web sites relevant to the topics in microbial biotechnology. Microb Biotechnol 13:1284–1285. https://doi.org/10.1111/1751-7915.13610
Wada OZ, Vincent AS, Mackey HR (2022) Single-cell protein production from purple non-sulphur bacteria-based wastewater treatment. rev Environ Sci Biotechnol 21:931–956https://doi.org/10.1128/aem.01159-16

Whittaker JA, Johnson RI, Finnigan TJA, Avery SV, Dyer PS (2020) The biotechnology of quorn mycoprotein: past, present and future challenges. In: Nevalainen H (ed) Grand challenges in fungal biotechnology. Springer, pp 59–79

Wiegand T, Hoffmann FT, Walker MWG, Tang S, Richard E, Le HC, Meers C, Sternberg SH (2024) TnpB homologues exapted from transposons are RNA-guided transcription factors. Nature 631(8020):439–448. https://doi.org/10.1038/s41586-024-07598-4

World Bank (2022) Research and development expenditure (% of GDP)—sub-Saharan Africa by countries from 1980–2021. URL: https://data.worldbank.org/indicator/GB.XPD.RSDV.GD.ZS?locations=ZG

Worldometer (2024) World population. URL: www.worldometers.info

Wright AV, Nunez JK, Doudna JA (2016) Biology and applications of CRISPR systems: harnessing nature's toolbox for genome engineering. Cell 164:29–44

Xiao Y, Ng S, Nam KH, Ke A (2017) How type II CRISPR-Cas establish immunity through Cas1-Cas2-mediated spacer integration. Nature 550:137–141. https://doi.org/10.1038/nature24020

Xu H, Wang X, Chi G, Tan B, Wang J (2019) Effects of *Bacillus thuringiensis* genetic engineering on induced volatile organic compounds emission in maize and the attractiveness to a parasitic wasp. Front Bioeng Biotechnol 7:160. https://doi.org/10.3389/fbioe.2019.00160

Xu Y et al (2022) De novo biosynthesis of rubusoside and rebaudiosides in engineered yeasts. Nat Commun 13:1–12

Yunus F, Nadeem M, Rashid F (2015) Single-cell protein production through microbial conversion of lignocellulosic residue (wheat bran) for animal feed. J Inst Brew 121:553–557

Yan WX, Mzazadeh R, Garnerone S, Scott D, Schneider MW, Kallas T, Custodio J, Wernersson E, Li Y, Gao L, Federova Y, Zetsche B, Zhang F, Bienko N (2017) BLISS is a versatile and quantitative method for genome-wide profiling of DNA double-stranded breaks. Nat Commun 8.15058

Zaidi SS, Mahas A, Vanderschuren H, Mahfouz MM (2020) Engineering crops of the future: CRISPR approaches to develop climate-resilient and disease-resistant plants. Genome Biol 21:289. https://doi.org/10.1186/s13059-020-02204-y

Zepka LO, Jacob-Lopes E, Goldbeck R, Queiroz MI (2008) Production and biochemical profile of the microalgae *Aphanothece microscopica* Nägeli submitted to different drying conditions. Chem Eng Process Process Intensif 47:1305–1310

Zha X, Tsapekos P, Zhu X, Khoshnevisan B, Lu X, Angelidaki I (2021) Bioconversion of wastewater to single cell protein by methanotrophic bacteria. Bioresour Technol 320:124351

Zhang F, Neik TX, Thomas WJW, Batley J (2023) CRISPR-based genome editing tools: An accelerator in crop breeding for a changing future. Int J Mol Sci 24:8623. https://doi.org/10.3390/ijms24108623

Zhu H, Zhang H, Xu Y, Laššáková S, Korabečná M, Neužil P (2020) PCR past, present and future. BioTechniques 69:317–325. https://doi.org/10.2144/btn-2020-005

Part V
Conservation and Governance of Microbial Applications

Chapter 15
Microbial Contributions to Biodiversity Conservation in Agroecosystems

Adenike Eunice Amoo, Rebaona Reaobaka Molefe, Priyanka More, Bukola Rhoda Aremu, Olumayowa Mary Olowe, and Olubukola Oluranti Babalola

Abstract Microbial biodiversity is an integral part of a natural environment that delivers essential services for the proper functioning of agroecosystems and associated flora and fauna. Hence, continuous exploration of microbial biodiversity, function, resilience and biodiversity conservation while addressing threats to microbial diversity in agricultural settings becomes mandatory. This chapter highlights the multifaceted roles of microbes in agroecosystems, including nutrient cycling, plant-microbe interactions, and biocontrol of plant diseases. Discussions related to symbiotic relationships such as mycorrhizal associations and rhizobia-legume symbiosis, as well as the role of plant growth-promoting rhizobacteria (PGPR) and microbial-induced systemic resistance (ISR) in plant health are included. The chapter also investigates the link between microbial diversity and climate change resilience, including microbial contributions to carbon sequestration and greenhouse gas mitigation. Challenges to microbial biodiversity in agroecosystems, such as land-use changes and environmental contamination are also addressed here. Strategies for

A. E. Amoo (✉) · P. More
Department of Biology, Edge Hill University, Ormskirk, UK

R. R. Molefe
Food Security and Safety Focus Area, Faculty of Natural and Agricultural Sciences, North-West University, Mmabatho, South Africa

B. R. Aremu
Food Security and Safety Focus Area, Faculty of Natural and Agricultural Sciences, North-West University, Mmabatho, South Africa

Department of Chemistry and Biochemistry, University of Windsor, Windsor, ON, Canada

O. M. Olowe
Plant Pathology Unit, Department of Botany, University of Ibadan, Ibadan, Nigeria

O. O. Babalola (✉)
Department of Life Sciences, Imperial College London, Ascot, Berkshire, United Kingdom

Food Security and Safety Focus Area, Faculty of Natural and Agricultural Sciences, North-West University, Mmabatho, South Africa
e-mail: olubukola.babalola@nwu.ac.za

O. O. Babalola et al. (eds.), *Ecofriendly Frontiers*,
https://doi.org/10.1007/978-3-031-98700-7_15

preserving microbial biodiversity are proposed, including cataloguing initiatives and incorporating microbial diversity into agricultural management. The chapter concludes by emphasizing the potential of harnessing microbial contributions for sustainable agriculture and ecosystem resilience, highlighting the need for further research in microbial ecology and agroecosystem management.

Keywords Agroecosystem · Microorganisms · Biodiversity · Plant-microbe interactions · Conservation · Economically viable agricultural production

15.1 Introduction

Approximately 40% of the Earth's total land surface is composed of agroecosystems - cultivated ecosystems that have been modified by humans for agricultural purposes (Krause et al. 2022; Stokes et al. 2023). Just as in natural ecosystems, agroecosystems consist of organisms, including humans, interacting within a chemical and physical abiotic context (Case et al. 2020). Agroecosystems are essential contributors to global food production and the provision of essential ecosystem services (Rendon et al. 2020). They play vital roles in ensuring global food security by producing crops, livestock, and other agricultural goods that satisfy the dietary requirements of populations worldwide (Feledyn-Szewczyk et al. 2016). Beyond their essential contributions to food production, agroecosystems also provide a diverse array of ecosystem services that are crucial for environmental sustainability, including the maintenance of soil fertility, carbon sequestration, water regulation and biodiversity conservation. The sustainable management of agroecosystems is therefore of critical importance for ensuring food security and environmental sustainability on a global scale (Dubey et al. 2022; Isbell et al. 2017; Raj et al. 2021).

Microbial biodiversity is an integral component of the overall biodiversity in agroecosystems and plays a crucial role in ecosystem function and resilience. Soil microorganisms are responsible for a wide range of essential ecosystem processes in agroecosystems, such as nutrient cycling, soil fertility maintenance, and biological pest control (Amoo and Babalola 2017; Cheeke et al. 2012; Gupta et al. 2022; Xiong and Lu 2022). The level of biodiversity in agroecosystems is influenced by factors such as the diversity of vegetation, crop permanence, management intensity, and distance from natural vegetation (Altieri 1999; Amoo and Babalola 2019).

The key threats to microbial biodiversity in agroecosystems stem from the negative impacts of intensive agricultural practices, the detrimental effects of land use changes, and the lack of integration of microbial conservation into broader ecosystem management and biodiversity protection efforts (Amoo et al. 2021a; Raj et al. 2021; Seneviratne and Kulasooriya 2013; Thrupp 2004). The intense management practices focused solely on maximizing productivity in agroecosystems can compromise important ecosystem functions and services (Choruma and Odume 2019; Emmerson et al. 2016). These threats can lead to the collapse of soil microbial diversity, compromise essential ecosystem functions, and increase the sensitivity of agroecosystems to various abiotic and biotic stresses, ultimately undermining their resilience and productivity (Qiu et al. 2021; Seneviratne and Kulasooriya 2013; Xiong and Lu 2022).

The integration of microbial biodiversity conservation with profitable and productive agriculture poses a central challenge in modern sustainable agriculture. However, it is crucial for preserving genetic, species, and ecosystem diversity (Feledyn-Szewczyk et al. 2016; Mamabolo et al. 2020), ensuring the long-term sustainability and resilience of these human-modified landscapes (Gupta et al. 2022; Xiong and Lu 2022). Implementing sustainable, biodiversity-focused management practices within agroecosystems is of critical importance for ensuring the health of soils, contributing to climate change mitigation, and promoting the overall environmental sustainability of these coupled social-ecological systems (Feledyn-Szewczyk et al. 2016; Mamabolo et al. 2020; Raj et al. 2021).

While acknowledging the prominence of microbial biodiversity, this chapter demonstrates specific strategies and examples of biodiversity conservation practices and its successful integration in agroecosystems. It provides tangible insights into how these principles can be effectively applied on the ground for more sustainable and economically viable agricultural production.

15.2 Microbial Roles in Agroecosystems

Microorganisms play a central role in nutrient cycling within agroecosystems by actively producing and conserving soil organic matter (SOM). While it was once thought that SOM primarily came from plant root exudates, more recent research indicates that microbial biomass is crucial in this process, highlighting the importance of microbes in nutrient recycling (Ma et al. 2021). SOM is essential for supplying nutrients that stimulate microbial activity and promote plant growth. Understanding these microbial processes can lead to strategies for enhancing nutrient availability, increasing soil fertility, and ultimately boosting agricultural crop yields. The chapter highlights four major nutrient cycles where microbes play a critical role: carbon, nitrogen, phosphorus, and sulfur. These cycles are interconnected through soil microorganisms, underscoring their vital role in nutrient dynamics and the sustainability of agroecosystems (Fontaine et al. 2024).

15.2.1 Microbial Role in Carbon Cycling

Microbial mediated carbon turnover is a key process controlling the flux of carbon in the soil and the breakdown of SOM. Microorganisms such as bacteria, fungi, archaea, and protozoa produce enzymes that break down complex compounds like cellulose, lignin, and proteins into simpler forms that plants can absorb (Gupta et al. 2016). Specific examples include anaerobic bacteria like *Cellulomonas*, aerobic bacteria like *Clostridium thermocellum*, and fungi such as *Phanerochaete* and *Trichoderma*, which specialize in cellulose degradation. These microbial activities are essential for recycling carbon and nutrients and enhancing soil productivity. However, these activities are highly dependent on abiotic factors such as soil pH, temperature, and moisture, which affect enzyme activity, microbial metabolism, community composition, enzyme efficiency, oxygen availability, and nutrient availability. Microbial Carbon Use Efficiency (CUE) is a critical concept that describes the balance between carbon used for microbial growth and carbon lost through respiration (Hagerty et al. 2018). Growth rate and yield affect CUE. Slower microbial growth can lead to more carbon being allocated to biomass production, improving efficiency and aiding long-term carbon sequestration when the energy required to break down SOM exceeds the energy gained.

15.2.2 Microbial Role in Nitrogen Cycling

The microbial nitrogen cycle defines how nitrogen is cycled within biomes, involving numerous transformations through microbial action to make it available for plant uptake. Microbial nitrogen fixation, nitrification, and denitrification are crucial parts of this cycle, determining the productive capacity of ecosystems (Zhang et al. 2020). This process converts atmospheric nitrogen gas into soluble ammonia, which is usable by plants. This is largely carried out by diazotrophic bacteria including symbiotic *Rhizobium* (in legumes) and free-living bacteria like *Azotobacter* and *Clostridium*. Non-symbiotic fixers like *Azospirillum* also enhance soil nitrogen levels in the rhizosphere, and *Frankia* associates with actinorhizal plants for nitrogen fixation. Certain archaea can also fix nitrogen, particularly under anaerobic conditions (Jehani et al. 2023). The nitrogen-fixing enzyme, nitrogenase, is oxygen-sensitive, and microbes have developed strategies to protect it. Nitrification is a two-step process where ammonia is converted to nitrite by ammonia-oxidizing bacteria (AOB) and archaea (AOA), and then nitrite is oxidized to nitrate by nitrite-oxidizing bacteria, making nitrogen easily absorbed by plants. Soil pH and humidity affect the efficiency of nitrification, with AOB being sensitive to acidic soils (Wang et al. 2019). Denitrification removes excess nitrogen from the soil by converting nitrate into nitrogen gas, releasing it back into the atmosphere. This process is carried out by heterotrophic bacteria (e.g., *Alcaligenes*, *Bacillus*, *Paracoccus*) and denitrifying fungi (e.g., *Fusarium*). Denitrification is enhanced when oxygen levels are low and organic substrate is available (de Klein et al. 2017).

15.2.3 Microbial Role in Phosphorus Cycling

Although phosphorus is present in soil, its availability for plant uptake is often limited due to its insoluble form. Soil microorganisms play a vital role in phosphorus cycling by solubilizing and mobilizing phosphorus, making it accessible to plants

(Tian et al. 2024). Phosphorus deficiency can hinder the uptake of other nutrients and affect plant growth. Microorganisms break down inorganic and organic phosphates through phosphorus solubilization and phosphorus mineralization. Phosphorus solubilization involves bacteria (e.g., *Bacillus*, *Pseudomonas*) and fungi (e.g., *Hymenella*, *Neosartorya*) converting unavailable phosphorus into soluble forms by producing acids and enzymes (Emami-Karvani and Chitsaz-Esfahani 2021). Fungi like *Trichoderma* release organic acids to achieve this, leading to improved plant growth. Phosphorus mineralization involves the breakdown of organic phosphates into inorganic forms through extracellular enzymes like phytases and phosphatases, produced by various phosphate-solubilizing bacteria and endophytic fungi.

15.2.4 Microbial Role in Sulfur Cycling

Sulfur is a crucial plant nutrient, and many soils are deficient in it, leading to poorer agricultural yields. The sulfur cycle, driven by microbial processes, makes sulfur available to plants. Soil microorganisms oxidize and mineralize sulfur-containing organic matter, transforming it into plant-accessible sulfates. Bacteria and fungi break down sulfur-containing organic matter, and *Thiobacillus* species oxidize reduced sulfur compounds to sulfate (Gupta and Germida 2021). This process increases sulfate availability for plants, improving yield and secondary metabolite content. The rate of elemental sulfur oxidation depends on the size and activity of microbial biomass. Under anaerobic conditions, sulfate-reducing bacteria convert sulfate to hydrogen sulfide. Plants and microorganisms can also absorb sulfate and incorporate sulfur into organic molecules through assimilatory sulfate reduction (Zhou et al. 2025).

Beyond nutrient cycling, plant-microbe interactions are fundamental to the functioning of agroecosystems and have profound implications for agriculture and environmental sustainability. These interactions can range from beneficial symbioses to pathogenic relationships. Microbes and endophytes can adjust their metabolism to their plant hosts and confer resistance against various stresses. Key types of plant-microbe interactions include symbiotic relationships (e.g., mycorrhizal associations and rhizobia-legume symbiosis), plant growth-promoting rhizobacteria (PGPR), endophytic associations, and pathogenic interactions. Understanding these interactions is crucial for developing sustainable agricultural practices.

15.3 Beneficial Symbiotic Relationships

15.3.1 Mycorrhizal Associations

These are symbiotic relationships between plant roots and mycorrhizal fungi, enhancing the uptake of phosphorus and other nutrients, and potentially water. Plants provide fungi with photoassimilates (Rozpądek et al. 2019). There are two main classifications: endomycorrhizae (fungal hyphae colonize the inner root cortex) and ectomycorrhizae (fungal hyphae grow surrounding the root). Arbuscular mycorrhizae (AM), formed by Glomeromycetes, are common in over 80% of vascular plants, where fungi colonize root cortical cells and form arbuscules for nutrient exchange (Prasad et al. 2022). Ectomycorrhizae (ECM) form an external fungal mantle and a Hartig net. Mycorrhizal associations can improve nutrient uptake, stress tolerance, plant growth, and productivity. They can also provide physical and chemical barriers against pathogens and herbivores and induce beneficial phytochemicals.

15.3.2 Rhizobia-Legume Symbiosis

This is a well-known symbiotic relationship where rhizobia bacteria infect legume roots and form nodules, within which they convert atmospheric nitrogen into ammonia. This provides legumes with a vital nitrogen source, especially in nitrogen-deficient soils. In return, plants provide rhizobia with a microenvironment and organic carbon sources (Ledermann et al. 2021). This process involves complex signaling between the plant and bacteria, including the exchange of Nod factors.

15.3.3 Plant Growth-Promoting Rhizobacteria (PGPR) and Microbial-Induced Systemic Resistance (ISR)

PGPR are beneficial soil bacteria that colonize plant roots and enhance plant growth, health, and defence. Examples include *Pseudomonas*, *Bacillus*, and *Azospirillum*. PGPR promote plant growth through direct mechanisms like nitrogen fixation, phosphate solubilization, phytohormone production, and siderophore production, and indirect mechanisms like pathogen suppression and the induction of systemic resistance (Meena et al. 2020). Microbial-induced systemic resistance (ISR) is a state of enhanced plant defence capacity triggered by certain microorganisms, including PGPR. ISR involves priming of the plant's defence mechanisms, broad-spectrum effectiveness against pathogens, and activation of jasmonic acid and ethylene signalling pathways (Kamle et al. 2020). Plants recognize microbial elicitors, leading to enhanced expression of defence-related genes upon pathogen attack. The molecular communication in ISR shares similarities with other beneficial interactions like rhizobia-legume symbiosis (Tonelli et al. 2020). PGPR and ISR contribute to improved nutrient uptake, stress tolerance, pathogen suppression, enhanced defence responses, and ultimately improved plant productivity. They represent valuable tools for sustainable agriculture.

15.4 Microbial Biocontrol of Plant Diseases

Microorganisms are ubiquitous in our environment. Plant-associated microbes, such as bacteria, fungi, algae, and bacteriophages, can originate from a plant's native commensal microbiota, including those borne by seeds, or from external sources like soil, water, air, vectors, and other eukaryotic organisms (Teplitski et al. 2011). Depending on their habitat niches, plant microbes are categorized as epiphytic, endophytic, or rhizospheric. These microbes can be transferred horizontally between different plant species or vertically within the same plant species and its subsequent generations (Glassner et al. 2018; Johnston-Monje and Raizada 2011).

Plant microbes can have harmful, beneficial, or neutral effects on the form and physiology of both below-ground and above-ground parts of plants. Nearly 80–90% of plant pathogens are of bacterial and fungal origin, causing significant agricultural and economic losses (Naz et al. 2022). Phytopathogens can also pose risks to human health (Alshannaq and Yu 2017). These pathogens infect their host plants and manipulate their physiology through various mechanisms, including the synthesis of microbial phytotoxins, degrading enzymes, extracellular polysaccharides, and phytohormones (Prasannath 2013). Through metabiosis—a type of commensalism where one organism creates or prepares a suitable environment for another unrelated organism (McCullers 2014)—phytopathogenic bacteria and fungi facilitate the entry and internalization of pathogens in plants (Wade and Beuchat 2003).

Within agricultural microbial populations, there are beneficial associations that significantly influence host physiology and overall plant health. These beneficial microbes can be applied as biofertilizers and phyto-stimulators (Igiehon et al. 2019; Sayyed and Uarrota 2022). Additionally, they include a variety of antagonistic microbes known to restrict or suppress the harmful effects of phytopathogens. These agriculturally important microbes exhibit various biocontrol activities against plant pathogens, such as antibiosis, competition for colonization and nutrients, inactivation of pathogenic germination factors present in seed or root exudates, and neutralization or degradation of phytotoxins.

15.4.1 *Beneficial Microbes and Their Biocontrol Mechanisms to Suppress Plant Pathogens*

The use of microbes in agricultural biocontrol practices has the potential to naturally suppress pathogens, reducing the reliance on chemical, non-biodegradable contaminants. This approach aligns with the principles of organic farming, which emphasizes organic, naturally occurring, less toxic, cost-effective, and productive farming methods. Consequently, the application of naturally antagonistic microbes or their active biomolecules is becoming increasingly essential. The effects of these beneficial microbes and their antimicrobial compounds on plant pathogens have been described at various levels. In this discussion, we will explore some of the major antimicrobial mechanisms of agricultural importance at the molecular level.

15.4.1.1 Antibiotics

The production of secondary metabolites is a unique characteristic of certain microbes, enabling them to interact with their surrounding environment. Among these secondary metabolites, antibiotics are particularly notable for their ability to inhibit the growth of competing organisms. The biocontrol process mediated by these antimicrobial compounds is known as 'antibiosis.'

In the agricultural ecosystem, antibiotic-producing microbial strains such as *Pseudomonas*, *Bacillus*, *Pantoea*, *Stenotrophomonas*, *Burkholderia*, and *Rahnella* play an integral role. These microbes create a competitive environment that suppresses bacterial, fungal, nematodal, and oomycetal phytopathogens (Johnson et al. 2023; Meyer et al. 2009; Paulin et al. 2017, 2009; Pusey et al. 2011; Wilhelm et al. 1998; Yang et al. 2024). Through antibiosis, they restrict the growth of phytopathogens, disrupt their life cycles, suppress disease progression, and indirectly promote plant growth (Pusey et al. 2011).

Table 15.1 lists some of these antibiotic-producing microbes, along with the types of antibiotics they produce and their target phytopathogens.

2,4-Diacetylphloroglucinol (DAPG) is a phenolic secondary metabolite that serves as a biocontrol agent against phytopathogens (Fig. 15.1) and has been extensively studied for the past 40 years (Johnson et al. 2023). This broad-spectrum antibiotic effectively controls a wide range of plant pathogens, including bacteria, fungi, and nematodes, through direct inhibition (Johnson et al. 2023; Li et al. 2013; Meyer et al. 2009; Paulin et al. 2017, 2009). In natural environments, DAPG is typically produced by the rhizobacteria *Pseudomonas* strains (Brodhagen et al. 2004; Lanteigne et al. 2012; Ramette et al. 2011; Rezzonico et al. 2007). Additionally, genomic studies have indicated that *C. vaccinii* may also have the potential to produce DAPG (Johnson et al. 2023). This bacterium was previously noted for its nematicidal properties against *Globodera rostochiensis*, a phytopathogenic cyst nematode of significant agronomic importance (Dunne et al. 1998).

Tomato is an economically important agricultural crop worldwide, but its yields have been severely impacted by canker caused by the bacterial pathogen *Clavibacter michiganensis* subsp. *michiganensis* (CMM), leading to significant losses (Chang et al. 1992). As a result, this issue has garnered the attention of many plant researchers (Lanteigne et al. 2012). The application of DAPG-producing *Pseudomonas* spp. has been recognized in numerous studies as an effective biocontrol technique to eradicate CMM-induced cankers in tomato fields (Lanteigne et al. 2012). Umesha (2006) identified infected tomato seeds as a key source of CMM-induced canker outbreaks in tomato plants. Sowing CMM-infected seeds leads to reduced seed germination, shorter shoot and root lengths, and overall diminished plant health and yield. However, field trials involving the co-inoculation of infected seeds with pure

Table 15.1 Antibiotic producing microbes of agriculture importance

Antibiotic producer	Antibiotic type	Mode of action	Biocontrol action on (associated pathological condition)	References and study types
Pseudomonas brassicacearum LBUM300	DAPG	Bactericide	*Clavibacter michiganensis* subsp. *michiganensis* (bacterial canker on tomato plants)	In-vitro and In-planta assays (Paulin et al. 2017) In-vitro and In-planta assays (Lanteigne et al. 2012)
Pseudomonas fluorescens Pf-5	Pyoluteorin	Oomycide	*Pythium ultimum* (seed- and root rotting diseases)	In-vitro (Brodhagen et al. 2004)
Pseudomonas brassicacearum LBUM300	DAPG	Oomycide and fungicide	*Phytophthora cactorum* and *Verticillium dahliae* (rot and wilt diseases)	In-vitro (Paulin et al. 2009)
Pseudomonas fluorescens	DAPG	Nematicide	*Xiphinema americanum* (root diseases)	In-vitro (Meyer et al. 2009)
Pseudomonas spp.	DAPG	Oomycide and fungicide	*Pythium ultimum–cucumber* (damping-off and root rot of cucumber) *Fusarium oxysporum f. sp. radicislycopersici–*–(root and crown rot disease)	In-vitro or In-pot study (Rezzonico et al. 2007)
Pseudomonas putida, Pseudomonas fluorescens	DAPG	Fungicide	*Aspergillus niger* (collar rot), *Aspergillus flavus* (aflaroot) and *Sclerotium rolfsii* (stem rot)	In-vitro (Sherathia et al. 2016)
Pseudomonas fluorescens Pf-5	Pyoluteorin	Fungicide	*Pythium ultimum* (cotton damping-off)	In-vitro (Howell and Stipanovic 1980)
Chromobacterium vaccinii DSM 25150^T	DAPG, violacein and bioactive cyclic depsipeptide FR900359	Antimicrobial and insecticidal	–	In-vitro and genome study (Johnson et al. 2023)
Pseudomonas protegens CHA0^T	DAPG and pyoluteorin	Antimicrobial	–	In-vitro and genome study (Ramette et al. 2011)

(continued)

Table 15.1 (continued)

Antibiotic producer	Antibiotic type	Mode of action	Biocontrol action on (associated pathological condition)	References and study types
Pseudomonas putida BW11M1	Xantholysin A (cyclic lipopeptides)	Bactericide and fungicide	*Xanthomonas* spp., *Aeromonas*, *Burkholderia* and *Erwinia sp.* Several fungal spp. such as *Fusarium spp.*, *Alternaria longipes*, *Xanthomonas spp.*, *Ascochyta pisi* MUCL30164, *Botrytis cinerea* MUCL30158	In-vitro, net house, and field experiment (Li et al. 2013)
Pseudomonas chlororaphis Phz24, *Pseudomonas sp.* SH-C52	Phenazine antibiotics and lipopeptide surfactants thanamycin	Fungicide	*Sclerotium rolfsii* (stem rot disease on groundnut)	In-vitro (Le et al. 2012)
Stenotrophomonas maltophilia PD3533	Potential antibiotic producers	Bactericide	*Ralstonia solanacearum* (potato brown rot)	In-vitro and greenhouse (Messiha et al. 2007)
Pantoea agglomerans E325	Potential antibiotic producers	Bactericide	*Erwinia amylovora* (fire blight of apple trees)	In-vitro and orchard trials (Pusey et al. 2011)
Bacillus amyloliquefaciens NJN-6	Bacillomycin D and macrolactin	Fungicide and bactericide	*F. oxysporum f. sp. cubense* and *Ralstonia solanacearum*	In-vitro (Yuan et al. 2012)
Bacillus amyloliquefaciens XY-13, *Bacillus velezensis* XY-2-3 and GJ-1-15	Bacilysin, surfactin, yndj, fengycin, iturin, and bacillomycin D.	Fungicide	*Fusarium oxysporum* (soil-borne diseases in watermelon)	In vitro and pot experiments (Yang et al. 2024)

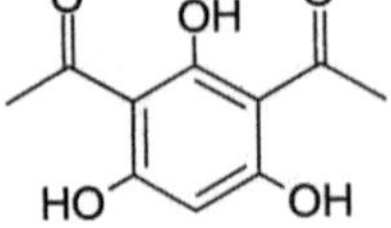

Fig. 15.1 Structure of 2,4-diacetylphloroglucinol (Brodhagen et al. 2004)

cultures and formulations of *Pseudomonas fluorescens* demonstrated a significant reduction in tomato canker incidences, increased seed germination, and improved crop yields. This suggests that *P. fluorescens* can be a reliable biological control strategy (Umesha 2006). Further in vitro and in planta assays have shown that DAPG produced by *Pseudomonas* spp. is a potent antimicrobial compound against CMM, highlighting its potential as a biocontrol agent for bacterial canker (Nandi et al. 2018; Paulin et al. 2017). Enhanced rhizospheric populations of DAPG-producing *Pseudomonas brassicacearum* and the overexpression of DAPG biosynthesis upon infection by *C. michiganensis* subsp. *michiganensis* demonstrate that this bacterium can augment rhizosphere colonization in the presence of CMM (Paulin et al. 2017).

The antifungal effects of DAPG production by *Pseudomonas* have been tested against various *Pythium* spp., revealing modifications in fungal plasma membranes, vacuolization, cell content fragmentation, and altered mycelial growth (de Souza et al. 2003). Stepanov et al. (2023) studied the effect of DAPG on the filamentous fungus *Aspergillus fumigatus*, a mycotoxin-producing, ubiquitous saprotrophic fungus that is both phytopathogenic in agricultural environments and capable of causing lethal clinical infections (Kang et al. 2022). The application of DAPG on *A. fumigatus* biofilms grown in polystyrene flat-bottom 96-well microtiter plates reduced the fungus's adaptive and competitive abilities by restricting the biosynthesis of fungal cell walls and mycotoxins. Controlling mycotoxin levels in agricultural soil through the application of DAPG has been proposed as a solution to improve the quality of agricultural produce (Stepanov et al. 2023). However, the study concluded that the biofilm status of *A. fumigatus* in natural environments may exhibit resistance to DAPG, necessitating further optimization and validation of this biological control activity in large-scale greenhouse and field conditions (Amkraz et al. 2010; Stepanov et al. 2023).

The effects of *Pseudomonas* spp. and their DAPG production have also been tested on broomrape (*Orobanche* and *Phelipanche* spp.), a difficult-to-control parasitic weed that affects the growth of many agriculturally important plants (Lurthy et al. 2023; Zermane et al. 2007). DAPG produced by *Pseudomonas ogarae* F113 has been successfully shown to inhibit broomrape germination without affecting the growth of the host plant, oilseed rape (*Brassica napus*) (Lurthy et al. 2023).

Microbes can also synthesize combinations of different antibiotics, demonstrating broad antimicrobial activity. For example, *Pseudomonas protegens* produces both 2,4-diacetylphloroglucinol and pyoluteorin (Ramette et al. 2011), while *Chromobacterium vaccinii* synthesizes 2,4-diacetylphloroglucinol, violacein, and the bioactive cyclic depsipeptide FR900359, which have antimicrobial and insecticidal properties (Johnson et al. 2023). *Pseudomonas chlororaphis* Phz24 produces phenazine antibiotics and lipopeptides, which have fungicidal properties (Le et al. 2012), and *Bacillus amyloliquefaciens* synthesizes bacillomycin D and macrolactin, known for their antifungal and antibacterial activities (Yuan et al. 2012). Additionally, *Bacillus velezensis* and *Bacillus amyloliquefaciens* produce a combination of bacilysin, surfactin, fengycin, iturin, and bacillomycin D, all of which have fungicidal properties (Yang et al. 2024). This underscores the broad antimicrobial potential of these microbes, warranting further exploration of their use as biocontrol agents in field studies.

15.4.1.2 Microbial Hydrogen Cyanide

Hydrogen cyanide (HCN) is a volatile secondary metabolite produced naturally by various bacteria, fungi, algae, and plants. In the rhizosphere, microorganisms utilize this chemical as a competitive and defensive tool against other organisms, thereby assisting agricultural plants in fighting against pathogens, and maintaining their health, growth, and productivity.

The biocontrol capability of bacterial HCN is primarily due to its ability to form strong bonds with metalloproteins. Once released in the rhizosphere, microbial HCN interacts with competing organisms by easily penetrating cellular and subcellular membranes of target cells, binding to their metal-containing enzymes, such as the iron-containing heme protein Cytochrome C Oxidase (CCO). The iron chelation effect of microbial HCN can hinder cellular growth, as iron becomes unavailable for the synthesis of heme and CCO in various organisms (Kumar and Padmanaban 1980; Saltzgaber-Müller and Schatz 1978).

An experiment by Kumar and Padmanaban (1980) demonstrated reduced levels of heme and CCO in *Neurospora crassa* when this mould was grown in an iron-deficient medium. Additionally, decreased protein synthesis on mitochondrial ribosomes was observed in these iron-deficient cultures, both in in vivo and in vitro measurements. CCO, a terminal enzyme in the mitochondrial electron transport chain, plays a crucial role in reducing molecular oxygen to water and producing ATP (Collins 2017). When CCO is inactivated by coupling with HCN, the electron transport chain is disrupted, leading to inhibited cellular oxygen utilization (histotoxic hypoxia), increased pyruvate reduction to lactic acid, and resulting acidosis.

Rhizosphere microbes produce HCN through a process called cyanogenesis, utilizing glycine—an amino acid found in plant root exudates (Bunch and Knowles 1980; Lesuffleur et al. 2007) and soil pore water (Fischer et al. 2007; Kielland 1995)—as a substrate. Glycine is decarboxylated to HCN by a microbial membrane-bound flavoenzyme known as HCN synthase (Fig. 15.2) (Wissing 1974; Laville et al. 1998; Bakker and Schippers 1987). Cyanide production by bacteria from glycine is further enhanced in the presence of another amino acid, methionine (Rodgers and Knowles 1978; Wissing 1968).

Microbial cyanogenesis is recognized as one of the most critical mechanisms in microbial biocontrol (Sehrawat et al. 2022). Studies conducted in vivo, in vitro, and through genome-based analyses have demonstrated the broad-spectrum suppressive effects of HCN-producing microbes against a variety of phytopathogens, including bacteria, fungi, insects, nematodes, oomycetes, termites, pests, and weeds (Gallagher and Manoil 2001). HCN is also known to have lethal effects on eukaryotic organisms, such as the depression of the central nervous system and myocardial functions (Anderson et al. 2010; Petrikovics et al. 2015). In situ, HCN produced by *Pseudomonas fluorescens* strain CHA0 has proven to be a potent biocontrol agent against the plant pathogenic root-knot nematode *Meloidogyne javanica*. When in direct contact with the nematode-infected root systems of tomato plants, HCN reduced nematode's egg hatching rates, inhibited their juvenile growth, and suppressed disease development, indicating the nematocidal activity of these

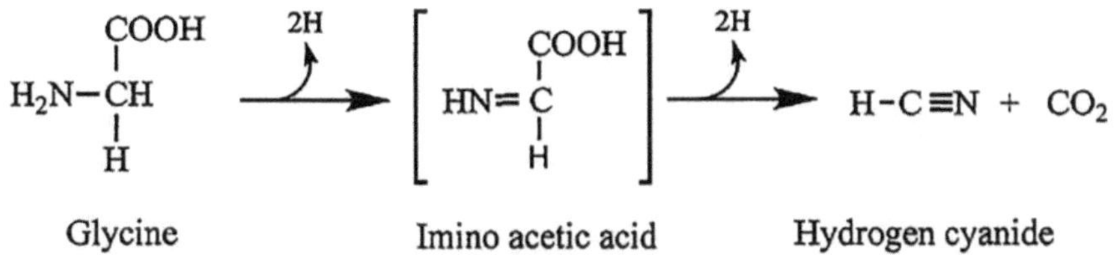

Fig. 15.2 Reaction of HCN synthase of proteobacteria. Glycine is decarboxylated to HCN with an intermediate product of iminoacetic acid (Wissing 1974)

HCN-producing rhizobacteria (Siddiqui et al. 2006). In vitro studies have also shown that microbial HCN inhibits Cytochrome C Oxidase (CCO) activity in the subterranean termite *Odontotermes obesus*, a significant plant pest that affects major agricultural crops and forest plantation trees. In this case, exposure of *O. obesus* to cyanide-producing isolates of *Rhizobium radiobacter* and *Alcaligenes latus* (Devi et al. 2007), as well as *Pseudomonas fluorescens* CHA0 (Devi and Kothamasi 2009), suggests the beneficial role of HCN producing microbes in termite control.

Cyanogenic *Pseudomonas* species have demonstrated the inhibition or killing of phytopathogenic fungi in various in vitro studies. Some of these bacterial isolates were agriculturally significant and directly retrieved from the rhizosphere of healthy crop plants (Sandikar 2018). For instance, an experiment by Sandikar (2018) showed that HCN-producing rhizosphere isolates *Pseudomonas aeruginosa* and *Pseudomonas fluorescens* were highly active against soil-borne phytopathogens such as *Pythium* and *Fusarium*. Additionally, Rezzonico et al. (2007) proposed the combined application of HCN and 2,4-diacetylphloroglucinol synthesized by *Pseudomonas spp.* could be a more robust and effective strategy for biocontrolling agricultural fungal phytopathogens like *Pythium* and *Fusarium*. The broad-spectrum activity of microbial HCN against various phytopathogens is further demonstrated and summarized in Table 15.2. Notably, these studies predominantly highlight microbial cyanogenesis in *Pseudomonas* species.

15.4.1.3 Bacterial and Fungal Siderophores of Agriculture Importance

Iron is one of the most abundant elements in the Earth's crust and is a crucial micronutrient required for the growth and development of all life forms. It predominantly exists in two oxidation states: ferrous (Fe^{2+}) and ferric (Fe^{3+}). While ferric iron (Fe^{3+}) is of greater biological importance, it is less soluble at neutral and alkaline soil pH, creating a challenge for its availability to organisms. Most biological systems operate at neutral pH, leading to a competitive environment for bioavailable iron among organisms. To survive in this iron-limited environment, some organisms synthesize siderophores, a process regulated by iron levels (Ghadamgahi et al. 2022; Karuppiah et al. 2022; Yang et al. 2024). Siderophores are low molecular weight compounds, typically between 500 and 1500 daltons (Hider and Kong 2010). Both plants and microbes synthesize these compounds, which have a high binding

Table 15.2 HCN producing biocontrol microbes and their broad-spectrum suppressing effects on phytopathogens

HCN producing microbes	Mode of action	Phytopathogen	References and study types
Pseudomonas spp.	Fungicide	*Pythium sp., Fusarium sp*	In-vitro (Rezzonico et al. 2007)
Pseudomonas fluorescens CHA0,	Fungicide	*Thielaviopsis basicola*	In-vitro (Ahl et al. 1986; Stutz et al. 1986; Voisard et al. 1989)
Pseudomonas chlororaphis PA23	Fungicide	*Sclerotinia sclerotiorum*	In-vitro (Athukorala et al. 2010)
Pseudomonas aeruginosa FG106	Fungicide, bactericide and oomycide	*Alternaria alternata, Clavibacter michiganensis subsp. michiganensis, Xanthomonas euvesicatoria pv. perforans, Phytophthora infestans, Phytophthora colocasiae, Rhizoctonia solani* and, *Botrytis cinerea*	In-vitro and greenhouse (Ghadamgahi et al. 2022)
Pseudomonas japonica strain NBRC 103040, *Bacillus megaterium* strain CtST3.5, *Pseudomonas sp.* strain Gamma-81, *P. tolaasii* strain ATCC 33618, *P. chlororaphis strain Lzh-T5,* and *P. mosselii strain CV25*	Nematicide and bactericide	*Meloidogyne incognita, Agrobacterium tumefaciens*	In-vitro and greenhouse (El-Rahman et al. 2019)

affinity for iron. Microbial siderophores, described as "molecular shuttles for metal cations" (Vijay et al. 2023), are evolved to specifically bind Fe^{3+} even in environments enriched with other metal ions (Hernlem et al. 1996). These siderophores solubilize iron from minerals like hematite, goethite, pyrite, and organic substrates, making it available to the target cell. The siderophores chelate insoluble Fe^{3+} and transport it into the intracellular space, where it is then reduced to Fe^{2+} by intracellular enzymes for use in various essential physiological processes (Bonneau et al. 2020; Brillet et al. 2012).

Siderophores are broadly classified into four main types based on their iron-chelating functional groups, ligand types, and structural characteristics: hydroxamates, catecholates (or phenolates), carboxylates, and mixed types that include two or more functional groups (Fig. 15.3) (Holden and Bachman 2015). Various microorganisms in nature produce a wide range of siderophores, with some species synthesizing mixtures of siderophores simultaneously. For instance, *Pseudomonas aeruginosa*, known for its antagonistic effects on plant-root pathogens and its ability to induce resistance against foliar pathogens, produces several types of siderophores including pyoverdine, pyochelin, pseudomonine, Pyridine-2, 6-dithiocarboxylic acid, quinolobactin, corrugatin, and nocardamine (Cornelis and Matthijs 2002). A study by Karuppiah et al. (2022) found that 20% of a consortium of plant growth-promoting rhizobacteria (PGPR), such as *Bacillus*, *Pseudomonas*, and *Achromobacter* species, were siderophore producers. Most of these organisms synthesized siderophores of the hydroxamate, catecholate, and carboxylate types, as

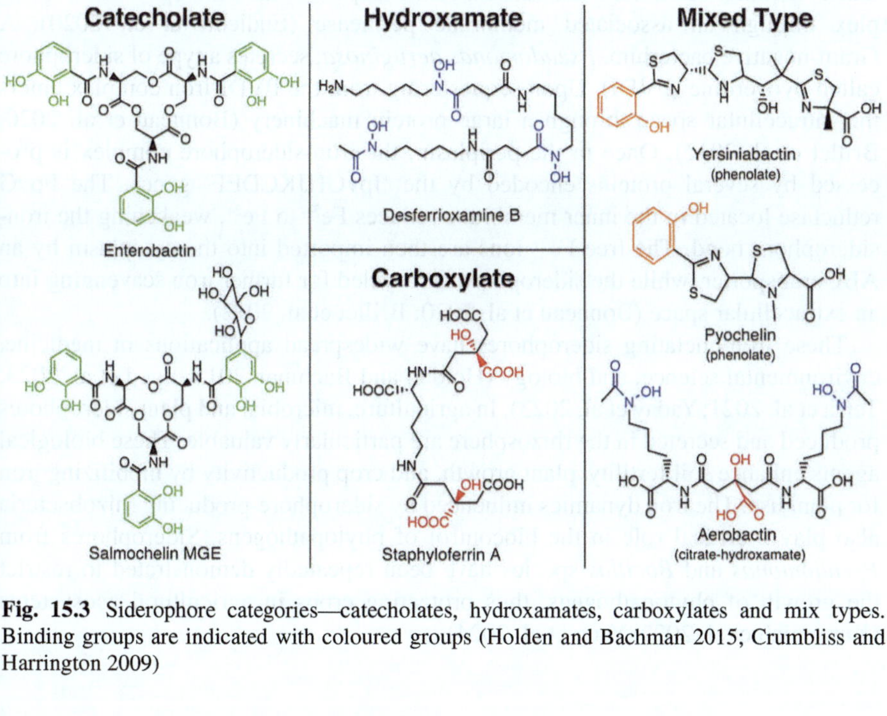

Fig. 15.3 Siderophore categories—catecholates, hydroxamates, carboxylates and mix types. Binding groups are indicated with coloured groups (Holden and Bachman 2015; Crumbliss and Harrington 2009)

well as mixed forms and pyoverdine types. These siderophore-producing organisms enhanced the bioavailability of iron for plant growth and inhibited the growth of *Fusarium* and *Sarocladium* species, which are fungal pathogens causing wilt in tomato plants. Among these, *Pseudomonas stutzeri* KRP8, which secretes hydroxamates, carboxylates, and pyoverdines, was identified as one of the most potent strains (Karuppiah et al. 2022).

Bacterial siderophores are typically extracellular, while in fungi, they can function both intra- and extracellularly, playing roles in the transport and storage of ferric iron (Winkelmann 2007). These compounds are predominantly produced under low iron conditions, with their biosynthesis linked to pathways involving non-ribosomal peptide synthetases (NRPS) and NRPS-independent synthetases (NIS). Upon synthesis, microbial extracellular siderophores bind Fe^{3+} ions in the environment, forming stable ferric iron-siderophore complexes. These complexes facilitate the transport and release of Fe^{3+} inside cells, though the specific mechanisms differ among microorganisms. In filamentous fungi and algae, Fe^{3+} is released from the siderophore complex in the extracellular space and enters the cytosol as a free Fe^{3+} ion (Ghosh et al. 2020). In contrast, in most bacteria, the entire Fe^{3+}-siderophore complex is transported across the cell membrane into the cytoplasm (Bonneau et al. 2020; Brillet et al. 2012; Crumbliss and Harrington 2009; Endicott et al. 2020; Ghosh et al. 2020). This process involves specific surface-exposed siderophore binding proteins (SBPs) or enzymes located on the bacterial cell membrane that act as shuttles for the Fe^{3+}-siderophore complex. For instance, in Gram-positive organisms, a catalytic SBP called "siderophore-dependent ferrichelatase" extracts ferric ions and facilitates the import of the siderophore-iron complex through an associated membrane permease (Endicott et al. 2020). A Gram-negative bacterium, *Pseudomonas aeruginosa*, secretes a type of siderophore called pyoverdine (PVDI). Upon sequestering iron, the PVDI-iron complex enters the intracellular space through a large protein machinery (Bonneau et al. 2020; Brillet et al. 2012). Once in the periplasm, the iron-siderophore complex is processed by several proteins encoded by the 'fpvGHJKCDEF' genes. The FpvG reductase located in the inner membrane reduces Fe^{3+} to Fe^{2+}, weakening the iron-siderophore bond. The free Fe^{2+} ions are then imported into the cytoplasm by an ABC transporter, while the siderophore is recycled for further iron scavenging into an extracellular space (Bonneau et al. 2020; Brillet et al. 2012).

These iron-chelating siderophores have widespread applications in medicine, environmental science, and biology (Holden and Bachman 2015; Syed et al. 2023; Terra et al. 2021; Yadav et al. 2023). In agriculture, microbial and plant siderophores produced and secreted in the rhizosphere are particularly valuable. These biological agents enhance soil fertility, plant growth, and crop productivity by mobilizing iron for plant use. The iron dynamics influenced by siderophore-producing rhizobacteria also play a critical role in the biocontrol of phytopathogens. Siderophores from *Pseudomonas* and *Bacillus* species have been repeatedly demonstrated to restrict the growth of phytopathogens, thus protecting crops in agricultural ecosystems (Karuppiah et al. 2022; Yang et al. 2024).

Table 15.3 Bacterial siderophores of agricultural importance

Siderophore producing microbes	Siderophore types	Mode of action	Phytopathogens	References and study types
Pseudomonas stutzeri KRP8	Hydroxamates, catecholates, carboxylates, pyoverdine types and mixed forms	Fungicide	*Fusarium oxysporum f. sp. lycopersici* MTCC10270, *Fusarium equiseti* and *Sarocladium sp.* SWL	In-vitro and in-pot (Karuppiah et al. 2022)
Bacillus velezensis	–	Fungicide	*Fusarium oxysporum sp.*	In-vitro and in-pot (Yang et al. 2024)
Pseudomonas aeruginosa FG106	Pyoverdine and pyochelin	Fungicide, bactericide and oomycide	*Alternaria alternata, Clavibacter michiganensis subsp. michiganensis, Xanthomonas euvesicatoria pv. perforans, Phytophthora infestans,* and *Phytophthora colocasiae*	In-vivo, in-vitro (Ghadamgahi et al. 2022)
Streptomyces sp. AzR-051	–	Fungicide	*Alternaria alternata*	In-vitro (Verma et al. 2011)
Pseudomonas sp. WCS417r, Pseudomonas putida WCS358r	Pseudobactins	Fungicide	*Fusarium oxysporum* f. *sp.dianthi*	In-vitro (Duijff et al. 1993)

The activity of bacterial HCN against various phytopathogens is summarized in Table 15.3. For example, the antagonistic effects of pyoverdine and pyochelin siderophores produced by *Pseudomonas aeruginosa* FG106 have been observed against various bacterial, fungal, and oomycete plant pathogens affecting crops like potato, tomato, and taro (Ghadamgahi et al. 2022). Siderophore-mediated competition for iron by *Pseudomonas sp.* WCS417r and *Pseudomonas putida* WCS358r has been shown to reduce the growth of *Fusarium oxysporum*, the causative agent of fusarium wilt in carnation roots. Purified pseudobactins, a siderophore from these *Pseudomonas* strains inhibited conidial germination and mycelial growth of *Fusarium oxysporum* (Duijff et al. 1993). Additionally, an in vitro antibiosis assay by Verma et al. (2011) demonstrated the antagonistic activity of an endophytic *Streptomyces* sp. against *Alternaria alternata*, the causative agent of early blight disease in tomato plants. The siderophore produced by this *Streptomyces* sp. exhibited a high Fe^{3+} complexing capacity, depriving the pathogen of iron and inhibiting its growth and survival (Verma et al. 2011).

15.4.2 *Integration of Microbial Biocontrol Strategies with Other Disease Management Practices*

In agricultural practices, protection against various phytopathogens is achieved through chemical, biological, cultural, mechanical, and monitoring tools. Biological tools, such as microbial formulations that exhibit plant disease control activities during in-vitro and greenhouse experiments, are eco-friendly and cost-effective. However, certain microbes in these formulations may struggle to survive and perform optimally when applied in the uncontrolled environmental and ecological conditions of agricultural fields (Brizuela et al. 2023; Meshram et al. 2022).

On the other hand, chemical pesticides are stable under field conditions and offer long-term activity against phytopathogens. However, their application is costly, and their toxic effects on non-target organisms can negatively impact flora, fauna, and human health. Moreover, the prolonged and excessive use of chemical pesticides and fertilizers has been reported to disrupt beneficial microbial diversity, increase the severity of plant diseases (Aremu et al. 2017; Daisley et al. 2022; Patyka et al. 2016; Zhang et al. 2024), and compromise the quality and safety of agricultural products (Carvalho 2006, 2017; Clarkson 1995).

These concerns have led to strong regulatory actions on agrochemical products and have accelerated the development of microbial bioformulations to be used as biofertilizers and biopesticides as a remedy in agricultural fields (Abawi and Widmer 2000; Sellitto et al. 2016). Sustainable and practical pest management approaches using carefully screened microbial bioformulations via an integrated pest management (IPM) method have been well documented through various field-based studies and are discussed in the following sections. IPM in plant pathology involves the application of a combination of traditional and modern plant management tools that offer targeted, long-term protection to plants from a wide range of infesting pests (Baker et al. 2020; Lindsey et al. 2020).

15.4.2.1 Microbial Consortia and Soil Amendments

Bioformulations combining intragenus and intergenus compatible microbes, whether applied alone or in combination, have demonstrated the capacity to enhance disease-suppressive activity, control single or multiple phytopathogens, and improve plant growth in both lab- and field-based agricultural studies (Tamiru and Muleta 2018). For instance, biocontrol isolates of *Trichoderma harzianum* and *T. asperellum* were tested against the fusarium wilt-causing pathogen *Fusarium oxysporum* f.sp. *lycopersici* (FOL). Each isolate individually revealed high activity of defence-related enzymes, reduced the soil population of FOL by 64–88%, and enhanced plant growth parameters and biomass production in tomato pot plants. The results with *T. harzianum* were comparable to those achieved with the fungicide carbendazim (Haque et al. 2023).

Consortia of bacterial and fungal biocontrol rhizosphere microbes, such as species of *Bacillus*, *Pseudomonas*, and *Trichoderma*, have shown extended plant growth promotion and diverse antagonistic activities against root pathogen *Fusarium oxysporum* and foliar pathogen *Botrytis cinerea* in tomato crops (Minchev et al. 2021) and in pea plants challenged by the soft-rot pathogen *Sclerotinia sclerotiorum* (Jain et al. 2012). Upon challenge with *Sclerotinia sclerotiorum*, a stronger activation of plant cellular defence responses was observed, with the interactions between the microbial consortium, plant host, and pathogen in experimental studies closely mimicking those in natural rhizosphere environments (Jain et al. 2012).

Moreover, long-term disease control and improved soil fertility can be achieved by introducing a consortium of beneficial microbes alongside soil amendments (Bamdad et al. 2022; Bonanomi et al. 2018). A meta-analysis of experimental studies examining the effects of organically amended soils on crop yield showed the most favourable impacts with farmyard manure (49%), followed by compost (17%) and straw (8%) (Luo et al. 2018). These amendments positively influenced soil organic carbon, total nitrogen, and soil microbial functionality (Liu et al. 2023). In one field study, sunflower plants inoculated with plant growth-promoting rhizobacteria, such as *Azotobacter* and *Azospirillum*, in combination with farmyard manure and the chemical fertilizer urea, exhibited enhanced plant biomass, grain production, and oil yield (Shoghi-Kalkhoran et al. 2018). Another field study involving co-culturing bioinoculants like *Frateuria aurantia*, *Rhizobium* spp., *Pseudomonas fluorescens*, and *Trichoderma viride* in different combinations with phosphocompost resulted in higher nutrient content and enhanced crop yield in fenugreek. In this study, *F. aurantia*, *Rhizobium* spp., *P. fluorescens*, and *T. viride* were applied as a potassium mobilizer, nitrogen source, growth promoter, and biocontrol agent, respectively (Biswas and Anusuya 2014).

In the arid regions of India, where alternative farming of cowpea and cumin is practiced in the same field, diseases such as charcoal rot of cowpea caused by *Macrophomina phaseolina* and wilt of cumin caused by *Fusarium oxysporum* f.sp. *cumini* are highly destructive. Field experiments by Singh et al. (2012) demonstrated a significant increase in disease control (96–99%) for these pathogens when *Bacillus firmus* coated cowpea seeds were sown in soil amended with farmyard manure, and *Aspergillus versicolor* was applied in neem compost-enriched cumin fields. The study also revealed compatibility and improved colonization between *B. firmus* and *A. versicolor* when applied together. In these studies, the integration of organic compost into agricultural soils provided essential growth substrates for beneficial microbial agents, enhancing their survival and biocontrol activity. Soil health, crop quality, and yield were further improved through microbe-induced nitrogen absorption, mineral dissolution, and phytohormone production, reducing the need for chemical fertilizers and pesticides (Abawi and Widmer 2000). These and similar experiments highlight the importance of applying biological agents through appropriate methods to achieve optimal disease management results (Dubey and Singh 2013; Dubey et al. 2013).

15.4.2.2 Microbial Consortia, Soil Amendments and Crop Rotation

Crop rotation, the practice of alternating different crop species on the same field, has proven to be a moderately to highly effective method for pest control compared to monoculture (Abawi and Widmer 2000). When combined with other agricultural practices, such as the application of microbial consortia and soil amendments, crop rotation has been employed to enhance soil health, manage pests, and improve agricultural yield. This integrated approach creates an environment less conducive to phytopathogens by limiting pests' access to their hosts, reducing their survival rates, lowering their inoculum levels in the soil, and fostering beneficial microbial growth at the expense of harmful pathogens (Abawi and Widmer 2000; Bailey and Lazarovits 2003; Roth et al. 2020). The success of this integrated pest management (IPM) strategy relies on the careful selection of rotation crops and biocontrol organisms (Larkin and Brewer 2020).

Meloidogyne incognita, a root-knot nematode (RKN), is known to form galls on plant roots, impairing water and nutrient uptake, and causing significant economic losses in crops like tomatoes (Nono-Womdim et al. 2002). Greenhouse and field studies have shown that rotating tomato crops with maize, combined with soil amendments and the application of the facultative nematode parasite *Pochonia chlamydosporia*, effectively managed RKN infestations. This approach reduced root galling in tomato plants and increased fruit yield by 63% (Luambano et al. 2015). A commercial product based on *P. chlamydosporia* (Pochar™, Microspore, Larino, Italy) is also available for use in irrigation systems to manage RKN populations (Sellitto et al. 2016).

15.4.2.3 Microbial Bioformulations with Fungicide

Global efforts to increase food production have led to the overuse of chemical pesticides on agricultural land, resulting in the emergence of resistant pathogens and environmental pollution. This has driven regulatory controls and even bans on pesticide usage in many regions (Lykogianni et al. 2021; Popp et al. 2013; Sparks 2013). The discovery of new pesticides has also become more challenging and costly. Consequently, sustainable pest control strategies have shifted towards minimizing chemical pesticide use and integrating it with alternative approaches, such as microbial biocontrol agents. Among these, *Bacillus*, *Pseudomonas*, and *Trichoderma*-based formulations have been widely studied for their effectiveness in managing crop diseases (Abd-El-Khair et al. 2019; Dubey et al. 2015, 2020; Jacobsen et al. 2004; Jain et al. 2012; Minchev et al. 2021; Woo et al. 2014). When used in combination with controlled amounts of fungicides, these microbes have demonstrated high levels of disease control, steady results, and enhanced fungicide resistance management under variable field conditions (Jacobsen et al. 2004; Ons et al. 2020).

For example, *Bacillus* species have shown antimicrobial activity through antibiotic production and stability due to their endospore formation. *Bacillus subtilis*

NJ-18, in combination with the fungicide jinggangmycin (produced by *Streptomyces*), improved the suppression of *Rhizoctonia solani*, the causal agent of rice sheath blight. Greenhouse and field trials revealed that rice seedlings treated with both *B. subtilis* NJ-18 and jinggangmycin exhibited 20–40% higher disease control efficacy compared to the biocontrol agent alone (Peng et al. 2014). Similarly, *Bacillus cereus* combined with the fungicide chlorothalonil significantly reduced early blight, late blight, and septoria leaf spot in tomato plants while increasing fruit yield (Silva et al. 2004). These studies suggest that integrating microbial agents with lower fungicide application frequencies can induce plant resistance and improve crop yield (Peng et al. 2014; Silva et al. 2004).

Trichoderma species have also demonstrated high efficacy in disease control, particularly when combined with chemical pesticides and other microbial agents. Field studies on crops such as mungbean, dry beans, chickpea, and tomato highlight *Trichoderma*'s ability to inhibit a wide range of phytopathogens (Abd-El-Khair et al. 2019; Dubey et al. 2011, 2013, 2015, 2020) (Table 15.4). For instance, *Trichoderma* spp. showed 90–99% inhibition of the fungal wilt pathogen *Fusarium oxysporum* and outperformed both *Bacillus subtilis* and commercial *Aspergillus niger* formulations (Charoenporn et al. 2010; Dubey et al. 2007). In mungbean, seed treatment with *Trichoderma virens* combined with thiamethoxam effectively managed wet root rot, cercospora leaf spots, and yellow mosaic virus, while increasing seed germination, shoot and root lengths, and grain yield (Dubey and Singh 2013).

Additionally, combining *Trichoderma virens* with thiophanate-methyl seed treatments significantly reduced damping-off and root rot/wilt in dry beans and enhanced phytoenzyme production, indicating systemic defence activation (Abd-El-Khair et al. 2019). In tomato, root dipping in *Trichoderma harzianum* and *T. viride* with carbendazim improved disease management and crop growth (Dubey et al. 2020). Similarly, seed dressing with *T. harzianum*, *Pseudomonas fluorescens*, *Mesorhizobium ciceri*, and Vitavax powder in chickpea farming significantly reduced *Fusarium* wilt and increased plant biomass and yield (Dubey et al. 2015, 2020).

These findings underscore the effectiveness of integrating biocontrol agents with chemical treatments in sustainable agricultural practices, offering enhanced disease control and crop productivity while minimizing the reliance on harmful pesticides.

15.5 Microbial Diversity and Climate Change Resilience

Microbes play a vital role in supporting ecosystem functions, and the loss of microbial taxa has been associated with the disruption of these essential processes (Philippot et al. 2021). Microbial diversity plays a crucial role in enhancing agroecosystem resilience to climate change and contributing to carbon sequestration and greenhouse gas mitigation (Bastida et al. 2021; Domeignoz-Horta et al. 2020; Vicena et al. 2022).

Table 15.4 *Trichoderma* species applied as a biocontrol agent in-combination with chemical pesticide and other microbial agents

Potential biocontrol agent	Fungicide	Phytopathogen (associated pathological condition)	Host plant	References and study types
T. virens	Carboxin	*Rhizoctonia solani* (wet root rot)	Mungbean (*Vigna radiata* L.)	Pot and field (Dubey et al. 2011)
T. virens	Thiamethoxam and carboxin, thiamethoxam and carbendazim	*Rhizoctonia solani* (wet root rot), *Cercospora canescens* and *Pseudocercospora cruenta* (cercospora leaf spots), and Yellow Mosaic Virus (yellow mosaic)	Mungbean (*Vigna radiata* L.)	Field (Dubey and Singh 2013)
T. virens	Thiophanate-methyl	Fusarium solani and *Fusarium oxysporum* (Damping-off and root rot/wilt)	Dry bean (*Phaseolus vulgaris* L.)	Greenhouse and field (Abd-El-Khair et al. 2019)
T. harzianum	Carboxin	*Fusarium oxysporum f. sp. ciceris (Padwick)* Matuo and *K. Sato)* (Fusarium wilt)	Chickpea (*Cicer arietinum* L.)	In-vitro, green house, and field (Dubey et al. 2007)
Trichoderma harzianum, T. viride	Carbendazim	*Fusarium oxysporum f. sp. lycopersici* (Fusarim Wilt)	Tomato	Pot and net house (Dubey et al. 2020)
Trichoderma harzianum Pseudomonas fluorescens, Mesorhizobium cicero	Vitavax	*Fusarium oxysporum f. sp. ciceris* (Fuarium wilt)	Chickpea	Pot and field (Dubey et al. 2015)

The resilience of microbial communities from disturbance is a crucial aspect in shaping the future of ecosystem functioning, particularly in the face of challenges posed by climatic changes and increasing anthropogenic activities (Allison and Martiny 2008). The resilience of microbial communities can manifest in their composition, function or both (Philippot et al. 2021).

15.5.1 Mechanisms by Which Diverse Microbial Communities Improve Resilience

Diverse microbial communities in agroecosystems contribute to resilience through several key mechanisms:

15.5.1.1 Enhancing Plant Stress Tolerance

Microbes produce osmolytes such as proline, glycine betaine, and trehalose, which help plants maintain cellular osmotic balance under abiotic stress conditions like drought or salinity. These compounds stabilize proteins and cell membranes, protecting plant cells from dehydration (Kaushal and Wani 2016). Furthermore, microbes synthesize antioxidants like glutathione and superoxide dismutase, which mitigate oxidative damage caused by environmental stresses, thereby improving plant resilience (Singh et al. 2015). Microbes also synthesize auxins, cytokinins, and other phytohormones that regulate plant growth and stress responses (Egamberdieva et al. 2019).

Certain beneficial microbes, such as plant growth-promoting rhizobacteria (PGPR), can trigger induced systemic resistance (ISR) in plants—a physiological state of enhanced defensive capacity in plants (Babalola et al. 2023). ISR differs from Systemic Acquired Resistance (SAR) in its elicitors and regulatory pathways, providing broad-spectrum protection against pathogens and herbivorous insects (Pieterse et al. 2014). ISR is mediated through signalling molecules like jasmonic acid and ethylene, which prime the plant's immune system to respond more effectively to biotic and abiotic stresses (Choudhary et al. 2007; Van Loon et al. 1998).

15.5.1.2 Improving Soil Structure and Water Retention

Microbial activities significantly enhance soil physical properties. Soil microbes produce extracellular polymeric substances (EPS), which are high-molecular-weight compounds that play a crucial role in soil structure formation (Or et al. 2007). These compounds bind soil particles together forming aggregates—the EPS act as a glue. Microbial activity contributes to the development of soil aggregates—clusters of soil particles held together by organic matter and microbial byproducts, thereby improving soil structure (Colica et al. 2014). Aggregates form a porous structure that enhances aeration, water infiltration, and root penetration (Ghezzehei et al. 2019). Well-aggregated soils are more resilient to drought and heavy rainfall, thereby strengthening the overall resilience of agroecosystems.

The improved soil structure resulting from microbial activity increases water retention. Increased porosity enhances water infiltration during rainfall, reducing surface runoff, while improved water retention ensures that crops have access to moisture during dry periods. This enhanced soil structure is essential for sustaining crop productivity amid fluctuating climatic conditions (Le Gall et al. 2021; Vu et al. 2022).

15.5.1.3 Accelerating Nutrient Cycling

Microbial diversity encompasses a range of functional groups that specialize in various nutrient transformations, including nitrogen fixation, phosphorus solubilization, and organic matter decomposition. This diversity ensures the continuous availability of essential nutrients to plants, even as environmental conditions fluctuate, thereby bolstering the resilience of agroecosystems (Delgado-Baquerizo et al. 2016; Wall et al. 2015).

Multiple microbial species can perform similar functions, ensuring nutrient cycling continues even if some species are lost. This redundancy acts as a buffer against disturbances, helping to maintain soil fertility and plant health (Allison and Martiny 2008). Different microbial groups work together, enhancing overall nutrient cycling efficiency (Jiao et al. 2018). For example, the collaboration between nitrogen-fixing bacteria and mycorrhizal fungi can enhance nutrient uptake by plants, promoting growth even in nutrient-deficient soils (Wagg et al. 2014).

15.5.1.4 Suppressing Plant Pathogens

Microbial diversity contributes to plant health by suppressing pathogens. Many microbes produce substances that inhibit pathogen growth (Compant et al. 2019). Certain soil microbes produce antimicrobial compounds, such as antibiotics and lytic enzymes, that directly inhibit or kill plant pathogens (Raaijmakers and Mazzola 2012). These compounds help lower the occurrence of plant diseases, leading to healthier crops and more resilient agroecosystems (Mendes et al. 2011). For instance, *Streptomyces* species produce antibiotics like streptomycin, which can effectively suppress fungal and bacterial pathogens (Hopwood 2007). Resource competition in the soil, driven by microbial diversity, results in beneficial microbes vying with pathogens for space and nutrients. This competition reduces the chances of infection, with competitive exclusion serving as a critical mechanism through which diverse microbial communities safeguard plants from diseases (Berendsen et al. 2012; Hibbing et al. 2010).

15.5.1.5 Induced Plant Defences

Beneficial microbes can trigger plant defence mechanisms, priming plants to respond more effectively to pathogen attacks (Pieterse et al. 2014). This process, akin to induced systemic resistance (ISR), may also involve the production of signalling molecules such as salicylic acid, which is linked to systemic acquired resistance (SAR) (Durrant and Dong 2004; Van Loon et al. 1998). By preparing plant defences, these microbes bolster the plant's ability to resist infections, thereby enhancing overall resilience (Bakker et al. 2007; Conrath et al. 2006).

15.5.2 Empirical Evidence Supporting the Diversity-Stability Relationship in Microbial Communities

Numerous studies have provided empirical evidence supporting the positive relationship between microbial diversity and ecosystem stability. These studies span various ecosystems and employ different methodologies, offering robust support for the importance of microbial diversity in maintaining ecosystem functions.

15.5.2.1 Functional Stability in Diverse Communities

Griffiths et al. (2000) conducted a seminal study examining the relationship between microbial diversity and functional stability in soil ecosystems. They manipulated soil microbial diversity through chloroform fumigation and heat treatments, then measured the stability of multiple soil functions. The study found that more diverse microbial communities exhibited higher functional stability when faced with perturbations such as copper contamination or heat stress. This research provided early experimental evidence for the diversity-stability hypothesis in microbial systems.

15.5.2.2 Buffering Against Environmental Fluctuations

Wittebolle et al. (2009) investigated how initial community evenness affects the functional stability of microbial communities under varying environmental conditions. Using denitrifying bacterial communities as a model system, they demonstrated that communities with higher initial evenness maintained their denitrification function more consistently across a range of salinity levels. This work highlighted how diversity, particularly evenness, can buffer ecosystem functions against environmental fluctuations.

15.5.2.3 Resilience in Soil Microbial Communities

A study by Tardy et al. (2014) examined the resilience of soil microbial communities to heat stress. They found that soils with higher microbial diversity recovered their microbial biomass and functional capabilities more quickly after heat stress, demonstrating greater resilience. This research underscored the importance of microbial diversity in maintaining ecosystem functions under climate change scenarios.

These studies collectively provide strong empirical support for the positive relationship between microbial diversity and ecosystem stability. They demonstrate that diverse microbial communities are better able to maintain their functions in the face of environmental perturbations, underlining the critical importance of preserving microbial diversity for ecosystem resilience in a changing world.

15.5.3 Microbial Contributions to Carbon Sequestration and Greenhouse Gas Mitigation

15.5.3.1 Microbial Role in Carbon Sequestration

Microbial communities play a crucial role in carbon sequestration through various mechanisms:

Soil Organic Matter Formation

Microbes contribute significantly to the formation of stable soil organic matter (SOM), a key component of long-term carbon storage in soils. Through the decomposition and transformation of plant residues and other organic materials, microbes produce compounds that can be resistant to further degradation, effectively sequestering carbon in the soil (Liang et al. 2017).

A significant portion of soil organic matter (SOM) originates from microbial necromass—the residues of dead microbial cells. These residues tend to be more stable and resistant to decomposition compared to plant-derived organic matter, thereby playing a crucial role in long-term carbon storage (Kallenbach et al. 2016). During the decomposition of organic matter, microbes produce various metabolites, some of which can form stable associations with soil minerals. These associations further enhance carbon sequestration, contributing to long-term soil carbon storage (Kögel-Knabner et al. 2008).

Mycorrhizal Fungi Networks

Arbuscular mycorrhizal fungi (AMF) form extensive networks in the soil, transferring carbon from plants to the soil. These fungi can store significant amounts of carbon in their biomass and contribute to the formation of stable soil aggregates, further enhancing carbon sequestration (Rillig et al. 2001). Arbuscular mycorrhizal fungi (AMF) can receive up to 20% of the carbon fixed by plants, with a substantial portion being transferred to the soil through fungal biomass turnover and exudation (Averill et al. 2014). Additionally, AMF produce glomalin, a glycoprotein that plays a crucial role in enhancing soil aggregate stability, thereby protecting occluded organic matter from decomposition (Wright and Upadhyaya 1998).

Microbial Carbon Pump

The microbial carbon pump refers to the microbial processes that transform labile organic carbon into recalcitrant forms, contributing to long-term carbon storage in soils and aquatic systems (Jiao et al. 2010). Microbes can transform simple, labile carbon compounds into more complex, recalcitrant forms that are resistant to further decomposition (Liang et al. 2017). The microbial carbon pump is particularly important in deep soil layers, where it contributes to the formation of stable, long-lived carbon pools (Rumpel and Kögel-Knabner 2011).

15.5.3.2 Microbial Mitigation of Greenhouse Gas Emissions

Microbes also play a significant role in mitigating greenhouse gas emissions:

Methane Oxidation

Methanotrophic bacteria in soils can oxidize methane, a potent greenhouse gas, into carbon dioxide, effectively reducing its global warming potential (Hanson and Hanson 1996). Well-aerated upland soils can act as methane sinks, with methanotrophs consuming atmospheric methane (Dutaur and Verchot 2007). In flooded rice paddies, methanotrophs in the rhizosphere can oxidize up to 90% of the methane produced before it reaches the atmosphere (Conrad 2007).

Nitrous Oxide Reduction

Certain soil microbes can reduce nitrous oxide, another potent greenhouse gas, to dinitrogen gas, mitigating its emission from agricultural soils (Hallin et al. 2018). Complete denitrifiers can reduce nitrous oxide to harmless dinitrogen gas, effectively mitigating emissions (Philippot et al. 2011). Promoting microbial communities capable of complete denitrification through soil management practices can significantly reduce nitrous oxide emissions from agricultural soils (Domeignoz-Horta et al. 2015).

Enhanced Nitrogen Use Efficiency

Microbial communities can improve nitrogen use efficiency in agricultural systems, reducing the need for synthetic fertilizers and consequently lowering nitrous oxide emissions (Ladha et al. 2016). Symbiotic and free-living nitrogen-fixing bacteria can provide plants with bioavailable nitrogen, reducing the need for synthetic fertilizers (Amoo et al. 2021b; Peoples et al. 2009). Diverse microbial communities promote efficient nutrient cycling, ensuring that more nitrogen is retained in the plant-soil system rather than lost as nitrous oxide (Jackson et al. 2008).

By leveraging these microbial processes through appropriate management practices, we can enhance carbon sequestration and mitigate greenhouse gas emissions in agricultural systems, contributing to climate change mitigation efforts.

15.5.4 Harnessing Microbial Diversity for Agroecosystem Resilience

Microbial diversity holds significant potential for enhancing agroecosystem resilience through various strategies:

15.5.4.1 Development of Diverse Microbial Inoculants

Microbial Inoculants as Soil Amendments

Microbial inoculants, which include beneficial bacteria and fungi, can improve soil health and plant growth by enhancing nutrient availability and supporting soil structure. They function similarly to probiotics in humans, promoting a healthy soil microbiome that supports plant health and productivity (Alori et al. 2017).

Mechanisms of Action

These inoculants work through various mechanisms, such as nitrogen fixation, phosphorus solubilization, and production of plant hormones. They can also enhance plant stress tolerance by producing compounds like ACC deaminase, which helps plants cope with abiotic stresses such as drought and salinity (Alori et al. 2017; Souza et al. 2015).

Challenges and Opportunities

While microbial inoculants offer numerous benefits, their effectiveness can be inconsistent due to environmental variability and interactions with native soil microorganisms. Understanding the ecological characteristics and colonization abilities of these inoculants is crucial for predicting their success and improving their application in the field (Alori and Babalola 2018; Li et al. 2023).

15.5.4.2 Management Practices Promoting Indigenous Diversity

Reduced Tillage and Crop Rotation

Practices such as reduced tillage and diverse crop rotations can enhance indigenous microbial diversity by creating a more stable and varied habitat for microbial communities. These practices help maintain soil structure and fertility, promoting resilient agroecosystems (Alori et al. 2017).

Integration with Sustainable Practices

Combining microbial inoculants with sustainable agricultural practices can reduce reliance on chemical fertilizers and pesticides, minimizing environmental impact and enhancing soil health (Alori and Babalola 2018).

15.5.4.3 Research Needs and Challenges

Predictable Enhancements in Resilience

Translating microbial diversity into predictable enhancements in agroecosystem resilience requires further research. This includes understanding the interactions between introduced microbial inoculants and native soil communities, as well as the long-term impacts on soil health and crop productivity (Li et al. 2023; Souza et al. 2015).

Ecological Frameworks

Incorporating ecological frameworks in the evaluation and application of microbial products can improve their consistency and effectiveness. This involves assessing the ability of inoculants to colonize new habitats and interact with native microbial communities (Li et al. 2023).

By leveraging microbial diversity through inoculants and sustainable management practices, we can enhance the resilience of agroecosystems to environmental stresses, ultimately contributing to more sustainable agricultural systems.

15.6 Challenges Facing Microbial Biodiversity Within Agroecosystems

15.6.1 The Consequences of Changes in Land Use

15.6.1.1 Deforestation and Habitat Fragmentation

High rates of deforestation and wide-scale agriculture expansion pose significant threats to microbial biodiversity in agroecosystems (Brown et al. 2020). The transformation of diverse forests and ecosystems into agricultural land leads to severe destruction and fragmentation eroding the intricate relationships of microbial networks (Sharma et al. 2018). These alterations are due to the involvement of intensive agricultural practices, such as monoculture farming, using large amount of pesticides and fertilizers, and mechanical cultivation, which make soil quality poor and results in change or loss of biodiversity (Mengist et al. 2021). Yang et al. (2022)

found that microbial diversity and community structure were altered by the conversion of natural grasslands to croplands in agriculture. However, the extent of land-use changes' impact on the richness and resilience of microbial co-occurrence networks under various natural ecosystem shifts remains unclear (Yang et al. 2022). Additionally, the soil erosion and agricultural expansion that frequently take place may cause soil carbon depletion and the loss of the nutrient retention ability of microbes which would further increase the challenges faced by microbial communities within agroecosystem (Tian et al. 2024).

Furthermore, the accumulation of chemicals, including pesticides and pollutants, which directly impacts beneficial microbes and contributes to the overall breakdown of the soil health and long-term sustainability of the agroecosystem (Mandal et al. 2020). Larsen et al. (2015) reported that the application of pesticide had an effect on the bacterial community structure and functions, ultimately impacting organic matter decomposition, phosphorus solubilization, iron, nitrogen fixation, nitrification, and additional functions, ultimately impacting plant nutrient supply and growth in agrosystems. These challenges can be overcome by the adoption of sustainable land management practices that focus on the conservation and restoration of natural habitats, promote agroecological practices and support the conservation of soil microbial biodiversity (Cárceles Rodríguez et al. 2022).

The destruction and disappearance of natural ecosystems brought on by human activity and climate change is mostly to blame for the sixth extinction event (Rybicki et al. 2020). Habitat fragmentation, the process of dividing a landscape into smaller parts, contributes to reduced microbial diversity and disrupts ecological connections (Mendes and Prevedello 2020). Consequently, ecological functions including carbon sequestration and nutrient cycling are hindered, and functional diversity decreases. Mendes and Prevedello (2020) conducted an experiment showing that habitat fragmentation influences species diversity, with larger environments supporting higher species diversity compared to smaller ones. Additionally, Vannette et al. (2016) noted microbial dispersion limitations, suggesting that certain microbial communities could be more vulnerable to human habitat fragmentation as a result of these restrictions. For example, the fungal diversity associated with roots was found to lower in smaller and less connected patches, whereas fungal community composition was more consistent in highly connected areas in a naturally fragment environment (Kiesewetter and Afkhami 2021). Grilli et al. (2014) indicated that plants cultivated with soils from larger fragments had higher mycorrhizal colonization.

15.6.1.2 Urbanization

The main determinants of the soil microbes abundance and diversity are increasingly understood and researched, especially in rural areas (Guilland et al. 2018). However, microbial communities in urban soils can only be considered poorly explored because specific features of its microbial soil ecology remain unclear (Guilland et al. 2018). According to Barberán et al. (2015), urbanization has effects which are not uniform among the taxonomic groups of microbials. Another study revealed that there is a decline in soil fungal richness in urbanized areas, while bacterial richness remains at the same level (Epp Schmidt et al. 2017). Additionally, Christel et al. (2023) reported that urban constructions disrupt the microbial carbon and nutrient cycling which affect the microbial community structure. However, the temporal dynamics of microbial communities in urban habitats remain poorly understood, particularly considering the variable levels of urbanization. Further research is needed to elucidate these complexities.

15.6.2 Climate Change Impact on Microbes

Microbial communities within ecosystems are highly susceptible to the impacts of climate change. Research has shown that climate change frequently alters the diversity, composition, and ecological functions of these microbial communities (Cavicchioli et al. 2019). Changes in temperature, precipitation patterns, and extreme weather events associated with climate change can prompt responses in microbial communities. These responses often lead to alterations in nutrient cycling, carbon sequestration, and soil fertility (Ballantyne et al. 2017). For example, modifications of microbial-mediated processes such as decomposition triggered by climate change will likely shift the balance of greenhouse gas emissions and potentiate feedbacks loops, those that contribute to a climate change (Abatenh et al. 2018). Weiman (2015) reported that bacteria, fungi, algal, and archaea speed up global warming through organic matter decomposition and enhanced carbon dioxide emissions in the atmosphere. In agriculture, an environment highly susceptible to environmental changes, shifts in climate can impact interactions among plants, microorganisms, and animals (Dubey et al. 2019). Temperature can disturb the symbiotic relationships between the hosts and their microbiomes which will upset the plants' health, make a negative influence on the animal production and undermine long-run sustainability of agriculture (Classen et al. 2015). Frater et al. (2018)'s study found that Arbuscular mycorrhizal fungi colonization increased parallel to temperature and pH in the Central United States, but decreased across the Mediterranean. Bintanja (2018) reported that the temperature in temperate forests increases by 5 °C, which changes the relative abundances of microorganisms like bacteria in the soil and raises the bacterial to fungal ratio, thereby altering respiration rates of soil microbes due to temperature-sensitive processes they mediate. Abdul Rahman et al. (2021) demonstrated that different temperatures have different effects on the major enzymes of N2-fixing bacteria. For example, vanadium nitrogenase works best at 5 °C, whereas molybdenum nitrogenase works best at 30 °C. These differences in effectiveness may encourage the emergence of novel bacteria (Abdul Rahman et al. 2021). Moreover, extreme climate instances such as drought and flood can affect soil microbial community enough that it compromises soil drainage and erosion; hence, these side effects dispose of soil function (Cárceles Rodríguez et al. 2022).

15.7 Strategies for Preserving Microbial Biodiversity

15.7.1 The Importance of Monitoring Microbial Diversity

The microbiome is essential for signalling ecosystem health and is extremely sensitive to environmental changes, including anthropogenic activity (Trivedi et al. 2022). It was shown that there was a reduction in the composition, overall abundance, and complexity of the microbiome, along with the number of global change factors such as microplastics, abiotic factors, resource availability, and toxic compounds. Also, Mukhtar et al. (2023) reported a shift in the response of soil microbiomes to project the impact of climate change, such as elevated temperatures, droughts, and floods. Loss of habitat resulting from changes in land use is one of the main anthropogenic stressors causing the degradation of ecosystems globally, and this also affects the soil microbiome. Pollution from oil spills, microplastics, chemicals, light, and heavy metals, and mine drainage from urban centres can potentially cause detectable changes in the environmental microbiomes of marine, terrestrial, and freshwater ecosystems (Sudan et al. 2024). One of the causes of global change is biological invasions by microorganisms, animals, and plants. These invasions cause changes in the microbiome of the native ecosystems.

Vibha Bhardwaj and Neelam Garg (2012) established the extraction of microorganisms from soil. They used them in industrial production for different purposes such as medicines, food production and processes, the development of biocontrol agents and biocides, and other natural products. Employing microorganisms as tools for exploring fundamental life processes becomes appealing following the recognition of a common set of biochemical life processes in higher forms of life and microorganisms; these include their ease of culture in both small and large quantities, rapid reproduction, and ease of manipulation of their growth through physical and chemical means (Ho et al. 2017). In the field of biotechnology, microorganisms play a dominant role. According to Vibha Bhardwaj and Neelam Garg (2012) report, microbial enzyme technology has grown to become a multimillion-dollar industry in recent times, and it is currently exploring microbial strain production to discover both enzymes with novel properties and the production of pharmaceutical products. Also, many novel genes, isolated from newly discovered microbes, have demonstrated phylogenetic similarities across various microorganism groups. Thus, this insight creates an avenue for the classification and analysis of microbes using new, advanced methods (Crits-Christoph et al. 2018). Between 1500 and 6000 genes exist in the ten most clinically significant human pathogenic bacteria. These genes are crucial for growth and survival, and they aid in the development of antimicrobial compounds that inhibit bacterial growth. It was reported that a novel program, Via Gene, has been developed to predict the essentiality of bacterial genes by utilizing a blend of DNA sequence descriptors and advanced pattern recognition techniques (Li et al. 2021). Phage therapy is another new area of discovery and interest employing bacteriophages for therapeutic measures in various countries, including those of the former Soviet Union. Phages could be employed biotechnologically to manage genetically modified microbes that pose threats to the environment (Harada et al. 2018).

Soil microbes have an important influence on global biochemical cycles, such as carbon and nitrogen. Physiological changes driven by climate can impact the rates of these cycles (Crowther et al. 2019). To better understand the temperature sensitivity of microbial respiration within ecosystems requires a shift from traditional climate models that primarily concentrate on the size of soil carbon pools. Emphasis is required in understanding the diversity of soil microbial community, composition, and functions for mechanistic insight into their responses to climate warming. Several studies suggest that soil microorganisms either undergo a modification of function of their existing populations or community composition shift in response to environmental changes, leading to temperature sensitivity alterations of soil organic carbon (SOC) decomposition (Adedayo and Babalola 2023; Kruse et al. 2013; Molefe et al. 2023). Higher temperatures increase microbial and metabolic activity rates, resulting in an increase in microbial respiration in the soils. However, the long-term microbial respiration-endogenous response to climate change will depend on the microbial resource capture (substrate) and the environment of the microbial ecosystem (Hagerty et al. 2018). Also, environmental changes can be influenced by shifts in microbial community composition, which may consequently affect rates of soil respiration. Microbial respiration rate and SOC decomposition will both drop if they store more carbon than they metabolize. However, a shift in the diversity of microbial communities that can degrade complex soil carbon will result in favourable emissions of CO2, decreased SOC decomposition, and a positive response to climate change (Chatterjee and Saha 2018). Thus, owing to their capability of SOC decomposition, it is important to identify functional groups of microbes to understand the feedback responses. The impact of the feedback responses across the ecosystems varies, and quantifying such responses is dependent on understanding which mechanism is prevalent and how microbial diversity affects the response.

15.7.2 Establishing Microbial Conservation Initiatives: The Microbiota Vault

Microbial diversity is important for balancing ecosystems and human activities increasingly threaten this. This is why there is a need to conserve microbes because such human activities greatly threaten microbial diversity. Microbiota vaults represent an innovative and crucial method for preserving invaluable microbial resources. These are repositories for isolated microbes obtained from different environments, such as air and soil (Chandel et al. 2021). To maintain the genetic integrity and long-term viability of the stored samples, microbial vaults are equipped with advanced technology. Microbial diversity is safeguarded by the microbial vaults, which are essential for the provision of ecosystem resilience and services. Conservation of microbial diversity requires specialized techniques to reclaim habitats that have been degraded (Bello et al. 2018). Microbiome preservation and samples that are intact are among the best approaches for potential culturomics and their effect on ecological perturbation on ecological shift and functionality. Intact microbiome preservation is needed for the dimensional structure of samples and the reconstruction of the spatial organization of microbial community alterations (Hartmann and Six 2023). The ecological function is negatively affected by the loss of microbial biodiversity, and this has made researchers focus on better ecosystem functioning by conserving the intact structure of the microbiome and restoring the natural microbial community. Primarily, the microbiota vault initiative focuses on preservation and is closely linked to research on several levels. It interacts with research efforts to collect microbial diversity; its long-term safe storage of microbial samples is a research question in itself, and most importantly, the initiative plays an important part in catalyzing research into the health relevance and diversity of microbiota.

Safeguarding microbial diversity through the microbial vaults initiative, provides a safe place for long-term storage for local and regional collections around the world. The Svalbard Global Seed Vault on the island of Svalbard was chosen, which allows the survival of the storage collection in the absence of human intervention (Smedsrud 2021). The Microbiota Vault has the possibility of using a specific storage mode, such as lyophilization. Bacteria specimens need ultralow temperatures to be preserved, which is similar to biobanks. This mode of operation would call for a more central, actively maintained, well-accessible, and politically stable location. Implementing parallel locations and independent safeguarding methodologies would result in increased chances of resuscitation and redundancy in terms of location. To assess different options, several factors, including cost, sample storage, overall risk, and time need to be considered.

15.7.3 Incorporating Microbial Diversity into Agricultural Management Strategies

Agriculture is the main source of employment and income in developing countries, especially sub-Saharan Africa, and is the backbone of the economy. Due to the adoption of agricultural practices that were faulty in the past, the agricultural industry's sustenance has become debatable (Sims and Kienzle 2017). The negative impact of excessive use of agrochemicals during various agricultural practices can be remedied by adopting beneficial microorganisms in the soil. Positive traits of agriculturally relevant soil microorganisms are employed for beneficial activities including nutrient cycling, biocontrol, biodegradation, and nutrient acquisition (Olowe et al. 2023a; Yadav et al. 2021).

Soil biodiversity may be disrupted by agricultural intensification in a way that the overall ecosystem is affected. Several agricultural practices, including the application of organic amendments, strongly affect soil microbial communities (Dornbush and von Haden 2017). Application of new technologies is needed to maintain healthy environments, especially inoculation of microbes into the soils. There are several approaches used to meet food requirements in a sustainable and environmentally friendly way; these include the use of microbial products, which play a crucial role in achieving food security, especially in a changing climate (de Graaff et al. 2019). Some microbial formulations provide alternatives to toxic agrochemicals such as chemical pesticides, and inorganic fertilizers are being used by crop producers to increase yield without affecting local natural resources such as the soil. Other possibilities for sustainable agriculture include indirect control of the effective microorganisms by rhizosphere engineering and other agricultural practices such as the adoption of appropriate cropping systems and irrigation regimes (Gupta et al. 2022). The use of new agricultural and advanced technologies is recommended to farmers to increase crop productivity. However, they will likely adopt more chemical pesticides and agrochemical-based inorganic fertilizers (Akinsorotan et al. 2023). This trend is mostly identified in policies relating to agriculture in the developing world including Nigeria.

Industrialisation as well as other man-made activities, especially those that are necessary for enhancing the development of the third-world countries like Nigeria, releases pollutants including carbon dioxide emissions, chemical discharge, and waste products; they gradually build up in the agricultural lands and the water bodies (Fayomi et al. 2019). Plants accommodate soils as important media for growth and depend on a diverse community of native microorganisms. The synergy between both plants and microorganisms is a basis for survival. Another challenge that is facing contemporary agricultural research is how to ensure sustainable economic and environmental problems even with quality produce and fairly high yields (Reisch et al. 2013). In the present context, an effective approach to addressing this challenge is exploiting soil microbial communities' agroecosystem services. Presently, attention is focused on developing green technologies that can break down pollutants that are toxic and that have infiltrated nature, reducing them to a level considered safe. Soil microbial communities management for crop production is one of the efficient and suitable ecological friendly ways of healthy and sustainable crop production as the biosphere is protected (Adeleke et al. 2023; Fasusi et al. 2021; Olowe et al. 2023b). Soil microorganisms in agricultural management are important to several processes that result in many beneficial effects on crop plants, amongst them being; soil quality acting as substrates for plant growth, decomposition of crop residue, and cycling of nutrients (Hirsch and Mauchline 2015). To support the ecological aspects of the agrotechnological applications, there are various strategies are being recommended for well-recognized low-input biotechnology and optimum utilization of beneficial microbial inputs. Since agronomic management and other ecological factors affect crop microbe relations, there should be considerations on how these environmental stressors affect interaction of crop microbes, especially in light of climate changes (Ullah et al. 2021). The productivity of both natural ecosystems and agricultural systems is affected by different stressors that are generated by intensive agricultural practices, which then restrict various ecosystem services. Drought, salinity, soil erosion and so on are some of the stress factors that prevail. Some of the microbes reside in the rhizoplane or rhizospheric soil and endophytes live within plant tissues and have many beneficial impacts on the overall performance of plants without showing their presence (Samad et al. 2019).

15.8 Conclusion: Reflections and Future Directions

This chapter has explored the crucial role of microbial contributions to biodiversity conservation in agroecosystems. As we reflect on the key points discussed, it becomes evident that microbial diversity is fundamental to the health, resilience, and productivity of agricultural ecosystems.

15.8.1 Summary of Key Points

- Microbes play essential roles in nutrient cycling, plant-microbe interactions, and biocontrol of plant diseases.
- Symbiotic relationships, such as mycorrhizal associations and rhizobia-legume symbiosis, significantly enhance plant growth and nutrient uptake.
- Plant growth-promoting rhizobacteria (PGPR) and microbial-induced systemic resistance (ISR) contribute to plant health and defence against pathogens.
- Microbial diversity is crucial for agroecosystem resilience to climate change and contributes to carbon sequestration.

15.8.2 Importance of Preserving Microbial Diversity

The microbiome is a sensitive indicator of ecosystem health and is highly responsive to environmental changes, including anthropogenic activities. Global change factors such as climate change, pollution, and land-use changes can significantly impact microbial communities. Crowther et al. (2019) noted that soil microbes have an important influence on global biogeochemical cycles, such as carbon and nitrogen, and that physiological changes driven by climate can impact the rates of these cycles.

Preserving microbial diversity is essential for maintaining ecosystem functions, enhancing agricultural productivity, and ensuring food security in the face of ongoing environmental challenges. As has been pointed out by various studies, conservation of microbial diversity requires specialized techniques to reclaim habitats that have been degraded.

15.8.3 Future Research Directions

1. **Advanced monitoring techniques:** Develop and implement advanced techniques for monitoring microbial diversity in agroecosystems, including high-throughput sequencing and metagenomic approaches.
2. **Climate change impacts:** Further investigate the effects of climate change on microbial communities and their functions in agroecosystems.
3. **Microbial biocontrol:** Explore the potential of beneficial microbes for sustainable pest and disease management in agriculture.
4. **Soil health indicators:** Develop microbial-based indicators of soil health to guide agricultural management practices.
5. **Microbiome engineering:** Investigate the potential for engineering beneficial microbial communities to enhance crop productivity and resilience.

15.8.4 Recommendations for Integrating Microbial Conservation

1. **Establish microbial conservation initiatives:** Support the development of initiatives like the Microbiota Vault to preserve microbial diversity for future research and applications.
2. **Policy integration:** Incorporate microbial conservation into broader biodiversity conservation policies and agricultural management strategies.
3. **Education and awareness:** Promote understanding of the importance of microbial diversity among farmers, policymakers, and the general public.
4. **Sustainable agricultural practices:** Encourage farming practices that support diverse and healthy microbial communities, such as reduced tillage, crop rotation, and minimal use of chemical inputs.

15.8.5 Final Thoughts

The potential of harnessing microbial contributions to promote biodiversity conservation and sustainable agriculture is immense. As we face ongoing environmental challenges, including climate change and food security issues, the role of microbial diversity in agroecosystems becomes increasingly critical. By recognizing the importance of microbial communities, investing in their conservation, and integrating this knowledge into agricultural practices and policies, we can work towards more resilient, productive, and sustainable agroecosystems.

The future of agriculture lies in our ability to work with nature, not against it. By embracing the power of microbial diversity, we can pave the way for a more sustainable and food-secure future, where agroecosystems thrive in harmony with the natural world.

References

Abatenh E, Gizaw B, Tsegaye Z, Tefera G (2018) Microbial function on climate change-a review. Environ Pollut Clim Change 2(1):1–6

Abawi G, Widmer T (2000) Impact of soil health management practices on soilborne pathogens, nematodes and root diseases of vegetable crops. Appl Soil Ecol 15(1):37–47

Abd-El-Khair H, Elshahawy I, Haggag HK (2019) Field application of *Trichoderma* spp. combined with thiophanate-methyl for controlling *Fusarium solani* and *Fusarium oxysporum* in dry bean. Bull Natl Res Cent 43:1–9

Abdul Rahman NSN, Abdul Hamid NW, Nadarajah K (2021) Effects of abiotic stress on soil microbiome. Int J Mol Sci 22(16):9036

Adedayo AA, Babalola OO (2023) The potential of biostimulants on soil microbial community: a review. Front Ind Microbiol 1:1308641

Adeleke BS, Muller D, Babalola OO (2023) A metagenomic lens into endosphere microbial communities, promises, and discoveries. Lett Appl Microbiol 76(2):ovac030

Ahl P, Voisard C, Défago G (1986) Iron bound-siderophores, cyanic acid, and antibiotics involved in suppression of *Thielaviopsis basicola* by a *Pseudomonas fluorescens* strain. J Phytopathol 116(2):121–134

Akinsorotan OA, Akinsorotan AM, Adewale RO, Akande AB (2023) Detrimental effects of agrochemical-based agricultural intensification on biodiversity: evidence from some past studies. In: One health implications of agrochemicals and their sustainable alternatives. Springer, pp 275–298

Allison SD, Martiny JB (2008) Resistance, resilience, and redundancy in microbial communities. Proc Natl Acad Sci 105(supplement_1):11512–11519

Alori ET, Babalola OO (2018) Microbial inoculants for improving crop quality and human health in Africa. Front Microbiol 9:2213

Alori ET, Dare MO, Babalola OO (2017) Microbial inoculants for soil quality and plant health. Sustain Agric Rev:281–307

Alshannaq A, Yu J-H (2017) Occurrence, toxicity, and analysis of major mycotoxins in food. Int J Environ Res Public Health 14(6):632

Altieri MA (1999) The ecological role of biodiversity in agroecosystems. Agric Ecosyst Environ 74(1–3):19–31

Amkraz N, Boudyach E, Boubaker H, Bouizgarne B, Ait Ben Aoumar A (2010) Screening for fluorescent pseudomonades, isolated from the rhizosphere of tomato, for antagonistic activity toward *Clavibacter michiganensis* subsp. *michiganensis*. World J Microbiol Biotechnol 26:1059–1065

Amoo AE, Babalola OO (2017) Ammonia-oxidizing microorganisms: key players in the promotion of plant growth. J Soil Sci Plant Nutr 17(4):935–947

Amoo AE, Babalola OO (2019) Impact of land use on bacterial diversity and community structure in temperate pine and indigenous forest soils. Diversity 11(11):217

Amoo AE, Delgado-Baquerizo M, Babalola OO (2021a) Forest plantations reduce soil functioning in terrestrial ecosystems from South Africa. Pedobiologia 89:150757

Amoo AE, Enagbonma BJ, Ayangbenro AS, Babalola OO (2021b) Biofertilizer: an eco-friendly approach for sustainable crop production. In: Food security and safety: African perspectives. Springer, pp 647–669

Anderson R, Roddam L, Bettiol S, Sanderson K, Reid D (2010) Biosignificance of bacterial cyanogenesis in the CF lung. J Cyst Fibros 9(3):158–164

Aremu BR, Alori ET, Kutu RF, Babalola OO (2017) Potentials of microbial inoculants in soil productivity: an outlook on African legumes. In: Microorganisms for green revolution: volume 1: microbes for sustainable crop production. Springer Singapore, pp 53–75

Athukorala SN, Fernando WD, Rashid KY, De Kievit T (2010) The role of volatile and non-volatile antibiotics produced by *Pseudomonas chlororaphis* strain PA23 in its root colonization and control of *Sclerotinia sclerotiorum*. Biocontrol Sci Tech 20(8):875–890

Averill C, Turner BL, Finzi AC (2014) Mycorrhiza-mediated competition between plants and decomposers drives soil carbon storage. Nature 505(7484):543–545

Babalola OO, Olanrewaju OS, Amoo AE, Ajilogba CF, Chukwuneme FC, Ojuederie OB, Omomowo OI (2023) The functionality of plant-microbe interactions in disease suppression. J King Saud Univ-Sci 35:102893

Bailey K, Lazarovits G (2003) Suppressing soil-borne diseases with residue management and organic amendments. Soil Tillage Res 72(2):169–180

Baker BP, Green TA, Loker AJ (2020) Biological control and integrated pest management in organic and conventional systems. Biol Control 140:104095

Bakker AW, Schippers B (1987) Microbial cyanide production in the rhizosphere in relation to potato yield reduction and Pseudomonas spp-mediated plant growth-stimulation. Soil Biol Biochem 19(4):451–457

Bakker PA, Pieterse CM, Van Loon L (2007) Induced systemic resistance by fluorescent *Pseudomonas* spp. Phytopathology 97(2):239–243

Ballantyne A, Smith W, Anderegg W, Kauppi P, Sarmiento J, Tans P, Shevliakova E, Pan Y, Poulter B, Anav A (2017) Accelerating net terrestrial carbon uptake during the warming hiatus due to reduced respiration. Nat Clim Chang 7(2):148–152

Bamdad H, Papari S, Lazarovits G, Berruti F (2022) Soil amendments for sustainable agriculture: microbial organic fertilizers. Soil Use Manag 38(1):94–120

Barberán A, Ladau J, Leff JW, Pollard KS, Menninger HL, Dunn RR, Fierer N (2015) Continental-scale distributions of dust-associated bacteria and fungi. Proc Natl Acad Sci 112(18):5756–5761

Bastida F, Eldridge DJ, García C, Kenny Png G, Bardgett RD, Delgado-Baquerizo M (2021) Soil microbial diversity–biomass relationships are driven by soil carbon content across global biomes. ISME J 15(7):2081–2091

Bello MGD, Knight R, Gilbert JA, Blaser MJ (2018) Preserving microbial diversity. Science 362(6410):33–34

Berendsen RL, Pieterse CM, Bakker PA (2012) The rhizosphere microbiome and plant health. Trends Plant Sci 17(8):478–486

Bintanja R (2018) The impact of Arctic warming on increased rainfall. Sci Rep 8(1):16001

Biswas S, Anusuya D (2014) Effect of bioinoculants and organic manure (phosphocompost) on growth, yield and nutrient uptake of Trigonella foenum-graecum L.(Fenugreek). Int J Sci Res 3:38–41

Bonanomi G, Lorito M, Vinale F, Woo SL (2018) Organic amendments, beneficial microbes, and soil microbiota: toward a unified framework for disease suppression. Annu Rev Phytopathol 56(1):1–20

Bonneau A, Roche B, Schalk IJ (2020) Iron acquisition in *Pseudomonas aeruginosa* by the siderophore pyoverdine: an intricate interacting network including periplasmic and membrane proteins. Sci Rep 10(1):120

Brillet K, Ruffenach F, Adams H, Journet L, Gasser VR, Hoegy F, Guillon L, Hannauer M, Page A, Schalk IJ (2012) An ABC transporter with two periplasmic binding proteins involved in iron acquisition in *Pseudomonas aeruginosa*. ACS Chem Biol 7(12):2036–2045

Brizuela AM, Gálvez L, Arroyo JM, Sánchez S, Palmero D (2023) Evaluation of *Trichoderma* spp. on *Fusarium oxysporum* f. sp. asparagi and *Fusarium* wilt Control in Asparagus Crop. Plants 12(15):2846

Brodhagen M, Henkels MD, Loper JE (2004) Positive autoregulation and signaling properties of pyoluteorin, an antibiotic produced by the biological control organism Pseudomonas fluorescens Pf-5. Appl Environ Microbiol 70(3):1758–1766

Brown G, Cooper M, Kobayashi M, Orgiazzi A, Domínguez A, Dias Turetta AP, Franco A, Zaitsev A, Winding A, Foereid B (2020) Threats to soil biodiversity-global and regional trends. In: State of knowledge of soil biodiversity - Status, challenges and potentialities, Report 2020. FAO

Bunch AW, Knowles CJ (1980) Cyanide production and degradation during growth of the snow mould fungus. Microbiology 116(1):9–16

Cárceles Rodríguez B, Durán-Zuazo VH, Soriano Rodríguez M, García-Tejero IF, Gálvez Ruiz B, Cuadros Tavira S (2022) Conservation agriculture as a sustainable system for soil health: a review. Soil Syst 6(4):87. https://www.mdpi.com/2571-8789/6/4/87
Carvalho FP (2006) Agriculture, pesticides, food security and food safety. Environ Sci Pol 9(7–8):685–692
Carvalho FP (2017) Pesticides, environment, and food safety. Food Energy Secur 6(2):48–60
Case BS, Pannell JL, Stanley MC, Norton DA, Brugman A, Funaki M, Mathieu C, Songling C, Suryaningrum F, Buckley HL (2020) The roles of non-production vegetation in agroecosystems: a research framework for filling process knowledge gaps in a social-ecological context. People Nat 2(2):292–304
Cavicchioli R, Ripple WJ, Timmis KN, Azam F, Bakken LR, Baylis M, Behrenfeld MJ, Boetius A, Boyd PW, Classen AT, Crowther TW, Danovaro R, Foreman CM, Huisman J, Hutchins DA, Jansson JK, Karl DM, Koskella B, Mark Welch DB et al (2019) Scientists' warning to humanity: microorganisms and climate change. Nat Rev Microbiol 17(9):569–586. https://doi.org/10.1038/s41579-019-0222-5
Chandel A, Mann R, Kaur J, Norton S, Edwards J, Spangenberg G, Sawbridge T (2021) Implications of seed vault storage strategies for conservation of seed bacterial microbiomes. Front Microbiol 12:784796
Chang R, Ries S, Pataky J (1992) Reductions in yield of processing tomatoes and incidence of bacterial canker. Plant Dis 76:805
Charoenporn C, Kanokmedhakul S, Lin F, Poeaim S, Soytong K (2010) Evaluation of bio-agent formulations to control Fusarium wilt of tomato. Afr J Biotechnol 9(36), 5836–5844
Chatterjee D, Saha S (2018) Response of soil properties and soil microbial communities to the projected climate change. In: Advances in crop environment interaction. Springer, pp 87–136
Cheeke TE, Coleman DC, Wall DH (2012) Microbial ecology in sustainable agroecosystems. Crc Press
Choruma DJ, Odume ON (2019) Exploring farmers' management practices and values of ecosystem services in an agroecosystem context—A case study from the eastern cape, South Africa. Sustainability 11(23):6567
Choudhary DK, Prakash A, Johri B (2007) Induced systemic resistance (ISR) in plants: mechanism of action. Indian J Microbiol 47:289–297
Christel A, Dequiedt S, Chemidlin-Prevost-Bouré N, Mercier F, Tripied J, Comment G, Djemiel C, Bargeot L, Matagne E, Fougeron A, Mina Passi J-B, Ranjard L, Maron P-A (2023) Urban land uses shape soil microbial abundance and diversity. Sci Total Environ 883:163455. https://doi.org/10.1016/j.scitotenv.2023.163455
Clarkson TW (1995) Environmental contaminants in the food chain. Am J Clin Nutr 61(3):682S–686S
Classen AT, Sundqvist MK, Henning JA, Newman GS, Moore JA, Cregger MA, Moorhead LC, Patterson CM (2015) Direct and indirect effects of climate change on soil microbial and soil microbial-plant interactions: what lies ahead? Ecosphere 6(8):1–21
Colica G, Li H, Rossi F, Li D, Liu Y, De Philippis R (2014) Microbial secreted exopolysaccharides affect the hydrological behavior of induced biological soil crusts in desert sandy soils. Soil Biol Biochem 68:62–70
Collins JF (2017) Copper: basic physiological and nutritional aspects. In: Molecular, genetic, and nutritional aspects of major and trace minerals. Elsevier, pp 69–83
Compant S, Samad A, Faist H, Sessitsch A (2019) A review on the plant microbiome: ecology, functions, and emerging trends in microbial application. J Adv Res 19:29–37
Conrad R (2007) Microbial ecology of methanogens and methanotrophs. Adv Agron 96:1–63
Conrath U, Beckers GJ, Flors V, García-Agustín P, Jakab G, Mauch F, Newman M-A, Pieterse CM, Poinssot B, Pozo MJ (2006) Priming: getting ready for battle. Mol Plant-Microbe Interact 19(10):1062–1071
Cornelis P, Matthijs S (2002) Diversity of siderophore-mediated iron uptake systems in fluorescent pseudomonads: not only pyoverdines. Environ Microbiol 4(12):787–798

Crits-Christoph A, Diamond S, Butterfield CN, Thomas BC, Banfield JF (2018) Novel soil bacteria possess diverse genes for secondary metabolite biosynthesis. Nature 558(7710):440–444
Crowther TW, Van den Hoogen J, Wan J, Mayes MA, Keiser A, Mo L, Averill C, Maynard DS (2019) The global soil community and its influence on biogeochemistry. Science 365(6455):eaav0550
Crumbliss AL, Harrington JM (2009) Iron sequestration by small molecules: thermodynamic and kinetic studies of natural siderophores and synthetic model compounds. Adv Inorg Chem 61:179–250
Daisley BA, Chernyshova AM, Thompson GJ, Allen-Vercoe E (2022) Deteriorating microbiomes in agriculture-the unintended effects of pesticides on microbial life. Microb Res Rep 1(1):6
de Klein JJ, Overbeek CC, Juncher Jorgensen C, Veraart AJ (2017) Effect of temperature on oxygen profiles and denitrification rates in freshwater sediments. Wetlands 37, 975–983
Delgado-Baquerizo M, Maestre FT, Reich PB, Jeffries TC, Gaitan JJ, Encinar D, Berdugo M, Campbell CD, Singh BK (2016) Microbial diversity drives multifunctionality in terrestrial ecosystems. Nat Commun 7(1):10541
Devi KK, Kothamasi D (2009) *Pseudomonas fluorescens* CHA0 can kill subterranean termite *Odontotermes obesus* by inhibiting cytochrome c oxidase of the termite respiratory chain. FEMS Microbiol Lett 300(2):195–200
Devi KK, Seth N, Kothamasi S, Kothamasi D (2007) Hydrogen cyanide-producing rhizobacteria kill subterranean termite *Odontotermes obesus* (Rambur) by cyanide poisoning under in vitro conditions. Curr Microbiol 54:74–78
Domeignoz-Horta LA, Spor A, Bru D, Breuil M-C, Bizouard F, Léonard J, Philippot L (2015) The diversity of the N2O reducers matters for the N2O: N2 denitrification end-product ratio across an annual and a perennial cropping system. Front Microbiol 6:971
Domeignoz-Horta LA, Pold G, Liu X-JA, Frey SD, Melillo JM, DeAngelis KM (2020) Microbial diversity drives carbon use efficiency in a model soil. Nat Commun 11(1):3684
Dornbush ME, von Haden AC (2017) Intensified agroecosystems and their effects on soil biodiversity and soil functions. In: Soil health and intensification of agroecosytems. Elsevier, pp 173–193
Dubey SC, Singh B (2013) Integrated management of major diseases of mungbean by seed treatment and foliar application of insecticide, fungicides and bioagent. Crop Prot 47:55–60
Dubey SC, Suresh M, Singh B (2007) Evaluation of *Trichoderma* species against *Fusarium oxysporum* f. sp. ciceris for integrated management of chickpea wilt. Biol Control 40(1):118–127
Dubey SC, Bhavani R, Singh B (2011) Integration of soil application and seed treatment formulations of *Trichoderma* species for management of wet root rot of mungbean caused by *Rhizoctonia solani*. Pest Manag Sci 67(9):1163–1168
Dubey SC, Tripathi A, Singh B (2013) Integrated management of Fusarium wilt by combined soil application and seed dressing formulations of Trichoderma species to increase grain yield of chickpea. Int J Pest Manag 59(1):47–54
Dubey SC, Singh V, Priyanka K, Upadhyay BK, Singh B (2015) Combined application of fungal and bacterial bio-agents, together with fungicide and Mesorhizobium for integrated management of Fusarium wilt of chickpea. BioControl 60:413–424
Dubey A, Malla MA, Khan F, Chowdhary K, Yadav S, Kumar A, Sharma S, Khare PK, Khan ML (2019) Soil microbiome: a key player for conservation of soil health under changing climate. Biodivers Conserv 28:2405–2429
Dubey SC, Tripathi A, Tak R, Devi SI (2020) Evaluation of bio-formulations of fungal and bacterial biological control agents in combination with fungicide in different mode of application for integrated management of tomato wilt. Indian Phytopathol 73(3):425–432
Dubey PK, Singh A, Merah O, Abhilash P (2022) Managing agroecosystems for food and nutrition security. Curr Res Environ Sustain 4:100127
Duijff BJ, Meijer JW, Bakker PA, Schippers B (1993) Siderophore-mediated competition for iron and induced resistance in the suppression of fusarium wilt of carnation by fluorescent Pseudomonas spp. Neth J Plant Pathol 99:277–289

Dunne C, Cronin D, Moënne-Loccoz Y, O'Gara F (1998) Biological control of phytopathogens by phloroglucinol and hydrolytic enzyme producing bacterial inoculants Bulletin OILB/SROP, 21(9), 19–25

Durrant WE, Dong X (2004) Systemic acquired resistance. Annu Rev Phytopathol 42(1):185–209

Dutaur L, Verchot LV (2007) A global inventory of the soil CH4 sink. Glob Biogeochem Cycles 21(4). https://doi.org/10.1029/2006GB002734

Egamberdieva D, Wirth S, Bellingrath-Kimura SD, Mishra J, Arora NK (2019) Salt-tolerant plant growth promoting rhizobacteria for enhancing crop productivity of saline soils. Front Microbiol 10:2791

El-Rahman A, Shaheen HA, El-Aziz A, Rabab M, Ibrahim DS (2019) Influence of hydrogen cyanide-producing rhizobacteria in controlling the crown gall and root-knot nematode, Meloidogyne incognita. Egypt J Biol Pest Control 29(1):1–11

Emami-Karvani Z, Chitsaz-Esfahani Z (2021) Phosphorus solubilization: mechanisms, recent advancement and future challenge. In Soil microbiomes for sustainable agriculture: functional annotation, pp. 85-131. Springer

Emmerson M, Morales MB, Oñate JJ, Batary P, Berendse F, Liira J, Aavik T, Guerrero I, Bommarco R, Eggers S (2016) How agricultural intensification affects biodiversity and ecosystem services. In: Advances in ecological research, vol 55. Elsevier, pp 43–97

Endicott NP, Rivera GSM, Yang J, Wencewicz TA (2020) Emergence of ferrichelatase activity in a siderophore-binding protein supports an iron shuttle in bacteria. ACS Central Sci 6(4):493–506

Epp Schmidt DJ, Pouyat R, Szlavecz K, Setälä H, Kotze DJ, Yesilonis I, Cilliers S, Hornung E, Dombos M, Yarwood SA (2017) Urbanization erodes ectomycorrhizal fungal diversity and may cause microbial communities to converge. Nat Ecol Evol 1(5):0123

Fasusi OA, Cruz C, Babalola OO (2021) Agricultural sustainability: microbial biofertilizers in rhizosphere management. Agriculture 11(2):163

Fayomi G, Mini S, Fayomi O, Oyeleke O, Omole D, Akinwumi I (2019) Overview of industrial pollution activities and its curbing mechanisms. IOP conference series: Mat Sci Eng 640:012097

Feledyn-Szewczyk B, Kuś J, Stalenga J, Berbeć AK, Radzikowski P (2016) The role of biological diversity in agroecosystems and organic farming. InTech

Fischer H, Meyer A, Fischer K, Kuzyakov Y (2007) Carbohydrate and amino acid composition of dissolved organic matter leached from soil. Soil Biol Biochem 39(11):2926–2935

Fontaine S, Abbadie L, Aubert M, Barot S, Bloor JM, Derrien D, Duchene O, Gross N, Henneron L, Le Roux X (2024) Plant–soil synchrony in nutrient cycles: Learning from ecosystems to design sustainable agrosystems. Global Change Biology 30, e17034

Frater PN, Borer ET, Fay PA, Jin V, Knaeble B, Seabloom E, Sullivan L, Wedin DA, Harpole WS (2018) Nutrients and environment influence arbuscular mycorrhizal colonization both independently and interactively in Schizachyrium scoparium. Plant Soil 425:493–506

Gallagher LA, Manoil C (2001) *Pseudomonas aeruginosa* PAO1 kills *Caenorhabditis elegans* by cyanide poisoning. J Bacteriol 183(21):6207–6214

Ghadamgahi F, Tarighi S, Taheri P, Saripella GV, Anzalone A, Kalyandurg PB, Catara V, Ortiz R, Vetukuri RR (2022) Plant growth-promoting activity of *Pseudomonas aeruginosa* FG106 and its ability to act as a biocontrol agent against potato, tomato and taro pathogens. Biology 11(1):140

Ghezzehei TA, Sulman B, Arnold CL, Bogie NA, Berhe AA (2019) On the role of soil water retention characteristic on aerobic microbial respiration. Biogeosciences 16(6):1187–1209

Ghosh SK, Bera T, Chakrabarty AM (2020) Microbial siderophore–A boon to agricultural sciences. Biol Control 144:104214

Glassner H, Zchori-Fein E, Yaron S, Sessitsch A, Sauer U, Compant S (2018) Bacterial niches inside seeds of *Cucumis melo* L. Plant Soil 422:101–113

de Graaff M-A, Hornslein N, Throop HL, Kardol P, van Diepen LT (2019) Effects of agricultural intensification on soil biodiversity and implications for ecosystem functioning: a meta-analysis. Adv Agron 155:1–44

Griffiths B, Ritz K, Bardgett RD, Cook R, Christensen S, Ekelund F, Sørensen SJ, Bååth E, Bloem J, De Ruiter P (2000) Ecosystem response of pasture soil communities to fumigation-induced microbial diversity reductions: an examination of the biodiversity–ecosystem function relationship. Oikos 90(2):279–294

Grilli G, Urcelay C, Longo MS, Galetto L (2014) Mycorrhizal fungi affect plant growth: experimental evidence comparing native and invasive hosts in the context of forest fragmentation. Plant Ecol 215:1513–1525

Guilland C, Maron P-A, Damas O, Ranjard L (2018) Biodiversity of urban soils for sustainable cities. Environ Chem Lett 16:1267–1282

Gupta A, Singh UB, Sahu PK, Paul S, Kumar A, Malviya D, Singh S, Kuppusamy P, Singh P, Paul D (2022) Linking soil microbial diversity to modern agriculture practices: a review. Int J Environ Res Public Health 19(5):3141

Gupta A, Gupta R, Singh RL (2016) Microbes and environment. In Principles and applications of environmental biotechnology for a sustainable future, pp. 43–84. Springer

Gupta V, Germida JJ (2021) Microbial transformations of sulfur in soil. In Principles and applications of soil microbiology, pp. 489–522. Elsevier

Hagerty SB, Allison SD, Schimel JP (2018) Evaluating soil microbial carbon use efficiency explicitly as a function of cellular processes: implications for measurements and models. Biogeochemistry 140:269–283

Hallin S, Philippot L, Löffler FE, Sanford RA, Jones CM (2018) Genomics and ecology of novel N2O-reducing microorganisms. Trends Microbiol 26(1):43–55

Hanson RS, Hanson TE (1996) Methanotrophic bacteria. Microbiol Rev 60(2):439–471

Haque Z, Pandey K, Zamir S (2023) Bio-management of *Fusarium* wilt of tomato (*Fusarium oxysporum* f. sp. *lycopersici*) with multifacial Trichoderma species. Discov Agriculture 1(1):7

Harada LK, Silva EC, Campos WF, Del Fiol FS, Vila M, Dąbrowska K, Krylov VN, Balcão VM (2018) Biotechnological applications of bacteriophages: state of the art. Microbiol Res 212:38–58

Hartmann M, Six J (2023) Soil structure and microbiome functions in agroecosystems. Nat Rev Earth Environ 4(1):4–18

Hernlem BJ, Vane LM, Sayles GD (1996) Stability constants for complexes of the siderophore desferrioxamine B with selected heavy metal cations. Inorg Chim Acta 244(2):179–184

Hibbing ME, Fuqua C, Parsek MR, Peterson SB (2010) Bacterial competition: surviving and thriving in the microbial jungle. Nat Rev Microbiol 8(1):15–25

Hider RC, Kong X (2010) Chemistry and biology of siderophores. Nat Prod Rep 27(5):637–657

Hirsch PR, Mauchline TH (2015) The importance of the microbial N cycle in soil for crop plant nutrition. Adv Appl Microbiol 93:45–71

Ho A, Di Lonardo DP, Bodelier PL (2017) Revisiting life strategy concepts in environmental microbial ecology. FEMS Microbiol Ecol 93(3):fix006

Holden VI, Bachman MA (2015) Diverging roles of bacterial siderophores during infection. Metallomics 7(6):986–995

Hopwood DA (2007) Streptomyces in nature and medicine: the antibiotic makers. Oxford University Press

Howell C, Stipanovic R (1980) Suppression of Pythium ultimum-induced damping-off of cotton seedlings by Pseudomonas fluorescens and its antibiotic, pyoluteorin. Phytopathology 70(8):712–715

Igiehon NO, Babalola OO, Aremu BR (2019) Genomic insights into plant growth promoting rhizobia capable of enhancing soybean germination under drought stress. BMC Microbiol 19:1–22

Isbell F, Adler PR, Eisenhauer N, Fornara D, Kimmel K, Kremen C, Letourneau DK, Liebman M, Polley HW, Quijas S (2017) Benefits of increasing plant diversity in sustainable agroecosystems. J Ecol 105(4):871–879

Jackson LE, Burger M, Cavagnaro TR (2008) Roots, nitrogen transformations, and ecosystem services. Annu Rev Plant Biol 59(1):341–363

Jacobsen B, Zidack N, Larson B (2004) The role of bacillus-based biological control agents in integrated pest management systems: plant diseases. Phytopathology 94(11):1272–1275

Jain A, Singh S, Kumar Sarma B, Bahadur Singh H (2012) Microbial consortium–mediated reprogramming of defence network in pea to enhance tolerance against Sclerotinia sclerotiorum. J Appl Microbiol 112(3):537–550

Jehani M, Singh S, Kumar D, Kumar G (2023) Azospirillum—A free-living nitrogen-fixing bacterium. In Rhizobiome, pp. 285–308. Elsevier

Jiao N, Herndl GJ, Hansell DA, Benner R, Kattner G, Wilhelm SW, Kirchman DL, Weinbauer MG, Luo T, Chen F (2010) Microbial production of recalcitrant dissolved organic matter: long-term carbon storage in the global ocean. Nat Rev Microbiol 8(8):593–599

Jiao S, Chen W, Wang J, Du N, Li Q, Wei G (2018) Soil microbiomes with distinct assemblies through vertical soil profiles drive the cycling of multiple nutrients in reforested ecosystems. Microbiome 6:1–13

Johnson ET, Bowman MJ, Gomes RP, Carneiro LC, Dunlap CA (2023) Identification of 2, 4-diacetylphloroglucinol production in the genus *Chromobacterium*. Sci Rep 13(1):14292

Johnston-Monje D, Raizada MN (2011) Conservation and diversity of seed associated endophytes in Zea across boundaries of evolution, ethnography and ecology. PLoS One 6(6):e20396

Kallenbach CM, Frey SD, Grandy AS (2016) Direct evidence for microbial-derived soil organic matter formation and its ecophysiological controls. Nat Commun 7(1):13630

Kamle M, Borah R, Bora H, Jaiswal AK, Singh RK, Kumar P (2020) Systemic acquired resistance (SAR) and induced systemic resistance (ISR): role and mechanism of action against phytopathogens. Fungal Biotechnology and Bioengineering, 457–470

Kang SE, Sumabat LG, Melie T, Mangum B, Momany M, Brewer MT (2022) Evidence for the agricultural origin of resistance to multiple antimicrobials in *Aspergillus fumigatus*, a fungal pathogen of humans. G3 12(2):jkab427

Karuppiah V, Natarajan S, Gangatharan M, Aldayel MF, Alsowayeh N, Thangavel K (2022) Development of siderophore-based rhizobacterial consortium for the mitigation of biotic and abiotic environmental stresses in tomatoes: an in vitro and in planta approach. J Appl Microbiol 133(6):3276–3287

Kaushal M, Wani SP (2016) Plant-growth-promoting rhizobacteria: drought stress alleviators to ameliorate crop production in drylands. Ann Microbiol 66:35–42

Kielland K (1995) Landscape patterns of free amino acids in arctic tundra soils. Biogeochemistry 31:85–98

Kiesewetter KN, Afkhami ME (2021) Microbiome-mediated effects of habitat fragmentation on native plant performance. New Phytol 232(4):1823–1838

Kögel-Knabner I, Guggenberger G, Kleber M, Kandeler E, Kalbitz K, Scheu S, Eusterhues K, Leinweber P (2008) Organo-mineral associations in temperate soils: integrating biology, mineralogy, and organic matter chemistry. J Plant Nutr Soil Sci 171(1):61–82

Krause SM, Bertilsson S, Grossart H-P, Bodelier PL, van Bodegom PM, Lennon JT, Philippot L, Le Roux X (2022) Microbial trait-based approaches for agroecosystems. Adv Agron 175:259–299

Kruse J, Simon J, Rennenberg H (2013) Soil respiration and soil organic matter decomposition in response to climate change. Dev Environ Sci 13:131–149

Kumar CC, Padmanaban G (1980) Role of heme in the synthesis of cytochrome c oxidase in Neurospora crassa. J Biol Chem 255(23):11130–11134

Ladha J, Tirol-Padre A, Reddy C, Cassman K, Verma S, Powlson D, Van Kessel C, de Richter BD, Chakraborty D, Pathak H (2016) Global nitrogen budgets in cereals: a 50-year assessment for maize, rice and wheat production systems. Sci Rep 6(1):1–9

Lanteigne C, Gadkar VJ, Wallon T, Novinscak A, Filion M (2012) Production of DAPG and HCN by *Pseudomonas* sp. LBUM300 contributes to the biological control of bacterial canker of tomato. Phytopathology 102(10):967–973

Larkin RP, Brewer MT (2020) Effects of crop rotation and biocontrol amendments on rhizoctonia disease of potato and soil microbial communities. Agriculture 10(4):128

Larsen J, Jaramillo-López P, Nájera-Rincon M, González-Esquivel C (2015) Biotic interactions in the rhizosphere in relation to plant and soil nutrient dynamics. J Soil Sci Plant Nutr 15(2):449–463

Laville J, Blumer C, Von Schroetter C, Gaia V, Défago GV, Keel C, Haas D (1998) Characterization of the hcnABC gene cluster encoding hydrogen cyanide synthase and anaerobic regulation by ANR in the strictly aerobic biocontrol agent Pseudomonas fluorescens CHA0. J Bacteriol 180(12):3187–3196

Le Gall S, Bérard A, Page D, Lanoe L, Bertin N, Doussan C (2021) Increased exopolysaccharide production and microbial activity affect soil water retention and field performance of tomato under water deficit. Rhizosphere 19:100408

Le C, Kruijt M, Raaijmakers J (2012) Involvement of phenazines and lipopeptides in interactions between *Pseudomonas* species and *Sclerotium rolfsii*, causal agent of stem rot disease on groundnut. J Appl Microbiol 112(2):390–403

Ledermann R, Schulte CC, Poole PS (2021) How rhizobia adapt to the nodule environment. Journal of Bacteriology 203. https://doi.org/10.1128/jb.00539–20

Lesuffleur F, Paynel F, Bataillé M-P, Le Deunff E, Cliquet J-B (2007) Root amino acid exudation: measurement of high efflux rates of glycine and serine from six different plant species. Plant Soil 294:235–246

Li W, Rokni-Zadeh H, De Vleeschouwer M, Ghequire MG, Sinnaeve D, Xie G-L, Rozenski J, Madder A, Martins JC, De Mot R (2013) The antimicrobial compound xantholysin defines a new group of Pseudomonas cyclic lipopeptides. PLoS One 8(5):e62946

Li J, Wang C, Liang W, Liu S (2021) Rhizosphere microbiome: the emerging barrier in plant-pathogen interactions. Front Microbiol 12:772420

Li J, Wang J, Liu H, Macdonald CA, Singh BK (2023) Microbial inoculants with higher capacity to colonize soils improved wheat drought tolerance. Microb Biotechnol 16(11):2131–2144

Liang C, Schimel JP, Jastrow JD (2017) The importance of anabolism in microbial control over soil carbon storage. Nat Microbiol 2(8):1–6

Lindsey APJ, Murugan S, Renitta RE (2020) Microbial disease management in agriculture: current status and future prospects. Biocatal Agric Biotechnol 23:101468

Liu W, Yang Z, Ye Q, Peng Z, Zhu S, Chen H, Liu D, Li Y, Deng L, Shu X (2023) Positive effects of organic amendments on soil microbes and their functionality in agro-ecosystems. Plants 12(22):3790

Luambano ND, Narla RD, Wanjohi WJ, Kimenju JW, Kerry BR (2015) Integrated management of root-knot nematodes in a tomato-maize crop system using the biocontrol fungus Pochonia clamydosporia. Crop Prot 71:45–50

Luo G, Li L, Friman V-P, Guo J, Guo S, Shen Q, Ling N (2018) Organic amendments increase crop yields by improving microbe-mediated soil functioning of agroecosystems: a meta-analysis. Soil Biol Biochem 124:105–115

Lurthy T, Perot S, Gerin-Eveillard F, Rey M, Wisniewski-Dyé F, Vacheron J, Prigent-Combaret C (2023) Inhibition of broomrape germination by 2, 4-diacetylphloroglucinol produced by environmental *Pseudomonas*. Microb Biotechnol 16(12):2313–2325

Lykogianni M, Bempelou E, Karamaouna F, Aliferis KA (2021) Do pesticides promote or hinder sustainability in agriculture? The challenge of sustainable use of pesticides in modern agriculture. Sci Total Environ 795:148625

Ma B, Stirling E, Liu Y, Zhao K, Zhou J, Singh BK, Tang C, Dahlgren RA, Xu J (2021) Soil biogeochemical cycle couplings inferred from a function-taxon network. Research 2021:7102769

Mamabolo E, Makwela MM, Tsilo TJ (2020) Achieving sustainability and biodiversity conservation in agriculture: importance, challenges and prospects. Eur J Sustain Dev 9(3):616–616

Mandal A, Sarkar B, Mandal S, Vithanage M, Patra AK, Manna MC (2020) Chapter 7 - impact of agrochemicals on soil health. In: Prasad MNV (ed) Agrochemicals detection, treatment and remediation. Butterworth-Heinemann, pp 161–187. https://doi.org/10.1016/B978-0-08-103017-2.00007-6

McCullers JA (2014) The public health policy implications of understanding metabiosis. Cell Host Microbe 16(1):3–4

Meena M, Swapnil P, Divyanshu K, Kumar S, Harish, Tripathi YN, Zehra A, Marwal, A, Upadhyay RS (2020) PGPR-mediated induction of systemic resistance and physiochemical alterations in plants against the pathogens: Current perspectives. Journal of Basic Microbiology 60:828–861

Mendes CB, Prevedello JA (2020) Does habitat fragmentation affect landscape-level temperatures? A global analysis. Landscape Ecol 35(8):1743–1756

Mendes R, Kruijt M, De Bruijn I, Dekkers E, Van Der Voort M, Schneider JH, Piceno YM, DeSantis TZ, Andersen GL, Bakker PA (2011) Deciphering the rhizosphere microbiome for disease-suppressive bacteria. Science 332(6033):1097–1100

Mengist W, Soromessa T, Feyisa GL (2021) Landscape change effects on habitat quality in a forest biosphere reserve: implications for the conservation of native habitats. J Clean Prod 329:129778. https://doi.org/10.1016/j.jclepro.2021.129778

Meshram S, Bisht S, Gogoi R (2022) Current development, application and constraints of biopesticides in plant disease management. In: Biopesticides. Elsevier, pp 207–224

Messiha N, Van Diepeningen A, Farag NS, Abdallah S, Janse J, Van Bruggen A (2007) *Stenotrophomonas maltophilia*: a new potential biocontrol agent of *Ralstonia solanacearum*, causal agent of potato brown rot. Eur J Plant Pathol 118:211–225

Meyer SL, Halbrendt JM, Carta LK, Skantar AM, Liu T, Abdelnabby HM, Vinyard BT (2009) Toxicity of 2, 4-diacetylphloroglucinol (DAPG) to plant-parasitic and bacterial-feeding nematodes. J Nematol 41(4):274–280

Minchev Z, Kostenko O, Soler R, Pozo MJ (2021) Microbial consortia for effective biocontrol of root and foliar diseases in tomato. Front Plant Sci 12:756368

Molefe RR, Amoo AE, Babalola OO (2023) Communication between plant roots and the soil microbiome; involvement in plant growth and development. Symbiosis 90(3):231–239

Mukhtar H, Wunderlich RF, Muzaffar A, Ansari A, Shipin OV, Cao TN-D, Lin Y-P (2023) Soil microbiome feedback to climate change and options for mitigation. Sci Total Environ 882:163412

Nandi M, Macdonald J, Liu P, Weselowski B, Yuan ZC (2018) *Clavibacter michiganensis* ssp. *michiganensis*: bacterial canker of tomato, molecular interactions and disease management. Mol Plant Pathol 19(8):2036–2050

Naz R, Khushhal S, Asif T, Mubeen S, Saranraj P, Sayyed R (2022) Inhibition of bacterial and fungal phytopathogens through volatile organic compounds produced by Pseudomonas sp. In: Secondary metabolites and volatiles of PGPR in plant-growth promotion. Springer, pp 95–118

Nono-Womdim R, Swai I, Mrosso L, Chadha M, Opena R (2002) Identification of root-knot nematode species occurring on tomatoes in Tanzania and resistant lines for their control. Plant Dis 86(2):127–130

Olowe OM, Akanmu AO, Ayangbenro AS, Fadiji AE, Bitire TD, Odhiambo JJ, Kutu FR, Babalola OO (2023a) Trenchant microbiological-based approach for the control of Striga: current practices and future prospects. Front Sustain Food Syst 7:1073339

Olowe OM, Ayangbenro AS, Akanmu AO, Kutu FR, Odhiambo JJ, Babalola OO (2023b) Comparative insights into the microbial diversity and community structure of Striga hermonthica-infested maize rhizosphere. Appl Sci 13(5):3260

Ons L, Bylemans D, Thevissen K, Cammue BP (2020) Combining biocontrol agents with chemical fungicides for integrated plant fungal disease control. Microorganisms 8(12):1930

Or D, Smets BF, Wraith J, Dechesne A, Friedman S (2007) Physical constraints affecting bacterial habitats and activity in unsaturated porous media–a review. Adv Water Resour 30(6–7):1505–1527

Prasad K, Khare A, Rawat P (2022) Glomalin arbuscular mycorrhizal fungal reproduction, lifestyle and dynamic role in global sustainable agriculture for future generation. Fungal Reproduction and Growth, 1–22

Patyka V, Buletsa N, Pasichnyk L, Zhitkevich N, Kalinichenko A, Gnatiuk T, Butsenko L (2016) Specifics of pesticides effects on the phytopathogenic bacteria. Ecol Chem Eng S 23(2):311–331

Paulin MM, Novinscak A, St-Arnaud M, Goyer C, DeCoste NJ, Prive J-P, Owen J, Filion M (2009) Transcriptional activity of antifungal metabolite-encoding genes phlD and hcnBC in *Pseudomonas* spp. using qRT-PCR. FEMS Microbiol Ecol 68(2):212–222

Paulin MM, Novinscak A, Lanteigne C, Gadkar VJ, Filion M (2017) Interaction between 2, 4-diacetylphloroglucinol-and hydrogen cyanide-producing *Pseudomonas brassicacearum* LBUM300 and *Clavibacter michiganensis* subsp. *michiganensis* in the tomato rhizosphere. Appl Environ Microbiol 83(13):e00073–e00017

Peng D, Li S, Wang J, Chen C, Zhou M (2014) Integrated biological and chemical control of rice sheath blight by *Bacillus subtilis* NJ-18 and jinggangmycin. Pest Manag Sci 70(2):258–263

Peoples M, Brockwell J, Herridge D, Rochester I, Alves B, Urquiaga S, Boddey R, Dakora F, Bhattarai S, Maskey S (2009) The contributions of nitrogen-fixing crop legumes to the productivity of agricultural systems. Symbiosis 48:1–17

Petrikovics I, Budai M, Kovacs K, Thompson DE (2015) Past, present and future of cyanide antagonism research: from the early remedies to the current therapies. World J Methodol 5(2):88–100

Philippot L, Andert J, Jones CM, Bru D, Hallin S (2011) Importance of denitrifiers lacking the genes encoding the nitrous oxide reductase for N2O emissions from soil. Glob Chang Biol 17(3):1497–1504

Philippot L, Griffiths BS, Langenheder S (2021) Microbial community resilience across ecosystems and multiple disturbances. Microbiol Mol Biol Rev 85(2). https://doi.org/10.1128/mmbr.00026-00020

Pieterse CM, Zamioudis C, Berendsen RL, Weller DM, Van Wees SC, Bakker PA (2014) Induced systemic resistance by beneficial microbes. Annu Rev Phytopathol 52(1):347–375

Popp J, Pető K, Nagy J (2013) Pesticide productivity and food security. A review. Agron Sustain Dev 33:243–255

Prasannath K (2013) Pathogenicity and virulence factors of phytobacteria. Sch Acad J Biosci 1(1):24–33

Pusey P, Stockwell V, Reardon C, Smits T, Duffy B (2011) Antibiosis activity of *Pantoea agglomerans* biocontrol strain E325 against *Erwinia amylovora* on apple flower stigmas. Phytopathology 101(10):1234–1241

Qiu L, Zhang Q, Zhu H, Reich PB, Banerjee S, van der Heijden MG, Sadowsky MJ, Ishii S, Jia X, Shao M (2021) Erosion reduces soil microbial diversity, network complexity and multifunctionality. ISME J 15(8):2474–2489

Raaijmakers JM, Mazzola M (2012) Diversity and natural functions of antibiotics produced by beneficial and plant pathogenic bacteria. Annu Rev Phytopathol 50(1):403–424

Raj A, Jhariya MK, Banerjee A, Khan N, Meena RS, Oraon PR, Yadav SK (2021) Agroecosystem service management and environmental sustainability. Springer Singapore

Ramette A, Frapolli M, Fischer-Le Saux M, Gruffaz C, Meyer J-M, Défago G, Sutra L, Moënne-Loccoz Y (2011) *Pseudomonas protegens* sp. nov., widespread plant-protecting bacteria producing the biocontrol compounds 2, 4-diacetylphloroglucinol and pyoluteorin. Syst Appl Microbiol 34(3):180–188

Reisch L, Eberle U, Lorek S (2013) Sustainable food consumption: an overview of contemporary issues and policies. Sustain Sci Pract Policy 9(2):7–25

Rendon P, Steinhoff-Knopp B, Saggau P, Burkhard B (2020) Assessment of the relationships between agroecosystem condition and the ecosystem service soil erosion regulation in northern Germany. PLoS One 15(12):e0234288

Rezzonico F, Zala M, Keel C, Duffy B, Moënne-Loccoz Y, Défago G (2007) Is the ability of biocontrol fluorescent pseudomonads to produce the antifungal metabolite 2, 4-diacetylphloroglucinol really synonymous with higher plant protection? New Phytol 173(4):861–872

Rillig MC, Wright SF, Nichols KA, Schmidt WF, Torn MS (2001) Large contribution of arbuscular mycorrhizal fungi to soil carbon pools in tropical forest soils. Plant Soil 233:167–177

Rodgers PB, Knowles CJ (1978) Cyanide production and degradation during growth of *Chromobacterium violaceum*. Microbiology 108(2):261–267

Roth MG, Webster RW, Mueller DS, Chilvers MI, Faske TR, Mathew FM, Bradley CA, Damicone JP, Kabbage M, Smith DL (2020) Integrated management of important soybean pathogens of the United States in changing climate. J Integr Pest Manag 11(1):17

Rozpądek P, Nosek M, Domka A, Ważny R, Jędrzejczyk R, Tokarz K, Pilarska M, Niewiadomska E, Turnau K (2019) Acclimation of the photosynthetic apparatus and alterations in sugar metabolism in response to inoculation with endophytic fungi. Plant, Cell & Environment 42, 1408–1423

Rumpel C, Kögel-Knabner I (2011) Deep soil organic matter—a key but poorly understood component of terrestrial C cycle. Plant Soil 338:143–158

Rybicki J, Abrego N, Ovaskainen O (2020) Habitat fragmentation and species diversity in competitive communities. Ecol Lett 23(3):506–517

Saltzgaber-Müller J, Schatz G (1978) Heme is necessary for the accumulation and assembly of cytochrome c oxidase subunits in Saccharomyces cerevisiae. J Biol Chem 253(1):305–310

Samad A, Brader G, Pfaffenbichler N, Sessitsch A (2019) Plant-associated bacteria and the rhizosphere. In: Modern soil microbiology, 3rd edn. CRC Press, pp 163–178

Sandikar BM (2018) Study of hydrogen cyanide production as a mechanism of antifungal activity of fluorescent *Pseudomonas* species against phytopathogenic *Pythium* and *Fusarium* species. World J Pharm Res, 7(14), 891–897

Sayyed RZ, Uarrota VG (2022) Secondary metabolites and volatiles of PGPR in plant-growth promotion. Springer

Sehrawat A, Sindhu SS, Glick BR (2022) Hydrogen cyanide production by soil bacteria: biological control of pests and promotion of plant growth in sustainable agriculture. Pedosphere 32(1):15–38

Sellitto VM, Curto G, Dallavalle E, Ciancio A, Colagiero M, Pietrantonio L, Bireescu G, Stoleru V, Storari M (2016) Effect of Pochonia chlamydosporia-based formulates on the regulation of root-knot nematodes and plant growth response. Front Life Sci 9(3):177–181

Seneviratne G, Kulasooriya S (2013) Reinstating soil microbial diversity in agroecosystems: the need of the hour for sustainability and health. Agric Ecosyst Environ 164:181–182

Sharma R, Nehren U, Rahman SA, Meyer M, Rimal B, Aria Seta G, Baral H (2018) Modeling land use and land cover changes and their effects on biodiversity in Central Kalimantan, Indonesia. Land 7(2):57

Sherathia D, Dey R, Thomas M, Dalsania T, Savsani K, Pal K (2016) Biochemical and molecular characterization of DAPG-producing plant growth-promoting rhizobacteria (PGPR) of groundnut (Arachis hypogaea L.). Legume Res Int J 39(4):614–622

Shoghi-Kalkhoran S, Ghalavand A, Modarres-Sanavy S, Mokhtassi-Bidgoli A, Akbari P (2018) Integrated fertilization systems enhance quality an d yield of sunflower (Helianthus annuus L.). J Agric Sci Technol 15:1343–1352

Siddiqui IA, Shaukat SS, Sheikh IH, Khan A (2006) Role of cyanide production by *Pseudomonas fluorescens* CHA0 in the suppression of root-knot nematode, *Meloidogyne javanica* in tomato. World J Microbiol Biotechnol 22:641–650

Silva HSA, Romeiro R, Carrer Filho R, Pereira J, Mizubuti E, Mounteer A (2004) Induction of systemic resistance by *Bacillus cereus* against tomato foliar diseases under field conditions. J Phytopathol 152(6):371–375

Sims B, Kienzle J (2017) Sustainable agricultural mechanization for smallholders: what is it and how can we implement it? Agriculture 7(6):50

Singh V, Mawar R, Lodha S (2012) Combined effects of biocontrol agents and soil amendments on soil microbial populations, plant growth and incidence of charcoal rot of cowpea and wilt of cumin. Phytopathol Mediterr 51(2):307–316

Singh RP, Jha P, Jha PN (2015) The plant-growth-promoting bacterium *Klebsiella* sp. SBP-8 confers induced systemic tolerance in wheat (*Triticum aestivum*) under salt stress. J Plant Physiol 184:57–67

Smedsrud SL (2021) What made the world's largest "final backup" facility possible? An analysis of the policy processes leading up to the establishment of the Svalbard Global Seed Vault.

Master's thesis, Norwegian University of Life Sciences (NMBU), Department of International Environment and Development Studies
de Souza JT, Arnould C, Deulvot C, Lemanceau P, Gianinazzi-Pearson V, Raaijmakers JM (2003) Effect of 2, 4-diacetylphloroglucinol on Pythium: cellular responses and variation in sensitivity among propagules and species. Phytopathology 93(8):966–975
Souza RD, Ambrosini A, Passaglia LM (2015) Plant growth-promoting bacteria as inoculants in agricultural soils. Genet Mol Biol 38:401–419
Sparks TC (2013) Insecticide discovery: an evaluation and analysis. Pestic Biochem Physiol 107(1):8–17
Stepanov AA, Vasilchenko AV, Vasilchenko AS (2023) Subinhibitory effects of 2, 4-diacetylphloroglucinol on filamentous fungus aspergillus fumigatus. J Appl Microbiol 134(12):lxad294
Stokes A, Bocquého G, Carrère P, Salazar RC, Deconchat M, Garcia L, Gardarin A, Gary C, Gaucherel C, Gueye M (2023) Services provided by multifunctional agroecosystems: questions, obstacles and solutions. Ecol Eng 191:106949
Stutz E, Défago G, Kern H (1986) Naturally occurring fluorescent pseudomonads involved in suppression. Phytopathology 76(2):181–185
Sudan A, Ghosh A, Verma M, Tomar P (2024) Impact of water pollution & perspective techniques to mitigate it: an overview. In: Handbook of water pollution. Wiley, pp 29–64
Syed A, Elgorban AM, Bahkali AH, Eswaramoorthy R, Iqbal RK, Danish S (2023) Metal-tolerant and siderophore producing *Pseudomonas fluorescence* and *Trichoderma* spp. improved the growth, biochemical features and yield attributes of chickpea by lowering Cd uptake. Sci Rep 13(1):4471
Tonelli ML, Figueredo MS, Rodriguez J, Fabra A, Ibanez F (2020) Induced systemic resistance-like responses elicited by rhizobia. Plant and Soil 448, 1–14
Tamiru G, Muleta D (2018) The effect of rhizobia isolates against black root rot disease of Faba bean (*Vicia faba* L) caused by F*usarium solani*. Open Agric J 12(1):131–147
Tardy V, Mathieu O, Lévêque J, Terrat S, Chabbi A, Lemanceau P, Ranjard L, Maron PA (2014) Stability of soil microbial structure and activity depends on microbial diversity. Environ Microbiol Rep 6(2):173–183
Teplitski M, Warriner K, Bartz J, Schneider KR (2011) Untangling metabolic and communication networks: interactions of enterics with phytobacteria and their implications in produce safety. Trends Microbiol 19(3):121–127
Terra L, Ratcliffe N, Castro HC, Vicente AC, Dyson P (2021) Biotechnological potential of Streptomyces siderophores as new antibiotics. Curr Med Chem 28(7):1407–1421
Thrupp LA (2004) The importance of biodiversity in agroecosystems. J Crop Improv 12(1–2):315–337
Tian J, Dungait JAJ, Hou R, Deng Y, Hartley IP, Yang Y, Kuzyakov Y, Zhang F, Cotrufo MF, Zhou J (2024) Microbially mediated mechanisms underlie soil carbon accrual by conservation agriculture under decade-long warming. Nat Commun 15(1):377. https://doi.org/10.1038/s41467-023-44647-4
Trivedi P, Batista BD, Bazany KE, Singh BK (2022) Plant–microbiome interactions under a changing world: responses, consequences and perspectives. New Phytol 234(6):1951–1959
Ullah A, Bano A, Khan N (2021) Climate change and salinity effects on crops and chemical communication between plants and plant growth-promoting microorganisms under stress. Front Sustain Food Syst 5:618092
Umesha S (2006) Occurrence of bacterial canker in tomato fields of Karnataka and effect of biological seed treatment on disease incidence. Crop Prot 25(4):375–381
Van Loon L, Bakker P, Pieterse C (1998) Systemic resistance induced by rhizosphere bacteria. Annu Rev Phytopathol 36(1):453–483
Vannette RL, Leopold DR, Fukami T (2016) Forest area and connectivity influence root-associated fungal communities in a fragmented landscape. Ecology 97(9):2374–2383

Verma V, Singh S, Prakash S (2011) Bio-control and plant growth promotion potential of siderophore producing endophytic *Streptomyces* from *Azadirachta indica* A. Juss. J Basic Microbiol 51(5):550–556

Vibha Bhardwaj VB, Neelam Garg NG (2012) Importance of exploration of microbial biodiversity. Int Res J Biol Sci 1(3):78–83

Vicena J, Ardestani MM, Baldrian P, Frouz J (2022) The effect of microbial diversity and biomass on microbial respiration in two soils along the soil chronosequence. Microorganisms 10(10):1920

Vijay K, Shibasini M, Sivasakthivelan P, Kavitha T (2023) Microbial siderophores as molecular shuttles for metal cations: sources, sinks and application perspectives. Arch Microbiol 205(9):322

Voisard C, Keel C, Haas D, Dèfago G (1989) Cyanide production by *Pseudomonas fluorescens* helps suppress black root rot of tobacco under gnotobiotic conditions. EMBO J 8(2):351–358

Vu E, Schaumann GE, Buchmann C (2022) The contribution of microbial activity to soil–water interactions and soil microstructural stability of a silty loam soil under moisture dynamics. Geoderma 417:115822

Wade W, Beuchat L (2003) Metabiosis of proteolytic moulds and Salmonella in raw, ripe tomatoes. J Appl Microbiol 95(3):437–450

Wang Z, Meng Y, Zhu-Barker X, He X, Horwath W R, Luo H, Zhao Y, Jiang X (2019) Responses of nitrification and ammonia oxidizers to a range of background and adjusted pH in purple soils. Geoderma 334, 9–14

Wagg C, Bender SF, Widmer F, Van Der Heijden MG (2014) Soil biodiversity and soil community composition determine ecosystem multifunctionality. Proc Natl Acad Sci 111(14):5266–5270

Wall DH, Nielsen UN, Six J (2015) Soil biodiversity and human health. Nature 528(7580):69–76

Weiman S (2015) Microbes help to drive global carbon cycling and climate change. Microbe Mag 10(6):233–238

Wilhelm E, Arthofer W, Schafleitner R, Krebs B (1998) *Bacillus subtilis* an endophyte of chestnut (*Castanea sativa*) as antagonist against chestnut blight (*Cryphonectria parasitica*). Plant Cell Tissue Organ Cult 52(1):105–108

Winkelmann G (2007) Ecology of siderophores with special reference to the fungi. Biometals 20:379–392

Wissing F (1968) Growth curves and pH-optima for cyanide producing bacteria. Physiol Plant 21(3):589–593

Wissing F (1974) Cyanide formation from oxidation of glycine by a Pseudomonas species. J Bacteriol 117(3):1289–1294

Wittebolle L, Marzorati M, Clement L, Balloi A, Daffonchio D, Heylen K, De Vos P, Verstraete W, Boon N (2009) Initial community evenness favours functionality under selective stress. Nature 458(7238):623–626

Woo SL, Ruocco M, Vinale F, Nigro M, Marra R, Lombardi N, Pascale A, Lanzuise S, Manganiello G, Lorito M (2014) Trichoderma-based products and their widespread use in agriculture. Open Mycol J 8(1):71–126

Wright SF, Upadhyaya A (1998) A survey of soils for aggregate stability and glomalin, a glycoprotein produced by hyphae of arbuscular mycorrhizal fungi. Plant Soil 198:97–107

Xiong C, Lu Y (2022) Microbiomes in agroecosystem: diversity, function and assembly mechanisms. Environ Microbiol Rep 14(6):833–849

Yadav AN, Kour D, Kaur T, Devi R, Yadav A, Dikilitas M, Abdel-Azeem AM, Ahluwalia AS, Saxena AK (2021) Biodiversity, and biotechnological contribution of beneficial soil microbiomes for nutrient cycling, plant growth improvement and nutrient uptake. Biocatal Agric Biotechnol 33:102009

Yadav G, Sharma N, Dabral S, Vaishnav A, Varma A, Kothari S, Choudhary D (2023) Functional insight of siderophore in reducing cadmium stress and inducing growth promotion in Solanum melongena. S Afr J Bot 158:479–494

Yang X, You L, Hu H, Chen Y (2022) Conversion of grassland to cropland altered soil nitrogen-related microbial communities at large scales. Sci Total Environ 816:151645

Yang D, Zhang X, Li Z, Chu R, Shah S, Wang X, Zhang X (2024) Antagonistic effect of Bacillus and Pseudomonas combinations against Fusarium oxysporum and their effect on disease resistance and growth promotion in watermelon. J Appl Microbiol 135(5):lxae074

Yuan J, Li B, Zhang N, Waseem R, Shen Q, Huang Q (2012) Production of bacillomycin-and macrolactin-type antibiotics by *Bacillus amyloliquefaciens* NJN-6 for suppressing soilborne plant pathogens. J Agric Food Chem 60(12):2976–2981

Zermane N, Souissi T, Kroschel J, Sikora R (2007) Biocontrol of broomrape (*Orobanche crenata* Forsk. and *Orobanche foetida* Poir.) by *Pseudomonas fluorescens* isolate Bf7-9 from the faba bean rhizosphere. Biocontrol Sci Tech 17(5):483–497

Zhang X, Ward BB, Sigman DM (2020) Global nitrogen cycle: critical enzymes, organisms, and processes for nitrogen budgets and dynamics. Chemical reviews 120, 5308–5351

Zhang L, Zuo Q, Cai H, Li S, Shen Z, Song T (2024) Fungicides reduce soil microbial diversity, network stability and complexity in wheat fields with different disease resistance. Appl Soil Ecol 201:105513

Zhou Z, Tran PQ, Cowley ES, Trembath-Reichert E, Anantharaman K (2025) Diversity and ecology of microbial sulfur metabolism. Nature Reviews Microbiology 23, 122–140

Chapter 16
Regulatory Frameworks and Safety Considerations for Microbial Applications

Ajayi Abraham and Smith Stella Ifeanyi

Abstract Microbes, amenable to genetic manipulation and capable of rapid growth and multiplication, have found applications in industry, health, agriculture, and the environment. With the emergence of modern biotechnology tools, the exploration of microbes for new applications is indisputable. However, safety concerns abound necessitating regulatory frameworks to guide the manipulation of microbes, their use in manufacturing, processing, treatment and storage of products. These frameworks also extended to metabolites or byproducts of microbes, such as enzymes, proteins, and toxins. Several countries and regions around the world have regulatory frameworks that evaluate, authorize and certify the application of microbes or their products for various purposes. The regulatory frameworks of most countries, such as those covering food, are broadly aligned with the recommendation of the global legislative framework, the Codex Alimentarius. As biotechnology evolves, there is a need for updated regulatory frameworks and circumspection in drafting laws to avoid regulations that could impede competitiveness and innovation.

Keywords Codex Alimentarius Comission · Biotechnology · Framework · Government · Genetically modified microorganisms

16.1 Introduction

Microbes are the most diverse of all living things on the planet earth and can be found virtually everywhere in the biosphere. Owing to their versatility, they thrive in normal and extremes of temperatures, pH, solar radiation and salinity (Maraz and Khan 2021; Thakura et al. 2022). Beyond their ubiquity, microbes are involved in a

A. Abraham · S. S. Ifeanyi (✉)
Molecular Biology and Biotechnology Department, Nigerian Institute of Medical Research
Yaba, Lagos, Nigeria

O. O. Babalola et al. (eds.), *Ecofriendly Frontiers*,
https://doi.org/10.1007/978-3-031-98700-7_16

plethora of beneficial activities that support our survival on earth. Similarly, they also create difficult situations when they cause diseases in humans, animals, and plants (Rappuoli et al. 2023).

Microorganisms have found applications in medicine, environment, agriculture, and industry (Vitorino and Bessa 2017). Very important feature of microbes that has been explored in their application is their amenability to genetic manipulation, rapid growth and multiplication (Frisk 2011). Thanks to modern and ever evolving biotechnological tools such as recombinant DNA technique, high-throughput sequencing and gene editing using the CRISPR-Cas9 (Zhao et al. 2020) new frontiers in the application of microbes are emerging. However, to fully maximize the potentials of the application of microbes in bettering the lives of animals, humans, plants and the environment, there must be defined scopes (boundaries) regarding safety, extent to which microbes can be manipulated and ethics. Several known risks and unintended consequences exist in the application of microbes in processes including their use in human therapy, in the production of food, fuel, feeds and chemicals or in the remediation of the environment (OECD 2015; Anomaly 2020; Xiong 2023). These risks and possible adverse consequences elicit ethical and safety concerns in various strata of society ranging from regulatory bodies to end-users.

In this chapter we will interrogate the various safety and ethical considerations for the application of microbes and discuss regulatory frameworks that should guide such considerations.

16.2 Application of Microbes and Safety Considerations

The application of microbes by man predates modern times. Evidence revealed that as far back as 7000 BC and 3000 BC the Babylonians and Egyptians respectively have applied microbes in the production of beer and cheese (Damerow 2012; Todaro et al. 2013). From this humble beginning, the application of microbes in various industries has seen tremendous expansion across the frontiers of agriculture, biofuel, bioelectronics, cosmetics, food, health, water, etc. (Fig. 16.1). Application of microbes in these industries have helped improve nutrition, treatment options for difficult health conditions, alternatives to fossil fuel as sources of energy, disease control through production of vaccines and many more. Although, there lies enormous benefits in the application of microbes, safety concerns abound in every application.

Beyond the threats and negative impacts associated with microbes as it relates to human and animal health, their usefulness in enhancing health and well-being has gained considerable traction over the years. Several areas/products not limited to prebiotics, probiotics, vaccines, antibiotics, and fermented foods have been largely explored in enhancing human health (Pathak et al. 2021). While emerging use of genetically modified microorganisms (GMMs) in the field of immunology, oncology and infectious diseases have shown huge promise. GMMs have demonstrated success in use as vehicles for the delivery of therapeutic genes and specific

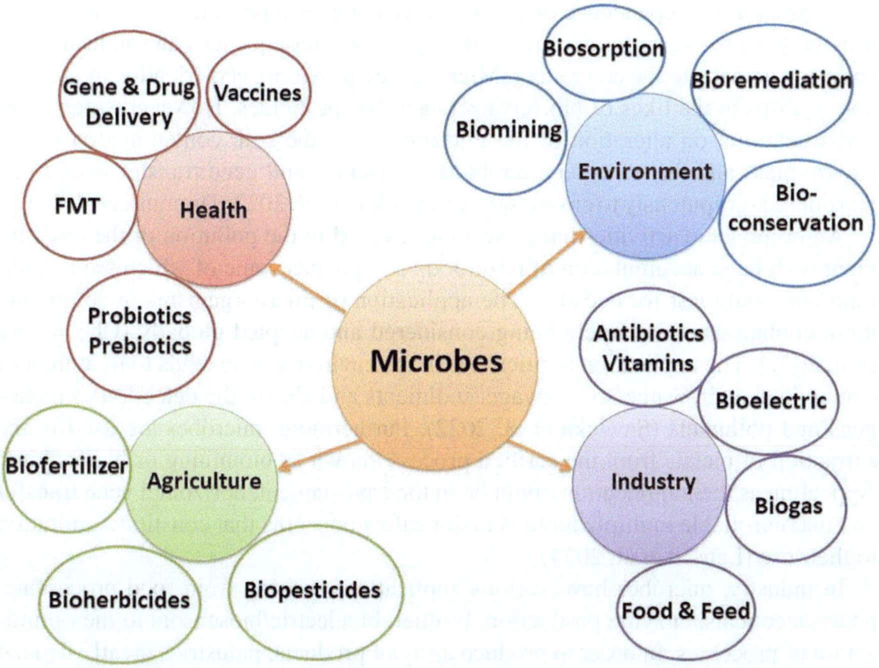

Fig. 16.1 Various applications of microbes in agriculture, environment, health, and industry.

molecules that triggers therapeutic effects in patients (Lensch et al. 2024). Similarly, leveraging on the beneficial bacteria present in the stool of healthy individuals, fecal microbiota transplantation (FMT) is used or considered to be used in the treatment/management of gastrointestinal disorders such as recurrent *Clostridioides difficile* infection (rCDI), inflammatory bowel disease (IBD), functional gastrointestinal disorders, obesity and metabolic syndrome (Gupta et al. 2016; Kim and Gluck 2019).

These itemized applications of microbes in enhancing health and well-being are not without safety concerns. Several safety concerns including resultant invasive infection, interference with drug metabolism, transfer of antimicrobial resistance genes, alteration of the gut microbiome, long-term colonization and possible physiological impacts on the host have been highlighted in the use of prebiotics and probiotics (Doron and Snydman 2015; Sherid et al. 2016; Merenstein et al. 2023). Although vaccines are useful and helpful, their microbial composition often raises safety concerns such as causing diseases they ought to protect against, lack of efficacy, allergy, and possible release of vaccine strains into the environment (Minor 2015; Khatoon et al. 2016). In the case of FMT there exist short- and long- term safety concerns which include immediate post FMT reactions like diarrhea, belching, abdominal cramping, fever, transfer of infectious diseases and activation of chronic diseases such as atherosclerosis, obesity, etc.

In agriculture, dependence on chemical fertilizers and pesticides in crop production has been the norm for decades with unpleasant consequences on soil health and enduring damage to the ecosystem. Microbes are providing eco-friendly and renewable options in the likes of bio-fertilizers and bio-pesticides. However, safety concerns boarding on alteration of the microbiome of the soil, contamination of the environment and farm workers, antibiotic resistance and gene transfer need to be considered scrupulously to ensure safety (Vassileva et al. 2022; Timmusk et al. 2017).

Anthropogenic activities have overtime resulted in the pollution of the environment with huge accumulation of hazardous compounds some of which are recalcitrant and could last for decades. The application of microorganisms in combating these contaminants is largely being considered and adapted globally (Chowdhary et al. 2022). The application of microbes in the environment extends to treatment of contaminated air, water, soil, sewage, sediments and also in the detection of pathogens and pollutants (Sawicka et al. 2022). Furthermore, microbes are used in the extraction of metals from the earth, a process known as biomining or bioleaching. Appealing as their application might be in the environment, horizontal gene transfer and uncontrollable multiplication is major safety concerns that constitute limitation to their use (Lensch et al. 2024).

In industry, microbes have various applications ranging from food processing, pharmaceuticals, enzyme production, biofuel, bioelectric/biosensors to the optimization of processes. In order to produce array of products, industries greatly depend on fermentation processes that are modulated by microbes. Some of these products such as beverages, cereals and products of vegetables, meat and dairy are often left with various proportions of the microbe, their enzymes or metabolic by products (Sharma et al. 2020; Kwang 2022). The health risk these could pose constitutes safety concern. For instance, β-amylase (4 α-D-glucan maltohydrolase; EC 3.2.1.2) derived from a genetically modified *Bacillus licheniformis* strain NZYM-JA which was considered a food enzyme was evaluated by a European Food Safety Authority (EFSA) panel on food contact materials, enzymes, flavorings and processing acids for safety concerns boarding on genotoxicity, systemic toxicity, mutagenicity and allergenicity before approval for use (ESFA et al. 2017).

16.3 Regulatory Frameworks in the Application of Microbes

To enable the efficient and safe application of microbes in any of the above discussed sectors, comprehensive and integrated regulatory frameworks are essential. Having robust and implementable frameworks in place can address known and in some cases; unforeseeable safety concerns. Such framework should clearly capture metadata of the microbe to be applied, for what purpose, genetic modification made and by what approach (non-targeted mutagenesis, genetic engineering, bioengineering, genome editing etc.), safety concerns, technical and laboratory analysis including trial experiments, labeling of products or formulations, etc. There seem not be a harmonized regulatory framework that cuts across every application of microbes for factors such as culture, geography, political priorities, implementation strategies, etc., that influence the formulation of regulatory frameworks (Hanlon and Sewalt 2021; Mukherjee et al. 2022). The lack of explicit and acceptable regulatory framework can hamper the judicious application of microbes. For example, a recent judgment by the European Union (EU) court of justice on gene editing technologies in manipulating microorganisms for use as genetically modified microorganisms (GMMs) is perceived to have a non-promising prospect for regulatory frameworks that will support the further application of GMMs (Wesseler et al. 2022).

There are diverse existing regional and country-wise regulatory frameworks (Table 16.1) that guide the application of microbes in industry, health, agriculture and the environment around the world which are either process-oriented or product-oriented (Rozas et al. 2022). Although they are diverse, the approach in formulating these frameworks is somewhat similar and some countries have come up with harmonized coordinated framework that enables proper oversight by regulatory agencies. For instance, in 2015, the United States government mandated primary regulatory agencies of biotechnology products to update the country's existing coordinated framework with the view to boosting the confidence of the public in the regulatory system and eliminate obstacles to competitiveness and future innovation (EPA, FDA and USDA 2017).

The EU have regulatory frameworks such as EU 2019/1009, EC 528/2012, EU 1107/2009, EU 2015/2283 among others that regulate the use of bacteria as biostimulants/biofertilizers, microbial biocides, use of microbial biological control in agriculture and novel foods/food ingredients consisting of microorganisms respectively. Similarly, other countries have regulatory frameworks that regulate the application of microbes in various fields as shown in Table 16.1. Most countries also draft regulatory framework in broad agreement with legislative frameworks of global organizations like the WHO and the United Nations Food and Agriculture Organization (FAO). For example, regulatory frameworks that cover foods of most countries are modeled in agreement with the recommendation of the Codex Alimentarius Commission (CAC) an appendage of the FAO and WHO (Mukherjee et al. 2022).

16.4 Conclusion

Microbes remain veritable tools needed in mitigating the myriads of challenges facing mankind. With the ever-growing world population and the climate change quagmire, the application of microbes in industry, health, agriculture and the environment cannot be relegated to the background. The need for robust and progressive regulatory frameworks is imperative to help drive the safe and effective application of microbes. Government and regulatory agencies or bodies must ensure that laws that birth regulatory frameworks are not inimical to competitiveness and innovation but are stringent enough to guaranty safety.

Table 16.1 Some regulatory frameworks currently being applied by regions and countries around the world

Country/ continent	Regulatory framework	Primary regulatory agency	Resource/product coverage	Reference
European Union	European Regulation (EU) 2019/1009		Use of bacteria as biostimulants and biofertilizers	Barros-Rodríguez et al. (2020)
	EU biocide regulation-regulation EC 528/2012		Authorizes the use of microbial biocides	Sundh and Eilenberg (2021)
	Regulation 1107/2009		Assessment of environmental safety risks for the use of microbial biological control agents (MBCAs) in agriculture	Wesseler et al. (2022)
	Regulation EU 2015/2283		Novel foods and food ingredients consisting of microorganisms	Laulund et al. (2017)
	Regulation (EC) 1331/2008		Establish regulation for the safety assessment and authorization procedure of food additives, food enzymes and food flavourings	EFSA et al. (2017)
United States of America	Coordinated regulatory framework for the regulation of biotechnology	U.S. Department of Agriculture (USDA), Federal Drug Administration (FDA), Environmental Protection Agency (EPA)	Regulation of organisms intended for agricultural use or other environmental release	https://usbiotechnologyregulation.mrp.usda.gov/eo14081-8b-plain-language.pdf
Peru	Law 27104, 29888, 29811, 31111		Regulation of GMOs (Regulation covers introduction, commercialization, research, transportation, storage, labeling, etc.)	Rozas et al. (2022)
Nigeria	National Biosafety Management Agency Act, 2015 (amended in 2019)	National Biosafety Management Agency	Application of modern biotechnology and handling of GMOs	Onwude et al. (2023)

Global	Codex Alimentarius	Food and Agriculture Organization (FAO)/World Health Organization (WHO)	Harmonize food and commodity standards. Provide guidelines that contribute to food quality and safety	Kocabaş and Grumet (2019)
Russia-Kazakhstan-Belarus	Technical Regulation (TRTS 033/2013	Customs Union	Regulation of safety of milk and dairy products, including the use of food culture	Laulund et al. (2017)
Argentina	Codigo Alimentario Argentino (Regulatory Decree 2126/71)	National Food Commission, Ministry of Health, Government of Argentina and the Ministry of Agriculture, Livestock and Fisheries	Regulate fermented, locally produced and imported foods	Mukherjee et al. (2022)
India	Essential Commodity Act (1955)	Ministry of Agriculture	Regulate biofertilizer implementation	Kumari et al. (2023)

References

Anomaly J (2020) The future of phage: ethical challenges of using phage therapy to treat bacterial infections. Public Health Ethics 13(1):82–88. https://doi.org/10.1093/phe/phaa003

Barros-Rodríguez A, Rangseekaew P, Lasudee K, Pathom-aree W, Manzanera M (2020) Regulatory risks associated with bacteria as biostimulants and biofertilizers in the frame of the European Regulation (EU) 2019/1009. Sci Total Environ 140239. https://doi.org/10.1016/j.scitotenv.2020.1402

Chowdhary P, Mani S, Shukla P, Raj A (2022) Microbes and environment. In: Chowdhary P, Mani S, Chaturvedi P (eds) Microbial biotechnology. Wiley, pp 1–28. https://doi.org/10.1002/9781119834489.ch1

Damerow P (2012) Sumerian beer: the origins of brewing technology in Ancient Mesopotamia. Cuneiform Digit Libr J 2012:2. https://cdli.mpiwg-berlin.mpg.de/articles/cdlj/2012-2

Doron S, Snydman DR (2015) Risk and safety of probiotics. Clin Infect Dis 60(S2):S129–S134

EFSA Panel on Food Contact Materials, Enzymes, Flavourings and Processing Aids (EFSA CEF Panel), Silano V, Bolognesi C, Castle L, Cravedi J-P, Fowler P, Franz R, Grob K, Gurtler R, Husøy T, Karenlampi S, Mennes W, Milana MR, Penninks A, Smith A, Tavares Pocas MF, Tlustos C, Wolfle D, Zorn H, Zugravu C-A, Chesson A, Glandorf B, Herman L, Jany K-D, Marcon F, Zeljezic D, Aguilera-Gomez M, Andryszkiewicz M, Diveki Z, Kovalkovicova N, Engel K-H (2017) Scientific opinion on the safety evaluation of the food enzyme b-amylase from genetically modified *Bacillus licheniformis* strain NZYM-JA. EFSA J 15(8):4896–4813. https://doi.org/10.2903/j.efsa.2017.4896

EPA, FDA USDA (2017) About the coordinated framework. https://usbiotechnologyregulation.mrp.usda.gov/sites/default/files/2017-coordinated-framework-update.pdf

Erkmen O, Bozoglu TF (2016) History and development of food microbiology. In: Food microbiology: principles into practice. Wiley, pp 3–12. https://doi.org/10.1002/9781119237860.ch1

Frisk ML (2011) Multifaceted microbes. Trends Biotechnol 29(2):51–52. https://doi.org/10.1016/j.tibtech.2010.12.005

Gupta S, Allen-Vercoe E, Petrof EO (2016) Fecal microbiota transplantation: in perspective. Ther Adv Gastroenterol 9(2):229–239. https://doi.org/10.1177/1756283X15607414

Hanlon P, Sewalt V (2021) GEMs: genetically engineered microorganisms and the regulatory oversight of their uses in modern food production. Crit Rev Food Sci Nutr 61(6):959–970. https://doi.org/10.1080/10408398.2020.1749026

Zhao J, Fang H, Zhang D (2020) Expanding application of CRISPR-Cas9 system in microorganisms. Synth Syst Biotechnol 5:269–276. https://doi.org/10.1016/j.synbio.2020.08.001

Khatoon S, Seida AA, Mahmood Z (2016) Bacterial vaccine: potential and safety concerns. EC Microbiol 3(2):438–439

Kim KO, Gluck M (2019) Fecal microbiota transplantation: an update on clinical practice. Clin Endosc 52(2):137–143. https://doi.org/10.5946/ce.2019.009

Kocabaş DS, Grumet R (2019) Evolving regulatory policies regarding food enzymes produced by recombinant microorganisms. GM Crops Food 10:191–207. https://doi.org/10.1080/21645698.2019.1649531

Kumari M, Swarupa P, Kesari KK, Kumar A (2023) Microbial inoculants as plant biostimulants: a review on risk status. Life 13:12. https://doi.org/10.3390/life13010012

Kwang J (2022) Importance of microbes in various industrial processes. Clin Microbiol 11:300. https://doi.org/10.35248/2327-5073.22.11.300

Laulund S, Wind A, Derkx PMF, Zuliani V (2017) Regulatory and safety requirements for food cultures. Microorganisms 5(2):28. https://doi.org/10.3390/microorganisms5020028

Lensch A, Lindfors HA, Duwenig E, Fleischmann T, Hjort C, Kärenlampi SO, McMurtry L, Melton E-D, Andersen MR, Skinner R, Wyss M, van Kranenburg R (2024) Safety aspects of microorganisms deliberately released into the environment. EFB Bioecon J 4:100061. https://doi.org/10.1016/j.bioeco.2023.100061

Maraz KM, Khan RA (2021) An overview on impact and application of microorganisms on human health, medicine and environment. GSC Biol Pharma Sci 14(01):089–104. https://doi.org/10.30574/gscbps.2021.16.1.0200

Merenstein D, Pot B, Leyer G, Ouwehand AC, Preidis GA, Elkins CA, Hill C, Lewis ZT, Shane AL, Zmora N, Petrova MI, Collado MC, Morelli L, Montoya GA, Szajewska H, Tancredi DJ, Sanders ME (2023) Emerging issues in probiotic safety: 2023 perspectives. Gut Microbes 15(1):2185034. https://doi.org/10.1080/19490976.2023.2185034

Minor PD (2015) Live attenuated vaccines: historical successes and current challenges. Virology 479–480:379–392. https://doi.org/10.1016/j.virol.2015.03.032

Mukherjee A, Gómez-Sala B, O'Connor EM, Kenny JG, Cotter PD (2022) Global regulatory frameworks for fermented foods: a review. Front Nutr 9:902642. https://doi.org/10.3389/fnut.2022.902642

OECD (2015) Biosafety and the environmental uses of micro-organisms: conference proceedings. OECD Publishing. https://doi.org/10.1787/9789264213562-en

Onwude J, Nwosu O, Ebegba R (2023) Biosecurity in Nigeria: emergence of the national biosecurity policy and action plan 2022–2026. GET J Biosecur One Health 2(1):19–22. https://doi.org/10.36108/GJOBOH/3202.20.0110

Pathak K, Saikia R, Gogoi U, Das A (2021) Potential application of microbes for human health and welfare. World J Pharma Pharma Sci 10(2):514–524. https://doi.org/10.20959/wjpps20212-18158

Rappuoli R, Young P, Ron E, Pecetta S, Pizza M (2023) Save the microbes to save the planet. A call to action of the International Union of the Microbiological Societies (IUMS). One Health Outlook 5:5. https://doi.org/10.1186/s42522-023-00077-2

Rozas P, Kessi-Perez EI, Martinez C (2022) Genetically modified organisms: adapting regulatory frameworks for evolving genome editing technologies. Biol Res 55:31. https://doi.org/10.1186/s40659-022-00399-x

Sawicka B, Pszczółkowski P, Barbaś P, Skiba D, Bienia B (2022) The role and importance of microorganisms in environmental sustainability. In: Chowdhary P, Mani S, Chaturvedi P (eds) Microbial biotechnology. Wiley, pp 107–133. https://doi.org/10.1002/9781119834489.ch7

Sharma R, Garg P, Kumar P, Bhatia SK, Kulshrestha S (2020) Microbial fermentation and its role in quality improvement of fermented foods. Fermentation 6:106. https://doi.org/10.3390/fermentation6040106

Sherid M, Samo S, Sulaiman S, Husein H, Sifuentes H, Sridhar S (2016) Liver abscess and bacteremia caused by lactobacillus: role of probiotics? Case report and review of the literature. BMC Gastroenterol 16:138. https://doi.org/10.1186/s12876-016-0552-y

Sundh I, Eilenberg J (2021) Why has the authorization of microbial biological control agents been slower in the EU than in comparable jurisdictions? Pest Manage Sci 77:2170–2178. https://doi.org/10.1002/ps.6177

Thakura N, Singh SP, Zhan C (2022) Microorganisms under extreme environments and their applications. Curr Res Microb Sci 3:100141. https://doi.org/10.1016/j.crmicr.2022.100141

Timmusk S, Behers L, Muthoni J, Muraya A, Aronsson A-C (2017) Perspectives and challenges of microbial application for crop improvement. Front Plant Sci 8:49. https://doi.org/10.3389/fpls.2017.00049

Todaro A, Adly FA, Omar OAH (2013) Traditional Egyptian Kariesh Cheese: a review. Food Sci Technol 1(1):1–6. https://doi.org/10.13189/fst.2013.010101

Vassileva M, Mocali S, Canfora L, Malusá E, García del Moral LF, Martos V, Flor-Peregrin E, Vassilev N (2022) Safety level of microorganism-bearing products applied in soil-plant systems. Front Plant Sci 13:862875. https://doi.org/10.3389/fpls.2022.862875

Vitorino LC, Bessa LA (2017) Technological microbiology: development and applications. Front Microbiol 8:827. https://doi.org/10.3389/fmicb.2017.00827

Wesseler J, Gijs K, Marthe M, Kai PP (2022) EU regulation of genetically modified microorganisms in light of new policy developments: possible implications for EU bioeconomy investments. Appl Econ Perspect Policy 1–21. https://doi.org/10.1002/aepp.13259

Xiong Y (2023) The role of microbial adaptation in biotechnology and its ethical considerations. Adv J Microbiol Res 17(3):001–002

Chapter 17
Ecofriendly Frontiers: The Concluding Section

Olubukola Oluranti Babalola, Olukayode Oladipo Amund, and Akinlolu Olalekan Akanmu

In this book, we explored the dynamic and vast role of microorganisms in reshaping agriculture and food systems. The transition from the conventional farming system that requires high chemical input into microbial-driven solutions that are characterized by sustainable agricultural systems represents a paradigm shift that has profound implications for environmental health, food security, and global sustainability. The microbial application offers a robust science-based pathway solution to the challenges of climate change, soil degradation, and food safety, including pest and disease resistance. We explored a range of products from plant growth-promoting microbes (PGPMs), biopesticides, microbial biostimulants, and fermentation technologies, exhibiting the roles of microbial strategies in redefining the frontiers of eco-friendly agricultural practices.

The significance of microbial applications extends beyond academic discourse into real-world impact. One of the major uses of the beneficial microbes is their application in soil health management where they enhance nutrient cycling, and reduce the huge dependence on synthetic fertilizers while fostering resilient cropping systems. Microbes as biofertilizers and biopesticides are environmentally safe alternatives to chemical pesticides that manage invading pests and diseases while

O. O. Babalola (✉)
Department of Life Sciences, Imperial College London, Ascot, Berkshire, UK

Food Security and Safety Focus Area, Faculty of Natural and Agricultural Sciences, North-West University, Mmabatho, South Africa
e-mail: olubukola.babalola@nwu.ac.za

O. O. Amund
Department of Microbiology, Faculty of Science, University of Lagos, Akoka, Lagos State, Nigeria

A. O. Akanmu
Food Security and Safety Focus Area, Faculty of Natural and Agricultural Sciences, North-West University, Mmabatho, South Africa

O. O. Babalola et al. (eds.), *Ecofriendly Frontiers*,
https://doi.org/10.1007/978-3-031-98700-7_17

preserving the soil biodiversity. Bioprocessing techniques and microbial fermentation contribute to food security, they enhance food preservation and nutritional quality, while ensuring the safety from foodborne pathogens. Hence, these approaches collectively form a holistic strategy for achieving sustainable food production and security in the twenty-first century.

17.1 Addressing Challenges and Scaling Up Microbial Innovations

The successful integration of microbial technologies into agriculture requires addressing barriers to product acceptability. Some of these key hurdles include:

(a) **Regulatory and policy constraints**: Regulations of microbial applications in agricultural and food manufacturing industries vary from one region to another. Therefore the need for standardization, quality assurance, and a streamlined process of product approval to achieve timely commercialization.
(b) **Farmers' awareness and adoption**: Many farmers in the developing region do not have sufficient knowledge, and lack awareness about microbial-based solutions. Hence, there is a need for more extension services, training programs, and demonstration farms, to efficiently bridge this knowledge gap.
(c) **Technological and infrastructural barriers**: Leveraging the recent developments in microbial biotechnology such as synthetic biology, next-generation sequencing, and bioinformatics has granted researchers a more in-depth understanding of the plant-microbe interaction. This has enhanced the optimization of microbial applications. However, despite their importance, such advanced technological tools are not available in resource-scarce settings and therefore impair development in such regions.
(d) **Scalability and market viability**: Microbial products must be cost-effective, have market viability, and potential to scale up across different agroecological systems. Strategic investments toward research development for commercialization activities will ensure the economic viability of the products.

The fundamental lesson learned from this book is that microbial application in food security needs to be multidisciplinary. This should entail the collaboration of experts from the fields of agronomy, microbiology, data science, and biotechnology while the policy regulation frameworks are important for translating research into impactful solutions. Similarly, academic institutions, governmental agencies, private sector stakeholders, and local communities need to work together to achieve successful implementation of microbial innovations. Support for microbial research requires funding programs and sustainable agricultural incentives as well as regulatory measures that allow proper utilization of microbial technologies. Governmental support for sustainable agriculture can be channelled through subsidies and educational programs as well as infrastructural development. These interventions prove vital for the development of sustainable farming practices. The Food and Agriculture

Organization (FAO), and the United Nations Environment Programme (UNEP) possess the capability to unite stakeholders on an international scale for strengthening microbial solutions that protect food security.

Microbial applications offer a promising resilience-building strategy, against climate change which continues to pose significant threats to global food security. This is because, beneficial microbes can mitigate climate-related stressors by improving soil carbon sequestration, enhancing drought tolerance, and reducing greenhouse gas emissions. Integration of climate-smart agricultural practices with microbial innovations will be key to sustaining food production in the face of increasing environmental uncertainties. In addition, the development of consortia of microbial communities together with microbiome engineering represents a transformative capability to redesign crops and soils that exhibit superior resilience. More so, researchers can develop tailored microbial consortia by leveraging cutting-edge genomic tools that optimize plant health, nutrient uptake, and stress tolerance. These advancements will be instrumental in shaping the future of sustainable agriculture and ensuring long-term food security.

17.2 Call to Action: Embracing Microbial Solutions for a Sustainable Future

It becomes essential to implement microbial applications as a cornerstone of sustainable agriculture and food security. This book has presented a comprehensive overview of microbial innovations, barriers, and prospects, emphasizing the need for a collective commitment to fostering microbial-based solutions. There is a need to foster collaboration among the stakeholders across the academia, industry, and policymakers to:

(a) Advanced Research and Development: Increasing the investment in microbial research will drive the discovery of novel microbial species and applications for food security.
(b) Promote Knowledge Dissemination: By bridging the knowledge gap through extension services and farmer education, workshops, and community-driven initiatives to facilitate widespread adoption.
(c) Encourage Sustainable Practices: Sustainable microbial applications need policy incentives to ensure long-term ecological balance and food production resilience.
(d) Leverage Technology for Microbial Innovations: The integration of artificial intelligence, big data analytics, and synthetic biology will revolutionize microbial applications in agriculture and food systems.

The natural world has existed with microorganisms shaping terrestrial ecosystems since billions of years ago yet the potential for these organisms to solve the modern-day problem of food insecurity is still vast. By harnessing the power of microbes, we can develop sustainable agroecosystems, with secure food production systems and promote environmental sustainability. The progress toward microbial agricultural transformation requires scientific findings and supportive government strategies alongside community-backed operations. This book lays a critical framework for upcoming research efforts and the execution steps that will guide society toward an environmentally friendly and food-secure sustainable system. The time to act is now. By embracing microbial applications as the base for sustainable food systems we will collectively build ecologically-friendly agricultural communities that promote harmony with nature ensuring food security for generations to come.

Index

O. O. Babalola et al. (eds.), *Ecofriendly Frontiers*,
https://doi.org/10.1007/978-3-031-98700-7

MIX
Papier aus verantwortungsvollen Quellen
Paper from responsible sources
FSC® C105338

If you have any concerns about our products,
you can contact us on
ProductSafety@springernature.com

In case Publisher is established outside the EU,
the EU authorized representative is:
Springer Nature Customer Service Center GmbH
Europaplatz 3, 69115 Heidelberg, Germany

Printed by Libri Plureos GmbH
in Hamburg, Germany